Biointerfaces
Where Material Meets Biology

RSC Smart Materials

Series Editors:
Hans-Jörg Schneider, *Saarland University, Germany*
Mohsen Shahinpoor, *University of Maine, USA*

Titles in this Series:

How to obtain future titles on publication:
A standing order plan is available for this series. A standing order will bring delivery of each new volume immediately on publication.

For further information please contact:
Book Sales Department, Royal Society of Chemistry, Thomas Graham House, Science Park, Milton Road, Cambridge, CB4 0WF, UK
Telephone: +44 (0)1223 420066, Fax: +44 (0)1223 420247
Email: booksales@rsc.org
Visit our website at www.rsc.org/books

Biointerfaces
Where Material Meets Biology

Edited by

Dietmar Hutmacher
Queensland University of Technology, Kelvin Grove, Australia
Email: dietmar.hutmacher@qut.edu.au

Wojciech Chrzanowski
University of Sydney, Sydney, Australia
Email: wojciech.chrzanowski@sydney.edu.au

THE QUEEN'S AWARDS
FOR ENTERPRISE:
INTERNATIONAL TRADE
2013

RSC Smart Materials No. 10

Print ISBN: 978-1-84973-876-7
PDF eISBN: 978-1-78262-845-3
ISSN: 2046-0066

A catalogue record for this book is available from the British Library

Published by The Royal Society of Chemistry,
Thomas Graham House, Science Park, Milton Road,
Cambridge CB4 0WF, UK

Registered Charity Number 207890

For further information see our web site at www.rsc.org

Printed and bound by CPI Group (UK) Ltd, Croydon, CR0 4YY

Preface

Biointerfaces: Surfaces that Guide Biology!

This book brings together scientists and engineers that work in diverse fields of physics, chemistry, biology, biotechnology, materials science, diagnostics, and medicine, with a broad focus on biointerface science. The engineered biomolecular interfaces described and discussed in the book are designed to recapitulate the functions of their natural counterparts. By uniting physical science and the engineering of surfaces with biology, fresh prospects are opened up and bridge the gap between fundamental research and development of application-oriented technologies and products.

The rapidly ageing population, maintaining high levels of activity well into retirement, together with the rising incidence of obesity- and diabetes-related dysfunction of the musculoskeletal system, results in an increasing number of surgeries that require implantable devices. There are now millions of cases each year requiring implantable fixation devices (over 6 million fractures in the USA alone; 11 fractures every minute). Regeneration of tissues can be aided with new classes of materials, *e.g.*, bioceramics, hydrogels or composites, which are emerging as the most promising approaches in many clinical applications. Despite many advances in materials synthesis and device design, the implantation of biomaterials almost universally leads to the device being enveloped in avascular fibrotic tissue that walls off the device from the body. This can be detrimental to the function of the implant/device, reducing its lifetime, and often necessitates repeated surgery and device replacement.

The implantation of a material/device into body is followed by the immediate interactions of body fluids, cells and tissues with the surface. These interactions are critical for obtaining the desired functionality of the material/device and, subsequently, they determine the success of the specific medical procedure. The communication between biological structures

RSC Smart Materials No. 10
Biointerfaces: Where Material Meets Biology
Edited by Dietmar Hutmacher and Wojciech Chrzanowski
© The Royal Society of Chemistry 2015
Published by the Royal Society of Chemistry, www.rsc.org

(*i.e.*, cells and tissues) and implanted physical objects (*i.e.*, device or material) occurs at the interface (termed the biointerface), and the surface of the implanted material is a key element that determines the outcomes of this communication. Surface properties are of special significance when biological and microbiological interactions are considered.

There are several, surface-related cues presented to biological milieu, which includes surface texture (also topography), mechanical properties, chemical composition, surface charge and stability of the surface (degradability). These cues can be presented to cell and tissues as two- and three-dimensional structures. It is well established that each of the surface parameters can modulate first interactions with the body fluids, and then cellular responses such as cell attachment, migration and differentiation. Although each of the individual parameters has been shown to be effective in controlling some of the biological functions, it is well known that there is a complex interplay between these parameters that may strengthen (cumulative effects) or weaken specific design modalities. Therefore, by tailoring the mechanical properties of materials, chemical composition, topography, and charge it is possible to design functional surfaces, which effectively control biological and microbiological responses. For example, osseointegration is often a key aim, which can be achieved by nanopatterning of the surface or attaching signalling molecules such as peptides.

This book describes the most recent advances in the manufacture and characterisation of biointerfaces that aim to provide the desired and predicted biological responses. It covers a wide range of materials and techniques used in the investigations of biomaterials, with a focus on musculoskeletal applications. Importantly, the book includes several examples of translational research and contains many 'how to do' or 'walk the talk' sections, which makes it suitable for student readers to gain a new knowledge how to design, fabricate and characterise a new generation of technologies and products based on biointerface science.

D. W. Hutmacher
W. Chrzanowski

Contents

RSC Smart Materials No. 10
Biointerfaces: Where Material Meets Biology
Edited by Dietmar Hutmacher and Wojciech Chrzanowski
© The Royal Society of Chemistry 2015
Published by the Royal Society of Chemistry, www.rsc.org

**Chapter 2   Additive Manufacturing and Surface Modification
of Biomaterials using Self-assembled
Monolayers   30**
*Jayasheelan Vaithilingam, Ruth D. Goodridge,
Steven D. R. Christie, Steve Edmondson and
Richard J. M. Hague*

Chapter 3   Probing Biointerfaces: Electrokinetics   55
*Ralf Zimmermann, Jérôme F. L. Duval and
Carsten Werner*

Section C   Multi-functional Biointerfaces

Chapter 7   Interfaces in Composite Materials   153

Ensanya A. Abou Neel, Wojciech Chrzanowski and Anne M. Young

Chapter 8   Bioactive Conducting Polymers for Optimising the Neural Interface   192

Josef Goding, Rylie Green, Penny Martens and Laura Poole-Warren

Section A
Biomolecular Interfaces for Cell Modulations

CHAPTER 1

Protein-based Biointerfaces to Control Stem Cell Differentiation

JORGE ALFREDO UQUILLAS PAREDES, ALESSANDRO POLINI AND WOJCIECH CHRZANOWSKI*

Faculty of Pharmacy, University of Sydney, Sydney, Australia
*Email: wojciech.chrzanowski@sydney.edu.au

1.1 Modification of Biomaterial Surfaces with Proteins

It is already well-established that cellular responses to foreign materials when implanted *in vivo* are guided mainly by the surface characteristics. The modulation of characteristics such as roughness, porosity, chemical and biological composition allows the regulation of material integration within the body as well as the guidance of specific responses, *e.g.*, cell adhesion, cell detachment, cell proliferation, differentiation, or metabolic activity. Material integration within the body is critically important for orthopedic implantable devices. Desired integration results in successful single surgery, thus reduces costs and complications related to adverse reactions and implant revisions. In this chapter, we described the methods of surface modification and functionalization, which enable the effective regulation of stem cells behavior for orthopedic applications. In particular, the chapter focuses on immobilization of different signaling molecules such as proteins, peptides, growth factors on the surface and, in three dimensions, to modulate cell

RSC Smart Materials No. 10
Biointerfaces: Where Material Meets Biology
Edited by Dietmar Hutmacher and Wojciech Chrzanowski
© The Royal Society of Chemistry 2015
Published by the Royal Society of Chemistry, www.rsc.org

adhesion and to trigger and regulate osteogenesis, osteointegration, and osteoregeneration. Furthermore, examination methods for the immobilization of biomolecules on surfaces are described in detail to explain the advantages and disadvantages of each method.

1.1.1   Introduction

Implantation of exogenous materials into the body triggers common body responses, and in some cases adverse inflammatory reactions. The body tries to ward-off the foreign object and protect from its 'perceived' negative influence on healthy cells and tissues. As a consequence the exogenous materials are encapsulated in a vascular, fibrous sacks which often limit the functionality of the implanted material. In the last decades research has been focused on the possibility of addressing this problem and creating surfaces that mimic the body's environment, thus enabling the host cells to accept and interact with the foreign object. Such interactions are facilitated by different cues incorporated into the surface, which includes topographical, mechanical, structural, chemical and biochemical signals. These are capable of stimulating specific cellular responses which include cell adhesion, proliferation, differentiation and, ultimately, cell death. The design of specific cues, single or combined, is guided by the application and tailored for specific needs.

Historically, one of the most successful approaches, which can be called biomimetic, was a deposition of hydroxyapatite coatings on the surface of implants that were placed within the bone environment. Hydroxyapatite, a natural component of bone tissues, when present on surfaces of metallic or polymeric devices allows for significant improvement of implant integration within the body. The most significant limitation is the adhesion of the coating to the substrate. It has to be noted that some of the implants are heavily bent (pre-operatively) and 'hammered' into the host tissue. This creates significant sheer and bending stresses that can lead to failure of the coating. Nevertheless, hydroxyapatite coatings have been clinically used and verified.[1]

The cross-talk between a bioactive layer and the tissue can similarly be achieved using different types of mineral, ceramic, glass–ceramic materials that include tri-calcium phosphate, bioglasses and phosphate glasses. These materials are coated onto the surface or are used as a base component to produce implants. The primary advantage of these materials is that they can be degradable and compounded with different ions (Zn, Ag, Co, Sr), which provides additional biochemical cues for cells at the interface. Zn, Co and Ag have been reported to have antimicrobial activity,[2] while Sr and Zn have been demonstrated to support bone formation and are used clinically in the treatment of osteoporosis.[2-4] Furthermore, each of these materials interacts with body fluids differently and can guide the adsorption of proteins and growth factors and other components from the body fluids and blood. Subsequently, these biomolecules modulate interactions with cells and tissues. Hence, by a careful selection of the material it is possible to stimulate cellular responses through the structural and chemical composition, and also to guide the adsorption of desired biomolecules that stimulates biological responses, *e.g.*, adsorption of cell adhesive proteins to hydroxyapatite.[2,3]

1.1.2 Influence of Modified Surfaces on Cell Adhesion

The communication of cells with surfaces is mediated by pre-adsorbed protein layers. The composition of this layer is critical to obtain specific responses such as cell adhesion/repulsion, or differentiation. It is well established that specific proteins, peptides, growth factors and drugs can stimulate cells.[5-10] These biomolecules are used to enhance the implant integration and also treat some dysfunction/diseases when released from the surface.[11-13] The most common approach to enhance cell adhesion to the surface of implantable materials using biomolecules is to incorporate the cell binding motif Arg-Gly-Asp (RGD).[13-15] RGD, together with the integrins that serve as receptors for them, constitute a primary recognition system for cell adhesion. The RGD sequence is the cell attachment site of many adhesive extracellular matrix, blood, and cell surface proteins, and nearly half of the over 20 known integrins recognize this sequence in their adhesion protein ligands.[16] The integrin-binding activity of adhesion proteins can be achieved by engineering short synthetic peptides that contain the RGD sequence. Interestingly, it was reported that such peptides promote cell adhesion when immobilized onto a surface, and inhibit it when presented to cells in solution.[16] Importantly, integrin-mediated cell attachment regulates further cell migration, growth, differentiation, and apoptosis.[14,15] The ability to target cell adhesion receptors, which are responsible not only for cell–matrix adhesion but also for signaling bidirectionally across the cell membrane, provides an opportunity to design new drugs (Figure 1.1). Because integrins are involved in many biological processes such as angiogenesis, thrombosis, inflammation, osteoporosis and cancer, these drugs, which are based on the RGD structure, can be used for the treatment of diseases such as thrombosis, osteoporosis and cancer[17-20] (Figure 1.1).

It is well established that tethering of protein and peptides facilitates communication with the cell. In fact, adhesion of cells is a prerequisite for the subsequent proliferation and differentiation of cells.[7,21] Recent studies by Webster demonstrated that adsorbed proteins including fibronectin, laminin and vitronectin play an important role in limiting bacteria colonization.[22] Hence fibronectin-functionalized surfaces can be considered as multi-functional with a dual purpose: preventing bacteria adhesion and enhancing cell adhesion.

While fibronectin and vitronectin receive the most attention to enhance cell adhesion to surfaces of biomaterials, there are several types of proteins and peptides that have been attempted to use for the same purpose. The use of fibronectin and vitronectin is primarily dictated by their composition, in

RGD-grafted PLGA-nanoparticle loaded with Paclitaxel (PTX)

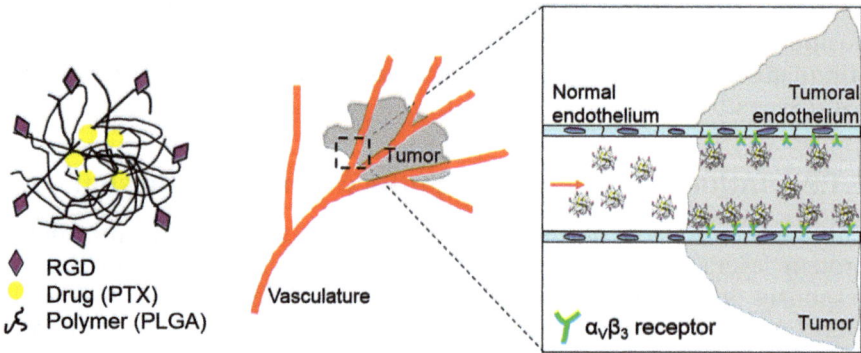

Figure 1.1 Functionalized drug carriers for active targeting of cancer cell.[20] Immobilization of the RGD sequence allows effective active binding to (targeting) cancer cells and then localized delivery of the drug cargo that is released upon degradation of the polymeric carrier.

which the RGD sequence plays a critical role.[23] RDG is recognized and bind *via* integrins ($\alpha_5\beta_1$, $\alpha_V\beta_1$, $\alpha_V\beta_3$, $\alpha_V\beta_5$, $\alpha_V\beta_6$, $\alpha_V\beta_8$ and $\alpha_{IIb}\beta_3$) cell adhesion receptors that bind to the extracellular matrix (ECM) proteins.[16,18,23–25] It has been demonstrated that laminins and collagens also contain RGD sequences but these are inaccessible, thus are not typically used for surface modifications.

Many integrins are expressed in various tissues; however there is some population which are expressed only in a certain type of cell or tissues. The integrin receptors, which were found in human osteoblasts and are typically used to regulate bone cell responses, are the fibronectin receptor ($\alpha_5\beta_1$),[16,18,23–27] vitronectin receptor ($\alpha_V\beta_3$),[28] and the type I collagen receptor ($\alpha_2\beta_1$).[16,18,23–27] For this reason modification—functionalization—of implant surfaces with biomolecules/protein to mediated cell adhesion to substrates *via* integrins has recently become one of the most interesting approaches in the development of new biomaterials including drug carriers that are capable of targeting specific sides.[15,29–33]

It has been already suggested that the RGD peptide, which interacts with the $\alpha_V\beta_3$ and $\alpha_V\beta_5$ integrin sub-units, commonly associated with vitronectin, increases biointegration of implants. Matsuura also demonstrated that RGD contributes to the osteoconductive effect of hydroxyapatite more than titanium.[34] This phenomenon is associated with higher affinity and adsorption capability of protein to hydroxyapatite surface than to titanium surface.

The positive effects of the RGD peptide on regulation of cell adhesion have been confirmed in many *in vitro* as well as *in vivo* studies. Elmengaard[35] has shown that RGD peptide-coated porous-coated titanium implant significantly increased bone formation on and around the implant. In this study a cyclic RGD (Figure 1.2), which interacts with both $\alpha_V\beta_3$ and $\alpha_V\beta_5$ integrin sub-units has been immobilized on unloaded press-fit titanium implants.

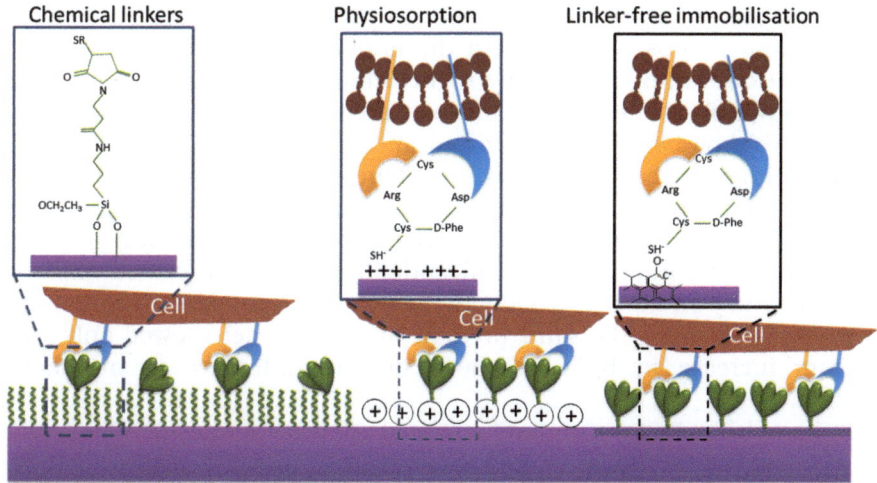

Figure 1.2 Mechanism of protein immobilization on the surface *via* three different mechanisms: chemical linking,[36] physiosorption and the linker-free (*via* active radicals) approach.[13,37,38]

Not only was the increase in bone formation observed, but also a significant reduction in presence of fibrous tissue around the implant was evidenced.[35] Because the RGD coating can be relatively easily applied to the surface it offers a cost-effective way to enhance the early osseo-integration of press-fitted clinical implants.

One of the interesting features of this approach was the use of cyclic peptide. Cells typically show higher integrin binding affinity to the cyclic RGD sequence and previous studies have shown that cyclic peptides are more stable when immobilized to complex and three-dimensional substrates.[39,40] Another important feature of cyclic RGD peptides is their high selectiveness toward the α_V integrin, which regulates activation of the osteoblast-specific transcription factor (core binding factor alpha-1/runt related transcription factor 2 during osteogenesis).[41] These peptides show also more resistance to enzymatic cleavage,[42] and may also promote a more stable cell–ligand bond. The integrin affinity and specificity to the RGD peptide is affected by both steric conformation and the amino-acid sequences flanking the RGD peptide.[43,44] Mooney[40] has also demonstrated that a linear RGD peptide promoted osteogenic differentiation of pre-osteoblasts (MC3T3-E1) but did not induce differentiation of hBMSCs or D1 stem cells. At the same time matrices that presented the higher-affinity cyclic form of this adhesion ligand enhanced osteoprogenitor differentiation in three dimensions.

In vivo studies by Elmengaard[35] on press-fit titanium alloy implants coated with cyclic RGD and its effect on bone ongrowth showed that after 4 weeks, cyclic RGD coating significantly stimulated bone formation directly at the interface. A two-fold increase in bone growth was found for RGD-coated implants compared to the uncoated. Importantly, RGD-coated implants showed less fibrous tissue around the implant, which is considered very positive. However, a significantly higher amount of bone for RGD coated implants was observed mostly at the interface (0–100 μm), and the difference in bone volume between both coated and uncoated groups gradually decreased with distance from the surface. There were no significant differences between both groups at a distance of 750 μm from the implant surface. Furthermore, an increase in the mechanical fixation was also observed; apparent shear stiffness was significantly higher for RGD-coated implants. A similar study on the osteoconductive hydroxyapatite coating with titanium alloy implants using the same press-fit implant model showed only a significant increase in bone ongrowth but no difference in mechanical fixation.[45]

To immobilize RGD on amino-functionalized glass surfaces Dechantsreiter *et al.*[43] suggested the use of isothiocyanate-terminated peptides. It was demonstrated that RGD peptides were fused to an isothiocyanate anchor during synthesis and bound to amino-terminated surfaces. Importantly, two types of linear peptides and one cyclic peptide were investigated and were shown to enhance cell spreading and induce the formation of focal adhesions in murine fibroblasts. The authors also concluded that formed adhesions were specific because cells did not recognize the corresponding negative control peptides and did not spread in the presence of soluble H-RGDS-OH peptide. This coupling method was shown to be effective also for patterning, where cells selectively recognized areas coated with RGD-containing peptides.

Better integration and deposition of the bone on the surface is directly linked to pullout strength of the implants. O'Toole suggested the use of bone sialoproteins (BSPs) to improve the pullout strength.[46] In his approach two scenarios were considered: (1) BSPs were coated on the surface and (2) BSPs were not placed directly on the surface but BSP-containing gelatin was used as a plug where the implant was placed. Results failed to demonstrate the benefits of the BSP. In general, BSP-coated acid-etched implants perform more poorly, mechanically, than do uncoated implants. It was concluded that BSP forms an insulating barrier on the implant surface, preventing the direct apposition of the bone on the implants. At the same time, histology tests showed that the BSP coating failed at the BSP–implant interface but osteoinductive behaviors were confirmed at the BSP–bone interface. It was postulated that a better method of coating is required to allow the formation of bone at the interface with the implant. Similar results were found for the second group where BSP-containing gelatin was used. These implants performed poorly both histologically and mechanically. Histologically, osteoid, osteoblasts and osteocytes are stimulated by the presence of BSP, but it was not observed at the surface. This distant osteoinduction does not correspond with better mechanical performances when implants are subjected to pullout testing analysis.

With respect to proteins, peptides benefit from a lower immunogenic activity as well as from the ability to be synthesized and handled.

In addition, over the last decade, highly active and $\alpha_v\beta_3$- and $\alpha_v\beta_5$-integrin selective cyclic pentapeptide ligands such as cyclo(-RGDfX) have been developed.[13] It was thus demonstrated that in addition to the RGD binding sequence a D-amino acid, especially D-Phe following the Asp residue in the cycle, is essential for high activities and α_v selectivity. Cyclic pentapeptides with D-amino acid in other positions and/or a non-hydrophobic amino acid following Asp as well as linear peptides have lower activity and are less selective towards α_v integrins. Our results clarify that osteoprogenitor cells exhibit a differential binding to RGD peptides displaying a specific conformation (linear RGDC and cyclo-DfKRG). Such a behavior has to be related to the different signal transduction pathways implied by both peptides. Indeed, some of us[14] have shown that linear GRGDSPC peptide interacts preferentially with $\alpha_\chi\beta_1$ integrins while cyclo-DfKRG peptide interacts with $\alpha_v\beta_3$ and $\alpha_v\beta_5$ integrins. Consequently, different cell adhesion at 24 h seeding may be linked to the different cellular activity, as extracellular proteins synthesis, implied by the signal transduction pathways that both RGD-containing peptides induce, respectively. In addition, cell adhesion behavior has to be related to the accessibility of integrins receptors by peptides displaying conformation. However, both cyclo-DfKRG and linear RGDC appear to be good candidates for developing hybrid biomaterials made of titanium alloys and human osteogenic cells.

1.1.3   Influence of Modified Surfaces on Cell Proliferation and Differentiation

Besides the improvement of cell adhesion by means of adhesive proteins or peptides, huge efforts are currently directed toward the development of osteoconductive/osteoinductive materials. Osteo-conductive materials are able to enhance the formation of new bone, *i.e.*, accelerating the bone development or improving it from a physiological point of view. Osteoinductive materials trigger the osteogenic process without external stimuli (*e.g.*, supplements in osteogenic media) and are therefore more promising for a rapid healing process. To assess these properties *in vitro*, stem cells are commonly cultured in basal and osteogenic medium and several markers of differentiation are evaluated. For *in vivo* studies, they are tested by implantation in ectopic and orthotopic bone site and assessing the bone formation. Several directions can be followed for obtaining an osteoinductive material. Few research groups have successfully modified the physical and chemical properties of an inert material, such as composition, elasticity and topography, in order to provide the inert material with osteoinductive capabilities.[47–56] Another approach is based on the delivery of biomolecules, mainly growth factors, integrated into the scaffold and able to control osteogenesis, bone tissue regeneration and ECM formation through specific cellular pathways in the implant site.[57] For example, insulin-like growth factors (IGFs) and transforming growth factor-β (TGF-β) affect the migration and recruitment of different bone cells involved in bone healing, respectively.[58,59] Bone morphogenetic protein (BMP) influences the osteogenic differentiation of progenitor bone cells, and vascular endothelial growth factor (VEGF) enhances the formation of functional blood vessels. Recently, small peptides resembling the functionality of physiologically active proteins have grown an increasing interest. Therapeutic peptides overcome many drawbacks of macromolecules, such as long synthesis time, high production costs, low stability, limited long-term efficiency, or risk of unforeseen side effects. Small peptides carrying the active site of several growth factors have already been developed. For example, Tanihara and co-workers found that a 20 amino acid peptide sequence from the 'knuckle' epitope of BMP-2 shows the biological activity of the full-length BMP-2 protein.[60] Similarly, a VEGF-mimetic peptide has been synthesized and has shown retaining the biological activity of VEGF protein.[61] Moreover, several peptides able to inhibit bone resorption have been synthesized after the identification of osteoprotegerin (OPG) and RANKL, important elements in the bone remodeling pathway.[62]

During the last few years orthopedic researchers aiming to regenerate and repair bone have realized that the appropriate function of engineered bone substitutes cannot rely on the sole diffusion of nutrients and oxygen.[63] The lack of a functional vascular network results in the formation of necrotic cores within the bone substitute.[64,65] Therefore, engineered bone scaffolds with clinically relevant sizes require a vascular network for the successful engraftment and survival of the bone substitute. However, host-derived vascularization of implanted constructs is largely limited by the capacity of host cells to invade and form *in situ* capillaries.[66] Creating functional and perfusable microvascular systems, replicating the structure, biological properties, heterotypic cell interactions and biomechanics of native microvascular environment, represents nowadays an important bottleneck of bone tissue engineering strategies. Below we describe important progress in strategies to create vascularized tissues using biological cues.

Biological techniques for microvessel formation rely on heterotypic cell–cell interactions and the cross-communication of cells with encapsulated growth factors in the extracellular microenvironment.[63] Endothelial cells co-cultured with human mesenchymal cells (hMSCSs) or perivascular cells have resulted in the formation of microvascular networks.[63,66–71] This tendency becomes more pronounced when endothelial and progenitor cells are encapsulated in ECM-like hydrogels. These types of hydrogels can be chemically modified to allow homing mechanisms to cells and facilitate the cross-talk of cell and the encapsulating matrix. Recently, Chen developed vascularized methacrylated gelatin hydrogels by encapsulating blood-derived endothelial colony forming cells (ECFCs) and bone marrow-derived MSCs in a hydrogel matrix.[71] The results demonstrated an extensive formation of capillary-like networks with lumens within the hydrogel structure. Pre-vascularized hydrogels have the potential to be combined with osteo-conductive materials to trigger angiogenesis in bone substitutes.

Presenting angiogenic growth factors and signaling markers such as VEGF can trigger the microvessel formation process. D'Andrea *et al*. reported the development of a synthetic short peptide of 15 amino acids called QK peptide which contained a VEGF binding domain to activate VEGF receptors in endothelial cells. This peptide was used to induce endothelial proliferation and VEGF signaling mechanisms on a Matrigel substrate.[61] Synthetic small peptides have been further modified to incorporate acrylate motifs to facilitate the functionalization of VEFG-like peptides in PEG hydrogels.[72] Additionally, Chiu *et al*. guided endothelial cell proliferation in a microfabricated chitosan–collagen hydrogel rich in Tβ4, an angiogenic and cardioprotective peptide that enhances cardiomyocyte survival.[73,74] This approach enabled capillary-mediated anchorage and anastomosis of arteries and veins.[73] In a recent study, Kim *et al*. integrated a microfluidic chip perfused with a fibrin gel to perform a comprehensive angiogenesis study by endothelial cell-mediated formation of perfusable microcapillaries.[75] Also, microvessels generated in a microfluidic chip presented morphological and biochemical cues replicating those in native blood vessels and capillaries niches.[75] The combination of these technologies offers the potential to create bone replacement scaffolds in the future in a cost-effective and reproducible manner. Biological techniques for pre-vascularization of bone substitutes hold a great promise for fabrication of functional large bones *in vitro*. However, there are some challenges that have limited the progress of these methods. Challenges associated with the fabrication of vascular networks are the inability to form three-dimensional (3D) biomimetic vascular networks in long bones, and the corresponding slow processing time of these networks.

1.2 Methods of Protein Immobilization

Over the years, many functionalization strategies have been developed and optimized in order to add new functionalities to bio-inert materials and broaden their applicability while overcoming any important material deficiency and preserving their qualities. Approaches for protein immobilization can generally be divided in two groups: (1) surface functionalization, when only the surface is involved in the modification process, commonly through a post-processing step; and (2) bulk functionalization, when the material is homogenously modified, usually before or during the processing. Here, we will focus on these two approaches for introducing biomolecules (proteins and peptides) to inert materials, having the creation of innovative bio-active materials as final goal. Protocols for the visualization and immobilization of proteins are given in Section 1.3.

1.2.1 Surface Functionalization

Surface functionalization strategies aim to introduce new biologically relevant properties to inert materials without undermining its bulk material properties. Surface properties are able to affect the overall performance of materials, control the first phase of material–cells interaction and trigger specific biological responses.

Generally, surface functionalization approaches can be classified into three main categories (physical, chemical and biological). However, this distinction is not always strict and multiple modifications can be employed. The choice of the right route is determined by either the material chemistry or the envisioned application and every choice presents advantages and drawbacks. Here, we will describe biological functionalization methods (physisorption and covalent bonding), overlooking physical (*e.g.*, polishing, grinding, laser treatment) and chemical (*e.g.*, ion bombardment, acid/alkaline treatment, sol–gel, small molecule grafting) approaches.

1.2.1.1 *Physisorption*

Physical immobilization is the simplest bio-functionalization method and consists just in dipping the material in a solution containing the target biomolecules. Physisorption (*i.e.*, physical adsorption) relies on electrostatic interactions, van der Waals forces, hydrogen bonds and/or hydrophobic interactions. It can be applied to most types of surfaces and, in general, does not require any surface pre-treatment. Although it is considered a gentle method (*i.e.*, non-destructive), the biomolecule–surface randomly-oriented interaction can lead to conformational changes and subsequent loss of functionality. In addition, due to the presence of weak interactions the binding stability of adsorbed molecules is dramatically affected by environmental conditions, such as pH, ionic strength and biomolecule concentration.[76]

Recently, a few approaches have been developed to introduce biomolecules by physisorption in a more controlled way. For example, Messersmith's group at Northwestern University introduced a fast, reproducible, versatile method to coat many different organic and inorganic surfaces by exploiting alkaline oxidative polymerization of dopamine.[77] Having a thickness ranging from few nanometers (nm) to >100 nm, the polydopamine (pD) coating does not affect the pristine material properties. The coating can be exploited to introduce biomolecules through a single step,[78] *i.e.*, with the biomolecules entrapped in the pD layer, or two separated steps, with the molecules attached on top of the pD layer.[79]

Another interesting physisorption-based approach for ceramic materials employed as bone substitutes involves the use of short (10–30 amino acids) modular peptides having two domains: (1) a domain able to bind strongly the surface through electrostatic interactions; and (2) a domain that introduces a biologically relevant function (*e.g.*, cell adhesive peptides or growth factor-derived moieties). The mechanism behind the surface functionalization mimics the natural mechanism by which human bone extracellular matrix proteins bind hydroxyapatite (HA) into the body. Different sequences have indeed been identified as responsible of this interaction in osteonectin and bone sialoprotein,[80,81] osteocalcin[82] and salivary statherin.[83] Moreover, following this approach, phage display technology has been employed for discovering new peptides able to bind ceramic surfaces.[84,85] Modular peptides have been reported for functionalizing several calcium phosphate-based materials, such as HA,[80,86] HA–titanium,[87] β-TCP,[88] HA–polymer composites,[89] native bone grafts[90] and allografts.[91]

In a series of recent studies, Rohanizadeh and his team investigated protein adsorption onto hydroxyapatite (HA) surface, factors affecting this phenomenon and techniques to modulate it. Using bovine serum albumin (BSA) and cytochrome *c* as model proteins, the effect of HA crystallinity on its protein adsorptive capacity was investigated by this team.[1] They concluded that regardless of the total surface charge of protein, the adsorption of proteins onto HA was directly influenced by HA crystallinity, where higher crystallinity resulted in a lower protein adsorption rate. The crystallinity of HA also affected proteins release profile, in which HA particle with higher crystallinity showed lower protein release kinetics. In addition, the study demonstrated that the surface charge of HA particles influence protein adsorption, where BSA, an acidic protein, was adsorbed at higher rate on a negatively charged surface compared to cytochrome *c*, a basic protein.

In another study, the team altered the HA surface charge by immobilization of different amino acids during HA precipitation.[2] Four amino acids with different isoelectric points were used in this study: neutral (serine, Ser; and asparagines, Asn), acidic (aspartic acid, Asp) and basic (arginine, Arg). In order to evaluate the affinity of amino acid-functionalized HA (AA-HA) to proteins, BSA and lysozyme were used respectively as an acidic and basic model protein. The protein adsorption onto the surface of AA-HA depended on the surface charges of HA particles, whereby BSA demonstrated higher affinity towards positively charged Arg-HA. Alternatively, lysozyme, a positively charged protein showed greater tendency to attach to negatively charged Asp-HA. The AA-HA particles that had higher proteins adsorption demonstrated a lower protein release rate. Thus, the results demonstrated that selective immobilization of amino acid onto HA could provide a strategy to tailor the adsorptive capacity of surface to a specific protein.

The team further investigated the effects of side chain length of amino acids on protein adsorption onto amino acid-treated HA.[3] The results showed that immobilization of amino acids with longer side chains decreased the crystallinity and increased the negative value of the surface charge of HA particles. This could be due to the steric hindrance of a 'bulky' side chain, disturbing the arrangement of HA crystals, due to lowered crystallinity of AA-HA. Furthermore, manipulation of external factors, namely pH and ionic strength, during proteins adsorption can significantly improve HA binding affinity to proteins. The protein adsorption rate in AA-HA increased with decreasing the pH, while reverse trend obtained in unmodified HA.

HA consisting of carboxyl rich (COO−) groups on its surface was also synthesized by Boccaccini an his team.[4] HA particles were precipitated in presence of different concentrations of citric acid (CA). Results showed that CA-HA displayed significantly higher affinity towards lysozyme. Using the optimized parameters obtained from lysozyme adsorption (as a model protein for BMP-2), the results demonstrated that HA particles immobilized with COO– significantly increased the HA loading capacity for bone morphogenic protein-2 (BMP-2) and prolonged BMP-2 release profile. The *in vitro* results showed that the prepared CA-HA was not toxic towards human osteoblasts and indeed citric ions present on HA surface shifted the surface charge towards negative value, which promoted cell proliferation on HA.

The pros and cons of physisorption are:

- Pros
  - Applicable to the broad range of natural or synthetic materials
  - No pre-treatment of the target molecules is necessary
  - Mild conditions
  - Usually a very quick and easy approach
- Cons
  - The coating is not stable for long-term application and molecules detach quickly from the surface with an unpredictable profile
  - Molecules can lose their functionality upon binding

1.2.1.2 Covalent Bonding

Covalent immobilization of biomolecules is usually chosen when possible over physisorption, as the attachment is controlled and stable, and the kinetics behind biomolecule release can be predicted and calculated. Numerous immobilization protocols have been optimized for different materials and different cross-linker molecules have been used to graft biomolecules to specific, surface functional groups. The most popular routes utilize glutaraldehyde (GA) or carbodiimide (EDC) chemistry to immobilize proteins and peptides as well as oligonucleotides.[76] GA can be used to activate the target surface and attach biomolecules irreversibly on the activated surface through an amino group (biomolecule)–aldehyde (surface) reaction. Due to prevalent cytotoxicity effects reported for GA,[92] EDC chemistry has become very common due to its 'zero-length property'. In other words, EDC facilitates the reaction between a carboxylic acid group and an amine group, leaving no residues on the new bond.

In polymers, the functional groups required for GA/EDC coupling reactions[93] can be exposed either naturally or by quick surface modification (*e.g.*, oxygen plasma treatment). For ceramic materials, pre-treatments are necessary to introduce such functional groups to the surface calcium phosphate surface. Silanization is the most widely used strategy to introduce functional groups on inorganic surfaces.[76] Silanization is extremely versatile in terms of materials to be used with and does not require extreme conditions (*i.e.*, mild pH, room temperature) or expensive equipment. In general, silane molecules are activated by hydrolysis and then condensation between Si–OH groups of the silanol and the OH present on the surface occurs. This activation leads to the exposure of functional groups (from the silanol molecules) on the ceramic surface. In this way amine, carboxylic acid or sulfur groups are introduced on the surface material and are often used for further functionalization reactions through GA or EDC. Though simple, several problems can occur such as layer stability, silane reactivity towards water, possible toxicity of free silanes or leachable products, complete hiding of the bulk material, and potential problem for bioactive ceramics.[94] Besides silanization, the formation of a pD coating layer can be a successful strategy for introducing functional amine groups, useful for other functionalization steps.

Overall, covalent immobilization procedures require laborious multi-functionalization work, can produce potentially toxic by-products, and the stability of linkers and bioconjugates during and after the immobilization reactions can be an issue.[95]

The pros and cons of the covalent bonding method are:

- Pros
  - Applicable to a broad range of natural or synthetic polymers
  - Stable functionalization over time and usually a predictable release profile
- Cons
  - Some polymers need to be activated (*e.g.*, grafting reactive groups on the polymer chain) before functionalization
  - By-products from the functionalization step can remain and show cytotoxicity

1.2.1.3 'Linker-free' Approaches

Simplicity and cost-effectiveness, besides biological efficacy, are two of the factors that remain critically important in the development of new functionalization technologies. The immobilization of biomolecules without chemical linkers—linker-free—addresses common problems related to physisorption and chemical linking. Linker-free immobilization was developed by Kondyurin and is currently used by producers of implantable devices (LfC Z.o.o).[37,38,96] This approach is particularly useful for polymeric devices and also has been translated to metallic devices by Chrzanowski.[13] In this approach biomolecules are immobilized *via* radicals created at the interfaces by energetic ion-implantation. Implanted ions disrupted the primary bonds of the polymer and results in the formation of a carbonized interface that contains highly reactive radicals, species with unpaired electrons.[97] The radicals are stabilized by delocalization on π-electron clouds of the condensed aromatic structures. The stability increases with larger conjugated areas in the aromatic structure. In addition to the chemical activity that they bring to the surface, the unpaired electrons typically react with oxygen in the environment[98] and create an elevated negative potential on the surface. This effect generates a strong double layer in solution and attracts charged protein molecules to the surface. The radicals in the interface also allow the covalent binding of biomolecules without the need for any additional linkers.[38] It is believed that the reactive radicals act as anchors to immobilize biomolecules through the substitution of mobile hydrogen atoms bonded to carbon, nitrogen or oxygen atoms in amino acids on the outer surface of the protein molecule. At the same time, the radicals and surface oxygen groups interact with water molecules due to strong hydrogen bonds providing a hydrophilic surface. It is known that an adsorbed water layer prevents direct physical contact of the protein molecule with the surface creating a water shell around the protein and maintaining the natural conformation of the protein molecule despite the covalent bonding with the surface.[99] This approach has been proven effective for both polymers and metals and immobilization of different types of proteins, peptides and yeast has been demonstrated.

The pros and cons of the linker-free approach are:

- Pros
  - Single step method
  - Applicable to the broad range of biomolecules
  - Applicable for all types of polymers
  - Significantly increases the amount of immobilized protein
- Cons
  - Not applicable for 3D structures
  - Activity of the surface decreases with time
  - Requires a specialized facility

1.2.2 Determination of Successful Functionalization of Proteins, Growth Factors or Peptides

1.2.2.1 *Determination of Successful Functionalization of Proteins, Growth Factors or Peptides in Three Dimensions*

The determination of successful covalent functionalization of growth factors or peptides in a 3D microenvironment (synthetic or naturally derived scaffolds) is not easy. As explained above, growth factors and small peptides can elicit specific cellular responses in encapsulated cells to recapitulate a particular tissue, and promote their proliferation or differentiation. Additionally, growth factors and peptides show short half-lives when injected *in vivo*, and their functionalization within a 3D matrix extends their functionality and protects them from degradation enzymes in the host environment.[100,101] In practice, we have found that the assessment of covalent functionalization of growth factors or peptides in naturally derived scaffolds made of type I collagen, gelatin, alginate, hyaluronic acid, or silk fibroin cannot be done using traditional analytical tools such as ¹H nuclear magnetic resonance (¹H NMR) or Fourier transform infrared spectroscopy (FTIR). Thus, the functionalization of fluorescent moieties to growth factors and small peptides is a practical approach to determine the extent of covalent functionalization of small molecules in a (hard) scaffold or hydrogel. When choosing the fluorescent tag, the researcher has to take into consideration the molecular mass of the fluorescent tag to attach to the growth factor or peptide. Fluorescent tags such as FITC (5,6-fluorescein isothiocyanate, MW = 389.38) or N-hydroxysuccinimide (NHS)–fluorescein (MW = 473.39) could be used as they are small compared to proteins, peptides, and growth factors, thus minimizing the impact in the biological function of the growth factor or peptide.[102]

A brief description of the process of determining the functionalization of growth factors and peptides in naturally derived hydrogels or scaffolds is outlined below. We need to clarify that no covalent functionalization is 100% effective and part of the growth factor or peptide will eventually leach out of the hydrogel or scaffolds. Thus, we present a simple and straightforward method to characterize the diffusivity of the growth factor or peptide out of the matrix. We will use FITC as the model fluorescent tag attached to a short peptide such as the QK peptide, mentioned previously, functionalized to a type I collagen hydrogel.

The pros and cons of the fluorescence method are:

- Pros
  - Applicable for 3D structures
  - Applicable to the broad range of natural or synthetic hydrogels
  - Applicable for all types of polymers
- Cons
  - The use of large fluorescent tags is not recommended
  - Activity of the functionalized 3D structure may decrease with time
  - Functionalization will be limited by diffusivity of species into hydrogels

A   Appendix: Protocols for the Visualization and Immobilization of Proteins

A.1   Protocol for the Visualization of Protein on a Surface Using Atomic Force Microscopy

Visualization of the presence of proteins on a surface and the evaluation of their density is a significant challenge due to the small size of the proteins (a few nanometers to tens of nanometers). Typically, surfaces have complex topography with roughness often much greater than a few nanometers, and small protein are difficult to be distinguished from surface features. Hence to visualize proteins on the surface a very flat substrate such as a silicon wafer or a mirror-finished surface are required.

> **NOTE:** *In such situations it will not be possible to evaluate effects of the surface topography on proteins conformation.*

> **TIP:** *If a silicon surface is not the preferred substrate and polishing is not trivial, layers of different materials of interest can be easily deposited on flat substrate. For examples to test embolization of protein to polymers such as polystyrene and PLA they can be spin coated on silicon wafer. This will provide flat surface that can allow distinguishing features below 1 nm.*

A protocol to visualize proteins on a surface is as follows:

1. Prepare a flat substrate with the roughness below the smallest protein (biomolecule) you plan to visualize.

   > **TIP:** *Typically, a silicon wafer offers the desired roughness/flatness.*

2. Modify the surface as required.
3. Scan the surface using atomic force microscopy. (For AFM settings, see below.)
4. Immobilize the protein of interest. Typically, immobilization is carried out by immersing the sample in the solution that contains protein.
5. Rinse the sample with ultra-pure water or phosphate-buffered saline (PBS) to remove unbound and loosely attached proteins.
6. Scan the surface after immobilization.

   > **NOTE:** *Scanning can be done both in air and in liquid. When you scan the surface in air the surface must be dry.*

7. Present the results. Typically, the height (topography) of the image both in 2D and 3D are used to present the proteins present on the surface. In addition, line profiles can be used to quantify the amount of proteins (Figure 1.3).

The scanning settings are:

- *Scan mode:* AC (tapping, non-contact mode); probe with a spring constant 40 N m^{-1} (a softer tip will also work; occasionally depending on the protein length softer probes might be required). Free amplitude of the tip can be in the few hundred nanometers to 1 μm.

  > **TIP:** *If the proteins are not well presented on the surface, the drive amplitude can be decreased, followed by a decrease of the set point. This will reduce the amplitude and should result in sharper images of proteins.*

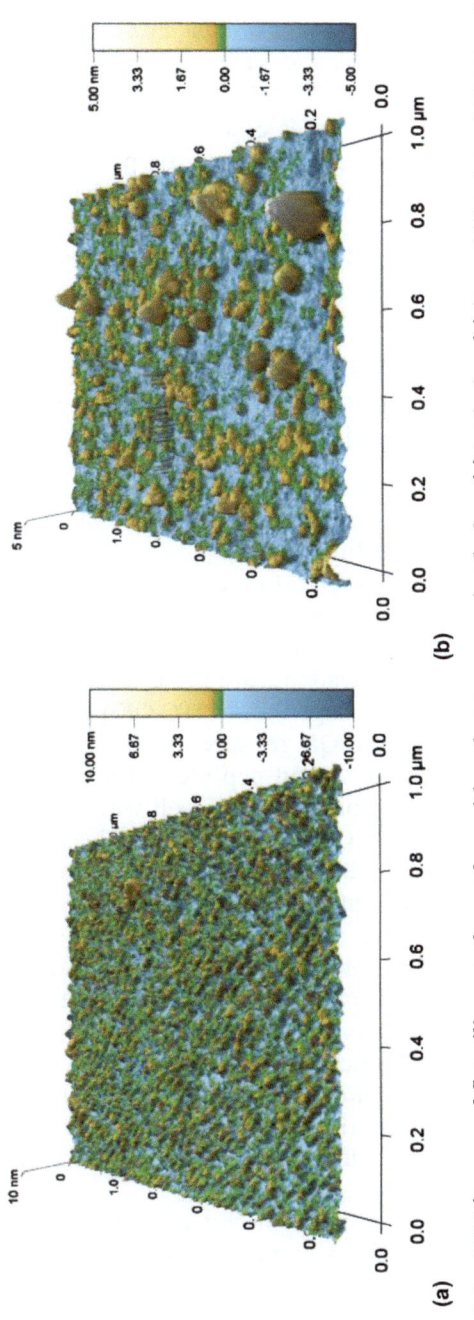

Figure 1.3 AFM images of flat silicon wafer surface with polystyrene coating before (a) and after (b) immobilization of fibronectin.

> **TIP:** *It is advisable to conduct 'a soft' approach, with a set point of 90% of free amplitude. When the surface is detected by the tip, further 'touching' of the surface should be done by decreasing the set point until the topography can be seen on the scope.*

- *Scan size*: Typically, scan sizes of 1×1 µm, 500×500 nm and 250×250 nm are sufficient to visualize most of the proteins.
- *Scan rate*: A slow scan rate of 0.2–0.4 Hz is advisable.

> **TIP:** *When proteins 'stick' to the tip and drag along the surface (lines are visible on the surface) then the scan rate can be increased (1 Hz).*

A.2 A Protocol to Determine the Degree of Functionalization of Growth Factors and Small Peptides in Naturally Derived Type I Collagen Hydrogels

1. Prepare 200 µL of a 25 µM solution of FITC-QK solution and bring it to 80 °C for 30 min to ensure solubilization of the peptide in Dulbecco's Phosphate Buffered Saline (DPBS).
2. Quench the solution to 37 °C in an ice-water bath and added 100 µL of acid-soluble type I collagen.
3. Neutralize the mixture solution with 0.1 N sodium hydroxide, and add 20 µL per well in a 48-well culture plate. Incubate at 37 °C for 2 h to allow gelation of the collagen and FITC-QK peptide.
4. Wash the collagen/FITC-QK gel several times, using 200 µL of chilled DPBS to remove any unbound peptide to the collagen matrix.
5. After the wash step, replenish each well with 200 µL of DPBS and incubate the plate at 37 °C for the FITC-QK release experiment.
6. Obtain four aliquots of 50 µL of the DPBS incubation bath every 12 h to determine the amount of FITC-QK peptide that is leaching out from the collagen hydrogel, and run the aliquots in a fluorescence microplate reader (Gemini EM; Molecular Devices, Sunnyvale, CA, USA). The fluorescence intensities are measured at 518 nm emission and 494 nm excitation wavelengths.
7. Replenish the DPBS incubation solution, wait for another 12 h and repeat Step 6 (above). Keep measuring the amount of FITC-QK in the incubation solution until the signal obtained from the FITC-QK incubation solution is below the detection limit of the fluorescence microplate reader.
8. The fluorescence readings will be converted to the corresponding concentration by using a standard calibration curve generated from known concentration of FITC-QK peptide in incubation solution at 37 °C.
9. Three possible functionalization scenarios (Figure 1.4) may occur depending on the characteristics of the probe and the 3D scaffold.

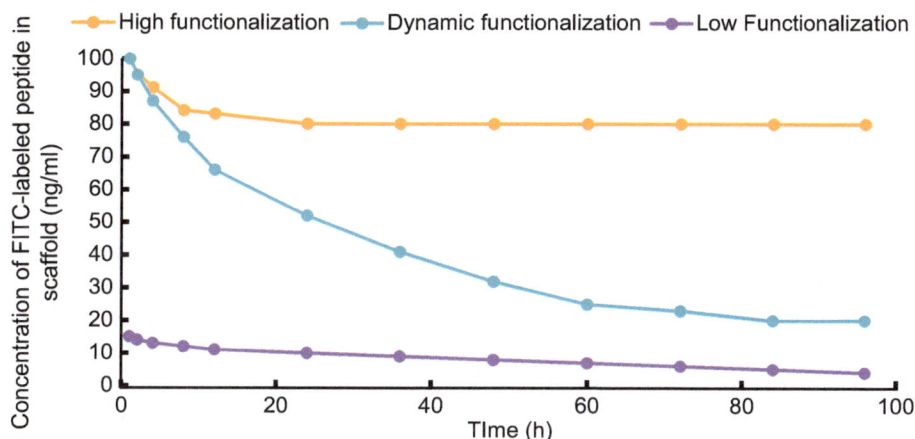

Figure 1.4 Three functionalization scenarios of fluorescently-labeled peptide in a hydrogel scaffold. High functionalization indicates a high concentration of peptide effectively bound to the hydrogel backbone. Dynamic functionalization indicates that the peptide can be slowly released to the microenvironment in a controlled manner over time. Low functionalization indicates that the peptide is loosely bound to the hydrogel backbone due to unspecific binding chemistry.

The fluorescently labeled probe can be highly functionalized to the scaffold where a strong interaction between the probe and the scaffold will take place. The probe could 'leach out' out of the scaffold over time, a phenomenon that can be described as dynamic functionalization. Finally, the interaction of the probe with the scaffold could be very low, making it very difficult to detect the probe in the 3D system.

A.3 A Protocol to Immobilize Growth Factors and Small Peptides by 'Classic' Physisorption

1. Dissolve the biomolecule to be immobilized in coating buffer at a pH near the biomolecule isoelectric point. Common buffers include: (1) 10 mM sodium phosphate, 0.15 M NaCl, pH 7.4 (PBS); (2) 50 mM sodium borate, pH 8.5; (3) 50 mM sodium acetate, pH 3.6–5.6; (4) 25 mM MES [2-(*N*-morpholino)ethane sulfonic acid], pH 6.1; and (5) 50 mM sodium bicarbonate. The concentration of the biomolecule should be calculated in order to have an excess of about 3–10 over the theoretical monolayer density for that protein on the surface, depending on the biomolecule size.

> **TIP:** *A protein titration study can be performed using a range of protein concentrations and a fixed incubation time.*

2. Soak the scaffold in the biomolecule solution and mix gently for 1 h at room temperature (unless the biomolecule is extremely sensitive to this temperature).

> **TIP:** *A range of incubation times and a fixed biomolecule concentration can be used to optimize Step 2.*

3. Remove the solution; wash the scaffold with coating solution.
4. Determine the amount of adsorbed proteins on the surface by bifunctional chelating agents (BCA Protein Assay; Thermo Fisher). For peptides, it is convenient to use fluorescent tags, as mentioned above, and calculate the amount adsorbed on the surface by depletion (*i.e.*, analyzing the coating solution before and after the incubation step). As alternative technique, FTIR in an attenuated total reflectance (ATR) configuration can be used, although it requires a strong background to use this technique for quantification purposes.

A.4 A Protocol to Immobilize Growth Factors and Small Peptides by Polydopamine Coating-based Physisorption

A.4.1 One-step Protocol

1. Soak the scaffold in a 2 mg mL^{-1} dopamine containing the biomolecules to be immobilized (for the optimal concentration, preliminary studies using a range of concentration should be considered) in 10 mM Tris-HCl (pH 8.5) solution.
2. Stir gently for 18 h at room temperature. The coating layer formation will turn the sample dark.
3. Wash with 10 mM Tris-HCl (pH 8.5) solution. To quantify the amount of biomolecules adhered on the surface, the use of fluorescent labels is the best choice as the polydopamine molecules can give false positive results by other techniques.

A.4.2 Two-step Protocol

1. Soak the scaffold in a 2 mg mL^{-1} dopamine in 10 mM Tris-HCl (pH 8.5) solution.
2. Stir gently for 18 h at room temperature. The coating layer formation will turn the sample dark.
3. Wash with 10 mM Tris-HCl (pH 8.5) solution.
4. Incubate the sample with the biomolecule solution (for the optimal concentration, preliminary studies using a range of concentration should be considered) for 12 h.

> **TIP:** *Step 4 has been proved to effective both with and without the use of cross-linking agents (EDC; Thermo Fisher) for coupling the biomolecules to the polydopamine layer. For experimental instructions on the use of EDC, refer to the manufacture's protocol.*

5. Wash with 10 mM Tris-HCl (pH 8.5) solution.

A.5 Protocol to Immobilize Growth Factors and Small Peptides by Silanization-based Physisorption

1. Prepare a 95% ethanol solution in water and adjust the pH to 4.5–5.5 with acetic acid.
2. Dissolve 3-aminopropyltriethoxysilane (APTS) at 5% vol in the above-described solution, allowing hydrolysis to occur for 5 min at room temperature.

> **TIP:** *In Step 2, APTS will introduce amine groups to the target surface; other silane molecules can be considered to introduce different functional groups (e.g., carboxyethylsilanetriol for carboxylic groups).*

3. Incubate the scaffolds with the silane solution for 2 h (shorter or longer times change the APTS content and its layer thickness).
4. Wash in ethanol or ethanol–water and cure at 110 °C for 30 min.
5. Incubate the silanized scaffolds in the biomolecule solution (for the optimal concentration, preliminary studies using a range of concentration should be considered) for 2 h at room temperature, under gentle stirring. As cross-linking agent, EDC is one of the most effective to covalently immobilize the biomolecule to the introduced functional groups of the surface. Cross-linking alternatives are glutaraldehyde and genipin.
6. Wash in PBS or water.
7. Determine the amount of adsorbed proteins on the surface by bifunctional chelating agents (BCA Protein Assay; Thermo Fisher). For peptides, it is convenient to use fluorescent tags, as mentioned above, and calculate the amount adsorbed on the surface by depletion (*i.e.*, analyzing the coating solution before and after the incubation step). As an alternative technique, FTIR in an ATR configuration can be used, although it requires a strong background to use this technique for quantification purposes.

References

1. E. Mohseni, E. Zalnezhad and A. R. Bushroa, Comparative investigation on the adhesion of hydroxyapatite coating on Ti–6Al–4V implant: A review paper, *Int. J. Adhes. Adhes.*, 2014, **48**, 238–257.

2. N. J. Lakhkar, *et al.*, Bone formation controlled by biologically relevant inorganic ions: Role and controlled delivery from phosphate-based glasses, *Adv. Drug Delivery Rev.*, 2013, **65**(4), 405–420.

3. A. Hoppe, N. S. Güldal and A. R. Boccaccini, A review of the biological response to ionic dissolution products from bioactive glasses and glass-ceramics, *Biomaterials*, 2011, **32**(11), 2757–2774.

4. A. Simchi, *et al.*, Recent progress in inorganic and composite coatings with bactericidal capability for orthopaedic applications, *Nanomedicine*, 2011, **7**(1), 22–39.

5. D. G. Castner and B. D. Ratner, Biomedical surface science: Foundations to frontiers, *Surf. Sci.*, 2002, **500**(1–3), 28–60.

6. S. Ebara and K. Nakayama, Mechanism for the action of bone morphogenetic proteins and regulation of their activity, *Spine*, 2002, **27**, S10–S15.

7. L. Y. Liu, *et al.*, Reduced foreign body reaction to implanted biomaterials by surface treatment with oriented osteopontin, *J. Biomater. Sci., Polym. Ed.*, 2008, **19**(6), 821–835.

8. G. McPhee, *et al.*, Can common adhesion molecules and microtopography affect cellular elasticity? A combined atomic force microscopy and optical study, *Med. Biol. Eng. Comput.*, 2010, **48**(10), 1043–1053.

9. A. Schindeler, *et al.*, Rapid cell culture and pre-clinical screening of a transforming growth factor-beta (TGF-beta) inhibitor for orthopaedics, *BMC Musculoskeletal Disord.*, 2010, **11**(1), 105.

10. W. Chrzanowski, *et al.*, Bone bonding ability-how to measure it?, *RSC Adv.*, 2012, **2**(24), 9214–9223.

11. A. V. Kondyurin, *et al.*, Drug release from polyureaurethane coating modified by plasma immersion ion implantation, *J. Biomater. Sci., Polym. Ed.*, 2004, **15**(2), 145–159.

12. K. Numata and D. L. Kaplan, Silk-based delivery systems of bioactive molecules, *Adv. Drug Delivery Rev.*, 2010, **62**(15), 1497–1508.

13. W. Chrzanowski, *et al.*, Biointerface: protein enhanced stem cells binding to implant surface, *J. Mater. Sci.: Mater. Med.*, 2012, **23**(9), 2203–2215.

14. E. A. Cavalcanti-Adam, *et al.*, Lateral spacing of integrin ligands influences cell spreading and focal adhesion assembly, *Eur. J. Cell Biol.*, 2006, **85**(3–4), 219–224.

15. E. A. Cavalcanti-Adam, *et al.*, Cell Spreading and Focal Adhesion Dynamics Are Regulated by Spacing of Integrin Ligands, *Biophys. J.*, 2007, **92**(8), 2964–2974.

16. E. Ruoslahti, RGD and other recognition sequences for integrins, *Annu. Rev. Cell Dev. Biol.*, 1996, **12**, 697–715.

17. T. Luhmann, *et al.*, Bone targeting for the treatment of osteoporosis, *J. Controlled Release*, 2012, **161**(2), 198–213.

18. J. Domingo-Espín, *et al.*, RGD-based cell ligands for cell-targeted drug delivery act as potent trophic factors, *Nanomedicine*, 2012, **8**(8), 1263–1266.

19. T. Zako, *et al.*, Cyclic RGD peptide-labeled upconversion nanophosphors for tumor cell-targeted imaging, *Biochem. Biophys. Res. Commun.*, 2009, **381**(1), 54–58.
20. F. Danhier, *et al.*, Targeting of tumor endothelium by RGD-grafted PLGA-nanoparticles loaded with Paclitaxel, *J. Controlled Release*, 2009, **140**(2), 166–173.
21. B. D. Ratner and S. J. Bryant, Biomaterials: Where we have been and where we are going, *Annu. Rev. Biomed. Eng.*, 2004, **6**, 41–75.
22. D. J. Gorth, *et al.*, Decreased bacteria activity on Si(3)N(4) surfaces compared with PEEK or titanium, *Int. J. Nanomed.*, 2012, 7, 4829–4840.
23. S. L. Bellis, Advantages of RGD peptides for directing cell association with biomaterials, *Biomaterials*, 2011, **32**(18), 4205–4210.
24. A. Meyer, *et al.*, Targeting RGD recognizing integrins: drug development, biomaterial research, tumor imaging and targeting, *Curr. Pharm. Des.*, 2006, **12**(22), 2723–2747.
25. Y. Takada, X. Ye and S. Simon, The integrins, *Genome Biol.*, 2007, **8**(5), 215.
26. A. Rezania and K. E. Healy, Integrin subunits responsible for adhesion of human osteoblast-like cells to biomimetic peptide surfaces, *J. Orthop. Res.*, 1999, **17**(4), 615–623.
27. S. Rammelt, *et al.*, Coating of titanium implants with type-I collagen, *J. Orthop. Res.*, 2004, **22**(5), 1025–1034.
28. M. Horton, Vitronectin receptor: tissue specific expression or adaptation to culture?, *Int. J. Exp. Pathol.*, 1990, **71**(5), 741–759.
29. M. J. P. Biggs, *et al.*, Regulation of implant surface cell adhesion: Characterization and quantification of S-phase primary osteoblast adhesions on biomimetic nanoscale substrates, *J. Orthop. Res.*, 2007, **25**(2), 273–282.
30. D. Pan, J. L. Turner and K. L. Wooley, Folic acid-conjugated nanostructured materials designed for cancer cell targeting, *Chem. Commun.*, 2003, 2400–2401.
31. S. Kim, *et al.*, Engineered polymers for advanced drug delivery, *Eur. J. Pharm. Biopharm.*, 2009, **71**(3), 420–430.
32. Y. Teow and S. Valiyaveettil, Active targeting of cancer cells using folic acid-conjugated platinum nanoparticles, *Nanoscale*, 2010, **2**(12), 2607–2613.
33. S. Fuchs and C. Coester, Protein-based nanoparticles as a drug delivery system: chances, risks, perspectives, *J. Drug Delivery Sci. Technol.*, 2010, **20**(5), 331–342.
34. T. Matsuura, *et al.*, Diverse mechanisms of osteoblast spreading on hydroxyapatite and titanium, *Biomaterials*, 2000, **21**(11), 1121–1127.
35. B. Elmengaard, J. E. Bechtold and K. Soballe, In vivo study of the effect of RGD treatment on bone ongrowth on press-fit titanium alloy implants, *Biomaterials*, 2005, **26**(17), 3521–3526.
36. M. C. Porté-Durrieu, *et al.*, Cyclo-(DfKRG) peptide grafting onto Ti–6Al–4V: physical characterization and interest towards human osteoprogenitor cells adhesion, *Biomaterials*, 2004, **25**(19), 4837–4846.

37. A. Kondyurin, N. J. Nosworthy and M. M. M. Bilek, Attachment of horseradish peroxidase to polytetrafluorethylene (teflon) after plasma immersion ion implantation, *Acta Biomater.*, 2008, **4**(5), 1218–1225.
38. M. M. M. Bilek, *et al.*, Free radical functionalization of surfaces to prevent adverse responses to biomedical devices, *Proc. Natl. Acad. Sci. U. S. A.*, 2011, **108**(35), 14405–14410.
39. J. Zhu, *et al.*, Design and synthesis of biomimetic hydrogel scaffolds with controlled organization of cyclic RGD peptides, *Bioconjugate Chem.*, 2009, **20**(2), 333–339.
40. S. X. Hsiong, *et al.*, Cyclic arginine-glycine-aspartate peptides enhance three-dimensional stem cell osteogenic differentiation, *Tissue Eng., Part A*, 2009, **15**(2), 263–272.
41. G. B. Schneider, R. Zaharias and C. Stanford, Osteoblast integrin adhesion and signaling regulate mineralization, *J. Dent. Res.*, 2001, **80**(6), 1540–1544.
42. S. J. Bogdanowich-Knipp, *et al.*, Solution stability of linear vs. cyclic RGD peptides, *J. Pept. Res.*, 1999, **53**(5), 530–541.
43. M. A. Dechantsreiter, *et al.*, N-Methylated cyclic RGD peptides as highly active and selective alpha(V)beta(3) integrin antagonists, *J. Med. Chem.*, 1999, **42**(16), 3033–3040.
44. A. Howe, *et al.*, Integrin signaling and cell growth control, *Curr. Opin. Cell Biol.*, 1998, **10**(2), 220–231.
45. K. Soballe, *et al.*, Hydroxyapatite coating enhances fixation of porous coated implants. A comparison in dogs between press fit and non-interference fit, *Acta Orthop. Scand.*, 1990, **61**(4), 299–306.
46. G. C. O'Toole, *et al.*, Bone sialoprotein-coated femoral implants are osteoinductive but mechanically compromised, *J. Orthop. Res.*, 2004, **22**(3), 641–646.
47. M. J. Dalby, *et al.*, The control of human mesenchymal cell differentiation using nanoscale symmetry and disorder, *Nat. Mater.*, 2007, **6**(12), 997–1003.
48. A. J. Engler, *et al.*, Matrix elasticity directs stem cell lineage specification, *Cell*, 2006, **126**(4), 677–689.
49. A. K. Gaharwar, *et al.*, Bioactive silicate nanoplatelets for osteogenic differentiation of human mesenchymal stem cells, *Adv. Mater.*, 2013, **25**(24), 3329–3336.
50. A. Higuchi, *et al.*, Physical cues of biomaterials guide stem cell differentiation fate, *Chem. Rev.*, 2013, **113**(5), 3297–3328.
51. K. A. Kilian and M. Mrksich, Directing stem cell fate by controlling the affinity and density of ligand-receptor interactions at the biomaterials interface, *Angew. Chem., Int. Ed.*, 2012, **51**(20), 4891–4895.
52. M. Lanniel, *et al.*, Substrate induced differentiation of human mesenchymal stem cells on hydrogels with modified surface chemistry and controlled modulus, *Soft Matter*, 2011, 7(14), 6501.

53. V. L. Lapointe, *et al.*, Nanoscale Topography and Chemistry Affect Embryonic Stem Cell Self-Renewal and Early Differentiation, *Adv. Healthcare Mater.*, 2013, **2**(12), 1644–1650.

54. R. McBeath, *et al.*, Cell Shape, Cytoskeletal Tension and RhoA Regulate Stem Cell Lineage Commitment, *Dev. Cell*, 2004, **6**(4), 483–495.

55. A. Polini, *et al.*, Osteoinduction of human mesenchymal stem cells by bioactive composite scaffolds without supplemental osteogenic growth factors, *PLoS One*, 2011, **6**(10), e26211.

56. J. Wang, *et al.*, Virus activated artificial ECM induces the osteoblastic differentiation of mesenchymal stem cells without osteogenic supplements, *Sci. Rep.*, 2013, **3**, 1242.

57. S. Bose, M. Roy and A. Bandyopadhyay, Recent advances in bone tissue engineering scaffolds, *Trends Biotechnol.*, 2012, **30**(10), 546–554.

58. M. Pollak, The insulin and insulin-like growth factor receptor family in neoplasia: an update, *Nat. Rev. Cancer*, 2012, **12**(3), 159–169.

59. J. T. Buijs, K. R. Stayrook and T. A. Guise, The role of TGF-beta in bone metastasis: novel therapeutic perspectives, *Bonekey Rep.*, 2012, **1**, 96.

60. A. Saito, *et al.*, Activation of osteo-progenitor cells by a novel synthetic peptide derived from the bone morphogenetic protein-2 knuckle epitope, *Biochim. Biophys. Acta, Proteins Proteomics*, 2003, **1651**(1–2), 60–67.

61. L. D. D'Andrea, *et al.*, Targeting angiogenesis: structural characterization and biological properties of a de novo engineered VEGF mimicking peptide, *Proc. Natl. Acad. Sci. U. S. A.*, 2005, **102**(40), 14215–14220.

62. K. Aoki, *et al.*, Peptide-based delivery to bone, *Adv. Drug Delivery Rev.*, 2012, **64**(12), 1220–1238.

63. H. Bae, *et al.*, Building vascular networks, *Sci. Transl. Med.*, 2012, **4**(160), 160ps23.

64. J. Malda, *et al.*, Effect of oxygen tension on adult articular chondrocytes in microcarrier bioreactor culture, *Tissue Eng.*, 2004, **10**(7–8), 987–994.

65. J. S. Miller, *et al.*, Rapid casting of patterned vascular networks for perfusable engineered three-dimensional tissues, *Nat. Mater.*, 2012, **11**(9), 768–774.

66. M. P. Cuchiara, *et al.*, Integration of Self-Assembled Microvascular Networks with Microfabricated PEG-Based Hydrogels, *Adv. Funct. Mater.*, 2012, **22**(21), 4511–4518.

67. J. D. Baranski, *et al.*, Geometric control of vascular networks to enhance engineered tissue integration and function, *Proc. Natl. Acad. Sci. U. S. A.*, 2013, **110**(19), 7586–7591.

68. D. H. Nguyen, *et al.*, Biomimetic model to reconstitute angiogenic sprouting morphogenesis in vitro, *Proc. Natl. Acad. Sci. U. S. A.*, 2013, **110**(17), 6712–6717.

69. Y. Zheng, *et al.*, In vitro microvessels for the study of angiogenesis and thrombosis, *Proc. Natl. Acad. Sci. U. S. A.*, 2012, **109**(24), 9342–9347.

70. J. Rouwkema, N. C. Rivron and C. A. van Blitterswijk, Vascularization in tissue engineering, *Trends Biotechnol.*, 2008, **26**(8), 434–441.

71. Y. C. Chen, *et al.*, Functional Human Vascular Network Generated in Photocrosslinkable Gelatin Methacrylate Hydrogels, *Adv. Funct. Mater.*, 2012, **22**(10), 2027–2039.

72. J. E. Leslie-Barbick, *et al.*, The promotion of microvasculature formation in poly(ethylene glycol) diacrylate hydrogels by an immobilized VEGF-mimetic peptide, *Biomaterials*, 2011, **32**(25), 5782–5789.

73. L. L. Chiu, *et al.*, Perfusable branching microvessel bed for vascularization of engineered tissues, *Proc. Natl. Acad. Sci. U. S. A.*, 2012, **109**(50), E3414–E3423.

74. N. Smart, *et al.*, Thymosin beta4 induces adult epicardial progenitor mobilization and neovascularization, *Nature*, 2007, **445**(7124), 177–182.

75. S. Kim, *et al.*, Engineering of functional, perfusable 3D microvascular networks on a chip, *Lab Chip*, 2013, **13**(8), 1489–1500.

76. G. T. Hermanson, *Bioconjugate Techniques*, Academic Press, New York, 2nd edn, 2008.

77. H. Lee, *et al.*, Mussel-inspired surface chemistry for multifunctional coatings, *Science*, 2007, **318**(5849), 426–430.

78. Y. Sun, *et al.*, Peptide decorated nano-hydroxyapatite with enhanced bioactivity and osteogenic differentiation via polydopamine coating, *Colloids Surf., B*, 2013, **111**, 107–116.

79. S. M. Kang, *et al.*, One-Step Multipurpose Surface Functionalization by Adhesive Catecholamine, *Adv. Funct. Mater.*, 2012, **22**(14), 2949–2955.

80. R. Fujisawa, *et al.*, Attachment of osteoblastic cells to hydroxyapatite crystals by a synthetic peptide (Glu7-Pro-Arg-Gly-Asp-Thr) containing two functional sequences of bone sialoprotein, *Matrix Biol.*, 1997, **16**(1), 21–28.

81. D. Itoh, *et al.*, Enhancement of osteogenesis on hydroxyapatite surface coated with synthetic peptide (EEEEEEEPRGDT) in vitro, *J. Biomed. Mater. Res.*, 2002, **62**(2), 292–298.

82. J. S. Lee, *et al.*, Modular peptide growth factors for substrate-mediated stem cell differentiation, *Angew. Chem., Int. Ed.*, 2009, **48**(34), 6266–6269.

83. M. Gilbert, *et al.*, Chimeric peptides of statherin and osteopontin that bind hydroxyapatite and mediate cell adhesion, *J. Biol. Chem.*, 2000, **275**(21), 16213–16218.

84. M. D. Roy, *et al.*, Identification of a Highly Specific Hydroxyapatite-binding Peptide using Phage Display, *Adv. Mater.*, 2008, **20**(10), 1830–1836.

85. S. J. Segvich, H. C. Smith and D. H. Kohn, The adsorption of preferential binding peptides to apatite-based materials, *Biomaterials*, 2009, **30**(7), 1287–1298.

86. A. Polini, *et al.*, Stable biofunctionalization of hydroxyapatite (HA) surfaces by HA-binding/osteogenic modular peptides for inducing osteogenic differentiation of mesenchymal stem cells, *Biomaterials Science*, 2014, DOI: 10.1039/C4BM00164H.

87. Y. Lu, *et al.*, Coating with a modular bone morphogenetic peptide promotes healing of a bone-implant gap in an ovine model, *PLoS One*, 2012, 7(11), e50378.

88. D. Suarez-Gonzalez, *et al.*, Mineral coatings modulate beta-TCP stability and enable growth factor binding and release, *Acta Biomater.*, 2012, 8(3), 1117–1124.

89. J. S. Lee, J. S. Lee and W. L. Murphy, Modular peptides promote human mesenchymal stem cell differentiation on biomaterial surfaces, *Acta Biomater.*, 2010, 6(1), 21–28.

90. S. H. Brounts, *et al.*, High affinity binding of an engineered, modular peptide to bone tissue, *Mol. Pharmaceutics*, 2013, 10(5), 2086–2090.

91. B. K. Culpepper, *et al.*, Polyglutamate directed coupling of bioactive peptides for the delivery of osteoinductive signals on allograft bone, *Biomaterials*, 2013, 34(5), 1506–1513.

92. P. B. van Wachem, *et al.*, Biocompatibility and tissue regenerating capacity of crosslinked dermal sheep collagen, *J. Biomed. Mater. Res.*, 1994, 28(3), 353–363.

93. A. Polini, *et al.*, Collagen-functionalised electrospun polymer fibers for bioengineering applications, *Soft Matter*, 2010, 6(8), 1668.

94. A. Polini, H. Bai and A. P. Tomsia, Dental applications of nano-structured bioactive glass and its composites, *Wiley Interdiscip. Rev.: Nanomed. Nanobiotechnol.*, 2013, 5(4), 399–410.

95. J. Kalia and R. Raines, Advances in Bioconjugation, *Curr. Org. Chem.*, 2010, 14(2), 138–147.

96. A. Kondyurin and M. Bilek, Biological and medical applications, in *Ion Beam Treatment of Polymers*, Elsevier, Amsterdam, 2008, pp. 205–241.

97. E. A. Kosobrodova, *et al.*, Free radical kinetics in a plasma immersion ion implanted polystyrene: theory and experiment, *Nucl. Instrum. Methods Phys. Res., Sect. B*, 2012, 280, 26–35.

98. A. Kondyurin, *et al.*, Mechanisms for surface energy changes observed in plasma immersion ion implanted polyethylene: the roles of free radicals and oxygen-containing groups, *Polym. Degrad. Stab.*, 2009, 94(4), 638–646.

99. A. Kondyurin, N. J. Nosworthy and M. M. Bilek, Effect of low molecular weight additives on immobilization strength, activity and conformation of protein immobilized on PVC and UHMWPE, *Langmuir*, 2011, 27(10), 6138–6148.

100. Y. C. Huang and T. J. Liu, Mobilization of mesenchymal stem cells by stromal cell-derived factor-1 released from chitosan/tripolyphosphate/fucoidan nanoparticles, *Acta Biomater.*, 2012, 8(3), 1048–1056.

101. N. R. Johnson and Y. Wang, Controlled delivery of sonic hedgehog morphogen and its potential for cardiac repair, *PLoS One*, 2013, 8(5), e63075.

102. O. V. Stepanenko, *et al.*, Fluorescent proteins as biomarkers and bio-sensors: throwing color lights on molecular and cellular processes, *Curr. Protein Pept. Sci.*, 2008, 9(4), 338–369.

CHAPTER 2

Additive Manufacturing and Surface Modification of Biomaterials using Self-assembled Monolayers

JAYASHEELAN VAITHILINGAM,[*a] RUTH D. GOODRIDGE,[a] STEVEN D. R. CHRISTIE,[b] STEVE EDMONDSON[c] AND RICHARD J. M. HAGUE[a]

[a] Department of Mechanical, Materials and Manufacturing Engineering, The University of Nottingham, Nottingham, NG7 2RD, UK; [b] Department of Chemistry, Loughborough University, Loughborough, LE11 3TU, UK; [c] School of Materials, The University of Manchester, Manchester, M13 9PL, UK
*Email: epxjv1@nottingham.ac.uk

2.1 Introduction

Biomaterials are natural or synthetic materials (other than drugs) that are used to evaluate, treat, augment or replace any damaged tissues/organs/function of the body.[1] The field of biomaterials is of major importance to mankind since their use has played a significant role in improving life expectancy and quality of life.[2] The use of biomaterials is not new; Egyptians and Romans used linen for sutures, gold and iron for dental applications and wood for toe replacements.[3] After the Second World War, nylon, teflon, silicone, stainless steel and titanium were some of the other materials

RSC Smart Materials No. 10
Biointerfaces: Where Material Meets Biology
Edited by Dietmar Hutmacher and Wojciech Chrzanowski
© The Royal Society of Chemistry 2015
Published by the Royal Society of Chemistry, www.rsc.org

introduced for biomedical applications. With recent advancements in human healthcare systems and with the availability of improved diagnostic tools, biomedical implants have found applications in almost all body functions. It is expected that virtually every individual will have contact with biomaterials and/or biomedical implants at some point of time during his or her life. This contact may occur in several ways including (1) permanent implantation (such as heart valves and total joint replacement); (2) long-term applications (such as contact lenses, dental prostheses and fracture fixation devices); and (3) transient applications (including needles for vaccination, wound healing dressings/sutures and cardiac assist systems).

Biomedical implants range from simple wires and screws for fracture plate fixation to joint prostheses for hips, knees, shoulders and so on. Depending on the requirement and load conditions, implants are made of natural and/ or synthetic materials such as polymers, metals, ceramics or composites. For example, metals are generally preferred for load-bearing implants and internal fixation devices whereas ceramics are preferred for skeletal and hard tissue repair.[4] Polymers are preferred in applications where elasticity of the material would be an advantage.

Biomedical implants have significantly improved the quality of life for countless people; however, they still impose significant challenges on current manufacturing processes due to their need to function within a relatively harsh biological environment.[2] Customisation, fabrication of complex geometries and the reduction of post-implant complications including implant migration, fracture and poor biocompatibility are major challenges to overcome.

2.1.1 Issues with Current Biomedical Implants

2.1.1.1 Customisation and Fabrication of Complex Geometries

Most off-the-shelf implants are available in a selection of dimensions to meet the patient requirement; however, they are not customised. Customisation is key in conditions such as where the patient's anatomy deviates from the standard sizes or is affected by individual defects. The primary objective of customisation is to fit the unique anatomy of a particular patient, especially where the implant is not fixed. Personalised hearing aid shells for 'in-the-ear' and 'in-the-canal' devices are one example of a customised implant that has achieved superior fit compared to conventionally manufactured aids, increasing both comfort and device functionality.[5] Orthopaedic implants usually perform better when they match exactly the anatomy of the patient, through the distribution and normalisation of the stresses incurred in the remaining skeletal system and reduction of stress shielding, migration and failure.[6]

Migration of an implant is the movement of the implant from the actual or surgically positioned area and is a mechanically triggered rather than a biological process.[7] Forces higher than expected may deteriorate fixation and cause implant migration. Migration of an implant to a more harmful anatomical space can sometimes be life threatening. Implant migration has been reported for a number of devices including total hip and knee replacement, dental, maxillo-facial and vascular implants.[8–12] Some of the factors that determine the rate of migration other than mechanical forces are implant design and fixation methods.[7] Cementing, screw retention of the implant, expanding the implant to conform to the walls and press fitting of the implant to the diseased/damaged area are some of the common fixation methods. The advantages, complications and limitations of these methods are reported extensively in literature.[13–16]

Fixation of implants may sometimes be difficult depending on the site and anatomy of the diseased/damaged part. In such cases, customisation may render improved fixation with the patient's anatomy and may normalise stresses, thus preventing implant migration. For example, customised hip implants can be more durable, in particular for younger patients, by improving the stability of the femoral stem, allowing load transfer to the proximal part of the femur and restoration of normal geometry of the hip joint.[6] However, the ability to achieve customised and complex geometries of biomedical implants using traditional manufacturing methods (such as moulding, die casting and subtractive processes) are limited and are often time consuming.[17] This is because traditional methods have limitations in the geometries that they can produce. Moulding and die casting methods require dies or moulds to fabricate the part. Hence, extra time for the manufacturing process is required in addition to fabricating the actual part. In addition, it is expensive to make a die or mould for a one-off design.

2.1.1.2   *Post-implant Complications*

There are many reasons for the failure of an implant within the biological environment including manufacturing, chemical, mechanical, tribological and surgical failures.[3] The patient's health condition and the physician's experience could also be contributing factors. A variety of biomaterials (polymers, metals, ceramics and composites), fabrication techniques (such as compression moulding, die casting, bar stock milling and laser cutting) and surface modification techniques (biocompatible material coating, surface polishing, drug loading *etc.*) have been employed to improve the mechanical and biological properties of the implant; however, there are still constraints in achieving this.

Even when using a biomaterial with the most superior material and biological properties, there are possibilities for the failure of an implant due to faulty mechanical design or inappropriate application of the implant. Inadequate mechanical properties (*e.g.*, elastic modulus, yield strength, tensile strength) can result in fracture leading to implant failure.[8,18,19] Fracture of an implant mainly occurs when the implant is not capable of bearing the load exerted on it. Optimisation of the implant designs to bear specific loads has been researched widely in recent years.[20,21] Optimisation methods allow the prediction of areas of fracture/failure of implants at specific loading conditions. Thus, implants can be designed to withstand the required load conditions with less material usage.

Although the mechanical properties of the bulk material, implant design and manufacturing process are considered to be the major factors determining implant fracture, corrosion of implants within the biological environment is also a crucial factor.[3] Corrosion of implants is one of the most challenging problems after implantation of dental, cardiovascular, maxillofacial and load-bearing implants such as hip and knee replacements.[3] Implants corrode within the biological environment due to electrochemical attack by the electrolytes present in the hostile environment. Leaching of metal ions from the implant surface due to corrosion will lead to erosion. The rate of corrosion is accelerated by increased surface area and loss of protective oxide, later leading to fractures. Also, leaching of metal ions has been observed to induce acute and chronic effects. Release of nickel was observed to affect the skin; cobalt was observed to cause anaemia which inhibits the absorption of iron into the bloodstream; chromium was observed to cause ulcers and disturb the central nervous system; aluminium was reported to cause epileptic effects and Alzheimer's disease and leaching of vanadium in its elemental state was observed to be toxic.[22] Hence, corrosion products formed as a result of implant–host interaction have a significant effect on the long-term stability of the prosthesis and its biocompatibility.[23]

Biocompatibility, a requirement for all biomedical implants, is the property of a material to be compatible with the tissues in the human body by not producing a toxic, injurious or immunological response. Chronic inflammations, lesions and scarring are some of the adverse reactions that may take place in the site if the implant is not biocompatible. To achieve better biocompatibility, two routes are often employed: (1) using biocompatible materials; and (2) fabrication of implants using off-the-shelf materials with the application of a suitable biocompatible layer.[6]

Contamination of implants by harmful micro-organisms cause severe infections and often forces replacement of the medical device. By entering the host through contaminated surfaces, bacteria multiply in the host environment and interfere with the host defence system leading to host tissue damage and inflammation.[24] Staphylococci species, including *Staphylococcus aureus* and *Staphylococcus epidermidis* are the more common cause of implant-associated infections.[25] Contamination of implants such as stents, dental, hip and knee implants by harmful microbes has been widely reported.[26–29]

2.1.2 Scope of this Chapter

The key to addressing these challenges is to optimise the implant–host interface. It is envisaged that customised implant designs, due to their superior fit compared to traditional designs, can reduce post-implant complications by preventing implant migration. In addition, by modifying the surface chemistry of the implant with drug/protein molecules specific to the target site, biochemical interactions at the implant–host interface can be optimised and post-implant complications can be reduced. This chapter will discuss a novel methodology to produce customised and functionalised biomedical implants by integrating a metal-based additive manufacturing (AM) process to fabricate customised biomedical implants with surface functionalisation using self-assembled monolayers.

2.2   Additive Manufacturing

Additive manufacturing (AM) has developed from the early days of rapid prototyping to enable the production of end-use parts. According to the American Society for Testing and Materials (ASTM), AM is defined as[30]: 'The process of joining materials to make objects from 3D model data, usually layer upon layer, as opposed to subtractive manufacturing technologies'.

Although there are many terms used to describe AM, the most commonly used terms are free-form manufacturing, rapid manufacturing, additive fabrication, additive layered manufacturing and 3D printing.

AM was first developed for polymeric materials and now all types of materials including metals, ceramics and composites can be processed using these techniques to varying levels of success.[31] Currently there are many different AM processes available, most working on a layer-by-layer basis. However, they differ through the form of the starting material and the mechanism used to consolidate it. Some of the consolidation methods used include the use of a laser or electron beam to selectively fuse polymer or metal powdered material; curing of liquid resin with ultraviolet (UV) light; extrusion of molten polymers from traversing nozzles; jetting of droplets from an array of nozzles, similar to inkjet printing; and consolidation of sheets of materials using ultrasonic vibrations. Table 2.1 lists and defines the categories of AM processes in accordance with the ASTM International Committee F42 on Additive Manufacturing Technologies published in 2010.[30]

AM has gained considerable interest in recent years due to its ability to build parts of complex geometries from three-dimensional (3D) model data.[32] AM is capable of building parts of almost any geometry in a single step regardless of the complexity of the part whereas conventional methods often need further steps and time or it is impossible to do so.[33] Furthermore, the use of AM speeds up the whole development process since there is no need for dies/moulds and toolings. Hence, the use of AM to fabricate customised one-off parts is faster and cheaper than traditional methods. Some of the active industries for AM include aerospace, automotive, medical, architectural, games, military, art, sport, construction and education;

Table 2.1 Definitions of AM process categories in accordance with the ASTM International Committee F42 on Additive Manufacturing Technologies (2010).[30]

Category	ASTM definition
Vat photopolymerisation	An additive manufacturing process in which liquid photopolymer in a vat is selectively cured by light-activated polymerisation
Material jetting	An additive manufacturing process in which droplets of build material are selectively deposited
Binder jetting	An additive manufacturing process in which a liquid bonding agent is selectively deposited to join powder materials
Material extrusion	An additive manufacturing process in which material is selectively dispensed through a nozzle or orifice
Powder bed fusion	An additive manufacturing process in which thermal energy selectively fuses regions of a powder bed
Sheet lamination	An additive manufacturing process in which sheets of material are bonded to form an object
Direct energy deposition	An additive manufacturing process in which focused thermal energy is used to fuse materials by melting as they are being deposited

however, there has been particular interest for AM in aerospace and biomedical industries owing to the possibility for high performance parts.[32,34]

For biomedical applications, AM offers a closed process chain. For example, bone replacement materials can be fabricated by scanning the bone defect using magnetic resonance imaging (MRI) or computer tomography (CT), designing the implant structure, then directly manufacturing the individual implant. This closed process chain will have the potential to offer custom-fitting implants to the damaged part's anatomy.[35] Also the freedom of design offered by this technology will enable the fabrication of complex geometries of the implants, in terms of both external and internal morphology.

2.2.1 Selective Laser Melting

Selective laser melting (SLM) is an AM technique capable of fabricating metallic parts. Being an AM technique, increased geometrical freedom allows the fabrication of a part as envisioned without the manufacturing constraints prevailed with conventional techniques such as CNC machining, moulding and casting. Gold, stainless steel (316L and 17-4PH), tool steel, commercially pure titanium and its alloys (Ti6Al4V, Ti6Al7Nb and Ti24Nb4Zr8Sn), aluminium AlSi12, cobalt-chrome (ASTM) and Inconel 718 and 625 are some of the powdered materials that can be processed currently using SLM.[36–42] SLM has been identified as a route to fabricate parts with a high material utilisation rate without using expensive moulds/forgings.[43]

2.2.1.1   Generation of a CAD File for Selective Laser Melting

In SLM, parts are built directly from 3D model data. For biomedical applications, the CAD file of the implant to be fabricated is normally generated either by designing the part using CAD software packages or by reconstructing the structure of the damaged part from CT or MRI data. A set of CT/MRI data can be combined using commercially available software such as Mimics (Materialise NV) to convert the two-dimensional (2D) DICOM files to a 3D surface tessellated (STL) file. In some cases it can also be reverse engineered by laser scanning the damaged part.

SLM is capable of building net-shape parts with complex geometries; however, there might be some areas of the part that overhang and may need a support material during building. In such cases, supports should be generated in a way that they are easy to remove without damaging the part's quality during post-processing. For example, SLM 250 (Renishaw PLC) machines use the same material for supports as they do for fabricating the part. However, they use a lower laser power for the supports than they use to fabricate the actual part so that the supports can be easily removed. Also it is important to position supports in a way that they can be removed easily and they cover all major overhangs. Once the part is designed and supports are generated (where required), the CAD file is then exported in STL format. The machine interface software is used to slice the STL file into several layers according to the predefined layer thickness along the Z direction.

2.2.1.2   Equipment and Principle of Selective Laser Melting

A schematic of the SLM process is shown in Figure 2.1. A typical SLM machine consists of a scanning laser (usually a fibre optic laser with wavelength, $\lambda = 1070$ nm) a hopper attached to a wiper (re-coater), an elevator that lowers a build platform to adjust the layer thickness, and a lens that focuses the laser to the build area. Before starting the build, the chamber is made to be inert by pumping in an inert gas such as argon. The powder from which the part is to be fabricated is spread over the build platform from the hopper to a pre-defined layer thickness. Depending on the surface quality and fabrication speed requirements, this layer thickness can be fixed to be between 20 and 100 μm.

Shortly after a layer of powder has been spread, the laser beam scans the powder in the areas specified by the layer of the model file and fuses them. Once the scan is complete, the build platform moves downwards by a pre-defined layer thickness for a new layer of powder to be spread over the previously scanned layer and this process continues until the part is completed. Once fabrication is complete, the build platform is raised and the part is removed from the substrate. The un-fused material on the build area can be sieved and recycled.[31] The part is normally sonicated with deionised water for 30 min to remove non-sintered particles.

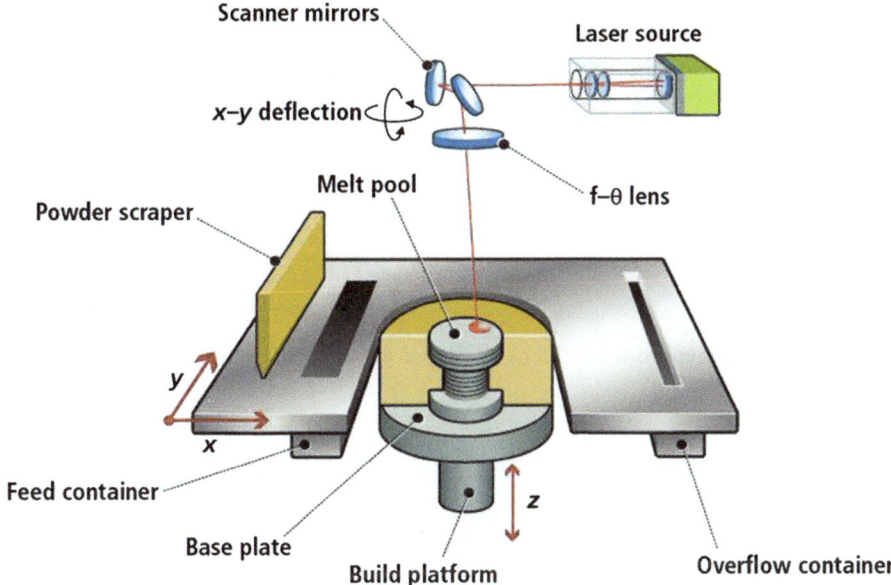

Scanner mirrors

Laser source

x–y deflection

Melt pool

Powder scraper

f–θ lens

y

x

Feed container

Base plate

z

Build platform

Overflow container

Figure 2.1 Schematic of the SLM process (image courtesy of Vision Systems (www.vision-systems.com)).

2.2.1.3 Fabrication of Medical Devices using Selective Laser Melting

The design freedom, potential for customisation, and direct fabrication of parts from 3D models offered by SLM provides great potential for the fabrication of metallic implants for biomedical applications. Research into the use of SLM to fabricate medical devices has considerably increased over the last decade. The Centre for Applied Reconstructive Technologies in Surgery (CARTIS, www.cartis.org) demonstrated the feasibility of the design and fabrication of a cranioplasty plate and the successful implant to restore the orbital floor and rim produced in titanium by SLM.[44] Wehmoller *et al.*[40] showed the ability of SLM to fabricate a lower jaw model (including teeth, the mandibular joint and the canal of the mandibular nerve), a spinal column model and a tubular bone femur. It was reported in their study that, to fabricate these implants in stainless steel, the production of the lower jaw required 24 h, the spinal column required 25 h and the bone femur took 26 h. Successful clinical use of a custom-made root analogue for a dental implant fabricated by a SLM process has also been reported in literature.[45] Currently, LayerWise, Renishaw PLC and 3T RPD are some of the companies actively involved in the production of biomedical implants for cranial, maxillo-facial, orthopaedic and dental applications.

Figure 2.2 shows different tracheobronchial stents with a strut thickness of 300 µm designed and fabricated using the SLM process. Figures 2.3 and

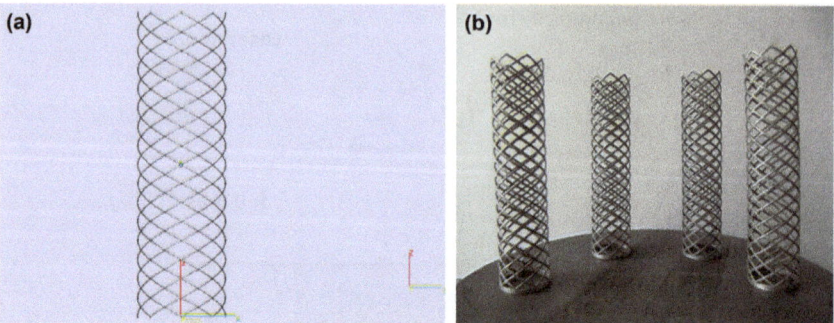

Figure 2.2   Tracheobronchial stent designed (a) and fabricated (b) using the SLM process.

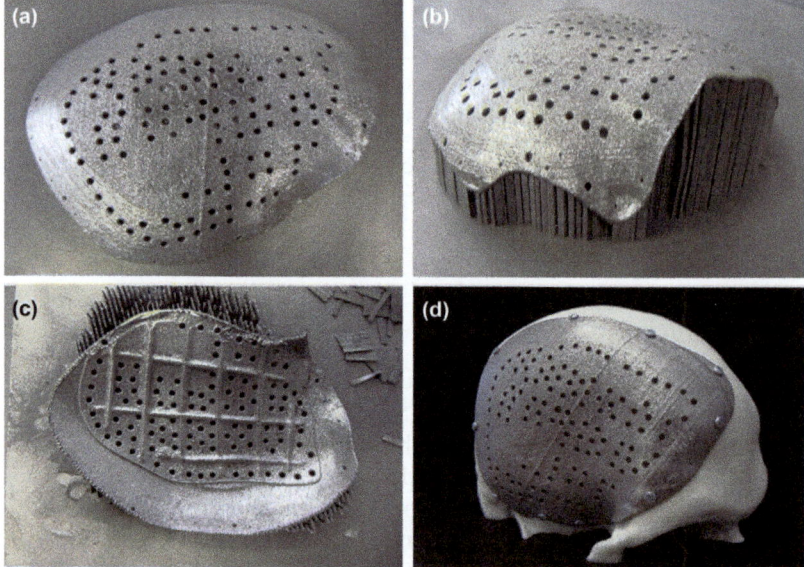

Figure 2.3   Customised cranioplasty implant fabricated using the SLM process. SLM fabricated implant (a); implant with support structures (b); implant removed from support structure showing inner geometry (c) and custom-fitting of the implant to the skull (d).
(Implant design and 3D model courtesy of CARTIS.)

2.4 show the demonstration cranioplasty implant and the implanted maxillofacial implant that were fabricated using the SLM process. The typical surface morphology of an implant fabricated using SLM 250 (Renishaw PLC) is shown in Figure 2.5. As can be observed from the figure, the as-fabricated surface is rough and porous due to partial sintering of particles. Several researchers have suggested that porosity on an implant surface could facilitate bone ingrowth and remodelling throughout the implant and

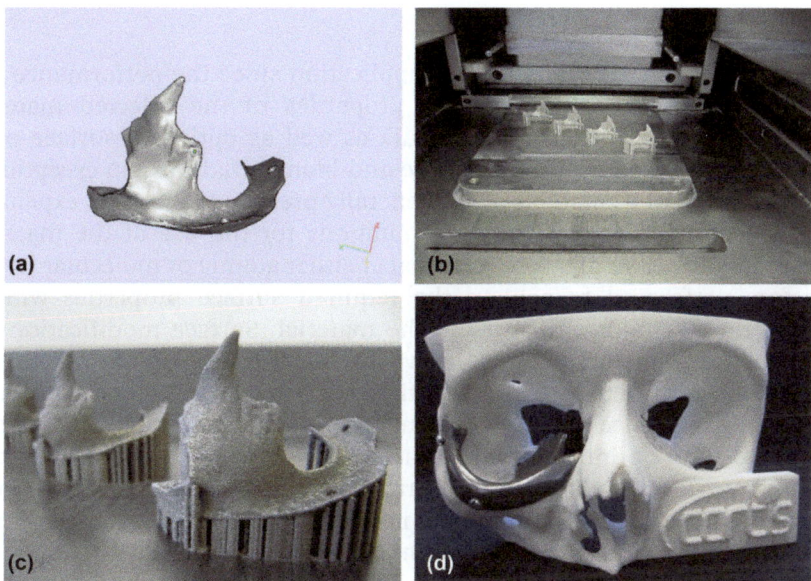

Figure 2.4 Customised maxillofacial implant fabricated using SLM. CAD design of a customised maxillofacial implant (a); implants fabricated in a SLM machine (b); implant with its support structure (c); custom fitting of implant to a replica model of the patient anatomy part (d). (Implant design and 3D model courtesy of CARTIS.)

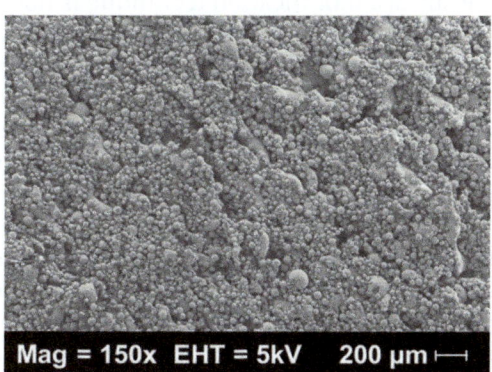

Figure 2.5 Typical surface morphology of a part fabricated using a Renishaw SLM 250 machine.

eventually improve bone-material contact.[46] Fabrication of porous scaffolds for bone tissue regeneration using SLM has been reported extensively in literature.[46–49] Although the SLM-produced surfaces will be highly advantageous in this context, care should be taken that the partially sintered particles do not detach from the surface since this can potentially affect the biocompatibility of the implant.

2.3 Surface Modification

Material selection is crucial for any application since the performance of a manufactured part will reflect the properties of the selected material. Finding a material with optimum bulk as well as optimum surface properties for any given application is rare and biomaterials are no exception.[50] Bulk alteration of a material is not generally preferred since it is expensive, time consuming and will impose limitations for the use of the material. Surface modification of a solid material at either atomic or molecular level is therefore performed to achieve the required surface properties without altering the key bulk properties of the material. Surface modification has wide applications including adhesive bonding enhancement, modifying wetting and non-wetting characteristics, creating micro-porous structures, altering the grain size and distribution, generating uniform 3D structures on the surface and optimisation of surface chemistries of an engineered part. Although surface modification is carried out in a number of engineering disciplines, its application in the biomedical field is unique since it directly affects the quality of life of the patient.

2.3.1 Issues with Current Surface Modification Techniques

Morphological, physico-chemical and biological modifications are the common types of surface modifications generally performed to improve the quality and properties of implants. However, they all have limitations. In most cases, a single surface modification technique is not enough to achieve the desired surface property and hence two or more surface modifications are performed to achieve better biological properties. For example, Kim *et al.* reported good adherence and spreading of human osteoblast cells for sand-blasted and acid-etched surfaces.[51] Though the use of two techniques provided improved results in terms of cellular attachments, the use of two or more techniques to modify the surface can be time consuming and expensive.

In recent years, surface modifications have been performed not just to improve the surface finish but also to make the implant biocompatible and a drug carrier. Bacterial infections due to implants are one of the main risk factors in orthopaedic surgery leading to surgical removal of implants.[8,52] Treatment measures available for bacterial infections include systemic, antibiotic-loaded cements and antimicrobial coatings.[52] However, the release of drugs in a systemic therapy prevails as one of the major problems when conventional coating techniques are used. Slower release of antibiotics for a longer period of time is a potential cause of antibiotic resistance in the human body.[53] For example, in drug eluting stents (DES), to reduce the restenosis rate, the stents are coated with polymers containing a drug using simple techniques including dip/spray/spin coating or solvent casting.[54,55] After implantation, the DES is expanded to conform to the vessel wall. During this expansion, fractures, peeling of the polymeric coating and uneven release of the drug material were reported as some of the major problems with surface modified DES.[56] Also, implants coated with conventional coating techniques have drawbacks including surface heterogeneity in the type and distribution of functional groups, hydrophilic and hydrophobic domains and surface roughness.[57] Thus a coating technique that offers precise control on the location and orientation of chemical groups/biomolecules on the surface is essential to cater for current needs.

Previous studies have shown the use of SLM to fabricate customised implants; however, there is not much research on modifying the surface of SLM fabricated parts with drug molecules. Surface modification and functionalisation of the SLM fabricated surface to deliver drugs can potentially reduce post-implant complications and deliver therapeutics on-site.

2.3.2 Surface Modification Using Self-assembled Monolayers

Self-assembled monolayers (SAMs) are ordered molecular assemblies with a head group, a spacer and a functional tail group as represented in Figure 2.6. Usually, the head group is bound to a surface and the tail group presents a chemical functional group. The head and tail groups are separated by a spacer which is generally an alkyl chain ($-CH_2$). SAM coatings are usually nano-sized adding only 1–10 nm thickness to the implant surface.[58] This organic interface provides well-defined thickness and acts as a physical barrier. SAMs can be designed at molecular level to be biologically inert or active by modifying the tail group with desired functionality.

SAMs are inexpensive and their ease of functionalisation with well-ordered arrays of SAMs makes them ideal systems for various applications such as the control of wetting and adhesion, chemical resistance and nanofabrication.[59–61] SAMs have wide applications in the healthcare industry

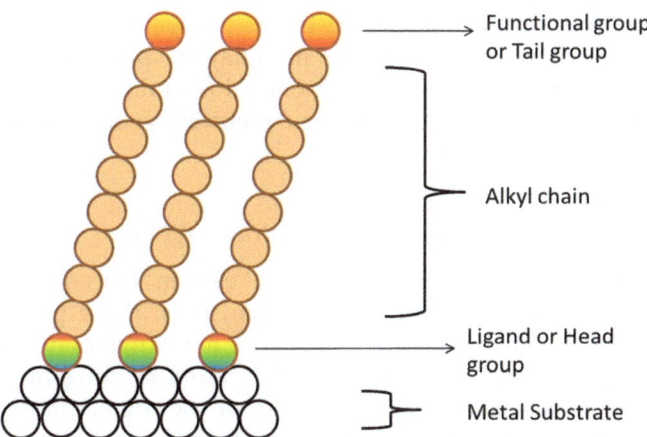

Figure 2.6   Schematic diagram of a single crystalline alkanethiol SAM formed on a metal surface.

including blood purification (algae removal from the bloodstream using SAM-coated nanoparticles), drug delivery, surface modification of implants and to study the cellular properties such as interfacial reactions between cells and organelles and/or proteins and implants.[62]

2.3.2.1   Types of Self-assembled Monolayers

Some of the current monolayer self-assembly chemistries include thiols, disulfides and thiols on gold; silanes on silicon dioxide and hydroxy functionalised surfaces; fatty acids and phosph(on)ates on metal oxides and isonitriles on platinum. Alkanethiol SAMs remain the most popular system despite being limited to noble metals.[63] Thiols on gold surfaces have been well studied and used for a variety of applications including corrosion protection, semi-conductors, biomaterial and biosensing applications.[64] Alkanethiol monolayers have also been formed on 316L stainless steel; however, alkanethiol SAMs formed on 316L SS are not as stable as they are on gold surfaces due to the complex surface chemistry of 316L SS.[58]

Similarly, the use of silane monolayers on silica surfaces is popular owing to their excellent grafting properties to silica surfaces.[63] The use of silane monolayers on metal oxide surfaces such as titanium and cobalt–chromium were also studied in recent years for drug delivery.[65,66] Mani *et al.* reported that the silane monolayers formed on a cobalt–chromium alloy surface was due to the covalent bonding of the SAMs with the alloy (Si–O–Cr and Si–O–W). However, their study reported that these monolayers remained ordered and bound to the alloy surface for only 7 days under *in-vitro* conditions.[66] Ajami *et al.* studied the functionalisation of silane monolayers with an antibacterial drug on an electropolished titanium surface.[65] This study concluded that the drug was released in a sustained manner over a 4-week period under *in-vitro* conditions. Although the release of a drug over 4 weeks is desirable, drug delivery from the implant surface for a longer period may be necessary in clinical application. In such cases, this system cannot be used. Apart from thiols and silane monolayers, phosphonic acids have gained considerable interest in recent years due to their ability to bond to a range of metal oxide surfaces and their relatively good hydrolytic stability under physiological conditions.[67,68]

2.3.2.2   Overview of Phosphonate Monolayers

Phosphonates have three oxygens and a carbon directly attached to a phosphorus molecule. The binding mechanism of these phosphonic acid monolayers to metal oxide surfaces can be a monodentate, bidentate or tridentate as represented in Figure 2.7.[61,69] This binding mechanism depends on both the surface and the nature of the organophosphorus compounds. Similar to thiol on gold, phosphonic acid monolayers adsorb on a metal surface with a tail-up orientation with a tilt angle of the hydrocarbon chains of about 30° with respect to the normal.[70] Phosphonic acid SAMs

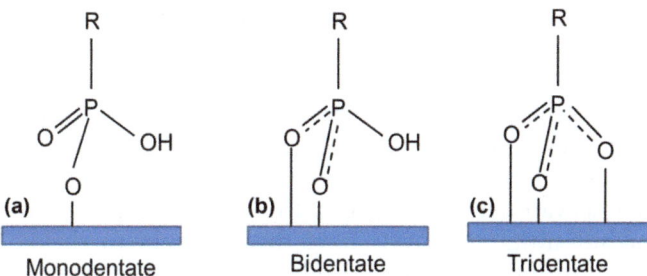

Figure 2.7 Schematic representation of the binding mechanism of phosphonic acid SAMs to a metal surface.

form a metal phosphonate bond when adsorbed to a metal oxide surface (Ti–O–P). Adsorption of long-chain alkyl phosphonic acid SAMs on metal oxides has been reported to offer densely packed and well-ordered SAMs.[71–74]

Various surfaces including silicon, titanium, cobalt–chromium, Nitinol and gold have been functionalised using phosphonic acid monolayers for biomedical applications.[72–75] Tosatti *et al.* studied the assembly of phosphate monolayers on rough and smooth titanium and titanium oxide coated glass surfaces.[76] The structure and order of phosphonic acid based monolayers on silicon was studied by Dubey *et al.*[77] Their studies concluded that alkyl organophosphate monolayers can be assembled on titanium surfaces. Bhure *et al.* investigated the stability of phosphonic acid monolayers on Co–Cr surfaces and showed that the monolayers were highly stable when the surfaces were exposed to atmospheric conditions.[78] Kaufmann *et al.* explored the *in-vitro* stability of phosphonic acid monolayers on an electropolished Co–Cr surface and showed that the monolayers gradually desorbed over the course of 28 days in Tris buffer solution.[73]

Mani *et al.*[52] have reported the use of phosphonic acid SAMs to immobilise flufenamic acid on gold and titanium surfaces and later immobilised paclitaxel, an anti-cancer drug, on a cobalt–chromium metal surface. The later study showed a sustainable release of paclitaxel from the monolayers. Torres *et al.* reported the use of phosphonic acid monolayers to deliver an antibacterial drug from the surface of biomedical implants to prevent bacterial infections.[79] Schwartz *et al.* demonstrated the functionalisation of a protein (RGD peptide) to phosphonate SAMs adsorbed on pure titanium and Ti6Al4V surfaces for osteoconductive surfaces.[80] The versatility of phosphonic acid monolayers to form densely packed, well-ordered and highly stable SAMs on metal oxide surfaces makes them ideal for these biomedical applications.

2.3.2.3 Formation of Self-assembled Monolayers

Deposition of monolayers on substrates offer one of the highest quality routes for preparing chemically and structurally well-defined surfaces.[81]

Adsorption of SAMs on to a surface is generally performed either in polar or non-polar solvents to attain a greater flexibility in the molecular design and control over the surface properties.[60] The adsorption of monolayers to a surface is due to the spontaneous chemical reaction at the interface, as the system approaches equilibrium. SAMs form a strong covalent bond on adsorption and the interactions between the sub-units (spacers) are non-covalent such as van der Waals forces. Due to this non-covalent interaction between the sub-units, the SAMs have the tendency to shrink and regain their original position. SAMs are adsorbed to a surface by various methods including spray deposition, electrochemical deposition or by simple immersion.[82] Annealing of the SAM coated specimens can be performed to obtain a strongly surface-bound film of monolayers.[78]

2.3.2.4   Assembly of Self-assembled Monolayers by Simple Immersion

The assembly of SAMs by a simple immersion method typically involves the immersion of the sample into a dilute solution (thiol, silane, phosphonic or other organic acids *etc.*), allowing molecules to assemble, then removal of the sample and rinsing. The attachment and assembly processes of SAMs are schematically represented in Figure 2.8a and b. The assembly process takes place once the sample is immersed into the solution. The amphiphilic SAM-forming molecules in solution come in contact with the surface in a few seconds leading to the organisation and packing of the molecules and form well-ordered SAMs in a few hours. As the monolayers continue to form, the non-covalent interaction between the hydrocarbon chain helps in packing molecules into a well-ordered crystalline layer.[68]

The degree of order of these monolayers is mainly determined by the length of the spacer (alkyl chain).[83] It has been reported that SAMs with short alkyl chains (fewer than 10 carbons) lack sufficient attraction between the alkyl chains to produce well-ordered monolayers. The other important factors that determine the order of SAMs include cleanliness of the substrate and purity of the solvent.[58,84] The important and problematic contaminants are hydrocarbons (oils which comes from pumps, skin *etc.*), impurities on the material's surface and some polymeric contaminants from the atmosphere.[85] The quality and assembly of SAMs also depend on several factors such as head group size, its properties and assembly time. A detailed explanation on the assembly of monolayers and how these factors affect the monolayer formation can be found in literature.[60]

2.3.3   Functionalisation of Self-assembled Monolayers

SAMs can be functionalised by modifying the tail group with a drug/protein molecule. This can be attained before or after the adsorption of SAMs to a surface. Although extensive literature is available for the modification of

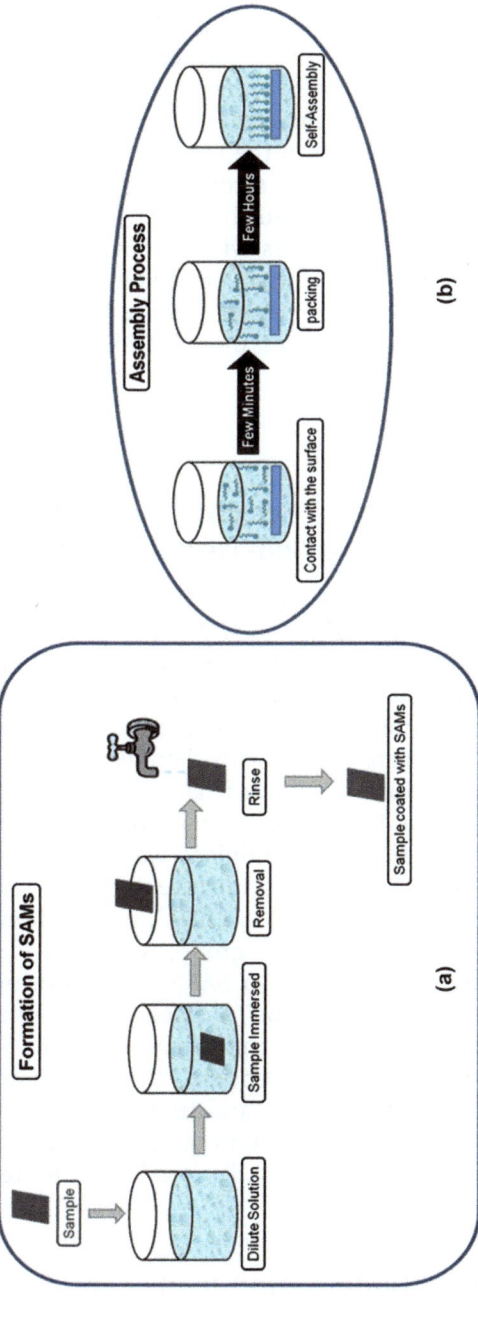

Figure 2.8 Steps involved in the adsorption (a) and assembly (b) of SAMs.

SAMs by both processes, functionalisation of SAMs after its adsorption to the surface is preferred. This is because, in certain cases, it will be difficult to control the reactivity of the head group towards the functional molecules.

There are a number of ways in which the functional group of SAMs can be modified. SAMs can be adsorbed to a surface and then dip coated with the active/functional ingredient. In other ways, the functional group of SAMs are made to react directly with the molecules to be attached. When performing this, there are possibilities for both the monolayers and the drug to possess the same functional groups such as alcohols, carboxylic groups *etc.* In these cases, the SAMs can be converted to their corresponding acid chlorides by reacting the surface adsorbed SAMs with thionyl chloride. Converting the functional group of SAMs to an acid chloride will enhance the reactivity and will effectively bind the desired molecule to the SAM. A case study on the additive manufacturing and surface modification of parts using SAMs is described below.

2.4 Case Study

2.4.1 Functionalisation of Selective Laser Melted Parts with Ciprofloxacin using Self-assembled Monolayers

Cuboidal samples with dimensions $10 \times 10 \times 3$ mm were fabricated from Ti6Al4V powder using a Renishaw SLM 250 machine. These samples were fabricated using previously optimised parameters.[86] The fabricated samples were mechanically polished using silicon carbide grits (200 μm, 400 μm, 600 μm, 800 μm and 1200 μm) and diamond pastes (6 μm and 1 μm). After polishing the samples were rinsed with deionised water and the samples were immersed in a mixture of sulphuric acid, hydrogen peroxide and water in the ratio of $1:1:5$ for 15 minutes to remove any surface contaminants. The samples were then rinsed in deionised water for 30 minutes (twice) to remove any residues of the chemical mixture.

16-Phosphanohexadecanoic acid (16-PhDA), a phosphonic acid SAM with carboxylic acid as its functional/tail group, was used to modify the SLM fabricated part. Assembly of 16-PhDA SAMs was performed using a previously optimised procedure.[73] In short, surface cleaned samples were rinsed with tetrahydrofuran (THF) and immediately immersed in 100 mL of 16-PhDA SAM solution (1 mM). After 24 hours, the samples were taken out and allowed to dry at room temperature and transferred to an oven at 120 °C. After 24 hours, the samples were taken out of the oven and allowed to cool before rinsing with THF and deionised water for 1 minutes each to remove any physisorbed molecules.

2.4.2 Surface Functionalisation using Ciprofloxacin

Functionalisation of the 16-PhDA SAMs adsorbed to the SLM fabricated Ti6Al4V surface was performed in an inert atmosphere. The apparatus included a round-bottomed flask (RBF) and a reflux condenser with a constant flow of nitrogen gas into the RBF to maintain an inert atmosphere. Using a flame gun, the apparatus was dried to remove any moisture. A sonicator was used to agitate the samples. The general reaction scheme to functionalise monolayers with a drug (Ciprofloxacin) is shown in Figure 2.9. Firstly, the SAM coated Ti6Al4V surfaces were transferred to the RBF with the SAM coated surface (mechanically polished surface) facing upwards. 3 mM of thionyl chloride ($SOCl_2$) in anhydrous THF was added to

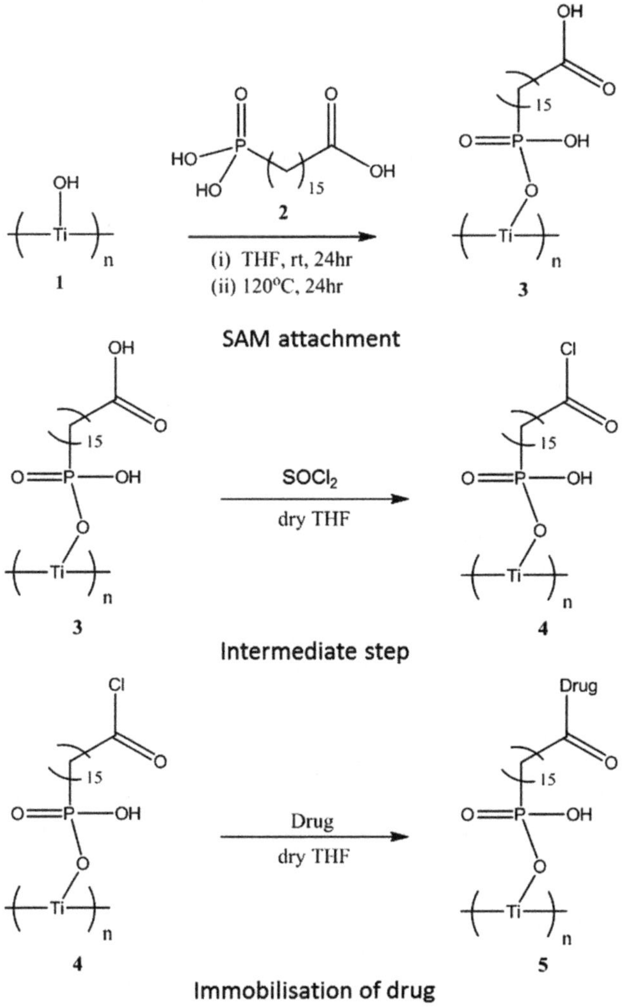

Figure 2.9 Reaction scheme for the surface modification and functionalisation of SAMs with Ciprofloxacin. Hydroxylated titanium surface (**1**); 16-phosphanohexadeconoic acid SAMs (**2**); SAMs assembled on titanium surface (**3**); intermediate compound with acid chloride as tail group (**4**) and drug attached to 16-PhDA SAMs (**5**).[86]

the RBF and the apparatus was sonicated for 1 hour. The reactive mixture should cover the samples in order to obtain a uniform reaction all over the surface. After 1 hour, the solution in the RBF was syringed out and the Ti6Al4V surfaces were sonicated three times for 3 minutes by adding an excess of anhydrous THF to the RBF each time. After this step, 10 mM of Ciprofloxacin in anhydrous THF was added to the RBF and sonicated for 1 hour. Finally, the Ciprofloxacin solution was syringed out and sonicated with an excess of anhydrous THF and deionised water (twice for 2 minutes each) to remove physisorbed molecules. Care should be taken so that the whole process is conducted in a nitrogen atmosphere to prevent the formation of HCl gas due to the use of $SOCl_2$.

Electron spectroscopy for chemical analysis (ESCA) (XPS; Thermo Scientific K-Alpha) was used to study the surface chemistry. Aluminium (Al) Kα monochromated radiation at 1486.6 eV was used and photoelectrons were collected at a take-off angle of 90°. The probing spot size used to obtain the measurement was 400 μm. Survey spectra were collected at a pass energy of 100 eV in a constant energy analyser mode. High resolution spectra were obtained at a pass energy of 20 eV. Peak deconvolution was performed using Gaussian–Lorentzian curves to determine the different chemical states. A built-in Thermo Advantage data system was used for data acquisition and processing. The NIST database and previously reported literature were used to identify the unknown spectral lines.

Figure 2.10 shows typical XPS spectra obtained for a Ti6Al4V surface before SAM attachment (a), after SAM attachment (b) and after the attachment of a drug (Ciprofloxacin). A phosphorus 2P peak would be expected after SAM coating since the monolayer used was a phosphonate group and an increase in carbon (C 1s) intensity would be expected due to the presence of a long alkyl chain. Similarly, the intensity of the base elements (Ti, Al, O) present on the surface of the sample would have been expected to decrease. This is because the intensity of photoelectron penetration to the base elements will be low and this will result in the decrease of base element's peak intensity after surface coating. After functionalisation, a fluorine peak (F 1s) would be expected with an increase in the nitrogen composition due to the presence of these elements in Ciprofloxacin.

It can be observed from the figure that the phosphorus peak was not observed initially in the control sample whereas after SAM attachment, introduction of a phosphorus 2P peak at 133.3 eV can be witnessed. An expected increase in the carbon (C 1s) intensity and decrease in the intensity of the titanium, aluminium and oxygen after surface modification with SAMs confirmed their attachment. Figure 2.11 shows the deconvoluted high-resolution spectra of fluorine 1s obtained using XPS for these control, SAM coated and drug coated samples. The appearance of fluorine 1s peak at the energy 687.1 ± 0.2 eV after the surface functionalisation shows the attachment of Ciprofloxacin. The introduction of the fluorine peak is attributed to its presence in the drug. The wettability of the samples before and after surface modification was studied using static water contact angle goniometry to confirm the attachment of SAMs and the drug. As can be

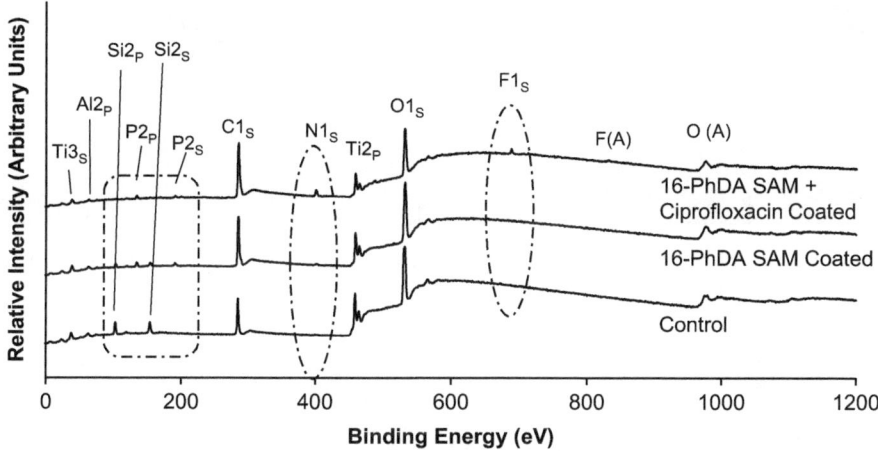

Figure 2.10   XPS survey spectra of a SLM fabricated Ti6Al4V surface before and after surface modification. The spectra clearly shows the introduction of a P 2p peak at 133.3 eV after surface modification with SAMs confirming its attachment. Similarly after drug attachment, a F 1s peak at 687.1 eV and N 1s at 400 eV were witnessed which confirms the functionalisation of SAMs with Ciprofloxacin.

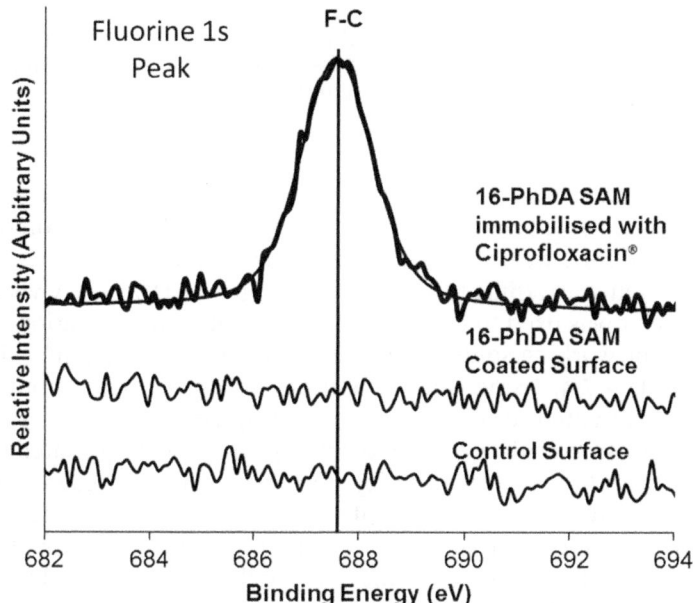

Figure 2.11   High resolution XPS spectra of Fluorine 1s peak confirming the immobilisation of Ciprofloxacin to 16-PhDA SAM coated Ti6Al4V surface.

observed from Figure 2.12, the change from hydrophilic to hydrophobic nature after drug attachment confirmed the functionalisation of SAMs with the drug.

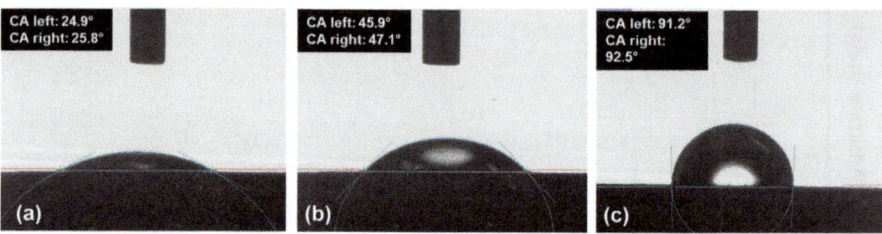

Figure 2.12   Surface wettability of SLM fabricated parts before SAM attachment (a); after SAM attachment (b) and after drug attachment (c).

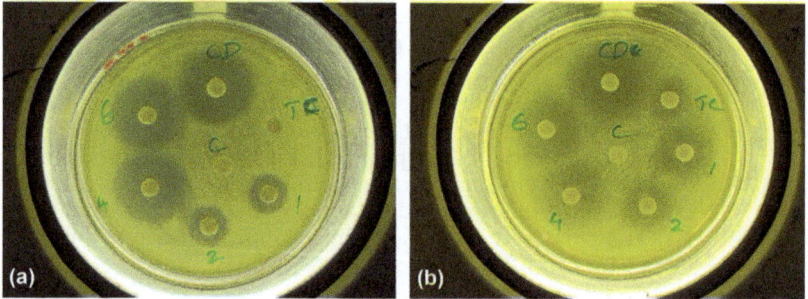

Figure 2.13   Steps antibacterial susceptibility test against (a) *S. aureus* and (b) *E. coli*. Label description: C, control disc with no drug; CD, control with drug on; TC, discs coated with Tris-HCl buffer; 1, 2, 4, 6, immersion time intervals (in weeks) of the samples in buffer solution.

Drug estimation was performed using a base hydrolysis method.[52] Drug coated samples ($n = 5$) were immersed in a 10 mM NaOH solution of pH $\sim$ 9 and incubated for 2 weeks to hydrolyse the drug attached to SAMs. The drug eluted solution was adjusted to pH 7.4 using 1 M HCl. A UV-visible spectrophotometer was used to estimate the drug eluted. A calibration curve was plotted for the working solution of ranges 1 ng mL^{-1} to 1 µg mL^{-1} in 10 mM NaOH of pH $\sim$ 7.4. Using this calibration curve, the total amount of drug coated was estimated to be approximately 1 µg cm^{-2}.

Tests to measure oxidative stability (by exposing the drug coated surface to atmosphere) and *in-vitro* stability (by immersing the drug coated surface in 10 mM of Tris-HCl buffer solution (pH $\sim$ 7.4) were performed. These studies revealed the drug coated surfaces were highly stable under oxidative condition whereas, under *in-vitro* conditions, the drug was observed to release in a sustained manner. After 6 weeks of immersion in Tris-HCl buffer, nearly 40% of the drug was observed to remain attached to the SAMs coated on the SLM fabricated surface. On performing an anti-bacterial susceptibility test with Gram-positive (*Staphylococcus aureus*) and Gram-negative bacteria (*Escherichia coli*), the discs coated with the drug showed anti-bacterial activity, confirming the drug was active upon its release (Figure 2.13).

2.5 Conclusion

The ability of additive manufacturing to build complex and customised parts for biomedical applications compared to conventional manufacturing methods and the unprecedented control of the surface chemistry rendered by SAMs offers a new route for biomedical implant fabrication and func-tionalisation. Surface modifying an additively manufactured part with de-sired bioactive substances has the potential to optimise the implant–host interface. Hence by optimising this biological interface, cell modulation can be achieved.

References

1. B. D. Ratner, A. S. Hoffman, F. J. Schoen and J. E. Lemons, *Biomaterials science: an introduction to materials in medicine*, Elsevier, London, 2nd edn, 2004.
2. H. C. Grillo, *Surgery of the Trachea and Bronchi*, BC Decker Inc, USA, 2004.
3. G. Manivasagam, D. Dhinasekaran and A. Rajamanickam, *Recent Pat. Corros. Sci.*, 2010, **2**, 40–54.
4. S. R. Paital and N. B. Dahotre, *Mater. Sci. Eng., R*, 2009, **66**, 1–70.
5. D. Eyers and K. Dotchev, *Assem. Autom.*, 2010, **30**, 39–46.
6. L. Hao and R. Harris, in *Bio-materials and prototyping applications in medicine*, ed. B. Bidanda and P. Bartolo, Springer, USA, 2008, pp. 79–108.
7. M. Taylor and K. E. Tanner, *J. Bone Joint Surg.*, 1997, **79–B**, 181–182.
8. L. Freitag, *Eur. Respir. Monogr.*, 2010, **48**, 190–217.
9. J. Bono and R. Scott, *Revision Total Knee Arthroplasty*, Springer, USA, 2005.
10. R. Freeman and P. Plante-bordeneuve, *J. Bone Joint Surg.*, 1994, **76**, 432–438.
11. M. H. Wholey, H. Ferral, R. Reyes, J. Lopera and W. Castan, *Cardiovasc. Inter. Rad.*, 1997, **20**, 477–480.
12. P. Galindo-Moreno, M. Padial-Molina, E. Sánchez-Fernández, P. Hernández-Cortés, H.-L. Wang and F. O'Valle, *Implant Dent.*, 2011, **20**, 400–405.
13. K. S. Hebel and R. C. Gajjar, *J. Prosthet. Dent.*, 1997, **77**, 28–35.
14. R. Shadid and N. Sadaqa, *J. Oral Implant.*, 2012, **38**, 298–307.
15. M. S. Chaar, W. Att and J. R. Strub, *J. Oral Rehabil.*, 2011, **38**, 697–711.
16. A. N. Natali, E. L. Carniel and P. G. Pavan, *Dent. Mater.*, 2009, **25**, 573–581.
17. R. Bibb, D. Eggbeer and P. Evans, *Rapid Prototyping J.*, 2010, **16**, 130–137.
18. D. Wood, in *Surgery of the Trachea and Bronchi*, ed. H. Grillo, BC Decker Inc, Hamilton, Ontario, 2004, pp. 763–790.
19. D. Stoeckel, A. Pelton and T. Duerig, *Eur. J. Radiol.*, 2004, **14**, 292–301.
20. S. Arabnejad Khanoki and D. Pasini, *J. Biomech. Eng.*, 2012, **134**, 031004.
21. G. Chahine, H. Atharifar, P. Smith and R. Kovacevic, *International Solid Freeform Fabrication Symposium*, 2009, pp. 631–640.

22. B. Aksakal, O. Yildirim and H. Gul, *J. Failure Anal. Prevent.*, 2004, **4**, 17–23.
23. R. Bhola, S. M. Bhola, B. Mishra and D. L. Olson, *Trends Biomater Artif. Organs*, 2011, **25**, 34–46.
24. Y. H. An, R. B. Dickinson and R. J. Doyle, in *Handbook of bacterial adhesion: principles, methods and applications*, eds. Y. H. An and R. J. Friedman, Humana Press Inc, New Jersey, 2002, pp. 1–27.
25. L. G. Harris and R. G. Richards, *Injury*, 2006, 37(suppl 2), S3–S14.
26. D. Pye, D. E. Lockhart, M. P. Dawson, C. Murray and J. Smith, *J. Hosp. Infect.*, 2009, **72**, 104–110.
27. B. a S. Knobben, J. R. van Horn, H. C. van der Mei and H. J. Busscher, *J. Hosp. Infect.*, 2006, **62**, 174–180.
28. J. W. Leung, *Self-Expandable Stents in the Gastrointestinal Tract*, Springer, New York, NY, 2013.
29. M. Noppen, D. Piérard, M. Meysman, I. Claes and W. Vincken, *Am. J. Respir. Crit. Care Med.*, 1999, **160**, 672–677.
30. ASTM-F2792, *Standard Terminology for Additive Manufacturing Technologies*, ASTM International, West Conshohocken, PA, 2003, 2010.
31. I. Gibson, D. W. Rosen and B. Stucker, *Additive Manufacturing Technologies*, Springer, Boston, MA, 2010.
32. E. Santos, M. Shiomi, K. Osakada and T. Laoui, *Int. J. Mach. Tool Manuf.*, 2006, **46**, 1459–1468.
33. H. Schleifenbaum, W. Meiners, K. Wissenbach and C. Hinke, *CIRP J. Manuf. Sci. Technol.*, 2010, **2**, 161–169.
34. N. Hopkinson, R. J. Hague and P. M. Dickens, *Rapid Manufacturing: An Industrial Revolution for the Digital Age*, John Wiley and Sons Ltd., West Sussex, UK, 1st edn, 2006.
35. M. Lindner, S. Hoeges, W. Meiners, K. Wissenbach, R. Smeets, R. Telle, R. Poprawe and H. Fischer, *J. Biomed. Mater. Res., Part A*, 2011, **97**, 466–471.
36. M. Badrossamay and T. Childs, *Int. J. Mach. Tools Manuf.*, 2007, **47**, 779–784.
37. K. Mumtaz and N. Hopkinson, *Rapid Prototyping J.*, 2010, **16**, 248–257.
38. I. Drstvensek, N. I. Hren, T. Strojnik, T. Brajlih and B. Valentan, *Int. J. Biol. Biomed. Eng.*, 2008, **2**, 29–38.
39. T. Sercombe, N. Jones, R. Day and A. Kop, *Rapid Prototyping J.*, 2008, **14**, 300–304.
40. M. Wehmoller, P. Warnke, C. Zilian and H. Eufinger, *Int. Congr. Ser.*, 2005, **1281**, 690–695.
41. M. Khan and P. Dickens, *Gold Bull.*, 2010, **43**, 114–121.
42. L. C. Zhang, D. Klemm, J. Eckert, Y. L. Hao and T. B. Sercombe, *Scr. Mater.*, 2011, **65**, 21–24.
43. I. Yadroitsev and I. Smurov, *Phys. Procedia*, 2010, **5**, 551–560.
44. R. J. Bibb, A. Bocca and P. Evans, *J. Maxillofac. Prosthet. Technol.*, 2002, **5**, 28–31.
45. M. Figliuzzi, F. Mangano and C. Mangano, *Int. J. Oral Maxillofac. Surg.*, 2012, **41**, 858–862.

46. G. M. de Peppo, A. Palmquist, P. Borchardt, M. Lennerås, J. Hyllner, A. Snis, J. Lausmaa, P. Thomsen and C. Karlsson, *Sci. World J.*, 2012, **2012**, 1–14.
47. D. K. Pattanayak, T. Matsushita, H. Takadama, a. Fukuda, M. Takemoto, S. Fujibayashi, K. Sasaki, N. Nishida, T. Nakamura and T. Kokubo, *Biceram. Dev. Appl.*, 2011, **2**, 1–3.
48. A. Fukuda, M. Takemoto, T. Saito, S. Fujibayashi, M. Neo, D. K. Pattanayak, T. Matsushita, K. Sasaki, N. Nishida, T. Kokubo and T. Nakamura, *Acta Biomater.*, 2011, **7**, 2327–2336.
49. P. H. Warnke, T. Douglas, P. Wollny, E. Sherry, M. Steiner, S. Galonska and S. Sivananthan, *Tissue Eng., Part C*, 2009, **15**, 115–124.
50. A. Kurella and N. B. Dahotre, *J. Biomater. Appl.*, 2005, **20**, 5–50.
51. H. Kim, S.-H. Choi, J.-J. Ryu, S.-Y. Koh, J.-H. Park and I.-S. Lee, *Biomed. Mater.*, 2008, **3**, 025011.
52. G. Mani, D. M. Johnson, D. Marton, M. D. Feldman, D. Patel, A. a Ayon and C. M. Agrawal, *Biomaterials*, 2008, **29**, 4561–4573.
53. K. Garvin and C. Feschuk, *Clin. Orthop. Relat. Res.*, 2005, **437**, 105 110.
54. R. Schmidmaier, M. Simsek, M. Baumann, P. Emmerich and G. Meinhardt, *Anti-cancer Drugs*, 2006, **17**, 621–630.
55. R. M. Stone, M. R. O'Donnell and M. a Sekeres, *Hematology Am. Soc. Hematol. Educ. Program*, 2004, 98–117.
56. Y. Levy, D. Mandler, J. Weinberger and A. J. Domb, *J. Biomed. Mater. Res., Part B*, 2009, **91**, 441–451.
57. N. Faucheux, R. Schweiss, K. Lützow, C. Werner and T. Groth, *Biomaterials*, 2004, **25**, 2721–2730.
58. A. Mahapatro, D. M. Johnson, D. N. Patel, M. D. Feldman, A. a Ayon and C. M. Agrawal, *Nanomedicine*, 2006, **2**, 182–190.
59. N. K. Chaki, M. Aslam, J. Sharma and K. Vijayamohanan, *J. Chem. Sci.*, 2001, **113**, 659–670.
60. D. K. Schwartz, *Annu. Rev. Phys. Chem.*, 2001, **52**, 107–137.
61. A. Ulman, *Chem. Rev.*, 1996, **96**, 1533–1554.
62. C. W. Yung, J. Fiering, A. J. Mueller and D. E. Ingber, *Lab Chip*, 2009, **9**, 1171–1177.
63. F. Durmaz, *Doctoral thesis (16942)*, Swizz Federal Institute of Technology, 2006.
64. L. Pardo and T. Boland, *J. Colloid Interface Sci.*, 2003, **257**, 116–120.
65. E. Ajami, K. Aguey-zinsou, *J. Colloid interface Sci.*, 2012, **385**, 258–267.
66. G. Mani, M. D. Feldman, S. Oh and C. M. Agrawal, *Appl. Surf. Sci.*, 2009, **255**, 5961–5970.
67. M. J. Pellerite, T. D. Dunbar, L. D. Boardman and E. J. Wood, *J. Phys. Chem. B*, 2003, **107**, 11726–11736.
68. F. Mastrangelo, G. Fioravanti, R. Quaresima, V. Raffaele and E. Gherlone, *J. Biomater. Nanobiotechnol.*, 2011, **2**, 533–543.
69. W. Gao, L. Dickinson, C. Grozinger, M. Frederick and L. Reven, *Langmuir*, 1996, **12**, 6429–6435.

70. R. Michel, J. W. Lussi, G. Csucs, I. Reviakine, G. Danuser, B. Ketterer, J. a. Hubbell, M. Textor and N. D. Spencer, *Langmuir*, 2002, **18**, 3281–3287.
71. G. Mani, N. Torres and S. Oh, *Biointerphases*, 2011, **6**, 33–42.
72. G. Mani, D. M. Johnson, D. Marton, V. L. Dougherty, M. D. Feldman, D. Patel, A. a Ayon and C. M. Agrawal, *Langmuir*, 2008, **24**, 6774–6784.
73. C. Kaufmann, G. Mani, D. Marton, D. Johnson and C. M. Agrawal, *J. Biomed. Mater. Res., Part A*, 2011, **98B**, 280–289.
74. S. Pawsey, K. Yach and L. Reven, *Langmuir*, 2002, **18**, 5205–5212.
75. H. Dadafarin, J. Katic, M. Metikos-Hukovic and S. Omanovic, *219th Electrochemical Society Meeting*, The Electrochemical Society, 2011, p. 120.
76. S. Tosatti, R. Michel, M. Textor and N. D. Spencer, *Langmuir*, 2002, **3858**, 3537–3548.
77. M. Dubey, T. Weidner, L. J. Gamble and D. G. Castner, *Langmuir*, 2010, **26**, 14747–14754.
78. R. Bhure, T. M. Abdel-Fattah, C. Bonner, F. Hall and A. Mahapatro, *Appl. Surf. Sci.*, 2011, **257**, 5605–5612.
79. N. Torres, S. Oh, M. Appleford, D. D. Dean, J. H. Jorgensen, J. L. Ong, C. M. Agrawal and G. Mani, *Acta Biomater.*, 2010, **6**, 3242–3255.
80. J. Schwartz, M. J. Avaltroni, M. P. Danahy, B. M. Silverman, E. L. Hanson, J. E. Schwarzbauer, K. S. Midwood and E. S. Gawalt, *Mater. Sci. Eng., C*, 2003, **23**, 395–400.
81. G. Shustak, A. J. Domb and D. Mandler, *Langmuir*, 2004, **20**, 7499–7506.
82. F. Schreiber, *Prog. Surf. Sci.*, 2000, **65**, 151–257.
83. G. a. Buckholtz and E. S. Gawalt, *Materials*, 2012, **5**, 1206–1218.
84. P. Thebault, E. Taffindegivenchy, F. Guittard, C. Guimon and S. Geribaldi, *Thin Solid Films*, 2008, **516**, 1765–1772.
85. B. O. Aronsson, J. Lausmaa and B. Kasemo, *J. Biomed. Mater. Res.*, 1997, **35**, 49–73.
86. J. Vaithilingam, S. Kilsby, R. D. Goodridge, S. D. R. Christie, S. Edmondson and R. J. M. Hague, *Appl. Surf. Sci.*, 2014, DOI: 10.1016/j.apsusc.2014.06.014.

CHAPTER 3

Probing Biointerfaces: Electrokinetics

RALF ZIMMERMANN,*[a] JÉRÔME F. L. DUVAL[b,c] AND
CARSTEN WERNER[a,d]

[a] Leibniz Institute of Polymer Research Dresden, Max Bergmann Center
of Biomaterials Dresden, Hohe Strasse 6, 01069 Dresden, Germany;
[b] Université de Lorraine, Laboratoire Interdisciplinaire des
Environnements Continentaux (LIEC), UMR 7360, Vandoeuvre-lès-Nancy
F-54501, France; [c] CNRS, LIEC, UMR 7360, Vandoeuvre-lès-Nancy, F-54501,
France; [d] Technische Universität Dresden, Center for Regenerative
Therapies Dresden, Tatzberg 47, 01307 Dresden, Germany
*Email: zimmermn@ipfdd.de

3.1 Introduction

The contact of biomaterials with aqueous environments very often involves
the formation of bulk and interfacial charges.[1–5] In turn, the structure and
function of the materials and their interfaces are largely governed by electrostatic interactions that depend on the degree of ionization of functional
groups and by ion-specific interactions within the molecular architecture of
the material.[6–9] Furthermore, electrostatic interactions were found to be
relevant for a number of interfacial processes such as wetting,[10] protein
binding to extracellular matrix,[11–13] proteins and nucleic acid stability in the
vicinity of/at biomaterial interfaces,[14] interaction kinetics between these
components and biosurfaces[14,15] and membrane permeability to proteins
and drugs.[16,17]

RSC Smart Materials No. 10
Biointerfaces: Where Material Meets Biology
Edited by Dietmar Hutmacher and Wojciech Chrzanowski
© The Royal Society of Chemistry 2015
Published by the Royal Society of Chemistry, www.rsc.org

The complex interplay between the processes that underlie biomaterial charge, ion interactions with ionized groups (including the formation of salt bridges), and interactions of the resulting material structure with (bio)molecules from the aqueous phase can hardly be derived from the primary material structure or from known ionization patterns of isolated functional groups. Therefore, revealing the fundamental principles of interfacial and bulk charge formation and related phenomena is of utmost importance for optimizing the performance of biomaterials in demanding products and technologies such as bioactive cell scaffolds and immunosorbent assays.

Electrokinetics is one of the oldest and most frequently applied methods to investigate charging processes at material interfaces of various shape and chemical nature.[18-20] The approach dictates that the interfacial charge of the investigated system be unambiguously derivable from a given measured electrokinetic quantity (*e.g.*, streaming current/potential for planar macroscopic surfaces or electrophoretic mobility for particulate systems). The fundamental theory for the interpretation of electrokinetic measurements at hard surfaces, *i.e.*, surfaces that are not permeable to water and electrolyte ions, was developed by Smoluchowski about a century ago and proven to provide valuable information on the impact of the electrical double layer on numerous interfacial and colloidal phenomena.[18,19]

The growing interest of the scientific community in complex systems and the emerging fields of micro- and nanofluidics has focused attention on basic interfacial phenomena associated with charge formation.[21] Advanced experimental techniques and theories have been developed in recent years that facilitate interpretation at a much higher level of sophistication.[21-25] The progress made in the quantitative interpretation of peculiar electrokinetic features of soft planar surfaces has its origin in recent theoretical developments, and the option to simultaneously access film surface conductivity and swelling. Measurements of the latter provide additional information on interfacial charging mechanisms and structure, which significantly constrains modeling film electrokinetic properties.

In this chapter, we provide an overview of state of the art experimental techniques and theory to analyze the charge and structure of soft planar (bio)polymeric systems. To demonstrate the options arising from these developments, we discuss strategies and recent results reported for stimuli-responsive polymer coatings, biohybrid hydrogels, and supported bilayer lipid membranes. Finally, we provide an outlook on potential future developments in this field.

3.2 Fundamental Principles and Theory

3.2.1 Streaming Current and Streaming Potential Measurements in Rectangular Microchannels

The electrokinetic phenomena streaming current and streaming potential are based on the transport of charge-compensating ions in the electrical double layer at solid–liquid interfaces. This transport is caused by an externally applied force (typically a pressure gradient) that leads to a tangential shift of the liquid phase in a capillary cell in relation to the solid under investigation. The convective ion transport can be detected by direct measurement of the electrical current, designated as streaming current, between two non-polarizable electrodes (*e.g.*, Ag/AgCl electrodes) using an amperemeter of sufficiently low internal resistance (Figure 3.1). Alternatively, an electrical potential, the streaming potential, can be measured if an electrometer of sufficiently high input resistance is connected to the two electrodes. This potential results from the steady-state balance between the streaming current and a back conduction current that originates from the electrical conductivity of the liquid phase in the capillary cell and from that of the analyzed material surface, *i.e.*, the surface conductivity. The latter results from mobile excess charge carriers located in the electrical double layer (excess with respect to the concentration of carriers present in bulk solution far from the interface).[18]

A typical set-up for performing streaming current and streaming potential measurements consists of a rectangular streaming channel formed by two identical coated sample surfaces of length L_0 and width ℓ (Figure 3.1).

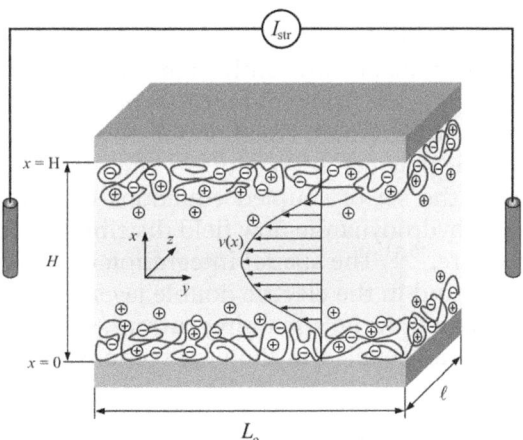

Figure 3.1   Schematic representation of a cell to measure the streaming current/ potential at charged soft polymer films supported by a hard carrier. The hydrodynamic flow field, $v(x)$, is indicated developing in the channel under application of a lateral pressure gradient along the y coordinate. The scheme details the nomenclature used in the theory section further. For the sake of simplicity, charges located at the carrier surface are not represented. Typical dimensions of the cell are: $L_0 = 20$ mm, $\ell = 10$ mm and $H = 5\ \mu m$ to $60\ \mu m$.

Reprinted from R. Zimmermann, S. S. Dukhin, C. Werner, J. F. L. Duval. On the use of electrokinetics for unravelling charging and structure of soft planar polymer films, *Curr. Opin. Colloid Interface Sci.*, 2013, **18**, 83–92, Copyright 2013, with permission from Elsevier.

The flow within the channel is driven by an externally applied pressure gradient ΔP. The system generally fulfils the condition that the separation distance between the two surfaces, H, is much smaller than the length and width of the surfaces, $(H \ll L_0$ and $H \ll \ell)$, which in turn minimizes the effects of side walls on the development of the hydrodynamic flow field within the channel.[26] Furthermore, the thickness d of the soft surface coating is much smaller than the channel height, *i.e.*, $d \ll H$. Typical dimensions of the experimental configuration are $L_0 = 20$ mm, $\ell = 10$ mm and $H = 30$ µm.[27] Under such conditions, laminar flow is achieved over the typical range of applied pressure used in practice ($\Delta P \sim$ 0–500 mBar).[28]

The fundamental equations for the streaming current under the conditions specified above were derived by Duval and collaborators.[29-33] In their formalism, the authors considered a soft interfacial (biopolymer) layer characterized by a given volume density of charges resulting from the ionization of acidic or basic groups distributed along the constitutive polymer chains. Likewise, a charge density or surface potential may be assigned to the hard substrate that supports the soft interfacial layer. The electrolyte solution is considered to comprise N ions of valence z_i with bulk concentrations c_i, with the index, i, running from $i = 1, \ldots, N$. The theory was derived for steady-state electrostatic and hydrodynamic fields, and it is applicable providing the separation distance between the two surfaces excludes any overlap of the electric double layers formed at the interfaces between the soft films and the electrolyte solution embedded in the channel. In addition, polymer films of sufficiently high water content are considered. For such hydrogels, only the first-order term in polymer volume fraction has to be considered in the Brinkman equation for the calculation of the friction force exerted by the polymer segments on the hydrodynamic flow. Procedures for solving the set of coupled equations determining the electrostatic potential and hydrodynamic flow field distribution across the channel are detailed elsewhere.[30,31] The spatial integration of these fields across the soft interfacial layer and in the electric double layer region outside the layer allows for the evaluation of the streaming current, streaming potential, and surface conductivity, as detailed in the next sections.

3.2.1.1 Hydrodynamic Flow Field

For the microchannel geometry described above, the flow field is oriented parallel to the sample surfaces (y dimension). For obvious reasons of symmetry, it is sufficient to consider the hydrodynamics at a single carrier/ soft film/solution interface. The magnitude of the velocity field, v, depends on the dimension perpendicular to the hard channel wall, x, according to the generalized Brinkman equation that may be written in the form:[30,34]

$$\frac{d^2 V(X)}{dX^2} - (\lambda_o H)^2 f(X) V(X) = -1 \tag{3.1}$$

where the spatial coordinate, x, and the velocity field, v, are normalized according to $X = x/H$ and $V(X) = v(X)/v_0$ respectively, with $v_0 = \Delta P H^2/\eta L_0$, where η is the dynamic viscosity of the electrolyte. The function f represents the distribution of polymer segment density across the soft interfacial layer and it may be derived from polymer theory (*e.g.*, for brushes) or experimentally determined by applying optical and scattering methods.[35–37] As extensively discussed by Duval and colleagues,[30] the following function allows for an appropriate modeling of soft interfaces:

$$f(x) = \frac{\omega}{2}\left[1 - \tanh\left(\frac{x - d}{\alpha}\right)\right] \qquad (3.2)$$

In eqn (3.2), the parameter α is the characteristic decay length of the segment density across the soft layer, and the scalar quantity ω ensures that the total amount of polymer segments within the soft film remains constant upon structural variations that result from changes in the electrolyte composition (pH, salt content) or temperature.[30] In the limit $\alpha/d \rightarrow 0$, f pertains to a film with homogeneous segment distribution (step-function like profile). The friction exerted by the polymer segments located at the position x on the fluid flow is determined by the term $(\lambda_0 H)^2 f(X)$ in the Brinkman equation. The parameter λ_0 stands for the hydrodynamic softness of the film. The quantity $1/\lambda_0$ is the Brinkman length that reflects the characteristic flow penetration within the soft film in case of homogeneous segment distribution, *i.e.*, $\alpha \rightarrow 0$ and $\omega \rightarrow 1$. The boundaries for the solution of eqn (3.1) are provided by the no-slip condition at the supporting hard carrier surface $(X = 0)$ and the symmetry of the flow field with respect to the position $X = 1/2$.[30]

3.2.1.2   Electrostatic Potential

The equilibrium electrical potential distribution, $\psi(x)$, across the substrate–soft polymer film–electrolyte solution interface is a function of the local charge density within the layer. The charge balance has two contributions, one stemming from the fixed charges within the layer and the other from mobile electrolyte ions that compensate the charge of the soft layer structure. These features are taken into account in the non-linear Poisson–Boltzmann equation that governs the position-dependent potential, $\psi(x)$, according to:[32]

$$\frac{d^2 y(X)}{dX^2} = -\frac{(\kappa H)^2}{\sum_{i=1}^{N} c_i z_i^2}\left\{\sum_{i=1}^{N} z_i c_i e^{-z_i y(X)} + \sum_{j=1}^{M} \frac{\rho_j/F}{1 + 10^{-\varepsilon_j(pK_j - pH)} e^{\varepsilon_j y(X)}} f(X)\right\} \qquad (3.3)$$

In eqn (3.3), y is the dimensionless potential $(y = F\psi/RT)$, κ is the reciprocal Debye length, F the Faraday constant, R the gas constant, and T the temperature. It is assumed that the film carries M types of ionizable groups. In the limits $\alpha \rightarrow 0$ and $\omega \rightarrow 1$ the charge source term, $\rho_{j=1,\ldots,M}/F$, corresponds to the total concentration of ionisable groups of type j within the film. The ionisation of the j-th type $(j = 1, \ldots, M)$ of the functional groups is characterized by its dissociation pK value, denoted as pK_j. The parameter $\varepsilon_{j=1,\dots,M}$ takes the value ± 1, with $\varepsilon_j = -1$ and $\varepsilon_j = +1$ corresponding to ionization of the groups due to deprotonation or protonation reactions as detailed elsewhere.[32] The boundary condition at the position $X=0$ can be expressed in terms of the surface potential or surface charge density of the hard carrier supporting the film.[31] For systems where the charge of the supporting surface does not significantly contribute to the streaming current and surface conductivity, the zero electric field condition may be imposed at $X=0$.[31] In line with the electroneutrality far from the film/solution interface, the additional required boundary condition for the potential is simply written, $y(X \rightarrow 1/2) = 0$, which is correct in absence of overlap of the electric double layers developed in the vicinity of the two samples in the microslit cell.

3.2.1.3 Streaming Current and Streaming Potential

To obtain the streaming current, I_{str}, the spatial integral over the hydrodynamic flow field, $V(X)$, [eqn (3.1)] times the charge density stemming from charged mobile carriers [first term in the right hand side of eqn (3.3)] must be solved. Using eqn (3.1) and (3.3), this integral can be written in the following form:[31]

$$\frac{I_{str}}{\Delta P} = \frac{2\ell F H^3}{\eta L_0} \int_0^{1/2} V(X) \sum_{i=1}^N z_i c_i e^{-z_i y(X)} dX \qquad (3.4)$$

Under conditions of low potentials (Debye–Hückel approximation) and homogeneous distribution of polymer segments in the soft layer, explicit analytical expressions may be derived for $I_{str}/\Delta P$.[30] For hard surfaces, the solution of eqn (3.4) is identical to the Smoluchowski equation that relates the streaming current to the zeta potential, the electrical potential at the 'plane of shear'.[18] Similar expressions can be obtained for the streaming potential, U_{str}, by converting the streaming current into a potential *via* Ohm's law,[18] where terms for the conductivity of the electrolyte and for the surface conductivity of the film are involved [see eqn (3.5)]. Measurement of the streaming current and potential is outlined in Appendix A.1.

3.2.2 Surface Conductivity

The excess conductivity, K^σ, is caused by the accumulation of mobile counter-ions inside and outside a soft surface layer and can be determined from streaming current and streaming potential measurements carried out at a single or varying channel heights using the following equation:

$$\frac{I_{str}/\Delta P}{U_{str}/\Delta P} \frac{L_0}{2\ell} = \frac{H}{2} K^B + K^\sigma \qquad (3.5)$$

where K^B is the conductivity of the bulk electrolyte. In eqn (3.5), the ratio $(I_{str}/\Delta P)/(U_{str}/\Delta P)$ represents the channel conductance that can be alternatively determined by applying a lateral electric field across the cell in the absence of flow and measuring the corresponding current.[29] If the electrokinetic or impedance measurements are performed at varying channel heights, K^σ corresponds to the intercept in the plot of the channel conductance *versus* the channel height.[28]

The surface conductivity, K^σ, consists of a migration component, K^σ_m, resulting from the ion transport in the tangential field, U_{str}/L_0, along the interface:

$$K^\sigma_m = \frac{HF^2}{RT} \int_0^{1/2} \sum_{i=1}^{N} |z_i| D_i c_i \beta_i(X) \left[e^{-z_i y(X)} - 1 \right] dX \qquad (3.6)$$

and a convective contribution, K^σ_{eo}, due to the electro-osmotic charge transport originating from the action of the field U_{str}/L_0 on mobile ions located within and outside the soft layer:[31]

$$K^\sigma_{eo} = \frac{H \varepsilon_0 \varepsilon_r RT (\kappa H)^2}{\eta} \int_0^{1/2} V_{eo}(X) \sum_{i=1}^{N} z_i c_i e^{-z_i y(X)} dX \qquad (3.7)$$

where D_i is the diffusion coefficient of ion i in the bulk electrolyte solution, $\beta_i(X)$ is the ratio between the mobilities of ion i at position X and in the bulk electrolyte solution, and $\varepsilon_0 \varepsilon_r$ is the dielectric permittivity of the medium. For soft films with sufficiently high water content, it is justified to assume $\beta_i(X) \sim 1$.[29] The electro-osmotic flow field, $V_{eo}(X)$, in eqn (3.7) is determined by the segment distribution at the interface and can be calculated applying the Brinkman equation [eqn (3.1)] after replacing the right hand side with a charge source term and making the substitution $V(X) \rightarrow V_{eo}(X)$.[31] The overall surface conductivity, K^σ, is then given by the sum $K^\sigma_m + K^\sigma_{eo}$. Simulations of electro-hydrodynamics at diffuse soft interfaces are explained in Appendix A.2.

3.3 Electrokinetic Analysis to Unravel Electrosurface Phenomena at Biointerfaces

3.3.1 Segment Distribution of Stimuli-responsive Soft Thin Films

Stimuli-responsive coatings on the basis of thermo-, light- or pH-sensitive polymers have gained widespread interest in numerous biomedical applications, because the properties of these 'smart' materials can be tuned by an external trigger.[38–41] Here we present an example that demonstrates how streaming current, surface conductivity, and swelling measurements over a broad range of pH and salt concentrations (pH 2.5–10, 0.1–10 mM KCl, constant temperature of 22 °C) were used to unravel the intertwined charging and structure of poly(*N*-isopropylacrylamide-*co*-carboxyacrylamide) (PNiPAAM-*co*-carboxyAAM) soft thin films.[31]

The low isoelectric point (IEP ∼ 2.9) and the negative streaming current measured for the PNiPAAM-*co*-carboxyAAM film over a broad range of pH (Figure 3.2a) was unambiguously attributed to the ionization of the carboxylic acid groups in the repeat unit of the copolymer. For a given electrolyte concentration, the streaming current exhibited a non-monotonous dependence on pH with the presence of a maximum at pH ∼ 6.4. This maximum was most pronounced in 0.1 mM KCl solution and gradually disappeared upon increase of the ionic strength.

To decipher the origins of this 'atypical' maximum in the streaming current *versus* pH plot, surface conductivity and film swelling were systematically investigated under the conditions adopted for the electrokinetic measurements.[31] Both streaming current and surface conductivity data could be consistently recovered by step-by-step modeling on the basis of the theory outlined above, taking into account the pH and salt concentration-dependent swelling of the PNiPAAM-*co*-carboxyAAM film. The electro-osmotic contribution, K_{eo}^{σ} [eqn (3.7)], to the overall surface

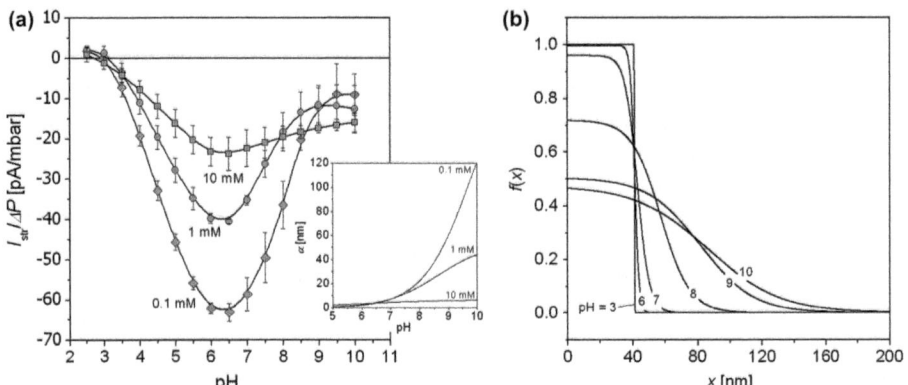

Figure 3.2   Ratio streaming current over applied pressure, $I_{str}/\Delta P$, as a function of pH for PNiPAAM-*co*-carboxyAAM soft thin films in 0.1, 1 and 10 mM KCl solution (a) and segment distribution in 1 mM KCl for different pH of the electrolyte (b). The inset on the left side shows the dependence of the softness parameter, α, on pH and salt concentration necessary for reproducing the experimental electrokinetic data in the pH range 2.5–10. Experimental data are shown by symbols. The corresponding solid lines represent reconstruction of the data by means of the theory.[30,31] The streaming current measurements were performed at a channel height of 30 µm.
Reprinted with permission from R. Zimmermann, D. Kuckling, M. Kaufmann, C. Werner, J. F. L. Duval. Electrokinetics of poly(*N*-isopropylacrylamide)-*co*-carboxyacrylamide soft thin-film. Evidence for diffuse segment distribution in swollen state, *Langmuir*, 2010, **26**, 18169–18181. Copyright 2010 American Chemical Society.

conductivity, K^σ, was verified to be insignificant for data analysis.[31] The density of the carboxyl groups in the collapsed film (low pH) was determined to be 255 ± 15 mM with a corresponding dissociation pK of 6.6. Furthermore, all data were reproduced using a single value of 0.53 ± 0.03 nm for the hydrodynamic penetration length, $1/\lambda_0$, of the collapsed film. The extremum in streaming current *versus* pH was unambiguously attributed to the formation of an interfacial gradient in polymer segment density upon swelling with increasing solution pH (leading to an increase of the parameter a, and to modulations of the segment distribution, as depicted in Figure 3.2). This expansion of the film increases the friction exerted by the outer tails of the polymer chains on the fluid flow. As a result, the streaming current decreases although the film charge and thus the number of counter-ions in the soft layer increases with pH. If the charging is considered alone, $|I_{str}/\Delta P|$ would gradually increase with increasing pH and reach a constant value reflecting a complete ionization of the carboxyl groups. At high electrolyte concentrations, the Debye length is sufficiently small for the electrokinetically active zone of the double layer to span the outmost region of the diffuse part of the film. The streaming current then becomes less sensitive to the hydrodynamic hindrance of the flow by the heterogeneous film structure, and the maximum in $|I_{str}/\Delta P|$ becomes thus less pronounced and ultimately disappears at high salt concentrations. Conversely, at low electrolyte concentrations, the double layer encompasses the entire heterogeneous region at the very interface between the film and bulk electrolyte. Under these conditions, the impact of α on $I_{str}/\Delta P$ becomes most significant, leading to the maximum in $|I_{str}/\Delta P|$ with increasing pH (or increasing α). Finally, it is emphasized that the data analysis indicated a difference of ΔpK ~ 0.9 between the pK values of the carboxyl groups in the film derived from streaming current and surface conductivity.[31] This difference was attributed to a position-dependent hydrophobicity across the film due to the preferred orientation of hydrophobic and hydrophilic polymer segments.[31]

Within the approximations underlying the applicability of the mean-field Poisson–Boltzmann equation,[18] the above example demonstrates how the coupled analysis of streaming current and surface conductivity allows for a quantitative characterization of structure and electric double layer properties of soft charged layers. Further examples covering the analysis of streaming current and surface conductivity data obtained for thermo-responsive films without ionisable groups and for highly charged polyelectrolyte layers can be found in literature.[30,32] In the latter case,[32] electro-osmosis may significantly contribute to the overall surface conductivity. K^σ then reflects not only the key electrostatic properties of the film but also its hydrodynamic permeability *via* the quantity $1/\lambda_0$ that impacts $V_{eo}(X)$.[32] Furthermore, for various systems of practical interest, *e.g.*, polyelectrolyte multilayers, the nature and composition of the films vary across the substrate–soft film–solution interface. An extension of the theory that captures the basics of the electrokinetics for such complex systems can be found in Duval *et al.*[33]

Further information regarding thin films for streaming current/potential measurements are given in Appendix A.3.

3.3.2 Composition and Structure of Biohybrid Hydrogels

Hydrogels, *i.e.*, three-dimensional matrices of physically or chemically cross-linked polymers, are widely used in tissue engineering.[42] The mechanical properties of these materials can be adapted to specific use in various tissues,[43] and hydrogels may further incorporate biomolecular constituents of extracellular matrices (ECM) to stimulate embedded cells *via* specifically orchestrated signals.[44] Biohybrid hydrogels, consisting of synthetic and biologically derived polymeric constituents, are increasingly being developed and employed for this purpose.[43] Among these systems, hydrogels containing polysaccharidic ECM components, glycosaminoglycans (GAGs), are particularly promising due to the effective binding, protection and sustained release of numerous growth factors.[5] The latter effect can be largely attributed to the sulfation pattern of the GAGs and, therefore, chiefly governed by electrostatics.

In order to analyze the charge and structure of biohybrid hydrogel films consisting of covalently linked star-shaped poly(ethylene glycol) (starPEG) and heparin,[44] Zimmermann *et al.*[45] developed and applied a mean-field approach for the numerical evaluation of surface conductivity data. The hydrogel film was treated as 3D meshwork, where ionizable groups are immobilized. The ionisation of these groups is related to an elevated ion concentration within the hydrogel film (due to the counter-ions) that, in turn, causes an electrical potential difference between the gel and the electrolyte. If the thickness of the film is much larger than the Debye screening length, interfacial effects are negligible and the potential difference between the gel and the bulk electrolyte can be described analogously to the potential difference across a semipermeable membrane by the Donnan potential.[29] For hydrogel films with a homogeneous distribution of the polymer segments and low to intermediate densities of ionisable groups, the surface conductivity, K^σ, is related to the Donnan potential, Ψ_D, between the hydrogel and an electrolyte composed of N monovalent ions according to the following equation:[29]

$$K^\sigma = \frac{F^2 d}{RT} \sum_{i=1}^{N} z_i c_i D_i e^{-z_i y_D} \tag{3.8}$$

where y_D is the dimensionless Donnan potential ($y_D = F\Psi_D/RT$). The dimensionless potential, y_D, can be calculated on the basis of eqn (3.3) after replacing $y(X)$ by y_D and the condition that the second derivative of the potential is zero with respect to position.

The surface conductivity of the starPEG–heparin hydrogel films was found to be strongly dependent on solution pH (Figure 3.3a). This property reflects the different ionization characteristics of the sulfate and carboxyl groups at

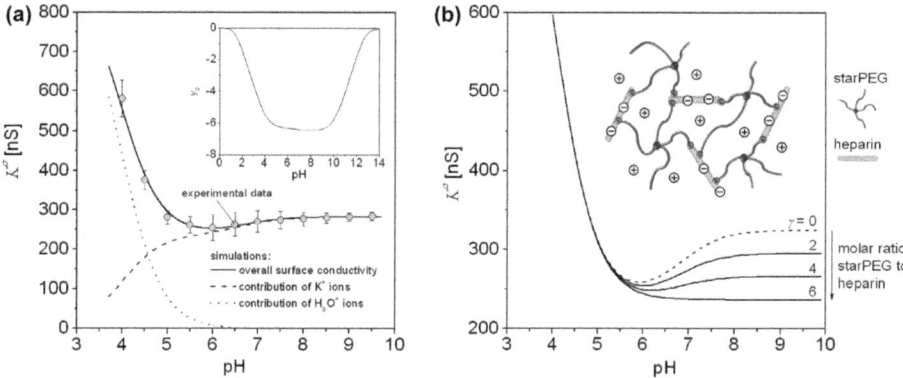

Figure 3.3 Surface conductivity, K^σ, and the dimensionless Donnan potential, y_D, of a starPEG–heparin hydrogel film (thickness $\sim$ 600 nm)[45] in 0.1 mM KCl solution at different pH values (a). The film was prepared at a molecular ratio starPEG to heparin of 3, *i.e.*, in the case of a quantitative reaction of the starPEG, a conversion of 12 of the 24 carboxyl groups of the heparin would occur. The experimental surface conductivity data were reproduced by the theory (solid line). The dashed and dotted lines show the contribution of K^+ ions and H_3O^+ ions to the overall surface conductivity. The simulation data on the right (b) illustrate the impact of the cross-linking degree on the surface conductivity of a starPEG hydrogel film under similar conditions. For further details see Zimmerman *et al.*[45]

Reprinted with permission from R. Zimmermann, S. Bartsch, U. Freudenberg, C. Werner. Electrokinetic analysis to reveal composition and structure of biohybrid hydrogels, *Anal. Chem.*, 2012, **84**, 9592–9595. Copyright 2012 American Chemical Society.

the heparin as well as the pH-dependent pattern of charge compensation. For pH < 5, the magnitude of K^σ is determined by the number and mobility of counter-ions that neutralize the charge of the strongly acidic sulfate groups. In the absence of the less acidic carboxyl groups, a plateau would arise at neutral and weak alkaline pH (see curve for the molar ratio of starPEG to heparin of $\gamma = 6$ in Figure 3.3b, corresponding to a complete conversion of the heparin carboxyl groups during the gel formation). Therefore, the increase of the surface conductivity above pH 6 un-ambiguously results from the ionization of non-converted carboxyl groups in the film. To illustrate the possible range of variations of K^σ with varying cross-linking degree, Figure 3.3b depicts simulation results for different molar ratios of starPEG and heparin. In the case of highly cross-linked hydrogels ($\gamma = 6$), all heparin carboxyl groups are involved in gel formation and, as mentioned above, K^σ remains nearly constant in the neutral pH range. In contrast, the dashed line in Figure 3.3b represents the upper boundary of K^σ increase due to the ionization of carboxyl groups. Weakly cross-linked hydrogel films[46] would show an increase in K^σ somewhat below this hypothetical curve. The surface conductivity further reflects the pH-dependent pattern of charge compensation in the hydrogel. The increase of K^{σ} below pH 5 can be attributed to the exchange of the K^+ ions, which predominantly compensate for the gel charge at neutral and alkaline pH, with H_3O^+ ions at lower pH values (see contributions of K^+ and H_3O^+ ions to K^{σ} in Figure 3.3a). H_3O^+ ions are about five times more mobile,[47] leading to the increase in K^{σ} observed.

The comparison between experimental data and simulation results (Figure 3.3a) revealed the concentration and pK values of the ionizable groups in the gel, *i.e.*, 53 mmol L^{-1} and p$K = 0.8$ for the sulfate groups and 10 mmol L^{-1} and p$K = 4$ for the carboxyl groups.[45] As an important prerequisite for the analysis of the gel composition and its cross-linking degree, heparin of known molecular weight and with known number of sulfate and carboxyl groups per molecule was employed for the formation of the hydrogel films.[45] As the sulfate groups are not involved in gel formation,[44] their concentration can be converted into the heparin concentration within the hydrogel. This calculation is straightforward and revealed a heparin concentration of 11.4 µg µL^{-1} for the hydrogel film, which agrees well with values obtained for similar macroscopic starPEG–heparin gels.[44] The heparin concentration obtained can be used further to derive information on the real cross-linking degree of the gel films. In excellent agreement with the theoretical value expected from the ratio starPEG to heparin, this analysis leads to an average number of 12.3 carboxyl groups per heparin molecule (*versus* 12 carboxyl groups expected for the molar ratio $\gamma = 3$).[45] The developed methodology has advantages over other analytical methods for the determination of the cross-linking degree *via* labeling of unreacted groups, because it is not prone to non-quantitative turnover or non-specific side effects.

The mean-field approach[45] can be applied further to quantify variations in the intrinsic sulfation pattern of the GAG components in the gel (*e.g.*, used to tune its interactions with signal molecules) and to investigate biomolecular interactions between the GAGs and soluble effectors. Altogether, the procedure introduced[45] provides new options for the interpretation of biophysical and biomolecular characteristics of GAG-based hydrogels for regenerative and sensoric applications. In particular, it enables the correlation of gel properties with the molecular transport and binding/release of cytokines, and thus the rational design of multibiofunctional polymer matrices.

3.3.3 Fluidity Modulation in Phospholipid Bilayers by Electrolyte Ions

Cell membranes separate the intracellular compartment from the extracellular medium and regulate important cellular processes such as signaling, organization of the cytoskeleton, protein sorting and apoptosis.[48,49] A key property of these membranes is their phase behavior. As lipid head groups consist of various acidic and/or basic moieties, the majority of lipids are charged in aqueous environments. The resulting electrostatic interactions within the lipid membranes, as well as between lipid head groups and ions from solution, are crucial for phase transition temperatures.[50] Here we present results that demonstrate how streaming current measurements were used in combination with measurements of the fluorescence recovery after photobleaching (FRAP) to analyze correlations between the charging and the fluidity of supported bilayer lipid membranes (sBLMs) prepared from 1,2-dioleoyl-*sn-glycero*-3-phosphatidylcholine (DOPC).[51,52]

The head group of DOPC contains both an acidic and a basic moiety that should charge up equally and in opposite directions at neutral pH. Therefore, on the basis of the molecular structure, one would expect an electrically neutral lipid membrane (no net charge) in the intermediate pH range.[53–55] However, streaming current measurements performed at different concentrations of the neutral background electrolyte KCl revealed a significant negative charge at neutral pH (Figure 3.4a). The isoelectric point was found to be about 4 at all salt concentrations in solution. Therefore, the authors of this study[51] concluded that the charging of the zwitterionic DOPC membranes is superposed by unsymmetrical water ion adsorption, *i.e.*, hydroxide ions (OH^-) are preferentially adsorbed as compared to hydronium ions

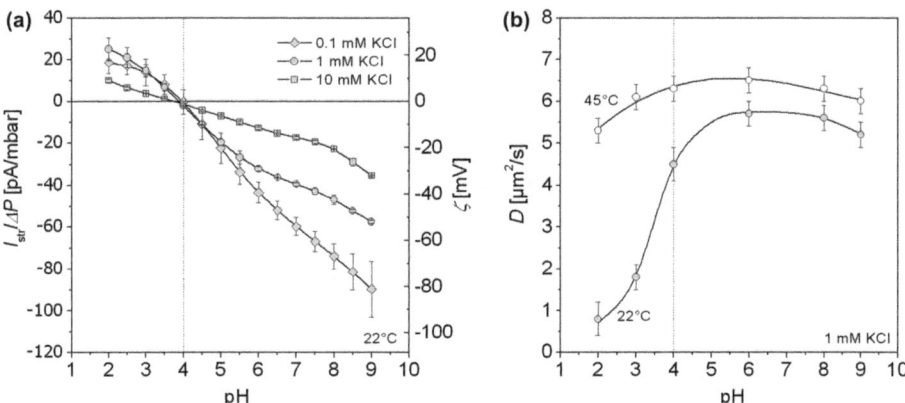

Figure 3.4 Ratio of streaming current over applied pressure, $I_{str}/\Delta P$, and zeta potential, ζ, of sBLMs prepared from DOPC on SiO_2 substrates as a function of pH for concentrations of 0.1 mM, 1 mM and 10 mM of the background electrolyte KCl (a) and diffusion coefficient of the DOPC membranes (doped with 1 mol% NBD-PE for FRAP) as a function of pH in 1 mM KCl solution at 22 °C and 45 °C (b). The streaming current measurements were performed at a channel height of 30 μm and a temperature of 22 °C.

(H_3O^+).[56] This phenomenon was found to be the origin of surface charge for different polymer materials without ionisable surface groups, but it is also proposed for explaining the origin of interfacial charges of dispersed organic liquids and air bubbles.[56] FRAP measurements performed under conditions similar to those for the electrokinetic measurements revealed a rather constant lipid mobility of (6 ± 1) μm^2 s^{-1} in the pH range from 9 to 6, a slight decrease at pH 4 to ~ 4.5 μm^2 s^{-1}, and a significant drop in mobility to about 1 μm^2 s^{-1} at pH 2. The observed behavior was found to be independent of solution ionic strength and vanished upon increasing the temperature to 45 °C (Figure 3.4b). The correlation between the sign reversal of the membrane charge at pH $\sim$ 4 in KCl and the significant reduction in lipid mobility under acidic pH conditions (at 22 °C) were interpreted in terms of a charge-induced transition from a fluid bilayer at neutral and alkaline pH to a gel/ordered bilayer phase at acidic pH. Obviously, the unsymmetrical adsorption of OH^- and H_3O^+ ions modulates the balance between attractive and repulsive forces in the DOPC membrane as well as between the membrane and the substrate and, thus, the phase transition temperature. The mechanisms underlying the observed transition are not directly obvious from the experiments and were discussed by Zimmermann *et al.*[51,52] with respect to ion-induced changes in the orientation and hydration of the lipid head groups and interactions of the membrane with the supporting SiO_2 substrate.

The experimental results reported for the supported DOPC membranes[51,52] were strikingly consistent with measurements of the electrophoretic mobility, dipole potential, and bending elasticity of SOPC vesicles.[57] In line with the streaming current and FRAP data for the supported DOPC membranes, Zhou and Raphael[57] reported an IEP of ~ 4 for vesicles prepared from the zwitterionic lipid 1-stearoyl-2-oleoyl-phosphatidyl-choline (SOPC). The elastic area compressibility modulus of the SOPC membranes was found to be unaffected between pH 3 and 9, but decreased by about 30% at pH 2, which was attributed to the presence of a gel phase.[57]

The results discussed so far refer to conditions where the measurements were performed in neutral background electrolytes. Taking these data as a reference, Zimmermann *et al.* further demonstrated how kosmotropic and chaotropic ions superimpose the charging by unsymmetrical water ion adsorption and how they influence the fluidity of DOPC membranes.[52] The IEP of the membranes was shifted towards the alkaline pH range in solutions containing kosmotropic cations, whereas lower IEPs were determined in the presence of chaotropic anions. The effects were more pronounced at higher solution concentrations for the more kosmotropic cations and chaotropic anions, respectively. Despite their different impact on membrane charging, both types of ions similarly influenced the fluidity of the bilayer under acidic pH conditions. Higher lipid mobilities were found in the acidic pH range in the presence of kosmotropic cations and chaotropic anions compared with FRAP measurements in KCl solutions. This observation was attributed to the adsorption of the ions at polar sites in the lipid head group region and to the resulting variations of the head group orientation and lipid packaging density. At the macroscopic scale, these effects manifest themselves in a variation (here a decrease) of the phase transition temperature.

In summary, the results discussed in this section demonstrate the application of electrokinetic measurements in combination with FRAP for analyzing the impact of charging and ion-specific interactions on phase transitions in sBLMs. The methodology can be analogously applied to practically more relevant membrane compositions and, thus, contribute to clarifying the influence of these processes on membrane formation, stability, structure, and fluidity.

3.4 Summary

Electrokinetic measurements provide a versatile tool for analyzing the charge and structure of soft biopolymer films in contact with aqueous solutions. Recent progress was achieved in interpreting the electrokinetics of complex systems by the development of sophisticated theories on the electrohydrodynamics of diffuse soft interfaces, and with merging of experimental techniques for the concomitant determination of surface conductivity and film swelling. The fruitful options arising from these various developments are illustrated here with selected case studies. In particular, we provide insights into the experimental strategies and theoretical procedures for investigating the interrelations between the electric charging and structure of stimuli-responsive coatings, and for evaluating the key ionization characteristics, composition, and cross-linking degree of biohybrid hydrogels. Finally, we demonstrate the suitability of combined electrokinetic and FRAP measurements to understand the variations in phospholipid bilayers fluidity through interactions with electrolyte ions. These examples cover only a small range of biomaterials and processes that can be studied by means of electrokinetics. By way of further illustration, and when combined with other analytical methods such as reflectometric interference spectroscopy,[58,59] reflectometry[60] or Fourier transform infrared spectroscopy, the methodology outlined here enables the *in situ* investigation of adsorption–desorption processes and the analysis of the impact of charge on the secondary structure of biopolymers at interfaces. Future developments will include the analysis of electrokinetics and surface conductivity of soft films under dynamic ion-transport conditions. Such measurements could extend our understanding on the binding and release of charged analytes, *e.g.*, proteins, nanoparticles or heavy metal ions, to/from soft polymer materials. With the increasing complexity of the systems studied, the field would also benefit from molecular approaches that take into account ion-specific characteristics, spatial location, and chemical environment of ionisable groups within the film and local fluctuation of polymer chains.

A Appendix: Methods

A.1 Measurement of Streaming Current/Potential

Various experimental conditions must be met to minimize the error in the electrokinetic experiment and ensure the applicability of the theory outlined in this chapter:

- Always ensure laminar flow in the measuring cell. This condition is met if the Reynolds number is smaller than 2300. Furthermore, the flow profile should be fully developed, *i.e.*, the hydrodynamic entrance length should be much smaller compared with the length of the channel.[28]
- Polarization of the electrodes can cause significant error in the measurement of the streaming current. To avoid polarization, use reversible (*e.g.*, Ag/AgCl or calomel) electrodes for the measurement of the streaming current.
- The magnitude of the streaming current/potential should be unambiguously associated with the charge and structure of the investigated system and not be superposed by asymmetry currents/potentials of the electrodes. To check the system for asymmetry, perform streaming current/potential measurements at various difference pressures across the streaming channel (in both flow directions). Asymmetry of the electrodes is negligible, if the plot of the streaming current/potential *vs.* the applied pressure yields a straight line with an intercept of zero. The contributions of the electrodes to the measured quantities (due to asymmetry) can be otherwise eliminated by calculating the streaming current/potential *versus* pressure gradient by linear regression. The slopes obtained can be used for the further evaluation of the experimental data, *e.g.*, for comparison with the theory as demonstrated in Section 3.3.

A.2 Simulations of the Electro-Hydrodynamics at Diffuse Soft Interfaces

The theoretical framework for the electro-hydrodynamics at diffuse soft interfaces deals with various spatial scales of significantly different magnitudes: the Debye length (from few nanometers to about 100 nm under practical conditions of electrolyte concentration), the spacing between soft layers in the electrokinetic cell (1 μm to 60 μm for the MICROSLIT set up used in this work), the Brinkman length (tens of nanometers at most), and the thickness of the soft surface layer (few nanometers for thin polymer films to several micrometers for thick hydrogels). Accordingly, finding accurate solutions of the governing equations often requires resorting to numerical algorithms other than those associated with standard finite element calculations. Instead, the use of the collocation method as developed by

Ascher *et al.*[38] in the COLSYS package is particularly powerful. Indeed, it efficiently approximates the searched electrostatic and hydrodynamic field distributions through spline-collocation at Gaussian nodes, and selects the relevant mesh subdivision following an autoadaptative strategy.

A.3 Thin Polymer Films for Streaming Current/Potential Measurements

The characterization of polymer films by streaming current/potential measurements requires chemical and mechanical stability (shear forces due to the hydrodynamic flow) of the films under the experimental conditions. If possible, covalently attach the polymers to the supporting surface (typically glass or silicon with a natural or thermal oxide layer) using reactive pre-coatings. In cases of polymers without reactive groups, the film stability can be improved by silanization of the carrier surface (for example, with hex-amethyldisilazane in the case of hydrophobic polymers such as Teflon® AF or polystyrene). Supported bilayer lipid membranes should be prepared directly in the measuring cell by injecting the vesicle solution into the channel formed by the supporting substrates.

The stability of the prepared films/membranes should be checked by re-peating the measurements in the same pH range at a given concentration of the background electrolyte. In the ideal case, the streaming current/potential *versus* pH plots are identical for all measurements for the same pair of sam-ples. Shifts in the IEP often indicate chemical variations of the polymer or delamination. Should stability problems arise, complementary methods like ellipsometry or photoelectron spectroscopy could help identify the source of the variation in the 'electrokinetic fingerprint' of the polymer films.

References

1. D. Myers, *Surfaces, Interfaces, and Colloids*, Wiley, New York, 2nd edn, 1999, ch. 5, pp. 79–96.
2. C. Dicke and G. Hähner, *J. Am. Chem. Soc.*, 2002, **124**, 12619–12625.
3. R. Zimmermann, S. S. Dukhin and C. Werner, *J. Phys. Chem. B*, 2001, **105**, 8544–8549.
4. A. Dickey and R. Faller, *Biophys. J.*, 2008, **95**, 2636–2646.
5. I. Capila and R. J. Linhardt, *Angew. Chem., Int. Ed.*, 2002, **41**, 391–412.
6. M. R. Yeaman and N. Y. Yount, *Pharmacol. Rev.*, 2003, **55**, 27–55.
7. P. Roach, D. Eglin, K. Rohde and C. C. Perry, *J. Mater. Sci.: Mater. Med.*, 2007, **18**, 1263–1277.
8. A. Rullo and M. Nitz, *Biopolymers*, 2010, **93**, 290–298.
9. T. Crouzier, T. Boudou and C. Picart, *Curr. Opin. Colloid Interface Sci.*, 2010, **15**, 417–426.
10. L. S. Puah, R. Sedev, D. Fornasiero and J. Ralston, *Langmuir*, 2010, **26**, 17218–17224.

11. A. Yayon, M. Klagsbrun, J. D. Esko, P. Leder and D. M. Ornitz, *Cell*, 1991, **64**, 841–848.
12. A. Y. Wang, S. Leong, Y.-C. Liang, R. C. C. Huang, C. S. Chen and S. M. Yu, *Biomacromolecules*, 2008, **9**, 2929–2936.
13. R. N. Prince, E. R. Schreiter, P. Zou, H. S. Wiley, A. Y. Ting, R. T. Lee and D. A. Lauffenburger, *J. Cell Sci.*, 2010, **123**, 2308–2318.
14. X. Liu, W. Farmerie, S. Schuster and W. Tan, *Anal. Biochem.*, 2000, **283**, 56–63.
15. A. Baerga-Ortiz, A. R. Rezaie and E. A. Komives, *J. Mol. Biol.*, 2000, **296**, 651–658.
16. M. M. Rohani and A. L. Zydney, *J. Membr. Sci.*, 2009, **337**, 324–331.
17. M. Inukai and M. Yonese, *Chem. Pharm. Bull.*, 1999, **47**, 1059–1063.
18. J. Lyklema, *Fundamentals of Interface and Colloid Science, Vol. II: Solid-Liquid Interfaces*, Academic Press, London, 1995.
19. Á. V. Delgado, F. González-Caballero, R. J. Hunter, L. K. Koopal and J. Lyklema, *J. Colloid Interface Sci.*, 2007, **309**, 194–224.
20. S. Wall, *Curr. Opin. Colloid Interface Sci.*, 2010, **15**, 119–124.
21. C. Werner and J. Lyklema, *Curr. Opin. Colloid Interface Sci.*, 2010, **15**, 117–118.
22. A. C. Barbati and B. J. Kirby, *Soft Matter*, 2012, **8**, 10598–10613.
23. J. F. L. Duval and F. Gaboriaud, *Curr. Opin. Colloid Interface Sci.*, 2010, **15**, 184–195.
24. H. Ohshima, *Curr. Opin. Colloid Interface Sci.*, 2013, **18**, 73–82.
25. R. Zimmermann, S. S. Dukhin, C. Werner and J. F. L. Duval, *Curr. Opin. Colloid Interface Sci.*, 2013, **18**, 83–92.
26. J. F. L. Duval, H. P. van Leeuwen, J. Cecilia and J. Galceran, *J. Phys. Chem. B*, 2003, **107**, 6782–6800.
27. R. Zimmermann, T. Osaki, R. Schweiss and C. Werner, *Microfluid. Nanofluid.*, 2006, **2**, 367–379.
28. C. Werner, H. Körber, R. Zimmermann, S. S. Dukhin and H.-J. Jacobasch, *J. Colloid Interface Sci.*, 1998, **208**, 329–346.
29. L. P. Yezek, J. F. L. Duval and H. P. van Leeuwen, *Langmuir*, 2005, **21**, 6220–6227.
30. J. F. L. Duval, R. Zimmermann, A. L. Cordeiro, N. Rein and C. Werner, *Langmuir*, 2009, **25**, 10691–10703.
31. R. Zimmermann, D. Kuckling, M. Kaufmann, C. Werner and J. F. L. Duval, *Langmuir*, 2010, **26**, 18169–18181.
32. J. F. L. Duval, D. Küttner, M. Nitschke, C. Werner and R. Zimmermann, *J. Colloid Interface Sci.*, 2011, **362**, 439–449.
33. J. F. L. Duval, D. Küttner, C. Werner and R. Zimmermann, *Langmuir*, 2011, **27**, 10739–10752.
34. H. C. Brinkman, *Appl. Sci. Res., Sect. A*, 1947, **1**, 27–34.
35. R. Toomey, D. Freidank and J. Rühe, *Macromolecules*, 2004, **37**, 882–887.
36. M. J. N. Junk, I. Anac, B. Menges and U. Jonas, *Langmuir*, 2010, **26**, 12253–12259.

37. S. Sanjuan, P. Perrin, N. Pantoustier and Y. Tran, *Langmuir*, 2007, **23**, 5769–5778.
38. U. Ascher, J. Christiansen and R. D. Russell, *ACM Trans. Math. Software*, 1981, **7**, 209–222.
39. Z. Ding, R. B. Fong, C. J. Long, P. S. Stayton and A. S. Hoffman, *Nature*, 2001, **411**, 59–62.
40. A. L. Cordeiro, M. E. Pettit, M. E. Callow, J. A. Callow and C. Werner, *Biotechnol. Lett.*, 2010, **32**, 489–495.
41. D. Schmaljohann, J. Oswald, B. Jorgensen, M. Nitschke, D. Beyerlein and C. Werner, *Biomacromolecules*, 2003, **4**, 1733–1739.
42. N. A. Peppas, J. Z. Hilt, A. Khademhosseini and R. Langer, *Adv. Mater.*, 2006, **18**, 1345–1360.
43. M. P. Lutolf and J. A. Hubbell, *Nat. Biotechnol.*, 2005, **23**, 47–55.
44. U. Freudenberg, J.-U. Sommer, K. R. Levental, P. B. Welzel, A. Zieris, K. Chwalek, K. Scheider, S. Prokoph, M. Prewitz, R. Dockhorn and C. Werner, *Adv. Funct. Mater.*, 2012, **22**, 1391–1398.
45. R. Zimmermann, S. Bartsch, U. Freudenberg and C. Werner, *Anal. Chem.*, 2012, **84**, 9592–9595.
46. U. Freudenberg, A. Hermann, P. B. Welzel, K. Stirl, S. C. Schwarz, M. Grimmer, A. Zieris, W. Panyanuwat, S. Zschoche, D. Meinhold, A. Storch and C. Werner, *Biomaterials*, 2009, **30**, 5049–5060.
47. D. R. Lide and H. P. R. Frederikse, *CRC Handbook of Chemistry and Physics*, CRC Press, Boca Raton, 78th edn, 1995, ch. 5, pp. 5-91–5-92.
48. K. Simons and D. Toomre, *Nat. Rev. Mol. Cell Biol.*, 2000, **1**, 31–39.
49. S. Munro, *Cell*, 2003, **115**, 377–388.
50. H. Träuble and H. Eibl, *Proc. Natl. Acad. Sci. U. S. A.*, 1974, **71**, 214–219.
51. R. Zimmermann, D. Küttner, L. Renner, M. Kaufmann, J. Zitzmann, M. Müller and C. Werner, *Biointerphases*, 2009, **4**, 1–6.
52. R. Zimmermann, D. Küttner, L. Renner, M. Kaufmann and C. Werner, *J. Phys. Chem. A*, 2012, **116**, 6519–6525.
53. R. Vacha, S. W. I. Siu, M. Petrov, R. A. Böckmann, J. Barucha-Kraszewska, P. Jurkiewicz, M. Hof, M. L. Berkowitz and P. Jungwirth, *J. Phys. Chem. B*, 2010, **114**, 9504–9509.
54. A. A. Gurtovenko and I. Vattulainen, *J. Phys. Chem. B*, 2008, **112**, 4629–4634.
55. J. J. Garcia-Celma, L. Hatahet, W. Kunz and K. Fendler, *Langmuir*, 2007, **23**, 10074–10080.
56. R. Zimmermann, U. Freudenberg, R. Schweiß, D. Küttner and C. Werner, *Curr. Opin. Colloid Interface Sci.*, 2010, **15**, 196–202.
57. Y. Zhou and R. M. Raphael, *Biophys. J.*, 2007, **92**, 2451–2462.
58. R. Zimmermann, T. Osaki, T. Kratzmüller, G. Gauglitz, S. S. Dukhin and C. Werner, *Anal. Chem.*, 2006, **78**, 5851–5857.
59. R. Zimmermann, T. Osaki, G. Gauglitz and C. Werner, *Biointerphases*, 2007, **2**, 159–164.
60. O. Theodoly, L. Cascão-Pereira, V. Bergeron and C. J. Radke, *Langmuir*, 2005, **21**, 10127–10139.

CHAPTER 4

Growth Factor Delivery Systems for the Treatment of Cardiovascular Diseases

NATALIA ZAPATA,[a] ELISA GARBAYO,[b]
MARIA J. BLANCO-PRIETO[b] AND FELIPE PROSPER*[a,c]

[a] Laboratory of Cell Therapy, Division of Oncology Foundation for Applied Medical Research, Clinica Universidad de Navarra, University of Navarra, Spain; [b] Pharmacy and Pharmaceutical Technology Department, School of Pharmacy, University of Navarra, Pamplona, Spain; [c] Hematology and Cell Therapy Service, Clinica Universidad de Navarra, University of Navarra, Spain
*Email: fprosper@unav.es

4.1 Introduction

According to World Health Organization (WHO), men and women around the world have gained little more than 10 years of life expectancy in general, but they spend more years living with injuries and illnesses. Heart disease and cancer have become the predominant causes of death and disability. Although a large proportion of cardiovascular disease (CVD) can be prevented, their incidence continues to rise. Almost half of the 36 million deaths from non-communicable diseases (NCDs) are caused by cardiovascular diseases, mainly due to insufficient preventive measures.[1]

Smoking, physical inactivity, unhealthy diet and alcohol abuse are major behavioral risk factors of CVD. These risk factors are common to other major

RSC Smart Materials No. 10
Biointerfaces: Where Material Meets Biology
Edited by Dietmar Hutmacher and Wojciech Chrzanowski
© The Royal Society of Chemistry 2015
Published by the Royal Society of Chemistry, www.rsc.org

non-communicable diseases such as cancer, diabetes and chronic respiratory diseases. Long-term exposure to behavioral risk factors leads to hypertension, increased blood glucose, dyslipidemia and obesity. Lipids are likely to deposit inside the arteries causing the inner surface of blood vessels to become irregular, so that it is difficult for blood to flow through, the disease process known as atherosclerosis. Eventually, fatty plaque can rupture, causing the formation of a blood clot in the heart resulting in coronary arterial disease.[2]

According to Mendis *et al.*, CVDs can be classified into two groups: (1) those due to atherosclerosis such as ischemic heart disease or coronary artery disease, cerebrovascular disease, diseases of the aorta and arteries, including hypertension and peripheral vascular disease; and (2) other CVDs, such as congenital heart disease, rheumatic heart disease, cardiomyopathies and cardiac arrhythmias. Ischemic heart disease and stroke have increased in the last two decades by 29% and 19%, respectively, with ischemic heart disease the first cause of disability-adjusted life years in the last regional ranking.[2]

When the blood flow to the heart is interrupted due to a ruptured atherosclerotic plaque, the decrease in the supply of oxygen and nutrients can damage the heart muscle. After the onset of myocardial ischemia histological cell death can occur and could lead to complete necrosis depending on the presence of collateral circulation to the ischemic zone, persistent or intermittent coronary arterial occlusion, the sensitivity of the cardiomyocytes to ischemia and the individual demand for oxygen and nutrients.[3]

During post-infarction Collagen fibers decrease in number and inflammatory cells infiltrate the necrotic tissue which leads to dilatation and wall thinning of the infarcted area, an increase in sarcomere length and a reduction in intercellular space; these changes are known as cardiac remodeling. This is a compensatory post-infarction mechanism that initially decreases wall stress and increases cardiac output and stroke volume. During this process cardiomyocytes suffer modifications in size and shape and in the molecular mechanisms involved in contraction and relaxation. At first, these adaptations seem to be beneficial; however, as the ventricle continues to dilate wall stress increases in the infarcted ventricle. Besides, matrix protein deposition and scar formation become evident constituting a maladaptive response leading to contractile dysfunction, arrhythmias and heart failure (HF).[4–6]

Percutaneous coronary intervention practice facilitates rapid relief of acute occlusion and pharmacological treatment constitutes the standard therapy but both could lead to HF and a posterior transplant requirement.[5,7] The application of regenerative therapies aim at preserving or restoring lost myocardial function which promotes long-term survival. These treatments are based on understanding the intrinsic cardiac repair process in order to modify inflammation in the ischemic zone, avoid cardiomyocyte apoptosis or stimulate neovascularization, cell survival, myocardial contraction and cardiomyogenesis. These novel methods include gene therapy and tissue engineering that comprises cell therapy and protein therapy.[5]

Nagai *et al.* extensively reviewed the use of cell therapy in the failing heart in preclinical and clinical studies and the use of pluripotent stem cells as source for generating cardiomyocytes.[8] However, efficacy, quality, effective translation of preclinical studies and clinical applicability of this therapy is still unresolved. Additionally, promotion of changes in cardiac performance after cell therapy in an infarcted heart seems to be due to molecules secreted by transplanted cells.[5,8]

Since the intrinsic process of cardiac repair is stimulated by both mechanical and biochemical factors like hormones and growth factors (GF), the secretome of transplanted cells may lead to reveal target molecules for novel strategies of treatment based on combination of cell, protein and gene therapy.

Currently much is known about the role of GF in myocardial infarction (MI; reviewed by Formiga *et al.*[9]). For instance placental growth factor (PlGF),[10] transforming growth factor-β (TGF-β) and vascular endothelial growth factor (VEGF) recognized pro-angiogenic factors, induce endothelial cell proliferation, migration, tube formation and vessel maturation.[11] In contrast, transforming growth factor-β1 (TGF-β1),[12] connective tissue growth factor (CTGF)[13] or insulin growth factor (IGF)[14] have been related to differentiation of cardiomyocytes and collagen deposition during cardiac remodeling post-infarction and platelet-derived growth factor (PDGF)[15] seems to induce smooth muscle cell proliferation. Some of this GF can act at different levels during this process; the family of fibroblast growth factor (FGF) are considered cardioprotective agents and differentiation cues for stem cells and are also implicated in inducing smooth muscle cell proliferation.[16,17] Some factors like hepatocyte growth factor (HGF) and granulocyte colony stimulating factor (G-CSF) have been related directly to inflammatory response;[5] the first one inhibits CD4 T-cell proliferation and the other increases proliferation and differentiation of neutrophils. Other GFs, such as angiopoietins,[18] are involved in several cardiovascular remodeling and vascular processes. Finally, yes-associated protein-1 (Yap-1)[19] is a known important regulator of cardiomyocyte proliferation and embryonic heart development while stromal-cell derived factor-1 (SDF-1)[20] may be involved in progenitor cell recruitment into the infarct zone as a result of its chemoattractant property. Recently, neuregulin-1 (Nrg1b) a cardioactive growth factor released from endothelial cells has attracted great attention.[21] This protein plays a crucial role in the adult cardiovascular system by inducing sarcomere membrane organization and integrity, cell survival and angiogenesis. Delivery of GFs into infarcted zones could enhance the intrinsic capacity of the heart to repair itself or regenerate after damage making it a promising treatment for MI.

GFs have been administered *via* intravenous, intracoronary, intramyocardial and perivascular routes in clinical studies. FGF-1 was first administrated *via* the intramyocardial (IM) route in 40 patients with heart disease and after 12 weeks a capillary network next to the FGF-1 injection area, as well as a local blood supply increase, were observed.[22] Nevertheless the latest clinical trial with FGF-2 did not show any difference between the placebo and treated groups.[23] VEGF has been injected *via* both the intracoronary and intravenous routes, showing an improvement in myocardial perfusion and probing the safety and tolerability of the treatment and a dose-dependent effect.[24,25] Despite these results Henry *et al.* did not find any significant difference between non-treated and VEGF-treated groups using a larger sample of patients, although they reported an improvement in quality of life. Another GF that has been administered to patients subcutaneously is G-CSF. The first trial promoted mononuclear CD34$^+$ cell mobilization, which correlated with better ventricular function preservation and less remodeling.[26] The following trials performed by Zohlnhöfer *et al.*[27] and Ripa *et al.*[28] also observed a significant stem cell mobilization but no impact on infarct size, left ventricular function or coronary restenosis.

The negative results on these trials have evidenced some limitations to overcome in future studies. Maintenance of GF half-life at the accurate concentration in the myocardium and the route of administration are the main challenges. These are labile molecules that are degraded in a very short period of time when directly administered to the organism, and that is why intracoronary injection was thought to be effective but low protein deposition was seen. Other methods, such as intrapericardial or intravenous administration, are less attractive in a post-cardiac surgery patient.[9]

Controlled drug delivery systems have been developed to overcome these limitations, maintaining precise concentrations of active GFs over days or weeks and drug protection emulating the innate extracellular matrix, since they can be made of different materials and they can incorporate two or more therapeutic proteins with different release profiles. Currently, these systems are at the preclinical stage of safety and efficacy evaluation.

In the following section we review the important criteria to design GF delivery systems, beginning with an overview of biomaterials and GF in order to deepen in DDS of GF for MI treatment, what has been done so far and the challenges faced by researchers in this field.

4.2 Drug Delivery Systems for Growth Factors

Once the GFs are synthesized they are released by cells for immediate signaling or are sequestered in the extracellular matrix (ECM) providing physical cues for cells through spatial presentation. ECM degradation coordinates the release of GFs in a controlled manner depending on cell requirements.[29]

Drug delivery systems (DDSs) using biomaterials that emulate natural ECM behavior might allow temporal distribution of specific GFs, at the right time and place and also would avoid the rapidly diffusion of these proteins from the injection site or their degradation by proteases present in inflammatory environments like infarcted tissue.[30,31]

4.2.1   Growth Factors

Once a GF interacts with its receptor it leads to changes in gene expression generating an integrated and multifaceted biological response on cells such as chemotactic, mitogenic, morphogenic, apoptotic, metabolic effects or the combination of several of these.[32,33]

The effect of a GF in directing the phenotype of both stem and differentiated cells depends on a variety of factors such as time of action, concentration, micro-environmental cues besides the presence of other GFs. Small amounts of a GF (picograms to nanograms) are needed to see an effect, which is why regulation of GFs occurs at different levels from inhibition of transcription to the degradation of the peptide.[33]

When thinking of GF molecules for therapy it is important to take into account that they are not conventional drugs. Herein we review some criteria when designing a DDS for GFs:

- The identification of the key GFs for delivery in a particular application should be based on an understanding of their biological developmental processes, their characterization and whether they are available in large quantities as recombinant proteins.[33]
- The GF or factors must target the chosen tissue when administered, and side effects should be avoided[33] by using local delivery or functionalization of the particles.
- The loading capacity of the DDS that corresponds to the amount of growth factor that can be included into the system should be assessed.[34]
- Often, fabrication of DDSs could cause degradation or inactivation of the peptide, changes in its metabolic half-life and immunogenicity. The strategies to preserve protein stability are based on an understanding of the degradation mechanisms and the effect of changes in the storage conditions and formulation conditions such as pH, ionic strength, temperature, and buffer composition.[9,33,35] A profound knowledge of the nature of the GF or factors, their degradation routes and micro-environmental conditions *in situ* would help in the choice of the most accurate DDS fabrication method. Adding carriers that contain the GF also help to preserve their bioactivity as well as their appropriate release.[34]
- The release profile of the GF from the DDS should be controlled temporally and spatially, allowing the appropriate dose of GF to reach the cells over a given period of time. DDS should be able to retain the GF or factors at the target tissue to see an effect. The range of effective concentration thresholds has to be considered. Spatial localization of the signaling molecule allows their direct effect within the tissue to be followed up.[9,34]
- The binding affinity defines how tightly the GF binds to the DDS; this must be sufficiently low to allow release, but high enough to prevent uncontrolled release.[34]
- To be economically viable, the DDS must be easy to manufacture and handle, and be cost-competitive.

4.2.2 Biomaterials

Biomaterials are synthetic or natural substances with unique properties of biodegradability and biocompatibility, with the capacity to transport nutrients and metabolites, the ability to regulate cell morphology and differentiation and have the presence of bioactive ligands for cell attachment. All of them can be chemically or physically modified.[30] The material of choice, whether it is natural or synthetic, will depend on the application. The most common biomaterials used in the cardiac repair context has been revised in Formiga *et al.*[9] and in Diaz-Herraez *et al.*[36]

Materials from natural sources, such as collagens, or naturally derived polymers, such as alginate, quitosan, gelatin or hyaluronic acid (HA), are biologically recognizable but are difficult to purify and could generate immunogenicity. In contrast, synthetic materials such as the polymers poly(lactic-*co*-glycolic) acid (PLGA), polycaprolactone (PCL), poly(ethylene glycol) (PEG), poly(vinyl alcohol) (PVA), poly(acrylic acid) (PAA), poly(propylene fumarate-*co*-ethylene glycol) [P(PF-*co*-EG)] and the polyketals present a stable composition.[34] Although these polymers are tough, their mechanical properties can be easily manipulated but they lack bioactive sites to interact with soluble proteins or cells, which could add complexity to the synthesis process in order to improve specificity.

Polymeric biomaterials are commonly and successfully used for administrating small doses of drugs or proteins like GF at defined dose rates directly to target cells. These DDSs generally contain an aqueous phase with the soluble molecule embedded in a degradable phase. They can also be designed to monitor their degradation rates shifting the molecular weight and co-polymerization ratio, present specific ligand-based signals, and/or control the release in response to the microenvironment.[30] These last are called smart delivery systems since they degrade in response to physiological or disease-specific signals with the advantage of reducing side effects and increasing efficacy of a variety of therapeutic agents.[37] Polymeric DDSs for GFs can be built as hydrogels, nano- and microparticles and scaffolds among others.

The use of lipids in drug delivery is a new trend for the delivery of poorly soluble drugs and for peptide and protein delivery. Materials used in lipid-based formulations include fatty acids, glycerides, phospholipids, sphingolipids, waxes and sterols among others. Several lipid-based DDSs have been developed over the last few years. Among them, liposomes and solid lipid particulate systems have been used successfully for GF delivery.[38]

Below we describe the most used vehicles explored so far for cardiac protein delivery (Figure 4.1), pitfalls and challenges of this area.

4.2.2.1 Hydrogels

Hydrogels are three-dimensional cross-linked polymeric networks which are able to absorb an aqueous solvent and release it in a controlled manner.

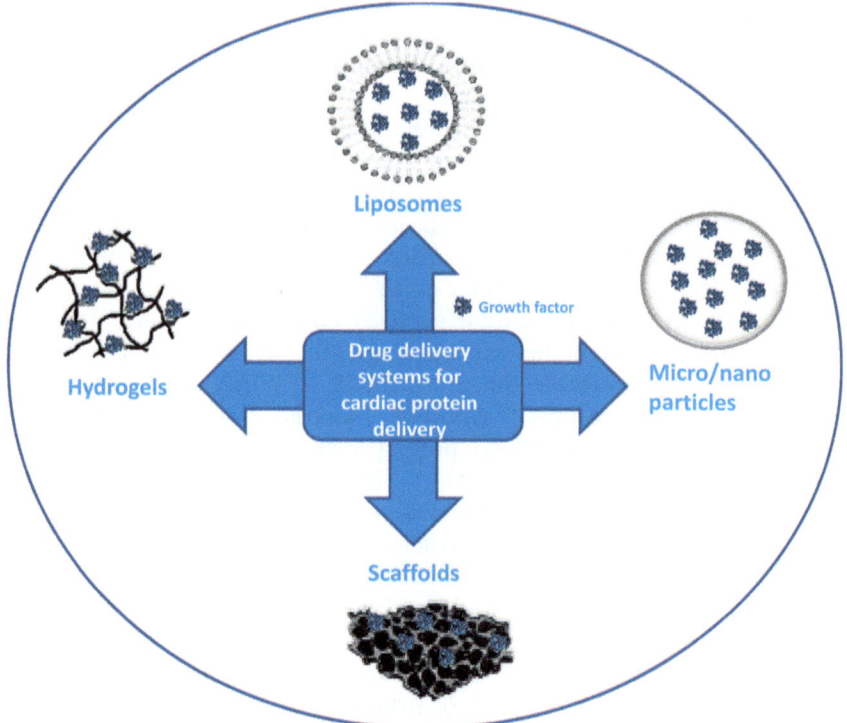

Figure 4.1 Systems for cardiac growth factor delivery: hydrogels, microparticles and nanoparticles, scaffolds and liposomes.

They can be classified as physical or chemical gels depending on the process by which the cross-links within the networks are produced. The synthesis method determines their properties. Physical hydrogels are less toxic which would be desired for biomedical use, however chemical gels have better mechanical properties but the agents involved in their synthesis could damage the molecules (*e.g.*, drugs, proteins) added in the solvent.[39]

GFs can be directly loaded into hydrogels showing a rapid burst release during the initial swelling phase, followed by the extended release of protein retained in the gel. It has been proved that the release profile can depend on the GF, since the interaction with the same biomaterial is different for each factor and the cross-linking density of the polymer networks can be changed in order to ensure the presence of the GF the needed time.[40] A synthetic hydrogel–DDS needs to be modified in order to induce release of the GF according to the cellular demands at the target site. This can be achieved by including strategies such as specific substrate–enzyme coating or inducing physical responsiveness in the target tissue.[41,42]

A large variety of hydrogels, both synthetic and natural were first directly loaded with GFs. Collagen type I hydrogel with the heparin-binding GFs was built to test their release profile *in vitro*;[43] FGF-2 was incorporated in a hydroxypropylcellulose (HPC) hydrogel for a clinical trial[44] as well as platelet derived growth factor-BB (PDGF-BB),[45] suggesting the necessity of improving direct loading. Variations on the cross-linking density,[46] the amount of cross-linking agent[47] or the charge of functional groups[43] could help to modify the releasing rate of proteins contained in hydrogels.

The incorporation of carriers within the hydrogels systems could help to extend the delivery period of time of proteins. The carriers are usually designed as particles containing the molecule of interest in the middle or as implants where the protein is dispersed within the polymeric network. Carries can be built with different biodegradable and non-biodegradable materials. The presence of a concentration gradient in the target site is mandatory when using a non-biodegradable carrier, in contrast to biodegradable carriers which release molecules along with the degradation of the hydrogel system.[48,49]

Proteins can also be covalently attached to hydrogels by the reaction of the different side chain functionalities of polymers with the amino acids of the GFs. For instance, PVA hydrogels have two neighboring hydroxyl groups which can be used to attach peptide sequences. On the other hand, PEG-based gels need a cell adhesion mediating peptide previous the incorporation of GFs. In the literature, it is also reported the covalent binding of PEG to epidermal growth factor (EGF), bFGF and TGF-b2.[34,50]

Hydrogels can be provided with enzymatic motifs to simulate ECM behavior allowing progressively release of GFs due to the matrix metalloproteinase (MMP)-mediated degradation. Zisch's group immobilized an adhesion peptide (RGD) and VEGF within a metallo-proteinase-sensitive synthetic hydrogel networks. An active VEGF liberation was confirmed from these hydrogels by incubation with MMP-2 or plasmin. Besides, when implanted subcutaneously in rats, these VEGF-containing matrices were completely remodeled into native vascularized tissue.[51]

A wide variety of hydrogels have been used to improve heart function after infarction. Gelatin hydrogels loaded with FGF-2 injected *via* the intramyocardial route in animal models of myocardial infarction seem to induce significant angiogenesis and improve left ventricular function. Moreover, alginate hydrogels alone or loaded with GFs have been demonstrated to be effective therapeutic systems when injected in the cardiac muscle after myocardial infarction.[6,32,52–54]

4.2.2.2 Microparticles and Nanoparticles

Microparticles can be classified as particles between 1 and 1000 μm while nanoparticles are those within a nanoscale (10–1000 nm). These systems seek to avoid the limitations of intravenous administration of therapeutic proteins. Same design parameters as liposomes need to be pursued here. Device biocompatibility and size are some of them.

Particles prepared with biodegradable polymers like PLGA have been widely used to deliver GFs to the heart due to its excellent biocompatibility and high safety profile. Various techniques for particle preparation have been reported being the double emulsion solvent evaporation method the most common method used to encapsulate proteins (Figure 4.2A). The main factors influencing the drug encapsulation efficiency using the water–oil–water solvent evaporation method are the viscosity of the internal water phase, the volume of the inner and outer water phase and the concentration of the polymer in the organic phase (Figure 4.2B). Other methods for preparing particles are coacervation, hot-melt technique and spray drying technique, among others.

A

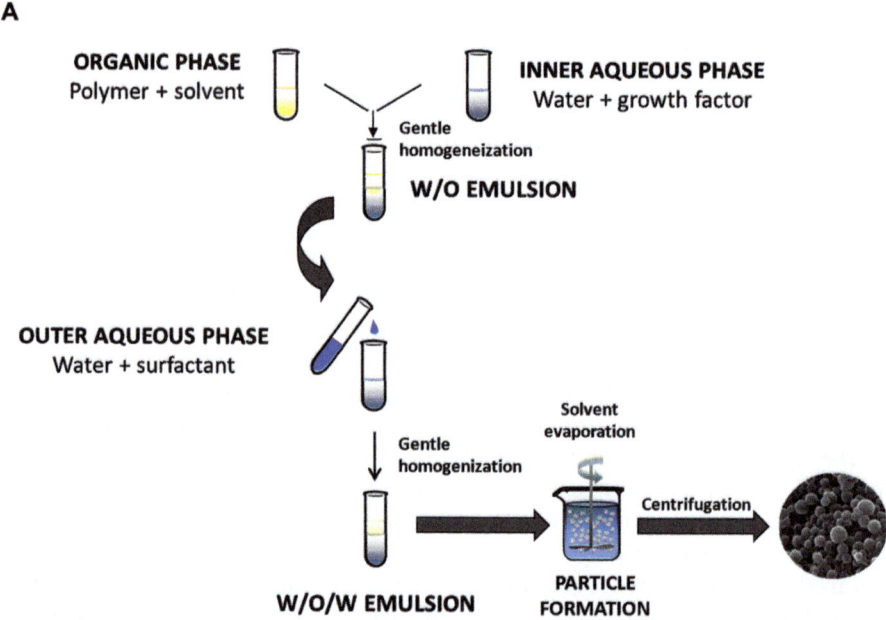

B

PARAMETERS	EFFECT
Increasing the volume of the inner water phase	-
Increasing the viscosity of the inner water phase	+
Increasing volume of the outer water phase	+
Increasing the viscosity of the organic phase	+
Increasing the amount of growth factor	+

Figure 4.2   (A) Schematic diagram of microparticle preparation by a water–oil–water (w–o–w) emulsion solvent evaporation method. (B) Factors affecting drug encapsulation efficiency using the w–o–w solvent evaporation method.

The combination of materials can allow more complex structures and patterns for GF encapsulation because it depends on hydrophobic–hydrophilic interactions among the molecules and polymers. As well as hydrophobicity, molecular weight distribution and porosity of microparticles can be manipulated in order to exercise control over GF delivery and prevent the loss of protein during manufacturing process.[55]

As well as for the other systems described above, a variety of synthetic polymers, including poly(α-hydroxy acids), poly(orthoesters), poly(anhydrides), poly(amino acids), dextrin, polyglycolic acid (PGA), poly(L-lactide) (PLA) and their co-polymers (PLG acid) have been used for GF encapsulation in micro and nanoparticles.[56] Also natural materials such as silk, keratin, collagen, gelatin, fibrinogen, elastin, chitosan, hyaluronic acid, cellulose and alginate have gained wide attention as drug carriers. Gelatin microspheres and combinations of glycidyl methacrylated dextran with gelatin are examples that have been successfully used to deliver diverse GFs.[57] The chosen material will depend on the application and will determine the method of fabrication.

4.2.2.3 Scaffolds

Polymer scaffolds can be fabricated into a variety of structures with the ultimate aim of promoting the viability and differentiation of stem cells seeded inside. This can be achieved based on both the intrinsic properties of the material and the incorporation of specific GFs. Biomaterials used for fabrication of scaffolds for DDSs may be natural or synthetic polymers whose properties can be varied to control protein adsorption. Natural polymers such as alginate, collagens, gelatin, fibrins, and albumin and synthetic such as polyvinyl PVA and PLA or PLGA have been extensively used in DDS scaffolds.[58] However, other novel degradable materials have been developed to meet the requirements of specific applications. For example polyketals have the unique property of degrading into neutral compounds not generating acid or basic residues that may induce inflammatory reactions. This is a potential advantage when treating diseases where inflammation plays a large role, such as cardiac dysfunction following MI.[31]

Several methods for the effective incorporation of bioactive molecules as GF into the scaffold have been developed. Direct lyophilized GF loading to the scaffold was proved to be highly ineffective due to protein retention in the device surface. An elevated GF amount can be eluted out due to the leaching step that takes place during the scaffold fabrication. Not to mention the possible loss of GF integrity due to the high temperatures used for instance in melt-fabrication method.[40] Innovative manufacturing techniques for polymer scaffolds have been developed in order to overcome these issues.[40,59]

Similarly to hydrogels, the addition to scaffolds of pre-encapsulated GFs into natural or synthetic carriers allows a more uniform distribution of the protein, keeps it from most of the chemical and physical excesses occurring during the manufacturing process, preserving its integrity and ensuring release kinetics controlled by degradation rate of the polymer used in the carriers.[59] Another method uses freeze-drying of an emulsion of the lipophilic polymer dissolved in methylene chloride and a water phase containing the different proteins used for encapsulation. The resulting polymer scaffolds exhibit quite different pore size distributions and morphologies although the emulsion could affect the stability of the protein.[60]

Electrospinning is a simple, versatile technique for generating nanofibers from a broad range of materials including polymers, composites, and ceramics.[61,62] These nanofibers have several advantages for tissue regeneration such as correct topography (*e.g.*, 3D porosity, nanoscale size, and alignment), encapsulation and local sustained release of growth factors, and surface functionalization in order to allow cell adhesion.[63] The electrospinning technique allows the production of DDS–nanofiber-based scaffolds when polymeric fibers and protein solution are collected on a plate that is oppositely charged to the solution involving the metal needle. Release studies have shown presence of bovine serum albumin for over 30 days, although its stability and biological activity need to be evaluated.[64]

The amino acids in proteins can form covalent bonds with polymer surfaces either by a natural mode or be forced to it by using a physiological method of protein fixation, such as the application of immobilized heparin to bind heparin-binding domains of certain proteins like FGF-2 or VEGF.[65,66] It is important to guarantee that the binding does not involve amino acids that are essential for the activity of the protein and that the receptors they are acting on are accessible within the target cell. The covalent immobilization of insulin[67] and the attachment of EGF[68] have proven that this method preserves the biological activity of GFs demonstrated by the stimulation in cell proliferation.

Finally, solid lipids can act as heat extractable porogen templates for scaffold manufacture. When the temperature increases, lipids melt forming pores, the polymer precipitates while a non-solvent solution that contains the proteins invades the new network. This process prevents excessive loss of encapsulated proteins, but the stability of the GF used at the chosen temperatures must be evaluated. An analogous method can be performed using gas bubble formation when high pressure carbon dioxide is used as a solvent.[40]

4.2.2.4 *Lipid-based Systems*

Liposomes are microscopic vesicles composed of lipid membranes surrounding discrete aqueous compartments. Their dimensions, composition, surface charge and structure can be modified to fix the requirements of the application. Natural and synthetic liposomes have been used both *in vitro* and *in vivo* to deliver proteins for cancer treatment, ocular inflammation, lung diseases, diabetes and infectious diseases.[69] Some innovative systems like inhalable liposomes to carry peptides for airway administration to treat lung diseases[70] or liposomes acoustically disrupted by ultrasound for cardiovascular applications,[31] resulted in localized protein delivery improving the outcome of the therapy.

Selection of an accurate size of liposomes when designing lipid-based DDSs is critical because, the larger the DDS, the higher the possibility to face various barriers within the host such as endothelial surfaces or the blood–brain or to become targets of the immune system.[69]

A wide range of methods has been employed in liposome manufacture. Each process has advantages and limitations that should be considered with respect to the individual application. Below we mention some of them:

- *Dried films*: these films are formed by roto-evaporation of a solvent mixture containing phospholipids. An aqueous phase containing a hydrophilic protein in solution is added to the dry film. This is the easiest process to fabricate lipo-based DDS, but usually the protein leaks out of the membrane preventing its controlled released.
- *Freeze-and-thaw liposomes*: submitting vesicles to freeze–thaw cycles or lyophilization seems to avoid protein leakiness and enhances encapsulation although it is important to evaluate the effects of temperature variations on the protein bioactivity.
- *Solvent-requiring methods*: the manufacturing process consists of an aqueous solution that is momentarily emulsified into an immiscible organic phase containing the required lipids; rapid removal of solvent induces formation of inverse micelle intermediates which upon hydration organize into bilayered liposomes. The protein of interest is present in the solvent.[71]

The release of drug from liposomes is related to both lipid composition and type of liposome constructed. DDSs made of liposomes can be administrated intramuscularly or subcutaneously, orally, through the eye, intranasally or topically and this versatility increases the number of clinical applications of liposomes.

Hedman *et al.* initiated a clinical trial in which a plasmid-encoding VEGF was delivered using liposomes during angioplasty and stenting for the treatment of chronic myocardial ischemia.[72] Currently, there are no liposomal formulations approved for human use for the treatment of cardiovascular disease nevertheless, liposomes have evolved as robust multifunctional platforms for drug delivery, with several innovative examples currently being examined in preclinical models of diverse diseases including infarction.

Solid lipid particulate systems appeared as alternative GF carriers to polymeric particles; however, there are only a few reports demonstrating its effectiveness as angiogenic proteins carriers for cardiac repair. Oh *et al.* reported the formation of a temperature-induced gel composed of core–shell nanoparticles for ischemic heart regeneration.[73] The particle core was composed of lecithin containing VEGF and the shell was composed of Pluronic-F127 (poly(ethylene oxide)–poly(propylene oxide)–poly(ethylene oxide) triblock co-polymer). VEGF-loaded core–shell nanoparticles and their gel improved heart function particularly with regard to ejection fraction and cardiac output. A minimal difference in neovascularization was observed between the core–shell nanoparticles and their gel.[73]

4.3 Growth Factor Drug Delivery Systems for Cardiac Applications

Several alternative strategies are being investigated to complement the current pharmacological therapies for myocardial diseases, including re-activation of cardiomyocytes (CMs) cell cycle activity and reduction of myocardial cell death. To improve regeneration of the damaged heart area two main strategies in the cell therapy field have been developed. The first aims at recruiting or promoting the homing of endogenous or circulating stem cells at the periphery or inside the damaged zone which improves the innate repair capacity of heart tissue. The second seeks for building a new tissue with a new population of functional and beating cardiomyocytes, with cell–cell communications, secreted factors and induction of neoangiogenesis. GFs play an essential role in both approaches and their administration together with cytokines and extracellular matrix molecules could help to reduce inflammation, promote cell survival, and promote recruitment and differentiation of progenitor cells towards cardiac lineages ensuring a successful result of this type of treatment.[74]

Below we review a group of GFs described as having a role in angiogenesis, remodeling, differentiation and proliferation processes within the heart after an infarction.[9,36] *In vitro* pre-clinical and clinical studies have been developed for almost all of them with different outcomes constituting an important field of research and a promising therapy for patients, especially for those who are not surgical candidates.

4.3.1 Angiogenesis

Infarcted myocardium is an ischemic and inflammatory environment hostile to most cells. New treatment strategies seek to enhance angiogenesis in order to increase survival of cardiac cells, remodeling of the tissue, stem cell engraftment and differentiation. GFs, by promoting the formation of new functional vessels help to decrease ischemia, improving cardiac performance after infarction.[30,75]

4.3.1.1 *Vascular Endothelial Growth Factor*

VEGF is a key regulator of angiogenesis but its systemic injection could lead to hyperpermeable vessels, hypotension, induction of tumor growth, and uncontrolled neovascularization in non-target tissues. DDSs of angiogenic factors localize and prolong the effect to a microenvironment, minimizing the negative side effects.[59,76-80]

For cardiac applications, VEGF has been successfully loaded into a wide variety of natural and synthetic materials using diverse methods. VEGF has been covalently incorporated into collagen gels scaffolds[81] and fibrin matrices,[82] encapsulated in gelatine[83] and PLGA microparticles, liposomes[84] or alginate gel matrices.[85] For most cases VEGF retained its bioactivity after DDS preparation and improved angiogenic performance *in vitro*.

VEGF-DDS efficacy has been widely evaluated in animal models. Some of those studies have proved the effect of this GF in the treatment of ischemic disease in mice using alginate matrices,[85] the improvement of vascularity and cardiac function in MI model[84] and the restoration of perfusion from hind limb ischemia, preventing severe necrosis in mice, when embedded in PLG scaffolds containing the microspheres with VEGF.[85] Also our group, demonstrated the improvement in vasculogenesis and tissue remodeling when PLGA and PEG-PLGA microparticles loaded with VEGF-A were delivered in a rat model of cardiac ischemia.[86,87]

4.3.1.2 Placental Growth Factor

PlGF is a member of the cysteine-knot family that can modulate the activity of VEGF. Gene inactivation experiments in mice showed that loss of PlGF correlated with impaired angiogenesis, but on its own PlGF cannot initiate endothelial cell stimulation and only enhances the effects of the available low concentrations of endogenous VEGF.[10] PlGF can stimulate the recruitment and activation of monocytes, which is an essential step in the process of formation of mature vascular networks. In animal models, systemic administration of recombinant PlGF-1 protected from induced ischemic lesions in the heart without affecting heart rate.[88,89]

PlGF is a potential candidate for DDS for angiogenesis; nevertheless, further *in vitro* and *in vivo* experiments need to be done with PlGF alone and in combination with VEGF which would probably be more effective in the treatment of cardiovascular diseases.

4.3.1.3 Platelet-derived Growth Factor

PDGF corresponds to a mitotic and motility signal for smooth muscle cells and pericytes during angiogenesis, as well as a chemo-attractant that induces VEGF expression in cardiac endothelial cells. Intramyocardial injection of PDGF has cardio-protective effect in mice, reducing the extent of myocardial injury following coronary occlusion.[90]

Several DDSs for PDGF have been examined, hydrogels that could be enzymatically degraded,[91] PLGA-nanoparticles[92] or porous scaffold devices made by entrapping PDGF-BB in PLG microspheres through a standard double-emulsion process.[93] Under optimal conditions, the encapsulation efficiency of PDGF-BB can reach 87% while maintaining bioactivity.

PDGF was used in an attempt to improve skeletal myoblast efficacy in a rat model of MI. While no statistical progress was seen in left ventricular function within the treatment groups, the sustained delivery PDGF did improve angiogenesis, cardiomyocyte survival, and long-term function.[15]

4.3.1.4   Angiopoietins

The best known proteins of this family are Ang-1 and Ang-2 and they seem to act as antagonists of each other in several cardiovascular remodeling and vascular processes. Ang-1, which is expressed by pericytes, may limit angiogenesis and its deletion leads to excessive angiogenesis and tissue fibrosis. In contrast, Ang-2 is expressed by endothelial cells and may regulate cell–matrix interactions in growing vessels.[18]

Human Ang-1 and VEGF were co-encapsulated in albumin nanoparticles for controlled delivery. The results indicated that the released proteins were biologically active and the combined application of both proteins demonstrated a significant highly proliferative and antiapoptotic effect on human umbilical vein endothelial cells (HUVECs) as compared with the effect demonstrated by each proteins alone.[94]

A further study demonstrated that a thermoresponsive hydrogel made of chitosan, conjugated with Ang-1 and mixed with collagen, allowed *in vitro* survival and maintenance of function of neonatal rat cardiomyocytes in comparison to unmodified chitosan–collagen hydrogels. Additionally, this group demonstrated that the hydrogel is suitable for subcutaneous cell injection and that it can be localized in the infarcted zone of mouse heart.[95] Recently, Anisimov *et al.* manufactured a VEGF–angiopoietin-1 chimera and confirmed its angiogenic potential by triggering, in a novel fashion, VEGF receptor-2. This chimera promoted less vessel leakiness, less tissue inflammation and better perfusion in ischemic muscle than VEGF alone.[96] Further development in angiogenesis will offer a better understanding of the role that angiopoietins play in response to ischemic insult and as potential therapeutic agents.

4.3.2   Direct Effect on Cardiac Cell Proliferation, Remodeling, Tissue Protection and Differentiation

The maintenance of cardiomyocyte number and growth of surviving cardiomyocytes are two important factors that modulate preservation of myocardial function in response to stress.

4.3.2.1   Fibroblast Growth Factor

Several members of this family have been used in DDS for cardiovascular diseases either to promote angiogenesis,[17,97,98] differentiation or a combination of both effects.[99] Under experimental conditions it has been proved that FGFs have both angiogenic and cardioprotective potential. These factors could promote resistance to injury (preconditioning), guard against secondary injury after an acute ischemic insult at the same that accelerate blood vessel formation which make them candidates for preclinical and clinical cardiovascular studies.[100]

Basic FGF corresponds to a mitotic signal for endothelial cells and Wang *et al.*, incorporating it into a PLGA scaffold could promote neovascular formation, enhanced blood-flow perfusion and improved myocardial function in a mini-swine animal model of MI.[101] Intramyocardial administration of bFGF microspheres injected in the peri-infarcted area can promote the growth of microvessels and improve left ventricular function and myocardial viability in the early period of acute myocardial infarction in dogs.[102] Composite fibrin matrices containing FGF-1 or FGF-2 were adsorbed onto Dacron meshes and then implanted subcutaneously in mice, showing an increment in the formation of microvessels.[103] On the other hand other member of the family FGF10 promoted CM differentiation from embryonic and induced pluripotent stem cells.[104]

These studies formed the basis for a further clinical trial, where 335 patients were administrated intravenous and intracoronary FGF2. It was concluded that a single intracoronary infusion of FGF2 does not improve exercise tolerance or myocardial perfusion but does show trends toward symptomatic improvement at 90 days.[23]

4.3.2.2   *Insulin Growth Factor*

IGF-1 is a small growth factor that is bound by IGF-binding proteins in serum; thus, delivery of the free protein is highly inefficient. Delivery of insulin-like growth factor-1 (IGF-1) enhanced both neonatal myocyte, as well as cardiac progenitor cell (CPC) efficacy.[14,105] Since, IGF-1 needs to be in the heart for more than 1 month to cause an effect DDSs are an ideal method to ensure this.[31]

IGF-I embedded in poly(DL-lactide-*co*-glycolide) microspheres was released from microsphere formulations for up to 13 days featuring a moderately pulsatile pattern, depending on the microsphere composition.[106] Biotinylated IGF-1 has been tethered to biotinylated self-assembling peptides using streptavidin as a linker and was evaluated *in vitro* and *in vivo*.[14] In cardiomyocytes, IGF-1 decreased activation of caspase-3, and increased expression of cardiac troponin I after experimental myocardial infarction in rats and an improvement in systolic function.[105] In a mouse model of MI, a single injection of nanoparticles loaded with IGF-1 was able to avoid apoptosis of cardiomyocytes, improving ejection fraction after 21 days of treatment.[107] Caplice *et al.* have an ongoing randomized trial called RESUS-AMI. This group, currently recruiting patients, is evaluating the safety and efficacy of a single low dose of intracoronary IGF-1 following percutaneous coronary intervention for ST-elevation acute MI[108] (clinicaltrials.gov-NCT01438086).

4.3.2.3 *Hepatocyte Growth Factor*

This factor induces the production of VEGF and PDGF, endothelial cell activation and motility and is a chemoattractant agent for pericytes.[109] Also, it has been demonstrated that, combined with IGF-1, it reduces arrhythmogenesis in chronic heart disease.[110]

HGF was encapsulated into the PLGA/PHBV [poly(3-hydroxybutyrate-*co*-3-hydroxyvalerate] composite microsphere with a core–shell structure, with sustained delivery of HGF maintaining its bioactivity for at least 40 days.[111] Ruvinov *et al.* used sulfated alginate to form electrostatically complexed HGF bound within alginate matrices. Following *in vivo* injection into ischemic myocardial tissue, 75% of the initially loaded HGF was lost from the injection site and subsequent release was slow; after 5 days, only $\sim 85\%$ of the HGF had been released.[112] HGFs have been electrostatically bound to the collagen but the manufacturing process reduced significantly HGF bioactivity.[113] A gelatin-based system to deliver HGF was capable of maintaining its bioactivity, increasing capillary density and enhancing blood flow into a hindlimb ischemic area in mice.[114] Recently, a clinical trial was open that aims to evaluate HGF as an early marker of myocardial injury and prognostic factor in patients with acute coronary syndrome[115] (clinicaltrials.gov-NCT01233336).

4.3.3 Other Proteins

Investigating the fundamental molecular mechanisms underlying CM growth and survival in response to stress allow researchers to identify new molecules with a potential use in protein therapy for MI.

4.3.3.1 *Neuregulin-1β*

Nrg1β regulates tissue organization during development through endocardial, endothelial, and cardiomyocyte interactions, as well as maintains cardiac function throughout life. After injury recombinant Nrg-1β improved cardiac systolic function *via* alterations in myocardial structure in dogs[21] and primates.[116]

Patients with systolic heart failure were treated for 10 days with Nrg-1β and after 90 days magnetic resonance imaging showed an improvement in cardiac structure and function at least at one dose studied. The analysis of acute hemodynamic effects demonstrated that there is a vasodilator effect along with an acute increase in cardiac output. Besides the nausea experienced by the treated patients, the dosing program used in these trials is the most important issue to tune up.[117] DDS with Nrg-1β would help to avoid the administration of protein that usually is prolonged (multi-hour), daily and for multiple days.

4.3.3.2 *Yes-associated Protein-1*

Yap-1 is a known important regulator of cardiomyocyte proliferation and embryonic heart development.[118] However, recently it has been elucidated its role in adult heart tissue. Del Re *et al.* demonstrated that, in response to stress, endogenous Yap-1 is a critical regulator of cardiomyocyte growth and survival in the adult mouse myocardium. They also found that a deficiency in Yap-1 intensifies injury after chronic MI, suggesting that specific targeting of Yap-1 activation in cardiomyocytes could serve to ameliorate the damage caused by myocardial ischemia. These findings highlight Yap-1 as a possible therapeutic target for driving pro-growth and pro-survival phenotype in a cell autonomous manner in cardiomyocytes of patients suffering a chronic heart disease.[19] *In vitro* and *in vivo* assays with Yap-1 direct injection or its encapsulation in DDSs for a release-controlled manner and their characterization has not yet been done.

4.3.3.3 Stromal-cell-derived Factor-1

SDF-1 may be involved in the recruitment of progenitor cells into the infarcted zone as a result of its chemo-attractant property. Few studies have been done using DDS for this protein in cardiovascular diseases. Segers *et al.* expressed SDF with a protease-resistant signal, followed by a self-assembling peptide sequence allowing it to be incorporated into a native self-assembling peptide scaffold. This system allowed the local and sustained delivery of SDF-1, resulting in an improved cardiac function due to the increase in stem cell recruitment and capillary density.[119] Another group covalently bound SDF-1 to a PEGylated fibrinogen patch for controlled release of the protein. The patch was placed on the surface of the infarct area of the left ventricle. SDF-1 delivery can improve the rate of c-kit(+) cell homing and improve LV function in a mouse AMI model.[120] Finally, Purcell *et al.* injected hydrogels containing SDF-1 into the ventricle wall of infarcted mice and showed sustained delivery and homing of bone marrow stem cells to the systemic circulation and to the injured area.[121]

4.3.4 Combined Therapies

4.3.4.1 Multiple Growth Factors

Simultaneous activation of diverse signaling pathways by over-expression of multiple growth factors ensures an appropriate angiogenesis and cause massive mobilization and homing of stem/progenitor cells from peripheral circulation, the bone marrow, and the heart. DDSs of multiple growth factors constitute a novel approach that could accelerate the repair of the infarcted myocardium.

Zieris *et al.* used the property of VEGF and FGF2 to bind strongly to heparin in order to fabricate functionalized PEG–heparin hydrogels for DDSs. Culture experiments with human umbilical vein endothelial cells (HUVECs) revealed that this system provides the correct cues to induce angiogenesis.[122]

Sequential delivery of VEGF and PDGF-BB, or IGF-1 and HGF, from alginate hydrogel systems induced the formation of mature vessels and improved cardiac function more efficiently than each factor separately, in animal models of MI.[112,123] More recently, Salimath et al. developed a PEG based protease-degradable hydrogel combining VEGF with HGF. It was demonstrated that dual factor release from a bioactive hydrogel was possible and had the capacity to significantly improve cardiac function in an ischemia/reperfusion rat model.[124]

Polymer microspheres fabricated from PLG containing pre-encapsulated PDGF with lyophilized VEGF were incorporated together in a single structural polymer scaffold. This process allows for sustained protein delivery and maintains the biological activity of incorporated and released growth factors. They were implanted into the subcutaneous pockets on the dorsal side of rats and histological analysis revealed a rapid formation of a mature vascular network due to the dual delivery of both growth factors with different kinetics.[93]

Gelatin microspheres have also been used for multiple GF delivery in the cardiac context. For instance, Cittadini et al. recently used gelatin microspheres for VEGF and IGF administration in a rat model of MI. Animals treated with microspheres containing both GFs showed remarkably better effects on infarct size and left ventricular volume reductions, heart function improvement, vascularization enhancement and apoptosis and inflammation reduction when compared with single GF microsphere administration.[125]

In a mouse model of hind-limb ischemia sequential delivery of VEGF and PDGF using an injectable, biodegradable alginate scaffold, resulted in stable and sustained improvement in perfusion.[126] Collagen scaffolds with a combination of FGF2 and VEGF displayed the highest density of blood vessels and most mature blood vessels when implanted subcutaneously in 3-month-old Wistar rats. These results indicated that the enhancement of early mature vasculature in an acellular construct could open prospects in conditions as ischemic heart disease.[127]

Kim et al. incorporated PDGF and FGF into self-assembling peptide nanofiber scaffolds that mimicked extracellular matrix porosity and gross structure and tested its efficacy in the rat model of MI. Animals treated with the system containing both GFs almost recovered cardiac function. This effect correlated with a decrease in cardiomyocyte apoptosis, capillary and arterial density recovery, with stable vessel formation, higher reduction in the infarction size and improvement in wall thickness.[128]

Recently, our group examined whether the administration of MP containing NRG1 and FGF1 in a rat model of MI promoted repair of heart tissue. Three months after treatment, cardiac function improvement was observed in rats treated with FGF1-MP, NRG1-MP or FGF1/NRG1-MP. A positive cardiac remodeling with a lower degree of fibrosis, smaller infarct size, and induction of tissue revascularization was also detected, together with cardiomyocyte proliferation and progenitor cell recruitment. A consistent synergistic in vitro or in vivo effect was not observed with the combination therapy involving the administration of both cytokines.[129] This important observation should be considered when designing new studies involving multiple GF administration.

4.4   Growth Factor Delivery Systems and Cell Therapy

The microenvironment of injured myocardium is characterized by high levels of oxidative stress and the presence of pro-inflammatory cytokines which directly affects the levels of cell engraftment following transplantation reducing the efficacy of cell therapy. After 24 h of delivery it has been calculated that the percentage of cell retention within the heart is between 1% and 20% depending on cell type and delivery method.[130] GF delivery would help to overcome these issues by providing cues for differentiation and proliferation controlling the environment of transplanted cells.

Different biomaterials, GFs and cellular sources have been tested in several studies to develop a combined therapy for the treatment of MI. The strategy has several advantages, such as active compounds are released over time and in close proximity to the transplanted cells, there are high local concentrations of the compounds, and fewer off-site effects. An extensive review on this field has been done recently.[74]

One example is the delivery of IGF-1 tethered in self-assembling peptide nanofibers together with cardiac stem cells. This led to a greater improvement of cardiac function after MI and to the development of more mature myocytes, compared to cells alone.[14] PLGA microspheres releasing HGF and IGF-1 commitmently induced stem cells into a cardiomyocyte phenotype. The HGF released could promote adult stem cell cardiomyocyte differentiation and local neoangiogenesis around the graft, while IGF-1 could improve survival of the grafted cells while, at the same time, inducing proliferation and survival of the cardiac progenitor cells.[74] This strategy has been used also to induce cardiac differentiation of pluripotent cells like embryonic stem cells (ESCs) and induced pluripotent stem cells (iPS). Although tough, there have been good results and it continues to be a challenging approach because of the specific temporal and spatial window in which these cells are responsive to differentiation.[30]

Bone marrow mesenchymal stem cells (BMSCs) were adhered into a bFGF-loaded polyglycolic acid scaffold impregnated with collagen I hydrogel and its efficacy in a rat model of MI was studied. The bFGF-BMSC-scaffold group obtained the highest density vessel formation and the best cardiac function, leading to a better improvement with respect to the BMSC-scaffold group 1 month after treatment.[131]

A porous collagen scaffold incorporate VEGF and FGF in combination with MSC. The system was tested in a rat model of MI. The group treated with the GF-MSC scaffold showed a higher angiogenic effect and better cardiac function compared to the MSC-scaffold group. A higher cell survival in the GF-scaffold could be responsible of this higher beneficial effect.[132]

VEGF was encapsulated in PLGA MP, which were then coated with fibronectin for MSC adhesion. The system was able to enhance *in vitro* cell proliferation and survival, but further studies are required to demonstrate its *in vivo* efficacy.[133] Lately our group encapsulated NRG in PLGA MP. The particles were next coated with collagen and/or poly-D-lysine for adipose derived stem cells (ADSC) adhesion and studied *in vivo* in a rat model of MI. The system was well integrated in the peri-infarcted area 2 weeks after treatment demonstrating its biocompatibility. This work, although preliminary, has yielded favorable results that require further studies that should focus in the efficacy of this treatment.[134]

Finally, there is an ongoing phase I clinical trial named ALCADIA that aims to evaluate the safety and efficacy of autologous human cardiac-derived stem cells (hCSCs) combined with a gelatin hydrogel sheet incorporating bFGF in severe refractory heart failure patients with chronic ischemic cardiomyopathy with left ventricular dysfunction ($15\% \leq LVEF \leq 35\%$)[135] (clinicaltrials.gov-NCT00981006).

4.5 Conclusions

Cardiovascular diseases (CVDs) remain the biggest cause of deaths worldwide, even though they are largely preventable. Occlusion of arteries, primarily as a result of atherosclerosis leads to a deprivation of oxygen and in tissues like the heart the ischemia produces an infarct that could result in permanent muscle damage and probably death.

After an ischemic event, the heart tries to heal and recover the muscle function but this process is highly inefficient, leading to heart failure. There are several treatment options for myocardial ischemia, including drugs, bypass grafting surgery or transplant. Many novel therapies include the delivery of different cellular types, proteins, like GFs or the genes that encodes for those factors.

The localized delivery of exogenous GFs is believed to be therapeutically effective for the replication of cardiac cellular components involve in both development and in the healing process, thus making them important factors for protein therapy. Different studies have helped to elucidate the principal GFs needed to promote heart regeneration, even though much detailed investigations on their signaling pathways, binding properties, half-life, dosage and timing of administration are needed.

DDSs of GFs for cardiac repair have been developed; however, all of them face major challenges, such as avoiding an initial burst of GF, achieving proper controlled release over time, degradation and loss of bioactivity of GF due to manufacturing methods, low efficiency of encapsulation, bioavailability of GF as recombinant proteins or high concentrations of GF loaded into the devices which normally exceed those seen in the body.

Cardiac repair is a complex process that implicates angiogenic signals along with cardiomyocyte proliferation, differentiation and remodeling, and thus involves different GFs and cell types whose participation is perfectly synchronized. In order to mimic the natural microenvironment of the heart, the development of systems with the ability to engraft different cell types and, at the same time, release numerous growth factors on multiple time-scales will be decisive for the success of these types of therapy.

Acknowledgements

This work was supported, in part, by funds from the ISCIII (RD12/0019/0031), MINECO (PLE2009-0116 and program INNPACTO Procardio), FP7 Program (INELPY) and the 'UTE project CIMA'.

References

1. C. J. L. Murray, T. Vos, R. Lozano, M. Naghavi, A. D. Flaxman, C. Michaud, M. Ezzati, K. Shibuya, J. A. Salomon, S. Abdalla, V. Aboyans, J. Abraham, *et al.*, *The Lancet*, 2012, **380**, 2197.
2. S. Mendis, P. Puska and B. Norrving, eds. Global Atlas on Cardiovascular Disease Prevention and Control. World Health Organization, Geneva, 2011.
3. K. Thygesen, J. S. Alpert, A. S. Jaffe, M. L. Simoons, B. R. Chaitman and H. D. White, *Eur. Heart J.*, 2012, **33**, 2551.
4. J. K. Gajarsa Robert, *Heart Failure Rev.*, 2011, **16**, 13.
5. K. Martin, C.-L. Huang and N. M. Caplice, *Curr. Pharm. Des.*, 2014, **20**, 1930.
6. B. Pelacho, M. Mazo, S. Montori, A. Simón-Yarza, M. Blanco-Prieto and F. Prosper, *Regenerative Medicine and Cell Therapy*, Springer, 2012.
7. M. T. Roe, J. C. Messenger, W. S. Weintraub, C. P. Cannon, G. C. Fonarow, D. Dai, A. Y. Chen, L. W. Klein, F. A. Masoudi, C. McKay, K. Hewitt, R. G. Brindis, E. D. Peterson and J. S. Rumsfeld, *J. Am. Coll. Cardiol.*, 2010, **56**, 254.
8. T. K. Nagai Issei, *Am. J. Physiol.: Heart Circ. Physiol.*, 2012, **303**, H501.
9. F. R. Formiga, E. Tamayo, T. Simón Yarza, B. Pelacho, F. Prósper and M. J. Blanco Prieto, *Heart Failure Rev.*, 2012, **17**, 449.
10. P. Carmeliet, L. Moons, A. Luttun, V. Vincenti, V. Compernolle, M. De Mol, Y. Wu, F. Bono, L. Devy, H. Beck, D. Scholz, T. Acker, T. DiPalma, M. Dewerchin, A. Noel, I. Stalmans, A. Barra, S. Blacher, T. VandenDriessche, A. Ponten, U. Eriksson, K. H. Plate, J. M. Foidart, W. Schaper, D.-S. Charnock Jones, D. J. Hicklin, J. M. Herbert, D. Collen and M. G. Persico, *Nat. Med.*, 2001, 7, 575.
11. R. Jain, *Nat. Med.*, 2003, **9**, 685.
12. W. Metzger, N. Grenner, S. Motsch, R. Strehlow, T. Pohlemann and M. Oberringer, *Tissue Eng.*, 2007, **13**, 2751.
13. M. S. Ahmed, E. Øie, L. Vinge, A. Yndestad, G. Øystein Andersen, Y. Andersson, T. Attramadal and H. Attramadal, *J. Mol. Cell. Cardiol.*, 2004, **36**, 393.
14. M. E. Padin Iruegas, Y. Misao, M. E. Davis, V. F. M. Segers, G. Esposito, T. Tokunou, K. Urbanek, T. Hosoda, M. Rota, P. Anversa, A. Leri, R. T. Lee and J. Kajstura, *Circulation*, 2009, **120**, 876.

15. G. Dubois, V. F. M. Segers, V. Bellamy, L. Sabbah, S. Peyrard, P. Bruneval, A. A. Hagège, R. T. Lee and P. Menasché, *J. Biomed. Mater. Res., Part B*, 2008, **87**, 222.

16. S. Sahoo, L. T. Ang, J. Cho-Hong Goh and S. L. Toh, *Differentiation*, 2010, **79**, 102.

17. J. C. Garbern, E. Minami, P. S. Stayton and C. E. Murry, *Biomaterials*, 2011, **32**, 2407.

18. P. A. Saharinen Kari, *J. Clin. Invest.*, 2011, **121**, 2157.

19. D. P. Del Re, Y. Yang, N. Nakano, J. Cho, P. Zhai, T. Yamamoto, N. Zhang, N. Yabuta, H. Nojima, D. Pan and J. Sadoshima, *J. Biol. Chem.*, 2013, **288**, 3977.

20. S. Dai, F. Yuan, J. Mu, C. Li, N. Chen, S. Guo, J. Kingery, S. D. Prabhu, R. Bolli and R. Rokosh, *J. Mol. Cell. Cardiol.*, 2010, **49**, 587.

21. X. Liu, X. Gu, Z. Li, X. Li, H. Li, J. Chang, P. Chen, J. Jin, B. Xi, D. Chen, D. Lai, R. M. Graham and M. Zhou, *J. Am. Coll. Cardiol.*, 2006, **48**, 1438.

22. B. Schumacher, P. Pecher, B. U. von Specht and Th. Stegmann, *Circulation*, 1998, **97**, 645.

23. M. Simons, B. H. Annex, R. J. Laham, N. Kleiman, T. Henry, H. Dauerman, J. E. Udelson, E. V. Gervino, M. Pike, M. J. Whitehouse, T. Moon and N. A. Chronos, *Circulation*, 2002, **105**, 788.

24. C. Gibson, R. Laham and F. Giordano, *J. Am. Coll. Cardiol.*, 1999, **33**, 384A.

25. R. C. Hendel, T. D. Henry, K. Rocha Singh, J. M. Isner, D. J. Kereiakes, F. J. Giordano, M. Simons and R. O. Bonow, *Circulation*, 2000, **101**, 118.

26. H. Ince, M. Petzsch, H. D. Kleine, H. Schmidt, T. Rehders, T. Körber, C. Schümichen, M. Freund and C. A. Nienaber, *Circulation*, 2005, **112**, 3097.

27. D. Zohlnhöfer, I. Ott, J. Mehilli, K. Schömig, F. Michalk, T. Ibrahim, G. Meisetschläger, J. von Wedel, H. Bollwein, M. Seyfarth, J. Dirschinger, C. Schmitt, M. Schwaiger, A. Kastrati and A. Schömig, *JAMA, J. Am. Med. Assoc.*, 2006, **295**, 1003.

28. R. S. Ripa, E. Jørgensen, Y. Wang, J. J. Thune, J. C. Nilsson, L. Søndergaard, H. E. Johnsen, L. Køber, P. Grande and J. Kastrup, *Circulation*, 2006, **113**, 1983.

29. M. Nomi, H. Miyake, Y. Sugita, M. Fujisawa and S. Soker, *Curr. Stem Cell Res. Ther.*, 2006, **1**, 333.

30. V. F. M. Segers and R. T. Lee, *Circ. Res.*, 2011, **109**, 910.

31. J. D. Sy Michael, *Journal of Cardiovascular Translational Research*, 2010, **3**, 461.

32. J. J. Rice, M. M. Martino, L. De Laporte, F. Tortelli, P. S. Briquez and J. A. Hubbell, *Adv. Healthcare Mater.*, 2013, **2**, 57.

33. R. M. Chen David, *Pharm. Res.*, 2003, **20**, 1103.

34. A. K. Andriola Silva, C. Richard, M. Bessodes, D. Scherman and O.-W. Merten, *Biomacromolecules*, 2009, **10**, 9.

35. J. L. Cleland, M. F. Powell and S. J. Shire, *Crit. Rev. Ther. Drug Carrier Syst.*, 1993, **10**, 307.

36. P. Diaz-Herraez, S. Pascual-Gil de Gomez, E. Garbayo, T. Simón-Yarza, F. Prósper and M. Blanco-Prieto, *Drug Delivery an Integrated Clinical and Engineering Approach*, Taylor & Francis Group, 2014, in press.
37. P. R. Wanakule Krishnendu, *Curr. Drug Metab.*, 2012, **13**, 42.
38. K. S. Oh, J. Y. Song, S. J. Yoon, Y. Park, D. Kim and S. H. Yuk, *J. Controlled Release*, 2010, **146**, 207.
39. A. Hoffman, *Adv. Drug Delivery Rev.*, 2002, **54**, 3.
40. J. G. Tessmar Achim, *Adv. Drug Delivery Rev.*, 2007, **59**, 274.
41. H. Zhao, L. Ma, J. Zhou, Z. Mao, C. Gao and J. Shen, *Biomed. Mater.*, 2008, **3**, 015001.
42. M. Leonard, M. R. De Boisseson, P. Hubert, F. Dalençon and E. Dellacherie, *J. Controlled Release*, 2004, **98**, 395.
43. Y. Tabata, M. Miyao, M. Ozeki and Y. Ikada, *J. Biomater. Sci., Polym. Ed.*, 2000, **11**, 915.
44. M. Kitamura, K. Nakashima, Y. Kowashi, T. Fujii, H. Shimauchi, T. Sasano, T. Furuuchi, M. Fukuda, T. Noguchi, T. Shibutani, Y. Iwayama, S. Takashiba, H. Kurihara, M. Ninomiya, J.-i. Kido, T. Nagata, T. Hamachi, K. Maeda, Y. Hara, Y. Izumi, T. Hirofuji, E. Imai, M. Omae, M. Watanuki and S. Murakami, *PLoS One*, 2008, **3**, e2611.
45. D. L. Steed, *J. Vasc. Surg.*, 1995, **21**, 71.
46. J. A. Cadée, C. J. de Groot, W. Jiskoot, W. den Otter and W. E. Hennink, *J. Controlled Release*, 2002, **78**, 1.
47. N. Bhattarai, H. R. Ramay, J. Gunn, F. A. Matsen and M. Zhang, *J. Controlled Release*, 2005, **103**, 609.
48. W. Mark Saltzman and S. P. Baldwin, *Adv. Drug Delivery Rev.*, 1998, **33**, 71.
49. M. Biondi, F. Ungaro, F. Quaglia and P. Netti, *Adv. Drug Delivery Rev.*, 2008, **60**, 229.
50. J. G. Tessmar Achim, *Adv. Drug Delivery Rev.*, 2007, **59**, 274.
51. A. H. Zisch, M. P. Lutolf, M. Ehrbar, G. P. Raeber, S. C. Rizzi, N. Davies, H. Schmökel, D. Bezuidenhout, V. Djonov, P. Zilla and J. Hubbell, *FASEB J.*, 2003, **17**, 2260.
52. A. Iwakura, M. Fujita, K. Kataoka, K. Tambara, Y. Sakakibara, M. Komeda and Y. Tabata, *Heart Vessels*, 2003, **18**, 93.
53. Y. Sakakibara, K. Tambara, G. Sakaguchi, F. Lu, M. Yamamoto, K. Nishimura, Y. Tabata and M. Komeda, *Eur. J. Cardiothorac. Surg.*, 2003, **24**, 105.
54. E. A. Silva and D. J. Mooney, *J. Thromb. Haemostasis*, 2007, **5**, 590.
55. K. Lee, E. A. Silva and D. J. Mooney, *J. R. Soc., Interface*, 2011, **8**, 153.
56. E. Wenk, A. J. Meinel, S. Wildy, H. P. Merkle and L. B. Meinel, *Biomaterials*, 2009, **30**, 2571.
57. F.-M. Chen, R. Chen, X. J. Wang, H. H. Sun and Z.-F. Wu, *Biomaterials*, 2009, **30**, 5215.
58. W. K. King Paul, *Adv. Drug Delivery Rev.*, 2012, **64**, 1239.

59. H. Naderi, M. M. Matin and A. R. Bahrami, *J. Biomater. Appl.*, 2011, **26**, 383.
60. K. Whang, T. K. Goldstick and K. E. Healy, *Biomaterials*, 2000, **21**, 2545.
61. D. Li and Y. Xia, *Adv. Mater.*, 2004, **16**, 1151.
62. Y. Dzenis, *Science*, 2004, **304**, 1917.
63. J. Xie and Y. Xia, *Mater. Matters*, 2008, **3**, 19.
64. Y. Z. Zhang, X. Wang, Y. Feng, J. Li, C. T. Lim and S. Ramakrishna, *Biomacromolecules*, 2006, **7**, 1049.
65. I. K. Kang, O. H. Kwon, Y. M. Lee and Y. K. Sung, *Biomaterials*, 1996, **17**, 841.
66. J. S. Bae, E. J. Seo and I. K. Kang, *Biomaterials*, 1999, **20**, 529.
67. E. J. Kim, I. K. Kang, M. K. Jang and Y. B. Park, *Biomaterials*, 1998, **19**, 239.
68. P. R. Kuhl and L. G. Griffith Cima, *Nat. Med.*, 1996, **2**, 1022.
69. M. L. Tan, P. F. M. Choong and C. R. Dass, *Peptides*, 2010, **31**, 184.
70. F. Hajos, B. Stark, S. Hensler, R. Prassl and W. Mosgoeller, *Int. J. Pharm.*, 2008, **357**, 286.
71. A. L. Weiner, *Adv. Drug Delivery Rev.*, 1989, **3**, 307.
72. M. Hedman, J. Hartikainen, M. Syvänne, J. Stjernvall, A. Hedman, A. Kivelä, E. Vanninen, H. Mussalo, E. Kauppila, S. Simula, O. Närvänen, A. Rantala, K. Peuhkurinen, M. S. Nieminen, M. Laakso and S. Ylä Herttuala, *Circulation*, 2003, **107**, 2677.
73. K. S. Oh, J. Y. Song, S. J. Yoon, Y. Park, D. Kim and S. H. Yuk, *J. Controlled Release*, 2010, **146**, 207.
74. J. P. Karam, C. Muscari and C. N. Montero Menei, *Biomaterials*, 2012, **33**, 5683.
75. A. G. Mikos, S. W. Herring, P. Ochareon, J. Elisseeff, H. H. Lu, R. Kandel, F. J. Schoen, M. Toner, D. Mooney, A. Atala, M. E. Van Dyke, D. Kaplan and G. Vunjak Novakovic, *Tissue Eng.*, 2006, **12**, 3307.
76. J. Zhang, L. Ding, Y. Zhao, W. Sun, B. Chen, H. Lin, X. Wang, L. Zhang, B. Xu and J. Dai, *Circulation*, 2009, **119**, 1776.
77. Y. Miyagi, L. L. Y. Chiu, M. Cimini, R. D. Weisel, M. Radisic and R.-K. Li, *Biomaterials*, 2011, **32**, 1280.
78. J. Wu, F. Zeng, X. P. Huang, J. C.-Y. Chung, F. Konecny, R. D. Weisel and R.-K. Li, *Biomaterials*, 2011, **32**, 579.
79. H. D. Guo, G.-H. Cui, J.-J. Yang, C. Wang, J. Zhu, L.-s. Zhang, J. Jiang and S.-j. Shao, *Biochem. Biophys. Res. Commun.*, 2012, **424**, 105.
80. Y. D. Lin, C. Y. Luo, Y.-N. Hu, M.-L. Yeh, Y.-C. Hsueh, M.-Y. Chang, D.-C. Tsai, J.-N. Wang, M.-J. Tang, E. I. H. Wei, M. L. Springer and P. C. H. Hsieh, *Sci. Transl. Med.*, 2012, **4**, 146ra109.
81. S. Koch, Ch. Yao, G. Grieb, P. Prével, E. M. Noah and G. C. M. Steffens, *J. Mater. Sci.: Mater. Med.*, 2006, **17**, 735.
82. A. H. Zisch, U. Schenk, J. C. Schense, S. E. Sakiyama Elbert and J. A. Hubbell, *J. Controlled Release*, 2001, **72**, 101.
83. Z. S. Patel, H. Ueda, M. Yamamoto, Y. Tabata and A. G. Mikos, *Pharm. Res.*, 2008, **25**, 2370.

84. R. C. Scott, J. M. Rosano, Z. Ivanov, B. Wang, P. L.-G. Chong, A. C. Issekutz, D. L. Crabbe and M. F. Kiani, *FASEB J.*, 2009, **23**, 3361.
85. R. R. Chen, E. A. Silva, W. W. Yuen, A. A. Brock, C. Fischbach, A. S. Lin, R. E. Guldberg and D. J. Mooney, *FASEB J.*, 2007, **21**, 3896.
86. F. R. Formiga, B. Pelacho, E. Garbayo, G. Abizanda, J. J. Gavira, T. Simon Yarza, M. Mazo, E. Tamayo, C. Jauquicoa, C. Ortiz-de-Solorzano, F. Prósper and M. J. Blanco Prieto, *J. Controlled Release*, 2010, **147**, 30.
87. T. Simón Yarza, E. Tamayo, C. Benavides, H. Lana, F. R. Formiga, C. N. Grama, C. Ortiz-de-Solorzano, M. N. Kumar, F. Prosper and M. J. Blanco Prieto, *Int. J. Pharm.*, 2013, **454**, 784.
88. D. Maglione, M. Battisti and M. Tucci, *Il Farmaco*, 2000, **55**, 165.
89. Z. M. Binsalamah, A. Paul, A. A. Khan, S. Prakash and D. Shum Tim, *Int. J. Nanomed.*, 2011, **6**, 2667.
90. J. M. Edelberg, D. Cai and M. Xaymardan, *Cardiovasc. Toxicol.*, 2003, **3**, 27.
91. B. Amsden, *Expert Opin. Drug Delivery*, 2011, **8**, 873.
92. I. D'Angelo, M. Garcia Fuentes, Y. Parajó, A. Welle, T. Vántus, A. Horváth, G. Bökönyi, G. Kéri and M. J. Alonso, *Mol. Pharmaceutics*, 2010, **7**, 1724.
93. T. P. Richardson, M. C. Peters, A. B. Ennett and D. J. Mooney, *Nat. Biotechnol.*, 2001, **19**, 1029.
94. A. A. Khan, A. Paul, S. Abbasi and S. Prakash, *Int. J. Nanomed.*, 2011, **6**, 1069.
95. L. A. Reis, L. L. Y. Chiu, Y. Liang, K. Hyunh, A. Momen and M. Radisic, *Acta Biomater.*, 2012, **8**, 1022.
96. A. Anisimov, D. Tvorogov, A. Alitalo, V.-M. Leppänen, Y. An, E. C. Han, F. Orsenigo, E. I. Gaál, T. Holopainen, Y. J. Koh, T. Tammela, P. Korpisalo, S. Keskitalo, M. Jeltsch, S. Ylä Herttuala, E. Dejana, G. Y. Koh, C. Choi, P. Saharinen and K. Alitalo, *Circulation*, 2013, **127**, 424.
97. M. Fujita, M. Ishihara, Y. Morimoto, M. Simizu, Y. Saito, H. Yura, T. Matsui, B. Takase, H. Hattori, Y. Kanatani, M. Kikuchi and T. Maehara, *J. Surg. Res.*, 2005, **126**, 27.
98. H. Wang, X. Zhang, Y. Li, Y. Ma, Y. Zhang, Z. Liu, J. Zhou, Q. Lin, Y. Wang, C. Duan and C. Wang, *J. Heart Lung Transpl.*, 2010, **29**, 881.
99. Y. Sakakibara, K. Tambara, G. Sakaguchi, F. Lu, M. Yamamoto, K. Nishimura, Y. Tabata and M. Komeda, *Eur. J. Cardiothorac. Surg.*, 2003, **24**, 105.
100. P. Cuevas, F. Carceller and G. Giménez Gallego, *J. Cell. Mol. Med.*, 2001, **5**, 132.
101. Y. Wang, X.-C. Liu, J. Zhao, X.-R. Kong, R.-F. Shi, X.-B. Zhao, C.-X. Song, T.-J. Liu and F. Lu, *Tex. Heart I. J.*, 2009, **36**, 89.
102. Y. Liu, L. Sun, Y. Huan, H. Zhao and J. Deng, *Eur. j. cardiothorac. surg.*, 2006, **30**, 103.
103. N. Fournier and C. J. Doillon, *Biomaterials*, 1996, **17**, 1659.

104. S. S. K. Chan, H. J. Li, Y. C. Hsueh, D. S. Lee, J.-H. Chen, S.-M. Hwang, C.-Y. Chen, E. Shih and P. C. H. Hsieh, *PLoS One*, 2010, **5**, e14414.
105. M. E. Davis, P. C. H. Hsieh, T. Takahashi, Q. Song, S. Zhang, R. D. Kamm, A. J. Grodzinsky, P. Anversa and R. T. Lee, *Proc. Natl. Acad. Sci. U. S. A.*, 2006, **103**, 8155.
106. L. Meinel, O. E. Illi, J. Zapf, M. Malfanti, H. Peter Merkle and B. Gander, *J. Controlled Release*, 2001, **70**, 193.
107. M.-Y. Chang, Y.-J. Yang, C.-H. Chang, A. C. L. Tang, W.-Y. Liao, F.-Y. Cheng, C.-S. Yeh, J. J. Lai, P. S. Stayton and P. C. H. Hsieh, *J. Controlled Release*, 2013, **170**, 287.
108. C. Noel, *A Randomised Trial Evaluating the Safety and Efficacy of a Single Low Dose of Intracoronary Insulin-like Growth Factor-1 Following Percutaneous Coronary Intervention for ST-Elevation Acute Myocardial Infarction (RESUS-AMI)*, University College Cork, 2013.
109. K. Kawaida, K. Matsumoto, H. Shimazu and T. Nakamura, *Proc. Natl. Acad. Sci. U. S. A.*, 1994, **91**, 4357.
110. L. Bocchi, M. Savi, G. Graiani, S. Rossi, A. Agnetti, F. Stillitano, C. Lagrasta, S. Baruffi, R. Berni, C. Frati, M. Vassalle, U. Squarcia, E. Cerbai, E. Macchi, D. Stilli, F. Quaini and E. Musso, *PLoS One*, 2011, **6**, e17750.
111. X. H. Zhu, C.-H. Wang and Y. W. Tong, *J. Biomed. Mater. Res., Part A*, 2009, **89**, 411.
112. E. Ruvinov, J. Leor and S. Cohen, *Biomaterials*, 2010, **31**, 4573.
113. M. T. Ozeki Yasuhiko, *J. Biomater. Sci., Polym. Ed.*, 2006, **17**, 163.
114. A. Marui, A. Kanematsu, K. Yamahara, K. Doi, T. Kushibiki, M. Yamamoto, H. Itoh, T. Ikeda, Y. Tabata and M. Komeda, *J. Vasc. Surg.*, 2005, **41**, 82.
115. Institute of Cardiology of Warsaw, *Hepatocyte Growth Factor as an Early Marker of Myocardial Injury and Prognostic Factor of Cardiovascular Events in Log-term Follow up in Patients With Acute Coronary Syndrome*, 2013.
116. J. Li, X.-h. Gu, J.-c. Duan, L. Zeng, Y. Li and L. Wang, *Journal of Sichuan University*, 2007, **38**, 105.
117. A. Jabbour, C. Hayward, A. M. Keogh, E. Kotlyar, J. A. McCrohon, J. F. England, R. Amor, X. Liu, X. Y. Li, M. D. Zhou, R. M. Graham and P. S. Macdonald, *Eur. J. Heart Fail.*, 2011, **13**, 83.
118. M. Xin, Y. Kim, L. B. Sutherland, X. Qi, J. McAnally, R. J. Schwartz, J. A. Richardson, R. Bassel Duby and E. N. Olson, *Sci. Signaling*, 2011, **4**, ra70.
119. V. F. M. Segers, T. Tokunou, L. J. Higgins, C. MacGillivray, J. Gannon and R. T. Lee, *Circulation*, 2007, **116**, 1683.
120. G. Zhang, Y. Nakamura, X. Wang, Q. Hu, L. J. Suggs and J. Zhang, *Tissue Eng.*, 2007, **13**, 2063.
121. B. P. Purcell, J. A. Elser, A. Mu, K. B. Margulies and J. A. Burdick, *Biomaterials*, 2012, **33**, 7849.
122. A. Zieris, S. Prokoph, K. Leventol, P. Welzel, M. Grimmer, U. Freudenberg and C. Werner, *Biomaterials*, 2010, **31**, 7985.

123. X. Hao, A. Månsson Broberg, T. Gustafsson, K. H. Grinnemo, P. Blomberg, A. J. Siddiqui, E. Wärdell and C. Sylvén, *Biochem. Biophys. Res. Commun.*, 2004, **315**, 1058.

124. A. S. Salimath, E. A. Phelps, A. V. Boopathy, P.-l. Che, M. Brown, A. J. García and M. E. Davis, *PLoS One*, 2012, 7, e50980.

125. A. Cittadini, M. Monti, V. Petrillo, G. Esposito, G. Imparato, A. Luciani, F. Urciuolo, E. Bobbio, C. Natale, L. Sacc and P. Netti, *Eur. J. Heart Failure*, 2011, **13**, 1264.

126. Q. Sun, E. A. Silva, A. Wang, J. C. Fritton, D. J. Mooney, M. B. Schaffler, P. M. Grossman and S. Rajagopalan, *Pharm. Res.*, 2010, **27**, 264.

127. S. T. M. Nillesen, P. J. Geutjes, R. Wismans, J. Schalkwijk, W. F. Daamen and T. H. van Kuppevelt, *Biomaterials*, 2007, **28**, 1123.

128. J. H. Kim, Y. Jung, S.-H. Kim, K. Sun, J. Choi, H. Kim and Y. Park, *Biomaterials*, 2011, **32**, 6080.

129. F. Formiga, B. Pelacho, E. Garbayo, I. Imbuluzqueta, P. Díaz-Herráez, G. Abizanda, J. Gavira, T. Simón-Yarza, E. Albiasu, E. Tamayo, F. Prósper and M. Blanco-Prieto, *J. Control. Release*, 2014, **10**, 132.

130. J. V. Terrovitis, R. R. Smith and E. Marbán, *Circ. Res.*, 2010, **106**, 479.

131. S. Fukuhara, S. Tomita, T. Nakatani, T. Fujisato, Y. Ohtsu, M. Ishida, C. Yutani and S. Kitamura, *Circ. J.*, 2005, **69**, 850.

132. K. Kang, L. Sun, Y. Xiao, S.-H. Li, J. Wu, J. Guo, S. L. Jiang, L. Yang, T. M. Yau, R. D. Weisel, M. Radisic and R.-K. Li, *J. Am. Coll. Cardiol.*, 2012, **60**, 2237.

133. C. Penna, M.-G. Perrelli, J.-P. Karam, C. Angotti, C. Muscari, C. N. Montero Menei and P. Pagliaro, *J. Cell. Mol. Med.*, 2013, **17**, 192.

134. P. Díaz Herráez, E. Garbayo, T. Simón Yarza, F. R. Formiga, F. Prosper and M. J. Blanco Prieto, *Eur. J. Pharm. Biopharm.*, 2013, **85**, 143.

135. H. Matsubara and N. Takehara, *Hybrid Biotherapy Involving Autologous Human Cardiac Stem Cell Transplantation Combined With the Controlled Release of bFGF Using a Gelatin Hydrogel Sheet to Treat Severe Refractory Heart Failure With Chronic Ischemic Cardiomyopathy*, 2013.

Section B
Structural Biointerfaces

CHAPTER 5

Titanium Phosphate Glass Microspheres as Microcarriers for In Vitro Bone Cell Tissue Engineering

NILAY J. LAKHKAR,[a,b] CARLOTTA PETICONE,[c]
DAVID DE SILVA-THOMPSON,[c] IVAN B. WALL,[c,d]
VEHID SALIH[a,e] AND JONATHAN C. KNOWLES*[a,d]

[a] Division of Biomaterials and Tissue Engineering, UCL Eastman Dental Institute, University College London, 256 Gray's Inn Road, London WC1X 8LD, UK; [b] Institute of Biomaterial Science and Berlin-Brandenburg Center for Regenerative Therapies, Helmholtz-Zentrum Geesthacht, Kantstrasse 55, 14513 Teltow, Germany; [c] Department of Biochemical Engineering, University College London, Torrington Place, London WC1E 7JE, UK; [d] Department of Nanobiomedical Science & BK21 Plus NBM Global Research Center for Regenerative Medicine, Dankook University, Cheonan 330-714, Republic of Korea; [e] Plymouth University Peninsula Schools of Medicine and Dentistry, C402, Portland Square, Drake Circus, Plymouth, Devon PL4 8AA, UK
*Email: j.knowles@ucl.ac.uk

5.1 Introduction

In the last decade, phosphate glasses, in general, and titanium phosphate glasses, in particular, have garnered increasing attention from researchers

RSC Smart Materials No. 10
Biointerfaces: Where Material Meets Biology
Edited by Dietmar Hutmacher and Wojciech Chrzanowski
Published by the Royal Society of Chemistry, www.rsc.org

investigating biomaterials that can help combat bone loss or bone deformities arising from trauma, disease or congenital defects.[1-4] In this context, it is worth mentioning at the outset that research on glasses for biomedical applications is not so recent; silicate glasses have been studied for a significantly longer period since Larry Hench developed 45S5 Bioglass in the late 1960s[5] and consequently far more literature exists on the properties and biomedical applications of silicate glasses in comparison with phosphate glasses. However, the amount of research on biological phosphate glasses suggests that they have now emerged as a distinct class of biomaterials in their own right.

The composition of the ternary P_2O_5–Na_2O–CaO glass system can be varied to control the rate at which the glass degrades; for example, an increase in the CaO content (at the expense of Na_2O) leads to a decrease in the degradation rate and *vice versa*. However, due to the relatively high dissolution rate (and hence poor biocompatibility) of most compositions within the ternary system above, most studies have focused on controlling the degradation rate of phosphate glasses by the addition of other metal oxides to the glass system to achieve their objective. To date, the metal oxides used for this purpose include Fe_2O_3,[6-8] CuO,[9-12] Al_2O_3,[13] TiO_2,[14-17] MgO,[18] ZnO,[19-23] Ag_2O,[24-28] Ga_2O_3[29-31] and SrO.[32-38] The purposes of metal oxide addition to the glass structure are two-fold. First, metal oxides serve as the most effective means to control the degradation rate by increasing the covalent nature of the bonds within the glass structure. Second, the metal ions released from such glasses can exert an antimicrobial effect and/or positively impact on cell proliferation and tissue regeneration.

The appeal of titanium phosphate glasses in biomedical research derives essentially from the strong inter-relationships between three sets of factors: (1) the glass chemistry and glass network structure, (2) the glass properties, and (3) the interactions between glasses and living cells or tissues. The link between the first two factors is that subtle changes in glass chemistry through variations in glass composition can bring about significant changes in glass properties such as the glass transition temperature, glass density or glass degradation. A glass not containing titanium oxide may disappear completely in a matter of a few hours when immersed in deionised water, whereas a glass containing 5 mol% titanium oxide may take several months to degrade completely. In turn, the glass properties greatly influence the ability of the glass surface to provide a stable substrate for cells to attach and proliferate and, in the case of stem cells, to possibly differentiate down a specific lineage, for example osteogenic or chondrogenic pathways. Using the same example as above, the rapid degradation of the former glass would lead to considerable fluctuations in the solution pH which would be detrimental to cells under *in vitro* cell culture conditions, while the latter glass would provide the stable surface required for the cells to flourish.

Thus, considering what is known about the strong links between structure, properties and cell–material interactions, a reasonably obvious question arises: what biomedical applications can we develop based on this knowledge? And this question leads us to a new challenge, which is to process the bulk glass into forms that would satisfy two seemingly disparate requirements: (1) to be suitable in terms of size and morphology for use either within the human body or within engineered devices that support an active biological environment, and (2) to be reproducible using rapid, inexpensive and easily scalable industrial techniques. Thus far, phosphate glasses have been processed in the form of monoliths, discs, granules, cloths and tubes in the macroscopic size range as well as powders and fibres in the micro- and nano-sized ranges.[39] The application described in this chapter is to use titanium phosphate glasses as a substrate material for the *in vitro* expansion (or scale-up) of bone cells and ultimately the formation of viable bone tissue for the treatment of critical sized bone defects (a few millimetres in size). The development of such substrate materials involves the investigation of the interactions between bone cells and glass systems firstly within static systems such as culture well plates and subsequently within dynamic systems such as spinner flasks and perfusion bioreactors.

For *in vitro* bone cell expansion, microspheres offer some unique advantages over other surfaces. For instance, in comparison with tissue culture plastic, they can provide a considerably larger surface area for cell proliferation. Furthermore, this surface area is easily and accurately quantifiable as opposed to the case of microscopic glass powders. Both factors assume significant importance in the context of industrial scale-up of the cell expansion process for providing a large quantity of cells that can be used in cell-based therapies or for high throughput screening.

In this chapter, the main focus is centred on three topics: (1) microsphere production, (2) microsphere degradation, and (3) cell–microsphere interactions. The production of glass microspheres using a simple, rapid and inexpensive technique is described. A novel time-lapse imaging method to visualise the degradation of microspheres and also obtain subsequent quantitative data is elaborated and this allows for correlation of the microsphere degradation rate with the glass composition. Subsequently, guidance on carrying out the culture of bone cells on microspheres and visualising cell–microsphere interactions *via* scanning electron microscopy is given. The experiments described in the chapter are easy to initiate and are also quite versatile; for instance, the time-lapse imaging technique can be used for visualising the degradation of other microscopic materials while the cell culturing techniques can be used for different cell types.

5.2 Production of Titanium Phosphate Glass Microspheres by Flame Spheroidisation

The low fusion temperatures and ease of glass formation without vitrification mean that phosphate glasses can be synthesised by simple conventional melt-quench routes. However, P_2O_5—the main component of these glasses—is highly hygroscopic; therefore, the inclusion of phosphorus in the glass structure without using excessive amounts of P_2O_5 becomes a significant technical challenge, and the presence of small amounts of water can cause crystallisation.[40]

TIP: *A good way to overcome this challenge is to (1) add some proportion of phosphorus in the form of a stable oxide that includes one of the modifying cations, e.g. $CaHPO_4$ or NaH_2PO_4, and then (2) add any remaining minimal amount as P_2O_5.*

Water uptake (and also P_2O_5 loss due to volatilisation) can be minimised by melting in ampoules or covered crucibles. A useful alternative to the conventional melt-quench process is the sol–gel process, which involves the mixing of inorganic alkoxide and metal chloride precursors to form a colloid through hydrolysis and polycondensation reactions and subsequent thermal treatment to yield a dense, porous, glassy material. This process has a high level of adaptability, which means that sol–gel glasses can be deposited as thin films on a substrate or drawn into macro- and micro-sized fibres or obtained as micrometre- or nanometre-sized particles. The sol–gel process can also be carried out at temperatures that are up to an order of magnitude lower than in the melt-quench process,[41,42] which facilitates the introduction of active biological ingredients such as proteins, antibiotics and chemotherapeutic molecules into the glass structure. It is worth noting, however, that the chemistry involved in phosphate sol–gel synthesis presents considerable technical challenges because alkyl phosphates undergo hydrolysis at considerably slower rates than alkyl silicates, and the synthesis and drying steps can require several weeks for completion. As a result, sol–gel synthesis of titanium phosphate glasses remains a relatively less explored area of research.

The production of microspheres from melt-quenched glasses takes place using a technique generally known as flame spheroidisation or flame spheronisation; both names belie the innate simplicity of the underlying principle, which is essentially the 'moulding' of glass microparticles into spherical form *via* surface tension forces whilst in the molten state.[43,44] Various designs for this procedure as well as apparatuses have been described in the literature, the basic components of which are: (1) a high-temperature jet, (2) a feed assembly, and (3) collectors, with particles travelling along the flame axis. The high-temperature jet can be a gas/oxygen flame, with the gas being acetylene or liquid petroleum gas, or it can be a radio-frequency plasma flame;[45–49] as mentioned in Section 5.2.2, a high-temperature flame can also be produced by using methylacetylene-propadiene propane (commercial name: MAPP gas) and can be used for heating, soldering, brazing and welding due to its high flame temperature of 2925 °C in air.[50] An alternative to the gas flame is a vertical tubular electric furnace, in which particles dropped at the top end of the furnace are transformed into microspheres as they fall through the furnace under the effect of gravity.[48,49,51] It is worth noting that the choice of the high-temperature jet depends on the required particle size; stronger jets generated at higher temperatures can spheroidise particles with a larger grain size. The particles to be spheroidised can be fed to the flame either manually or using an automated feed assembly. The microspheres can be collected over air or water in standard glass containers. Other techniques for producing glass microspheres are based on processing techniques that complement the sol–gel method, as described in the review by Park *et al.*[52] Silicate glasses are far more amenable to such processing than phosphate glasses, with tetraethyl orthosilicate (TEOS) being the silica precursor of choice, and various microsphere morphologies such as solid or hollow microspheres with or without surface porosity can be obtained,[53-58] as opposed to the flame spheroidisation process which is mostly used to produce non-porous solid microspheres.

5.2.1 Materials

All chemicals used for glass making have purities of >99% and can be sourced from VWR (Poole, UK).

- Calcium carbonate ($CaCO_3$; GPR Rectapur, catalog no. 22296)
- Phosphorus pentoxide (P_2O_5; BDH Prolabo, catalog no. 21411)
- Sodium dihydrogen orthophosphate (NaH_2PO_4; GPR Rectapur, catalog no. 28013)
- Titanium(IV) oxide (TiO_2; BDH Prolabo, catalog no. 20732)

5.2.2 Equipment

- Alcohol wipes (Azowipe bactericidal wipes may be used)
- Aluminium trough
- Connecting wires
- DC motor (15 800 rpm, 4.5–15 V, 35.8 mm diameter; RS Components, Corby, UK)
- Furnace capable of temperatures up to 1500 °C (RHF 1500; Carbolite, UK)
- Laboratory paddle blender (Stomacher®400 Circulator; Seward, UK)
- MAP-PLUS gas cylinder (453 g; Today's Tools, UK)
- Metal tongs
- Metal screws
- Micro spatula
- Microscope slides
- Milling machine (MM301; Retsch, Germany)
- Mortar and pestle
- Optical microscope
- Paddle blender bags (Stomacher®400 Classic BA6141)
- Parafilm
- Platinum crucible (Pt/10% Rh type 71040; Johnson Matthey, UK)
- Plastic container (capacity, 1000 mL)

- Polystyrene containers (Sterilin, UK)
- Programmable power supply (ISO-TECH IPS-405, RS Components, UK)
- Rectangular glass boxes (dimensions 275 mm×150 mm×60 mm)
- Retort stands
- Rothenberger Super Fire 2 gas torch (Rothenberger Werkzeuge GmbH, Germany)
- Sieves (Endecotts, UK)
- Sieve shaker (Fritsch Spartan, Germany)
- Stainless steel trowel spatula
- Steel plate
- Test connectors (4 mm banana plugs and alligator clips)
- Three-pronged clamps
- Weight balance (resolution of 0.1 g)

5.2.3 Calculations of the Precursor Amounts

As an example, consider a glass having the composition $0.5P_2O_5$–$0.4CaO$–$0.09Na_2O$–$0.01TiO_2$ (all values in molar fraction). The amounts of precursors used to manufacture this phosphate glasses can be calculated as follows.

5.2.3.1 Sodium Dihydrogen Orthophosphate (NaH_2PO_4)

	$2NaH_2PO_4$ $\rightarrow$	Na_2O +	P_2O_5 +	$2H_2O$
Mol. wt. (g mol^{-1})	240	62	142	36
Relative mol. wt. (g mol^{-1})		0.26	0.59	0.15

Here, the amount of NaH_2PO_4 to be added in order to obtain 9 mol% Na_2O:

$$= [(\text{mol. fraction}) \times (\text{mol. wt.})]/(\text{relative mol. wt.})$$

$$= (0.09 \times 62)/(0.26) = 21.46 \text{ g}$$

5.2.3.2 Phosphorus Pentoxide (P_2O_5)

The amount of P_2O_5 produced in the above reaction = (amount of NaH_2PO_4 produced)×(relative mol. wt. of P_2O_5)

$$= 21.46 \times 0.59 = 12.66 \text{ g}$$

The amount of P_2O_5 required to produce 50 mol% P_2O_5

$$= 0.50 \times 142 = 71.00 \text{ g}$$

Therefore, the actual amount of P_2O_5 to be added is:

$$= 71.00 - 12.66 = 58.34 \text{ g}$$

5.2.3.3 Calcium Carbonate (CaCO₃)

	$CaCO_3 \rightarrow$	CaO	$+$	CO_2
Mol. wt. (g mol^{-1})	100	56		44
Relative mol. wt. (g mol^{-1})		0.56		0.44

Amount of $CaCO_3$ to be added in order to obtain 40 mol% CaO:

$$= (0.4 \times 56)/0.56 = 40.00 \text{ g}$$

5.2.3.4 Titanium Dioxide (TiO₂)

Mol. wt. of Ti $= 47.90$, and the amount of TiO_2 to be added in order to obtain 1 mol% TiO_2:

$$= (47.90 + 32) \times 0.01 = 0.80 \text{ g}$$

> **TIP:** *The amounts of precursors actually used to make the glass will depend on the size of the Pt crucible. For smaller crucible sizes, half or one-third precursor quantities may be used.*

5.2.4 Preparation of Glass Microparticles

> **CAUTION!** *Appropriate laboratory safety apparel such as laboratory coats, goggles, nitrile/latex gloves and particulate respirator masks must be worn throughout the entire procedure.*

1. Switch on furnace and set the temperature to 700 °C.
2. Place an opened paddle blender bag into the 1000 mL plastic container with its edges folded over the container rim. Place this container on the weighing scale. Weigh the calculated amount of $CaCO_3$, remove bag from container and keep aside. Repeat these steps for NaH_2PO_4 and TiO_2.
3. Mix contents of all three bags into a separate bag. Place bag in paddle blender and mix at 200 rpm for 1 min.
4. Repeat Step 1 for P_2O_5 and then add P_2O_5 to bag containing the other precursors.

> **TIP:** *Since P_2O_5 is highly hygroscopic in nature, perform this step as quickly as possible to minimise water sorption. Repeat the mixing step.*

5. Pour the mixed precursors into the crucible and place the crucible in the preheated furnace using metal tongs

> **CAUTION!** *Furnace gloves must be worn while carrying out Step 5.*

6. Maintain the furnace at 700 °C for 30 min; then, increase the furnace temperature to 1300 °C and maintain at this temperature for 3 h.
7. Take the crucible out of the furnace using metal tongs and carefully pour the molten glass onto a steel plate.

> **CAUTION!** *Furnace gloves must be worn while carrying out Step 7. Also care must be taken to avoid spillage of molten glass anywhere outside the steel plate.*

8. Leave to cool overnight.
9. Crush glass into smaller fragments using a mortar and pestle. Carry out further crushing of fragments in a milling machine to obtain particles in the micrometre/nanometre size range.
10. Pass glass particles through sieves placed on a sieve shaker to obtain particles of the required size for making glass microspheres.

> **TIP:** *In order to reduce the time required to obtain a certain quantity of glass particles ready to be spheroidised, Steps 9 and 10 can be carried out in parallel; for example, at intervals of 2 h, the glass particles present in the uppermost sieve can be reintroduced into the milling machine for crushing and then placed back in the uppermost sieve for further sieving.*

5.2.5 Design of Flame Spheroidisation Apparatus

1. The flame spheroidisation apparatus comprises the following component assemblies: (1) blow torch, (2) feed, and (3) collectors (see Figure 5.1).
2. The blow torch assembly comprises the gas torch fitted to a MAPP-PLUS gas cylinder. The blow torch assembly is positioned using a retort stand and three-pronged clamps such that the flame of the gas torch is horizontal.
3. The feed assembly comprises an aluminium trough with dimensions of 200 mm×20 mm×30 mm, with the edge of one end positioned approximately 10 mm above the outlet of the torch at a slight angle to the horizontal by means of a retort stand and clamp. A DC motor (15 800 rpm, 4.5–15 V, 35.8 mm diameter; RS components, Corby, UK) is attached to the other end of the trough and is connected to an ISO-TECH IPS-405 programmable power supply (RS Components, Corby, UK) by wires fitted with the required connectors (4 mm banana test plug connectors for the power supply and alligator test connectors for the motor).

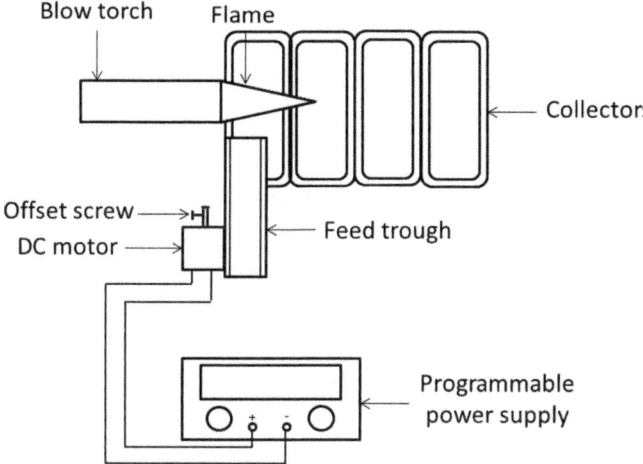

Figure 5.1   Schematic of flame spheroidisation set-up.

4. A metal screw connected to the axle of the motor serves as an off-set to generate vibration when the motor is in operation so that the particle feed for microsphere production is evenly distributed over the surface of the trough before the particles enter the flame. The output power from the programmable power supply may be varied during operation so as to obtain optimum particle distribution in the trough.
5. The collectors consist of four rectangular glass boxes with dimensions of 275 mm×150 mm×60 mm placed with their longer edges in contact with each other. The first collector is placed directly below the flame so as to collect particles that do not pass through the flame.

5.2.6   Preparation of Glass Microspheres

1. Place the glass microparticles to be spheroidised on the aluminium trough at the end away from the flame.

> **TIP:** *Prior to operation, examine the different components of the apparatus from appropriate positions to ensure proper alignment of (1) the trough end with the flame and (2) the flame with the collectors. This is important to ensure maximum possible microsphere recovery from the process.*

2. Switch on the blow torch and then switch on the power supply to the motor.
3. Monitor the particles as they travel down the trough. If required, a micro spatula can be used to aid distribution of particles over the trough surface before they fall into the flame, travel along the flame axis and then fall into one of the collectors.

4. Take a small portion of the powder in each collector on a microscope slide and examine under an optical microscope. Store powder from appropriate collectors in Sterilin vials and seal the lid with parafilm to minimise water uptake.

5.2.7 Expected Outcomes and Other Considerations

The powder obtained in the collectors will be a mixture of microspheres and non-spherical particles (Figure 5.2); depending on the ratio of spherical to non-spherical particles, the powder obtained in the collectors may be stored for further use, discarded or recycled in the flame spheroidisation process. The material present in the collector immediately below the flame may contain a significant proportion of non-spherical particles; this material can be discarded or recycled in the process to further increase the microsphere yield. The particle size distribution of the obtained particles will reveal that the microsphere size is independent of the glass composition, indicating that the microsphere size range can be varied significantly by suitably modifying the flame spheroidisation set up. The set-up described above yields microspheres in the size range of $\sim$ 10–210 µm. At sizes of $\sim$ 45 µm or lower, particle agglomeration in the feed apparatus and subsequent dispersal of the agglomerates in the flame considerably limits the number of glass microspheres obtained. This problem can possibly be overcome by dispersing the feed particles in a suitable liquid dispersant prior to introduction in the flame. At sizes of $\sim$ 100 µm or higher, the residence time of the particles in the flame is too small for achieving complete spheroidisation so that non-spherical particles are obtained. This problem can be resolved by generating a flame of greater length and higher temperature to provide a longer residence under hotter conditions. A larger blow torch connected to a gas–oxygen source such as acetylene–oxygen may be used for this purpose.

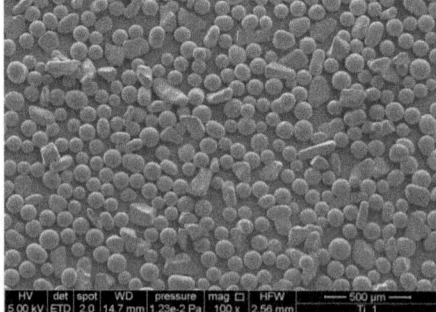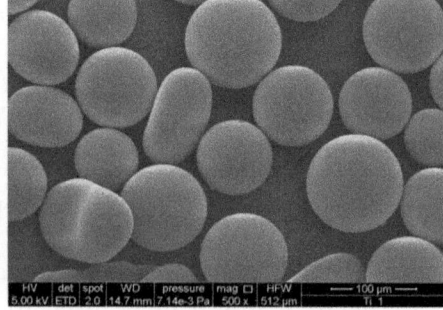

Figure 5.2 Scanning electron microscopy images of titanium phosphate glass microspheres at magnifications of ×100 (left) and ×500 (right).

5.3 Visualisation and Quantification of Microsphere Degradation by Time-lapse Imaging

Before going into the details of how microsphere degradation can be seen and quantitatively assessed, understanding how glass degradation occurs when glass specimens are immersed in deionised water is a worthwhile exercise. As already mentioned, phosphate glasses have the dual capability of being fully degradable and also easily modifiable to incorporate therapeutic ions, which means that these glasses are considered prime candidates as bone filling materials or as scaffolds for bone tissue engineering. The ions released in this process are mainly degradation products of the P_2O_5–Na_2O–CaO network skeleton, all of which are non-toxic in nature.[59] Compared to silicate glasses, weight loss and ion release in phosphate glasses are generally linear and constant over time.[60] As mentioned previously, the degradation rate can be altered by several orders of magnitude by engineering the glass structure,[12] thus making this material more suitable for clinical applications. This can be achieved by stabilising the glass network by doping with several metal oxides. Furthermore, these ions have been found to induce a biological response from human cells, thus enhancing the bioactivity of the scaffold material.

The main ions released from phosphate glasses are phosphate groups, which are known to play a role in enhancing tissue regeneration. The phosphate groups studied thus far include orthophosphate (PO_4^{3-}), linear species such as pyrophosphate ($P_2O_7^{4-}$) and tripolyphosphate ($P_3O_{10}^{5-}$) and cyclic species such as cyclic trimetaphosphate ($P_3O_9^{3-}$). Although higher phosphate species (P_4 and higher) may be released by these glasses, the absence of calibration standards containing these species complicates the identification of these ions by currently available measurement techniques. These techniques comprise chromatography techniques such as high-performance liquid chromatography and ion chromatography along with other approaches such as modified end group titration; ion chromatography is considered the simplest and most effective of these techniques.[61]

Phosphate glasses generally degrade and release ions at rates that increase with time until full glass dissolution. For instance, in ternary glass systems such as $0.45P_2O_5$–$xCaO$–$(0.55 - x)$ Na_2O where $x = 0.30$, 0.40 and 0.45 mol. fraction, all four phosphate species are released in increasing amounts over the study period.[62] Modifying the content of cationic oxides in the glass is considered to play a role in determining linear phosphate ion release, and the presence of single/multiple phases in the glass and the proximity of the glass composition to a eutectic point in phase diagrams may be contributing factors. The extent of cross-linking between the branched linear phosphates and the Ca^{2+} ions is proportional to the CaO content.[63] The degradation trends observed conform to the model by Bunker *et al.*[64] which proposes three types of reactions between the glass and the surrounding water: acid–base reactions, hydrolysis and hydration. The amount of cyclic $P_3O_9^{3-}$ released is generally the highest among all the investigated phosphate species. From this observation, it can be inferred that the cyclic species dominates the structure of these glasses.

Metal oxides are known to exert their own influence on glass degradation based on several factors such as metal ion valency and atomic radius.[65] When metal oxides with $\geq 3+$ valencies are incorporated in the glass structure, the overall trend is for the phosphate ion release to mirror the degradation profile so that glasses that degrade at slower rates release smaller amounts of all phosphate species. However, glasses containing divalent metal oxides may not show this trend; for instance, P_2O_5–CaO–Na_2O–ZnO glasses with different ZnO contents release phosphate ions at roughly similar rates, whereas in P_2O_5–CaO–Na_2O–SrO glasses, there is no particular consistency in terms of the overall trend of phosphate release.[20,35]

Most studies on titanium phosphate glass degradation have been carried out using macroscopic glass discs. The experimental technique used is a commonly used weight loss model.[66] However, for investigating the degradation of glass microspheres, we have used a novel time lapse imaging technique. This technique takes advantage of the microscopic size of these spheres to provide visual evidence of microsphere degradation while simultaneously providing quantitative data to correlate the microsphere degradation with the glass composition.

5.3.1 Materials

The following refers to microspheres made from a glass of known density.

5.3.2 Equipment

- Analytical balance (resolution of 0.1 mg)
- Alcohol wipes (Azowipe bactericidal wipes may be used)
- Beaker
- Brush
- Camera (TCA-10.0-N; Tucsen Image Technology Inc., China)

> **TIP:** *A wide range of cameras equipped with software to capture time-lapse images are available and can be used for this experiment.*

- Desktop PC loaded with µManager microscopy open source software (version 1.4; Ron Vale Lab, University of California – San Francisco, USA) and ImageJ (version 1.47; National Institutes of Health, USA)
- Micro spatula
- Microscope incubator system (Solent Scientific, UK)

> **TIP:** *The incubator system is used in order to carry out the time-lapse experiment at 37 °C. If such an incubator system is not available, the same experiment can be carried out at room temperature; our previous experience shows that the lower temperature will not drastically affect the experimental results.*

- Optical microscope (DMIRB microscope; Leica Microsystems, Germany)

> **NOTE:** *This microscope was procured together with the incubator system, but conventional optical microscopes having a camera port and appropriate adapters can be used.*

- Pipette with a capacity of 10 mL (Finnpipette; Thermo Scientific, UK)
- Pipette tips (10 mL; Thermo Scientific, UK)
- Tissue culture flask, 25 cm^2 (TPP; Helena Biosciences, UK)
- Type 1 ultra-pure water purification system capable of dispensing water with a resistivity of 18.2 MΩ cm^{-1} (Purelab UHQ-PS; Elga Labwater, UK)
- Weighing boats

5.3.3 Setting Up a Time-lapse Experiment

1. Switch on the incubator system about 2 h prior to setting up the experiment and set the temperature to 37 °C. Just before the experiment, switch on the desktop PC and load the μManager software.

2. Weigh 100 mg of glass microspheres on an analytical balance and pour into a T-25 cell culture flask. Add 10 mL of ultra-pure water to the flask using a pipette.

> **TIP:** *Prior to the addition of microspheres and water, the bottom inner surface of the flask can be scraped with a brush in order to provide surface roughness so that the microspheres remain reasonably stationary during the experiment. If this step is performed, the flask should then be washed two to three times with ultrapure water in order to wash away any debris resulting from the scraping action.*

3. Place the flask on the microscope stage and adjust the vertical stage position so that the microspheres are brought into focus on the software display screen (the Live View window, which can be opened by clicking on the Live option on the software interface). Adjust the horizontal stage position so that the display screen shows approximately eight to ten microspheres.

> **TIP:** *It is advisable to focus on microspheres that do not touch each other so that the image processing software will count the spheres as individual objects and provide diameter values for each sphere; this is because spheres touching each other may be interpreted by the software as a single object.*

4. To begin the time lapse sequence, click on Multi-D Acq. On the following window, tick the boxes for Time Points and Save Images (Figure 5.3). Below Time Points, list the number of images to be taken and the interval between images. Below Save Images, write the file pathway to be used to store the files. For Saving Format, click the option for Separate image files. Click on Acquire.

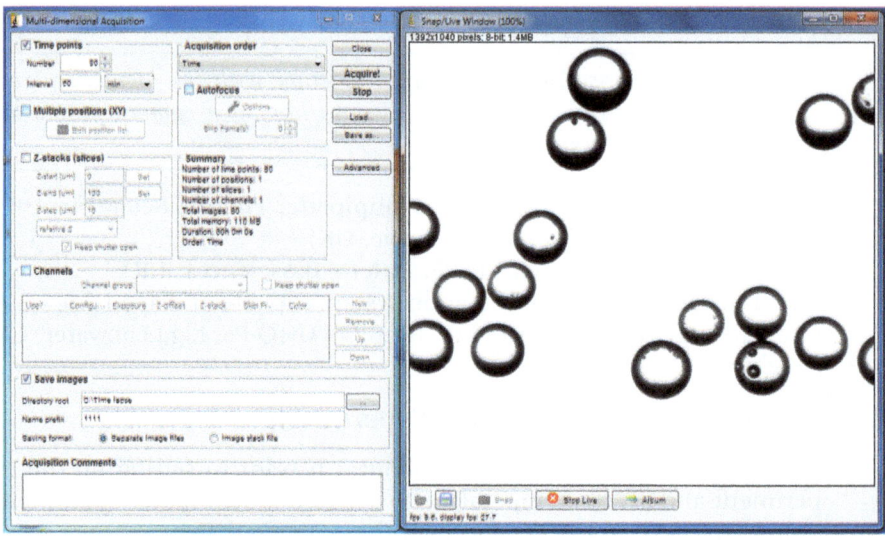

Figure 5.3 Screenshots of multi-dimensional acquisition and Snap/Live windows in μManager software.

5. Carry out the following steps to obtain microsphere diameter values from the stored time lapse images using ImageJ version 1.47.

    i. *Open the time lapse images*: Go to File | Import | Image Sequence. Go to the folder containing the images. Click on the first image from your sequence and press Open. Press OK on the window that appears.

    ii. *Binarise the images*: Go to Process | Binary | Make Binary and press OK on the window that appears.

    iii. *Select a rectangular area around a sphere*: Press on the rectangle icon and then draw a rectangle on the image wherever required by dragging the cursor (which looks like a '+' on the image) across the image with the left mouse button pressed. The position of the rectangle can be adjusted by keeping the cursor anywhere inside the rectangle (the '+' sign changes to a normal arrow) and dragging it over the image.

    iv. *Exclude all objects other than the ones inside the rectangular area*: Go to File | Edit | Clear Outside, and press OK on the window that appears. Remove the rectangular area by clicking anywhere on the image.

> **TIP:** *Before excluding all other spheres from each image, make sure that the bead of interest stays fully within the rectangle throughout all the images by moving the horizontal scroll bar from beginning to end. It is alright if parts of other spheres are visible inside the rectangular area since the software will exclude them during the analysis.*

    v. *Set the calibration scale*: Go to Analyse | Set Scale and input appropriate values for Distance in Pixels and Known Distance. Maintain Pixel Aspect Ratio = 1.0 and Unit of Length = micron. Make sure that the box next to Global is ticked. Press OK.

> **NOTE:** *If the calibration values are not known, use a micrometer calibration slide to measure a known distance in pixels. After proper focusing, draw a line on the screen between two divisions on the slide and measure the length in pixel units by going to Analyse | Measure. The output measurement is the Distance in Pixels and the distance between the divisions is the Known Distance in the Set Scale window.*

    vi. *Measure the diameters in the first and last images*: Measure the diameter of the sphere in the first image of your sequence by pressing on the circle icon next to the rectangular icon, repeating Step (iii) to draw a circle roughly around the sphere circumference, and going to Analyse | Measure. Make a mental note of the approximate area (say, 22788.789). Repeat the same step for the last image to get the approximate area (say, 20742.123). Approximations of both values are needed for the next step.

    vii. *Obtain the bead diameter*: Go to Analyze | Analyze Particles and use the following values: Size (micron^2) = 20 000–25 000, Circularity: 0.0–1.0, Show: Outlines (from drop-down menu). Tick the following boxes: Display Results, Summarize, Exclude on Edges, Include Holes, Record Starts. Press OK on the window that comes up. Three new windows will appear on the screen. Go to the Results window. Then go to Edit | Select All and Edit | Copy in order to paste the values in a suitable spread sheet software.

6. Measure the diameters of three or more microspheres by the above method and calculate the mass of the sphere by the equation:

$$m_{sphere} = \rho_{glass} \times v_{sphere} = \rho_{glass} \times 4\pi(r_{sphere})^3/3 \qquad (5.1)$$

where m_{sphere} is the mass of the sphere (µg); ρ_{glass} is the density of the glass (µg µm^{-3}); v_{sphere} is the volume of the sphere (µm^3); and r_{sphere} is the radius of the sphere (µm).

7. Calculate the percentage weight loss per unit surface area of the sphere at each time point using the following equation:

$$[(m_{sphere(0)} - m_{sphere(t)})/m_{sphere(0)}A_t] \times 100 \qquad (5.2)$$

where $m_{sphere(0)}$ is the mass of the microsphere at time $t = 0$ (µg); $m_{sphere(t)}$ is the mass of the microsphere at time t (µg); and A_t is the surface area of the microsphere at time t (µm^2).

5.3.4 Expected Outcomes and Other Considerations

The above procedure will yield a series of images which will provide visual evidence of microsphere degradation over the time period of the experiment. These images can be converted into time-lapse videos using software such as Microsoft Windows Movie Maker or QuickTime 7 Pro. Furthermore, quantitative data on sphere diameter reduction with time can be obtained which allows the correlation of microsphere degradation rates with the glass composition. It is worth noting that the procedure may not work for microspheres made of highly degradable glasses. To elaborate, consider the microspheres of the titanium-free phosphate glass (Ti0 in Figure 5.4). By $t = 24$ h, a hydrated layer begins to form on the outer surface of most of the microspheres and the microspheres continue to degrade inside this structure. As it degrades within the layer, each microsphere becomes increasingly irregular in shape and constantly changes its position inside the structure. In some cases, the hydrated layers of adjacent microspheres come in contact and fuse together, with the microspheres continuing to degrade inside the newly formed larger layer. By $t = 72$ h, almost all the microspheres have degraded completely, leaving behind only the transparent structures. Under

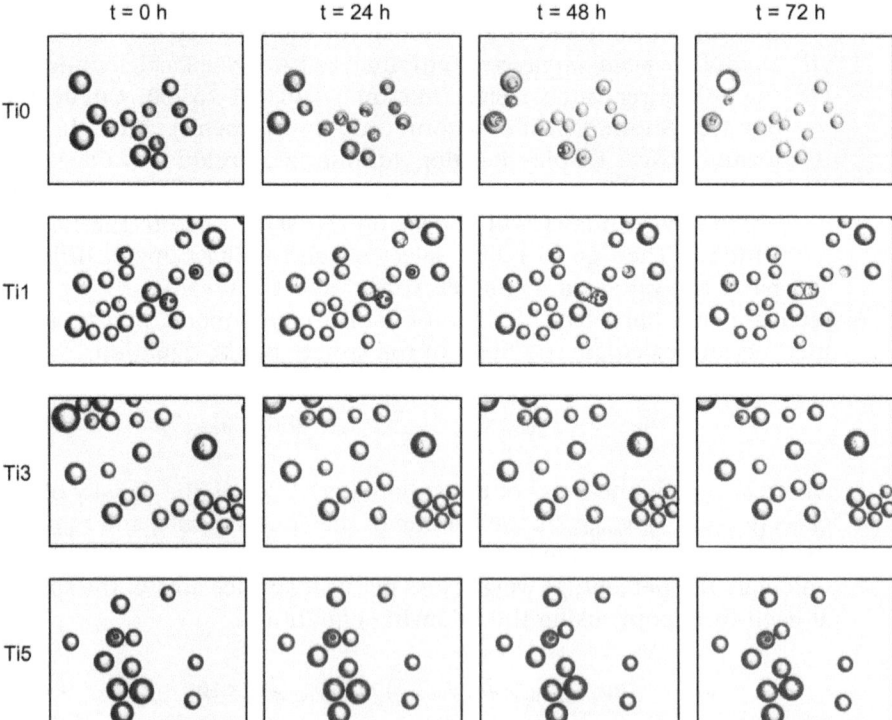

Figure 5.4 Time-lapse acquisition images showing the degradation of microspheres of glasses containing 0, 1, 3 and 5 mol% TiO_2 at 0, 24, 48 and 72 h for each composition.

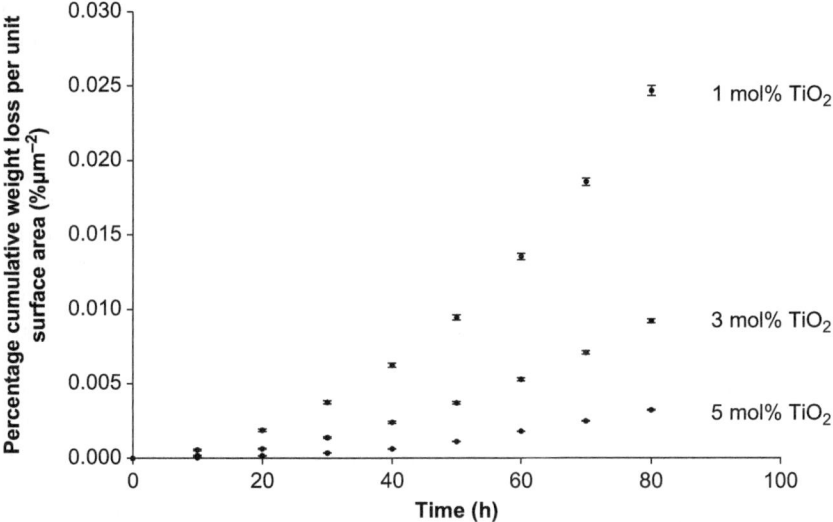

Figure 5.5 Degradation presented as per cent cumulative weight loss per unit surface area $(\%\mu m^{-2})$ as a function of time for titanium phosphate glass microspheres containing 1, 3 and 5 mol% TiO_2.

these circumstances, it is not possible for the image analysis software to correctly identify the microsphere boundary and measure the microsphere diameter. In contrast, microspheres made of titanium-containing glasses (Ti1, Ti3 and Ti5 in Figure 5.5) degrade at a much slower rate than those made of titanium-free glass. For most of the spheres, the sphere boundary can easily be identified by the software, which is then able to automatically measure the sphere diameter. At higher titanium concentrations, no differences in sphere size can be discerned with the naked eye from one image to the next.

Figure 5.5 shows the percentage cumulative weight loss per unit surface area of the microspheres as a function of the degradation time for titanium-containing phosphate glass microspheres. The graph shows that the microspheres undergo increased degradation with time. A clear correlation between the TiO_2 content and the degradation rate is observed, whereby as the TiO_2 content is increased, the degradation rate of the glass microspheres decreases. Of note is the small size of the error bars in the graphs, which indicates the reliability of the methodology.

5.4 Cell Culture of Microspheres with Bone Cells

When considering the effects of glasses on bone cells or tissues under *in vitro* or *in vivo* conditions, researchers have generally focused on three biological responses of significant clinical relevance: osteogenesis, angiogenesis and antimicrobial activity. From an osteogenesis point of view, some inorganic ions have been found to over-stimulate the expression of osteogenic markers, to increase osteoblast proliferation and/or to induce early osteoblasts towards an osteogenic differentiation pathway.[2,14,67] On the other hand, enhanced endothelial cell proliferation, the formation of cell tubules networks or the stimulated secretion of angiogenic growth factors (*i.e.*, vascular endothelial growth factor, VEGF) are some of the angiogenic responses that have been observed on cells exposed to trace elements such as Cu and Co.[68] Some metal ions were also found to present antimicrobial activity by altering the culture pH or by compromising cell integrity.[12,27–29,31,34–36] More generally, most metal ions play a key role in metabolic pathways, by interacting with proteins and in particular with enzymes.

When investigating the biocompatibility of phosphate glasses containing TiO_2, researchers have tended to use one or more of the following three approaches. The first and most common involves immersing glass specimens in artificially prepared fluids such as citric acid, Tris buffer, Tris-HCl, or simulated body fluid (SBF) for example, and then using techniques such as scanning electron microscopy (SEM) or X-ray diffraction (XRD) spectroscopy to detect the formation of apatites on the glass surface.[67,69] This approach is quite popular among biomaterials researchers but doubts exist regarding the ability of the approach to confirm whether a material that forms hydroxyapatite on its surface will elicit a favourable response from cells and tissues cultured on the material surface. Some materials are capable of directly bonding with bone tissue without apatite formation;[70,71] several titanium glass compositions have shown favourable results in cell culture studies with upregulation of a number of bone-related genes, and yet they have been found to show no evidence of apatite formation.[14,72,73]

Consequently, *in vitro* cell culture studies are a better means to investigate material biocompatibility since they yield qualitative and quantitative data on cell viability, adhesion, proliferation, differentiation and toxicity as well as gene expression, thereby providing us with an improved understanding of the ability of cells to adhere and proliferate on the material surface. In the case of titanium phosphate glasses, increasing the titanium oxide content has been found to improve cell viability, attachment, and proliferation, as confirmed by gene expression studies, which reveal significant upregulation of important bone cell markers such as COLIA1, ALP, osteonectin and Cbfa-1.[14–16,59,74,75] A threshold concentration of TiO_2, beyond which there is no further improvement in glass biocompatibility, has also been observed.[15] The morphology of the titanium glass surface may exert significant influence on cell proliferation, with rough and highly porous samples inhibiting cell proliferation over relatively short time-scales.[59]

There is a relative lack of *in vivo* implantation studies concerning titanium phosphate glasses. The few studies where *in vivo* implantation has been performed have used either murine or lapine models with the defect site being filled with glass granules of predefined size over periods of 5–12 weeks. Titanium glass solubility has been found to exert a significant effect on *in vivo* glass behaviour with glasses that are less soluble undergoing degradation to a much lesser extent in physiological environments. A combination of solution-mediated and cell-mediated processes is considered to contribute to glass degradation at the implant site.[76] Histology studies have revealed that less soluble titanium glasses are more likely to be fully surrounded by new bone tissue while more soluble Ti-free glasses show some empty spaces.

These studies have been carried out using macroscopic glass specimens or glass granules, but the culturing of cells on microspheres under *in vitro* conditions presents unique challenges. We focus on the use of Transwell® inserts to enable easier handling of microspheres inside the cell culture well and also to provide improved distribution of medium and nutrients within the system. These inserts are available in a variety of diameters, membrane types and pore sizes which can fulfil different research requirements.

5.4.1   Materials and Reagents

- MG63 osteosarcoma cell line

> **CAUTION!** *All cell culture studies must be carried out with prior ethics approval from the host institution.*

- Titanium phosphate glass microspheres
- Dulbecco modified Eagle's medium (DMEM; Gibco, catalog no. 31885-049)
- Ethanol (50%, 70%, 90% and 100%)
- Fetal bovine serum (FBS; Gibco, catalog no. 10106-169)
- Glutaraldehyde, 3% E.M. grade (Fisher Scientific, catalog no. R3281800150)
- Hexamethyldisilazane (HMDS; Agar Scientific, catalog no. AGR1228)
- Penicillin–streptomycin, 5000 U mL^{-1} (Gibco, catalog no. 15070-063)
- Phosphate-buffered saline (PBS; Lonza, catalog no. BE17-516F)
- Trypan Blue solution, 0.4% (Sigma-Aldrich, catalog no. T8154)
- Trypsin–EDTA solution, 1× (Gibco, catalog no. R-001-100)
- Virkon virucidal surface disinfectant (Dupont, catalog no. 12358667)

5.4.2   Equipment

- Aluminium foil
- Autoclave tape
- Carbon adhesive discs (Agar Scientific, UK)
- Cell culture well plates (TPP; Helena Biosciences, UK)
- Centrifuge tubes (TPP; Helena Biosciences, UK)
- Disposable scalpels (VWR, UK)
- Forceps
- Glass vials, 2 mL (Sigma-Aldrich, UK)
- Neubauer haemocytometer
- Phase contrast microscope
- Pipettes – 10, 200 and 1000 µL (Gilson, UK)
- Pipette controller (Easypet; Eppendorf, UK)
- Pipette tips – 10, 200 and 1000 µL (Star Labs, UK)

- Serological pipettes – 10 and 25 mL (Fisher Scientific, UK)
- SEM specimen stubs (Agar Scientific, UK)
- Tissue culture flasks, 75 cm^2 (TPP; Helena Biosciences, UK)
- Transwell® inserts (Corning Costar®, USA)

5.4.3 Seeding of MG63 Cells on Microspheres

1. The medium used for cell culture is standard low-glucose DMEM containing 10% FBS and 1% penicillin–streptomycin. To prepare this, pre-warm all the components in a water bath at 37 °C. In a sterile fume hood, remove 50 mL of medium from a 500 mL bottle of DMEM and add 50 mL of FBS and 5 mL of penicillin–streptomycin.

> **NOTE:** *An increasing number of laboratories are using antibiotic-free culture systems in order to avoid antibiotic resistance of contaminant microbes and also because of possible effects on cell function. If cell culture is carried out without antibiotics, extra care must be taken to ensure complete sterility of the cell culture environment and equipment used.*

2. The MG63 cells are grown in a 75 cm^2 cell culture flask containing the cell culture medium described in Step 1. The growth process is carried out in a 37 °C/5% CO_2 incubator till the cells reach up to 90% confluence on the flask surface.
3. At 1 day prior to cell seeding, weigh out the required quantity of microspheres.

> **NOTE:** *The weight of microspheres used per well will depend on the amount of microspheres required to cover the cell seeding surface with a thin layer of microspheres, which in turn depends on the surface area of the Transwell® insert mesh. This should be determined prior to the experiment.*

4. Add the microspheres to a glass vial (one vial per well) and seal the open end of the vial with aluminium foil and autoclave tape. Place the vials in a drying oven and sterilise the contents under dry heat at a temperature of 180 °C for 3 h before leaving to cool overnight.

> **TIP:** *If a drying oven is not available, sterilisation can be carried out by exposing the microspheres to UV light for 1 h. However, autoclaving of the microspheres must be avoided since the microspheres are water-soluble and will degrade under the effects of moisture and high temperature.*

5. On the day of the experiment, remove the medium from the flask in a sterile fume hood, wash the cells with 5 mL PBS and trypsinise them using 3 mL trypsin–EDTA to detach the cells from the flask. Ensure

proper detachment and floating of cells in the medium by visual observation using phase microscopy. Add 7 mL of fresh media to the trypsinised live cells to inhibit trypsinisation, transfer the contents of the flask to a 15 mL centrifuge tube and centrifuge the cells at 1000 rpm for 3 min, followed by careful removal of supernatant to avoid disturbing the cell pellet. Add 8 mL of media prior to cell counting using a Neubauer haemocytometer, the procedure for which is based on the trypan blue dye protocol.[77]

6. Remove Transwell® insert from its packaging in a sterile fume hood, and gently place it into the well either by hand or by using sterile forceps.
7. Remove autoclave tape and aluminium foil from the open end of the glass vial containing the microspheres and gently tip the vial contents onto the mesh of the insert, taking care to avoid spillage into the bottom of the well. Tap the bottom of the vial a few times to dislodge attached microspheres, if any, and empty into the insert. Repeat for all the Transwell® inserts to be used in the experiment.
8. Add pre-warmed media to both the well and the insert according to recommended volumes (listed in Table 5.1) and incubate for 1 h in a 37 °C/5% CO_2 incubator.

CAUTION! *Avoid trapping air under the insert as this will generate pressure, potentially forcing cells off the microspheres.*

---

9. Carefully add the cell suspension to the insert, again taking care to avoid spillage into the well outside the insert. Store the well plate in a 37 °C/5% CO_2 incubator.

NOTE: *The initial seeding density depends on the surface area available for cell attachment, which in turn depends on the microsphere dimensions. It also depends on the cell size and cell growth rate. We have found that for Transwell® inserts in a 24-well plate, 10 000 cells per well is an appropriate number for ensuring a sufficient degree of cell attachment while at the same time avoiding the problem of rapid confluence.*

Table 5.1 Physical specifications of the Transwell® insert.

Parameter	6-well	12-well	24-well
Effective diameter of membrane (mm)	23.1–24	10.5–12	6.4–6.5
Effective growth area of membrane (cm^2)	4.2–4.67	0.9–1.12	0.3–0.33
Insert height (mm)	17.2	17.2	17.5
Distance from membrane to the bottom of well (mm)	0.9	0.9	0.8
Suggested media in insert (mL)	1.5–2.5	0.4–1.0	0.1–0.35
Suggested media in well (mL)	2.6–3.2	1.4–2.3	0.6–0.9
Growth area in tissue culture plate well (cm^2)	9.6	3.8	2

10. At periodic intervals (*e.g.*, every 48 h), transfer the well plate to a sterile fume hood, remove media from the well in which the insert is placed using a 1000 μL pipette and replace with fresh medium. Transwell® inserts have three openings through which standard pipette tips can be inserted to reach the well bottom for medium replacement or sample removal.

5.4.4 Sample Preparation for Scanning Electron Microscopy

1. In a fume hood, remove the Transwell® insert from the cell culture plate and place in a new culture plate using forceps, taking care not to disturb the microsphere layer.
2. Add 3% glutaraldehyde fixative solution (in 100 mM sodium cacodylate buffer) to both the insert and the well using a Pasteur pipette. Leave undisturbed for 45 min at room temperature.
3. Replace the fixative in the insert and well with appropriate ethanol solution in a graded series of concentrations to dehydrate the samples. Dehydration should occur without the membrane of the insert drying out. The series is as follows:
   i. 50% ethanol: 10 min
   ii. 70% ethanol: 10 min
   iii. 90% ethanol: 10 min
   iv. 100% ethanol: 10 min (three times)
4. Replace the final concentration of ethanol in the insert and well with HMDS solvent or transfer the insert swiftly into a container with a sufficient quantity of HMDS to submerge the insert. Leave undisturbed for approximately 1–2 min.
5. Remove the HMDS and leave the plate containing the wells to dry overnight in a desiccator.
6. Take the insert out of the well and mount it carefully on an SEM specimen stub to which a carbon adhesive disc has been attached so that the membrane is in contact with the disc. Use a disposable scalpel to carefully cut around the outer edge of the membrane and remove it from the insert.

> **CAUTION!** *Avoid touching the surface of the membrane while cutting.*

7. Sputter coat the mounted sample with gold/palladium and examine with a scanning electron microscope at an operating voltage of 10 kV.

> **NOTE:** *Coatings of gold/palladium are preferable since they provide a better signal-to-noise ratio than carbon or gold alone.*

5.4.5 Expected Outcomes and Other Considerations

The above approaches should yield layers of cells on the glass microsphere surface that can be viewed by microscopic techniques such as SEM, confocal laser scanning microscopy, fluorescence microscopy and so on. The effect of the glass composition on the attachment and proliferation of cells on the microsphere surface is quite pronounced. For instance, when MG63 cells are cultured on titanium glass microspheres containing 0–5 mol% TiO_2 over a period of 4 days, cells are clearly visible on the microspheres containing 3 mol% TiO_2 or higher, whereas no cells can be found on those containing 0 or 1 mol% TiO_2 (Figure 5.6). In agreement with the degradation study results, the 0 mol% TiO_2 glass microspheres show the highest degradation in the cell culture medium and have almost completely disintegrated by day 4. The 1 mol% TiO_2 microspheres show less degradation than the Ti0 microspheres, yet considerable changes in surface morphology are observed in the form of irregular cracks on the surface, and no adherent cells can be discerned. In contrast, both the 3 and 5 mol% TiO_2 microspheres show significant numbers of cells on their surface. Most of the microspheres appear to be covered by a finite, countable number of cells that envelop the microsphere surface. In many cases, cells on different microspheres appear to be joined to each other by means of processes. Thus, the 3 and 5 mol%

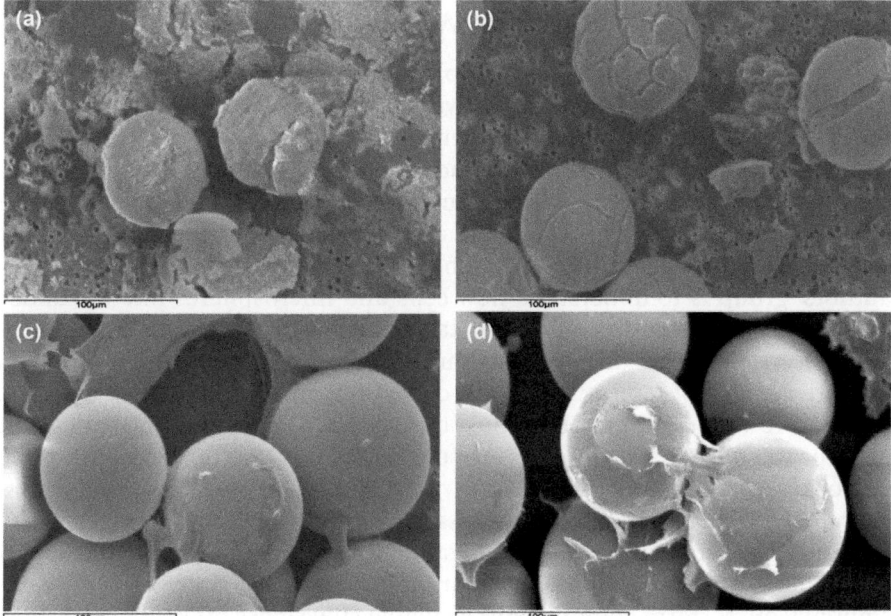

Figure 5.6 Scanning electron microscopy images of MG63 cells cultured on titanium phosphate glass microspheres containing (a) 0 mol% TiO_2, (b) 1 mol% TiO_2, (c) 3 mol% TiO_2 and (d) 5 mol% TiO_2 over a 4-day period. The magnification in all the images is 5005.

TiO$_2$ microspheres clearly support cell attachment and cell proliferation on their surface whereas the 0 and 1 mol% TiO$_2$ microspheres do not.

It is worth noting that considerable care is required when carrying out cell culture on glass microspheres because the various pipetting steps involved can lead to cells becoming detached from the microspheres. Furthermore, although the Transwell® inserts are a useful means to carry out cell culture because of the aforementioned advantages in terms of handling and nutrient distribution, they may not be particularly suited for experiments involving quantitative measurements such as cell proliferation or ELISA assays. This is because the membrane surface in the Transwell® insert is specially treated to encourage high levels of cell attachment and proliferation; thus, in the case of microsphere culture on such inserts, the cells could attach to either the microspheres or to the insert membrane and any assay results would indicate the combined effects of the microsphere surface and the membrane surface on the cells.

TIP: *Therefore, to quantify the effects of microspheres alone on the cells, it is advisable to use ultra-low attachment surfaces that prevent cellular attachment to the surface, so that the microspheres provide the only surface on which the cells can attach and proliferate.*

5.5   Summary: On-going Work and Future Directions

This chapter has presented titanium phosphate glass microspheres as viable microcarrier substrates for the growth of bone cells under *in vitro* conditions. Simple techniques have been used to illustrate how glass microspheres can be made, how the degradation of the microspheres in liquids can be visualised and quantified and how cells can be grown on the microsphere surface. The results obtained from these experiments present several directions for future research. One of the major objectives of the authors' group is to understand how cells and microspheres interact under dynamic culture environments. Bioreactors are a suitable tool for this purpose because they provide dynamic three-dimensional *in vitro* environments that are capable of replicating *in vivo* conditions more accurately than static cell culture well environments; thus, they provide valuable information not only for the *in vitro* growth of tissue substitutes but also for researching the responses of cells and tissues to various stimuli of mechanical and biochemical origin. The group has successfully used spinner flask bioreactors to culture MG63 cells on titanium phosphate glass microspheres, and it has been found that in terms of cell proliferation, these microspheres substantially outperform commercially available silica glass microspheres; thus, using the titanium phosphate glass microspheres in such bioreactor systems raises the possibility of rapid bone cell expansion for yielding large quantities of cells that can be used in cell-based therapies and high throughput screening applications. Another research avenue which is being pursued at present is to use the microspheres as substrates in perfusion bioreactors. These bioreactors could potentially allow the development of small-sized tissue fragments that would be useful for tissue replacement strategies. A major indicator of the success of these microspheres is whether the results obtained for MG63 cells can be replicated with stem cells, which offer remarkable possibilities for guided differentiation through particular lineages. However, the major challenges encountered in carrying out these cell culture approaches using stem cells are: (1) the significantly weaker attachment between the cells and the microsphere surface, and (2) the slow growth rates associated with these cells. It is anticipated that these challenges can be overcome by suitable biofunctionalisation of the microsphere surface and by the use of cell culture media that are more suited to rapid stem cell proliferation. Thus, the development of viable small-sized bone tissue to treat bone loss or bone defects caused by injury, disease or congenital deformities is now an achievable objective.

References

1. J. C. Knowles, *J. Mater. Chem.*, 2003, **13**, 2395–2401.
2. V. Rajendran, A. V. G. Devi, M. Azooz and F. H. El-Batal, *J. Non-Cryst. Solids*, 2007, **353**, 77–84.
3. E. A. Abou Neel, D. M. Pickup, S. P. Valappil, R. J. Newport and J. C. Knowles, *J. Mater. Chem.*, 2009, **19**, 690–701.
4. A. Kiani, N. J. Lakhkar, V. Salih, M. E. Smith, J. V. Hanna, R. J. Newport, D. M. Pickup and J. C. Knowles, *Philos. Trans. R. Soc., A*, 2012, **370**, 1352–1375.
5. L. L. Hench, *J. Mater. Sci.: Mater. Med.*, 2006, **17**, 967–978.
6. S. T. Reis, D. L. A. Faria, J. R. Martinelli, W. M. Pontuschka, D. E. Day and C. S. M. Partiti, *J. Non-Cryst. Solids*, 2002, **304**, 188–194.
7. M. Karabulut, E. Metwalli, D. E. Day and R. K. Brow, *J. Non-Cryst. Solids*, 2003, **328**, 199–206.
8. I. Ahmed, C. A. Collins, M. P. Lewis, I. Olsen and J. C. Knowles, *Biomaterials*, 2004, **25**, 3223–3232.
9. P. Y. Shih, S. W. Yung and T. S. Chin, *J. Non-Cryst. Solids*, 1998, **224** 143–152.
10. P. Y. Shih, S. W. Yung and T. S. Chin, *J. Non-Cryst. Solids*, 1999, **244** 211–222.
11. A. M. Mulligan, M. Wilson and J. C. Knowles, *Biomaterials*, 2003, **24**, 1797–1807.
12. E. A. Abou Neel, I. Ahmed, J. Pratten, S. N. Nazhat and J. C. Knowles, *Biomaterials*, 2005, **26**, 2247–2254.
13. R. Shah, A. C. M. Sinanan, J. C. Knowles, N. P. Hunt and M. P. Lewis, *Biomaterials*, 2005, **26**, 1497–1505.
14. E. A. Abou Neel, T. Mizoguchi, M. Ito, M. Bitar, V. Salih and J. C. Knowles, *Biomaterials*, 2007, **28**, 2967–2977.

15. E. A. Abou Neel, W. Chrzanowski and J. C. Knowles, *Acta Biomater.*, 2008, **4**, 523–534.

16. E. A. Abou Neel and J. C. Knowles, *J. Mater. Sci.: Mater. Med.*, 2008, **19**, 377–386.

17. E. A. Abou Neel, W. Chrzanowski, S. P. Valappil, L. A. O'Dell, D. M. Pickup, M. E. Smith, R. J. Newport and J. C. Knowles, *J. Non-Cryst. Solids*, 2009, **355**, 991–1000.

18. K. Franks, V. Salih, J. C. Knowles and I. Olsen, *J. Mater. Sci.: Mater. Med.*, 2002, **13**, 549–556.

19. K. Meyer, *J. Non-Cryst. Solids*, 1997, **209**, 227–239.

20. V. Salih, A. Patel and J. C. Knowles, *Biomed. Mater.*, 2007, **2**, 11–20.

21. J. C. Knowles, E. A. Abou Neel, L. A. O'Dell and M. E. Smith, *J. Mater. Sci.: Mater. Med.*, 2008, **19**, 1669–1679.

22. E. A. Abou Neel, L. A. O'Dell, W. Chrzanowski, M. E. Smith and J. C. Knowles, *J. Biomed. Mater. Res., Part B*, 2009, **89**, 392–407.

23. R. M. Moss, E. A. Abou Neel, D. M. Pickup, H. L. Twyman, R. A. Martin, M. D. Henson, E. R. Barney, A. C. Hannon, J. C. Knowles and R. J. Newport, *J. Non-Cryst. Solids*, 2010, **356**, 1319–1324.

24. A. M. Mulligan, M. Wilson and J. C. Knowles, *J. Biomed. Mater. Res., Part A*, 2003, **67**, 401–412.

25. I. Ahmed, D. Ready, M. Wilson and J. C. Knowles, *J. Biomed. Mater. Res., Part A*, 2006, **79**, 618–626.

26. I. Ahmed, E. A. Abou Neel, S. P. Valappil, S. N. Nazhat, D. M. Pickup, D. Carta, D. L. Carroll, R. J. Newport, M. E. Smith and J. C. Knowles, *J. Mater. Sci.*, 2007, **42**, 9827–9835.

27. S. P. Valappil, D. M. Pickup, D. L. Carroll, C. K. Hope, J. Pratten, R. J. Newport, M. E. Smith, M. Wilson and J. C. Knowles, *Antimicrob. Agents Chemother.*, 2007, **51**, 4453–4461.

28. S. P. Valappil, J. C. Knowles and M. Wilson, *Appl. Environ. Microbiol.*, 2008, **74**, 5228–5230.

29. S. P. Valappil, D. Ready, E. A. Abou Neel, D. M. Pickup, W. Chrzanowski, L. A. O'Dell, R. J. Newport, M. E. Smith, M. Wilson and J. C. Knowles, *Adv. Funct. Mater.*, 2008, **18**, 732–741.

30. D. M. Pickup, S. P. Valappil, R. M. Moss, H. L. Twyman, P. Guerry, M. E. Smith, M. Wilson, J. C. Knowles and R. J. Newport, *J. Mater. Sci.*, 2009, **44**, 1858–1867.

31. S. P. Valappil, D. Ready, E. A. Abou Neel, D. M. Pickup, L. A. O'Dell, W. Chrzanowski, J. Pratten, R. J. Newport, M. E. Smith, M. Wilson and J. C. Knowles, *Acta Biomater.*, 2009, **5**, 1198–1210.

32. P. Y. Shih and H. M. Shiu, *Mater. Chem. Phys.*, 2007, **106**, 222–226.

33. J. Lao, E. Jallot and J.-M. Nedelec, *Chem. Mater.*, 2008, **20**, 4969–4973.

34. N. J. Lakhkar, E. A. Abou Neel, V. Salih and J. C. Knowles, *J. Mater. Sci.: Mater. Med.*, 2009, **20**, 1339–1346.

35. E. A. Abou Neel, J. C. Knowles, W. Chrzanowski, D. M. Pickup, L. A. O'Dell, N. J. Mordan, R. J. Newport and M. E. Smith, *J. R. Soc., Interface*, 2009, **6**, 435–446.

36. N. Lakhkar, E. A. Abou Neel, V. Salih and J. C. Knowles, *J. Biomater. Appl.*, 2010, **25**, 877–893.
37. S. Hesaraki, M. Alizadeh, H. Nazarian and D. Sharifi, *J. Mater. Sci.: Mater. Med.*, 2010, **21**, 695–705.
38. M. D. O'Donnell and R. G. Hill, *Acta Biomater.*, 2010, **6**, 2382–2385.
39. T. Gilchrist, D. M. Healy and C. Drake, *Biomaterials*, 1991, **12**, 76–78.
40. R. K. Brow, R. J. Kirkpatrick and G. L. Turner, *J. Non-Cryst. Solids*, 1990, **116**, 39–45.
41. D. M. Pickup, R. J. Speight, J. C. Knowles, M. E. Smith and R. J. Newport, *Mater. Res. Bull.*, 2008, **43**, 333–342.
42. D. M. Pickup, K. M. Wetherall, J. C. Knowles, M. E. Smith and R. J. Newport, *J. Mater. Sci.: Mater. Med.*, 2008, **19**, 1661–1668.
43. J. E. White and D. E. Day, *Rare Elements in Glasses*, 1994, **vol. 94–99**, pp. 181–208.
44. W. Huang, *Shanghai Inst. Bldg. Mat.*, 1992, **4**, 347.
45. Y. W. Gu, A. U. J. Yap, P. Cheang and R. Kumar, *Biomaterials*, 2004, **25**, 4029–4035.
46. D. Cacaina, H. Ylanen, M. Hupa and S. Simon, *J. Mater. Sci.: Mater. Med.*, 2006, **17**, 709–716.
47. D. Cacaina, H. Ylanen, S. Simon and M. Hupa, *J. Mater. Sci.: Mater. Med.*, 2008, **19**, 1225–1233.
48. J. R. Martinelli, F. F. Sene, C. N. Kamikawachi, C. S. D. Partiti and D. R. Cornejo, *J. Non-Cryst. Solids*, 2010, **356**, 2683–2688.
49. F. F. Sene, E. Okuno and J. R. Martinelli, *J. Non-Cryst. Solids*, 2008, **354**, 4887–4893.
50. N. J. Lakhkar, J. H. Park, N. J. Mordan, V. Salih, I. B. Wall, H. W. Kim, S. P. King, J. V. Hanna, R. A. Martin, O. Addison, J. F. Mosselmans and J. C. Knowles, *Acta Biomater.*, 2012, **8**, 4181–4190.
51. J. M. Ward, Y. Q. Wu, K. Khalfi and S. N. Chormaic, *Rev. Sci. Instrum.*, 2010, **81**, 0731061–0731064.
52. J. H. Park, R. A. Perez, G. Z. Jin, S. J. Choi, H. W. Kim and I. B. Wall, *Tissue Eng., Part B*, 2013, **19**, 172–190.
53. M. Kawashita, S. Toda, H. M. Kim, T. Kokubo and N. Masuda, *J. Biomed. Mater. Res., Part A*, 2003, **66**, 266–274.
54. S. Y. Li, L. Nguyen, H. R. Xiong, M. Y. Wang, T. C. C. Hu, J. X. She, S. M. Serkiz, G. G. Wicks and W. S. Dynan, *Nanomedicine*, 2010, **6**, 127–136.
55. C. Wu, W. Fan, M. Gelinsky, Y. Xiao, P. Simon, R. Schulze, T. Doert, Y. Luo and G. Cuniberti, *Acta Biomater.*, 2011, **7**, 1797–1806.
56. C. T. Wu, Y. Xiao, Y. F. Zhang, X. B. Ke, Y. X. Xie, H. Y. Zhu and R. Crawford, *J. Biomed. Mater. Res., Part A*, 2010, **95**, 476–485.
57. H. Poorbaygi, S. M. R. Aghamiri, S. Sheibani, A. Kamali-asl and E. Mohagheghpoor, *Appl. Radiat. Isot.*, 2011, **69**, 1407–1414.
58. H. Poorbaygi, S. Aghamiri, A. Kamali-asl, S. Sheibani and E. Mohagheghpoor, *Intern. Med. J.*, 2011, **41**, 30.
59. D. S. Brauer, C. Russel, W. Li and S. Habelitz, *J. Biomed. Mater. Res., Part A*, 2006, **77**, 213–219.

60. N. J. Lakhkar, I. H. Lee, H. W. Kim, V. Salih, I. B. Wall and J. C. Knowles, *Adv. Drug Delivery Rev.*, 2013, **65**, 405–420.

61. E. S. Baluyot and C. G. Hartford, *J. Chromatogr. A*, 1996, **739**, 217–222.

62. I. Ahmed, M. P. Lewis, S. N. Nazhat and J. C. Knowles, *J. Biomater. Appl.*, 2005, **20**, 65–80.

63. I. Ahmed, M. P. Lewis, S. N. Nazhat and J. C. Knowles, *J. Biomater. Appl.*, 2005, **20**, 65–80.

64. B. C. Bunker, G. W. Arnold and J. A. Wilder, *J. Non-Cryst. Solids*, 1984, **64**, 291–316.

65. Y. C. Yang, J. Liu, C. Li, L. Fu, W. Huang and Z. L. Li, *Electrochim. Acta*, 2012, **72**, 94–100.

66. I. Ahmed, M. Lewis, I. Olsen and J. C. Knowles, *Biomaterials*, 2004, **25**, 491–499.

67. A. V. G. Devi, V. Rajendran and N. Rajendran, *Mater. Chem. Phys.*, 2010, **124**, 312–318.

68. I. H. Lee, H. S. Yu, N. J. Lakhkar, H. W. Kim, M. S. Gong, J. C. Knowles and I. B. Wall, *Mater. Sci. Eng., C*, 2013, **33**, 2104–2112.

69. T. Kasuga, M. Sawada, M. Nogami and Y. Abe, *Biomaterials*, 1999, **20**, 1415–1420.

70. T. Kokubo and H. Takadama, *Biomaterials*, 2006, **27**, 2907–2915.

71. K. Y. Lee, M. Park, H. M. Kim, Y. J. Lim, H. J. Chun, H. Kim and S. H. Moon, *Biomed. Mater.*, 2006, **1**, R31–R37.

72. R. C. Lucacel, M. Maier and V. Simon, *J. Non-Cryst. Solids*, 2010, **356**, 2869–2874.

73. H. A. ElBatal, E. M. A. Khalil and Y. M. Hamdy, *Ceram. Interfaces*, 2009, **35**, 1195–1204.

74. M. Navarro, M. P. Ginebra and J. A. Planell, *J. Biomed. Mater. Res., Part A*, 2003, **67**, 1009–1015.

75. A. G. Dias, M. A. Lopes, A. T. T. Cabral, J. D. Santos and M. H. Fernandes, *J. Biomed. Mater. Res., Part A*, 2005, **74**, 347–355.

76. A. G. Dias, M. A. Lopes, J. D. Santos, A. Afonso, K. Tsuru, A. Osaka, S. Hayakawa, S. Takashima and Y. Kurabayashi, *J. Biomater. Appl.*, 2006, **20**, 253–266.

77. W. Strober, *Current Protocols in Immunology*, John Wiley & Sons, Inc., 2001.

CHAPTER 6

Biointerfaces Between Cells and Substrates in Three Dimensions

ADAM S. HAYWARD,[a] NEIL R. CAMERON*[b] AND
STEFAN A. PRZYBORSKI*[a]

[a] School of Biological and Biomedical Science, Durham University, South Road, Durham, DH1 3LE, UK; [b] Department of Chemistry, Durham University, South Road, Durham, DH1 3LE, UK
*Email: stefan.przyborski@durham.ac.uk; n.r.cameron@durham.ac.uk

6.1 The Structural and Physiological Relevance of Three-dimensional Biointerfaces

In order to successfully replicate native cell and tissue function in the laboratory, the artificial growth environment needs to be structurally and physiologically relevant. In the body, cells adopt a complex three-dimensional (3D) architecture that involves anchorage to adjacent cells and a supporting extracellular cellular matrix (ECM). Tight junctions and other protein complexes bind cells together across multiple surfaces allowing constant 3D communication across the tissue. In parallel, cell surface receptors continually receive biochemical cues from the ECM as to when to proliferate, differentiate or initiate a specific biological process. This complex 3D interaction between cells and the ECM plays a critical role in maintaining normal cell and tissue function *in vivo*.

RSC Smart Materials No. 10
Biointerfaces: Where Material Meets Biology
Edited by Dietmar Hutmacher and Wojciech Chrzanowski
© The Royal Society of Chemistry 2015
Published by the Royal Society of Chemistry, www.rsc.org

Traditionally, plastic Petri dishes and welled plates have been used as the artificial growth environment for most *in vitro* cell studies, offering only a two-dimensional (2D) interface for cell attachment and growth (along the *XY* plane). Whilst this approach has proved an extremely practical and accessible method of cell growth in the past, it is now regarded as highly unrealistic compared to the complex 3D scenario described above.[1,2] Cells in a 2D monolayer are unable to form multi-directional tight junctions with their neighbours and become polarised, given that the majority of the cell surface is spread out along the plastic interface. In addition, a cell's cytoskeleton and delicate organelles are forced into a severely flattened shape when cultured in two dimensions. This can cause unnecessary stress, making it artificially vulnerable to external stimuli and mechanical forces. Unsurprisingly, many primary cells, such as hepatocytes, cultured at the 2D interface rapidly lose their native phenotype and viability after just a few days in culture.

Recognising the limitations of 2D culture substrates there is now a strong demand for materials that can offer cells a 3D interface for growth (*XYZ* plane).[3] By adding the third dimension to the growth substrate, cells can more readily approximate their native 3D organisation and cell density. This in turn encourages cell communication and anchorage, making cells less susceptible to trauma and thus prolonging a more representative *in vitro* phenotype. Indeed, the advantages of 3D interfaces for cell growth are already evident. 3D biointerfaces are now helping to progress our understanding of the molecular mechanisms involved in cancer.[4,5] They are also opening up exciting opportunities in the fields of tissue engineering and regenerative medicine, where 3D constructs grown *in vitro* or *ex vivo* can later be implanted into the body for organ repair.[6,7] This chapter will therefore briefly describe some of the materials used to create 3D interfaces for cells and tissues, with a particular emphasis on materials used in 3D cell culture applications. Some general methodologies used to examine cells in the 3D microenvironment will also be discussed, with the intention of merely highlighting to the reader the techniques available, rather than providing detailed experimental conditions or specific procedures.

6.2 Using Natural and Synthetic Materials to Create Three-dimensional Biointerfaces

There is now a variety of different materials that can offer a 3D interface for cell and tissue growth, with the choice of material depending on the application required. Natural-based materials such as protein hydrogels or alginate scaffolds are particularly attractive for tissue engineering or regenerative medicine applications, as the biocompatibility and biodegradability of these materials makes them suitable for eventual implantation into the body. These materials are also useful for studying cell–ECM interactions, given that natural-based materials are more likely to mimic specific carbohydrate motifs or peptide anchorage points present on the native ECM. In contrast, applications such as high-throughput laboratory models as a replacement for 2D plastic plates will more likely require synthetic substrates that are reproducible, chemically stable and easy to use. Many natural-based materials are often too variable in composition and structure for this application. Their biodegradability also adds an unwanted time-dependent variable into laboratory experiments.

Whilst it is not possible to list every natural and synthetic material used for 3D cell growth, broad categories of materials have been described below. Within each category, a brief description of the technology is available, along with any important advantages and disadvantages.

6.2.1 Sandwiching Cells with Extracellular Matrix Proteins

Extracellular matrix (ECM) proteins such as collagen or the sarcoma-derived protein mixture Matrigel™ (BD Biosciences) have long been used to coat Petri dishes and welled plates to improve the physiological relevance of the 2D interface and thus promote cell adhesion and function. Taking this further, researchers in the field of liver biology found that applying a second layer of ECM protein directly above the monolayer, to create a sandwich of protein–cell–protein, helped to keep cells in a more upright morphology during the culture period (Figure 6.1). As a result, cells were less flattened and could sustain a polarised state that led to enhanced functional capacity.[8] Consequently this approach is a popular method of culturing hepatocyte cells, particularly to monitor hepatobiliary mechanisms *in vitro*.[9]

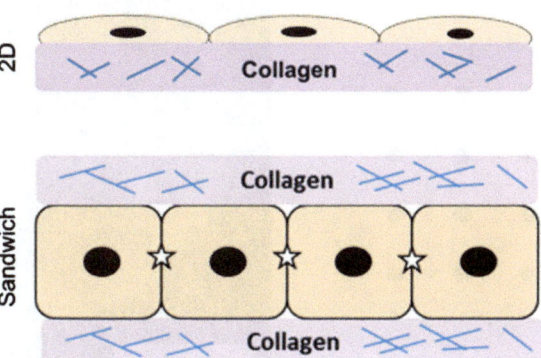

Figure 6.1 Schematic of a collagen sandwich culture compared to a conventional 2D culture. Top: conventional 2D culture where cells flatten out along the plastic surface. Bottom: collagen sandwich culture where the collagen above and below the cells helps to maintain a more appropriate upright morphology. Notice the reduction in cell–cell contact in the 2D culture whilst in 3D sandwich culture there is greater opportunity for intercellular interactions (stars).

6.2.2   Hydrogels

Hydrogels derived from natural or synthetic polymers can be used to im-mobilise cells within a 3D gel matrix that mimics certain physical, mech-anical and/or biochemical aspects of the native ECM. Within the gel, cells aggregate into 3D structures that promote cell contact and interaction across multiple surfaces (Figure 6.2). Many hydrogels are made from biodegradable polymers, such as natural proteins or synthetic polypeptides, enabling good suitability for tissue engineering or regenerative medicine applications.[10] Also, being approximately 90% water, hydrogels can display some of the mechanical properties of soft tissue, which for certain organs such as the bladder, is an important factor in maintaining normal cell function.

Whilst hydrogels do go a long way to mimic several components of the ECM and are therefore an attractive choice for many researchers, there are some notable limitations. Mass-transfer issues have long been a concern for hydrogels, in that the diffusion of nutrients and waste through the gel may be too slow to accommodate dense 3D structures. Another limitation is gel formation. Many hydrogels require some form of external switch to gel, such as a change in temperature, pH or exposure to UV radiation. Hence if cells are dispersed into the material prior to gelation, which is common practice, care needs to be taken to ensure that the gelation process does not have any detrimental effect on cell viability.

6.2.3   Electrospun Polymer Fibres

Electrospinning is a technique that converts polymeric liquids (natural and synthetic) into micro- or nanoscale fibres creating a 3D mesh topography.

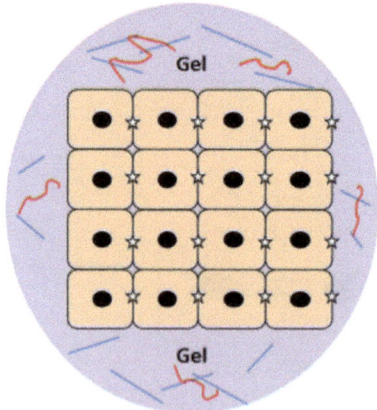 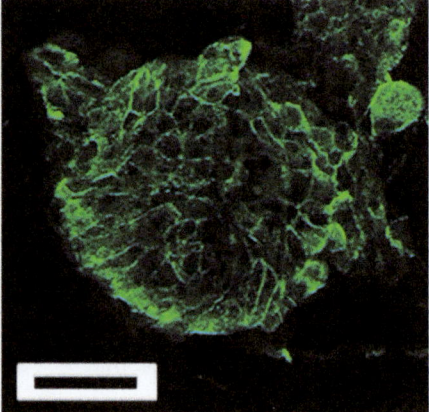

Figure 6.2   3D cell growth in hydrogels. Left: schematic showing how a gel will encapsulate a 3D aggregate of cells. Right: glioma (brain cancer) cells growing in a 3D hydrogel. The green fluorescent dye reflects the cyto-skeletons of the cells. Scale bar is 50 mm.
Image taken from www.medgadget.com (original image courtesy of Brendan Harley).

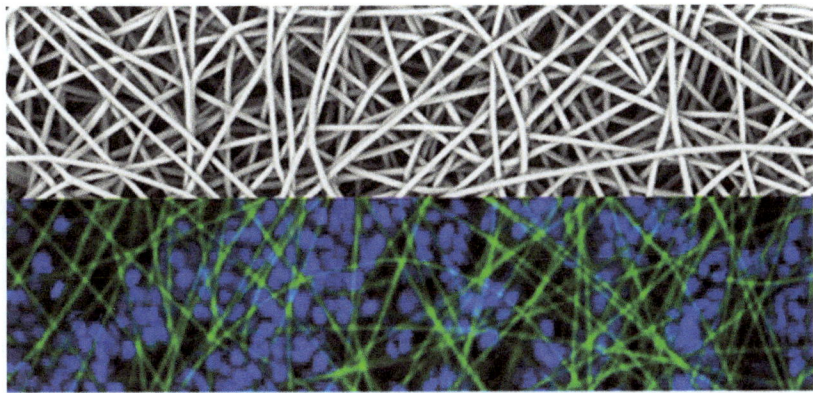

Figure 6.3 Cell culture in the electrospun scaffolds. Top: SEM micrograph of an electrospun scaffold. Bottom: breast cancer cells (blue) growing in an electrospun scaffold (green).
Images taken from the Electrospinning Company, UK (www.electrospinning.co.uk).

The technique is hugely versatile enabling a range of different fibrous compositions and architectures with different diameters and porosities to be formed. As a result the technique has attracted significant attention for 3D cell growth applications, with cells being able to migrate into the mesh structure and grow in a 3D manner.[11] Furthermore, electrospun materials are now available as commercial scaffolds for 3D cell growth (Mimetix™; Electrospinning Company). Figure 6.3, taken from the Electrospinning Company website (www.electrospinning.co.uk), illustrates how cells are able to grow in a 3D manner within these materials. There are also exciting new developments with aligned electrospun fibres, which may have important applications for cells that require a specific orientation for function, such as peripheral nerves and tendons. The materials are also inexpensive to produce, once the initial electrospinning equipment has been purchased.

However, the main limitation of electrospun materials is that 3D growth is generally restricted to the nodes where fibres overlap, as large holes separating individual fibres can prevent extensive organisation. Consequently, 3D growth tends to be as individual pockets dispersed throughout the material, and very rarely will high cell densities be packed into the structure. The mechanical properties of such materials are also quite poor, significantly reducing their suitability for routine *in vitro* applications.

6.2.4 Three-dimensional Printed Scaffolds

The development of 3D printing and selective laser sintering (SLS) has opened up new opportunities for a range of versatile 3D interfaces which are suitable for 3D cell growth, particularly for hard tissues such as bone.[12] Here digital prototypes are used to construct 3D structures of a material one layer at a time. As the materials are digitally designed, there is excellent control

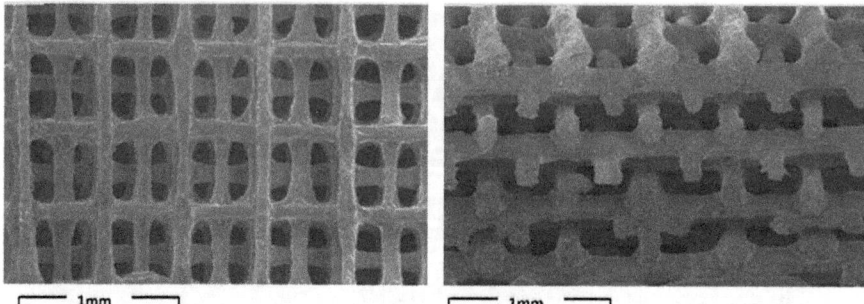

Figure 6.4   A 3D printed cell scaffold designed for 3D cell growth applications. Left: top view. Right: cross-sectional view.
The image represents a cropped section of a larger image taken, with permission, from *Acta Biomaterialia*, 2013, **9**(3), 5521.

over morphology and porosity (material reproducibility). However, a significant disadvantage of these materials is cost, in that expensive digital 3D printers must be purchased in order to fabricate these materials. The architecture of these materials is also limited to simpler structures (Figure 6.4), often only allowing individual compartments of 3D growth rather than extensive growth throughout the material.

6.2.5   Porous Polymeric Scaffolds

With a versatile range of morphologies and fabrication techniques available, porous polymers are attractive materials as substrates for 3D cell growth. Some of the first porous polymers to be used as 3D scaffolds were based on polylactic acid (PLA), polyglycolic acid (PGA) and polylactic-*co*-glycolic acid (PLGA) for tissue engineering applications, as the degradation products of these materials are metabolised and excreted by the body. Since then a range of natural and synthetic porous polymers have been developed and commercialised, such as AlgiMatrix™ (Invitrogen), a polysaccharide based scaffold (alginate) derived from brown algae.

One of the main fabrication methods for polymeric scaffolds is solvent casting/particle leaching. Here, a polymer–solvent mixture is cast over a 3D mould containing crystalline salts, which can leach into the polymer mixture. Upon solvent evaporation, the polymer–salt composite can then be washed to remove the crystalline salts, rendering a 3D porous structure. With this method high porosities can be achieved (90%); however, the technique does not really enable good control over scaffold morphology and reproducibility, given variability in salt crystal size and distribution. Consequently, several other fabrication methods have been employed in an attempt to improve the reproducibility of porous polymeric scaffolds for 3D cell growth applications. One such method is known as emulsion templating.

6.2.6 Emulsion Templating and Alvetex®

Emulsion templating uses emulsion internal phase droplets to template voids into a material polymerising in the emulsion external phase. Typically, water is used as the non-polymerising internal phase and organic monomers are used as the continuous external phase. A non-water soluble surfactant such as Span80™ is also included to create a stable emulsion. When high internal phase emulsions (HIPEs) are employed, defined as when the internal droplet phase occupies greater than 74% of the emulsion volume, the resulting polymer (or polyHIPE) can display a highly open-cell morphology that renders porosities above 90%. Cells are able to enter and migrate around these voids to create extensive 3D organisation in the material. Importantly, these materials are highly controllable, enabling excellent scaffold reproducibility.[13] By manipulating the initial HIPE characteristics (surfactant choice, temperature, introducing stabilising salts) the distribution and diameter of the internal phase water droplets can be controlled, thus controlling the final polymer morphology. This morphological control provides a key advantage over many other materials used for 3D cell growth.

When fabricating polyHIPE it is useful to keep the following points in mind:

- The polyHIPE void diameter can be increased *via* de-stabilisation of the initial HIPE.[13] Increasing emulsion temperature is one approach to emulsion de-stabilisation and thus increasing void diameter. Another approach is to include small organic molecule additives such as methanol or tetrahydrofuran into the emulsion aqueous phase.
- Functional polyHIPEs can be obtained by employing functional co-monomers in the initial HIPE emulsion.[14] For example, pentafluorophenyl acrylate, a hydrophobic ester-containing acrylate, will happily reside in the organic phase of styrene–water HIPEs to render pendent ester functionality in the final polymer.[15] Alternatively, hydrophilic functional co-monomers such as acrylic acid can be included in the aqueous phase of styrene–water HIPEs to render pendent acid functionality in the final polymer.[16]
- The choice of polymerisation initiator is an important consideration for polyHIPE fabrication. For example, potassium persulfate (KPS) is typically dissolved in the aqueous phase of the HIPE and thus favours polymerisation of those monomers close to the emulsion interface. Conversely, azobisisobutyronitrile (AIBN) is more soluble in the organic phase of the emulsion and thus favours polymerisation of those monomers not at the emulsion interface.

Recently, a polyHIPE derived from cross-linked polystyrene has been developed as a 3D cell scaffold designed for routine *in vitro* use (Alvetex®;

Reinnervate).[17] A simplified fabrication protocol for this type of scaffold has been reported previously[13,18] and is summarised below:

1. An oil phase consisting of the monomers styrene, divinylbenzene and 2-ethylhexyl acrylate, and the surfactant Span80™ is placed in a 250 mL three-necked round-bottom flask and stirred continually at 350 rpm using a PTFE paddle connected to an overhead stirrer.
2. An aqueous phase consisting of deionised water and 1 wt% potassium persulfate is heated to 85 °C and then transferred into a dropping funnel. This solution is then slowly added to the above organic phase over a period of 2 min with constant stirring.
3. The mixture (which should be a HIPE) is then transferred into a 50 mL falcon tube and placed in a 60 °C oven for 24 h to thermally polymerise.
4. The resulting polyHIPE is then washed in acetone for 24 h using soxhlet recirculation.

The structure of the Alvetex® scaffold can accommodate a range of cell types and has already been shown to significantly enhance cell function compared to conventional 2D models.[19–21] The material is chemically stable and mechanically strong and can support large cell numbers as well as survive most laboratory processing techniques. It can also be engineered into thin 200 µm membranes that fit conveniently into the bottom of tissue culture plasticware and well inserts, offering the potential for high-throughput applications (Figure 6.5).

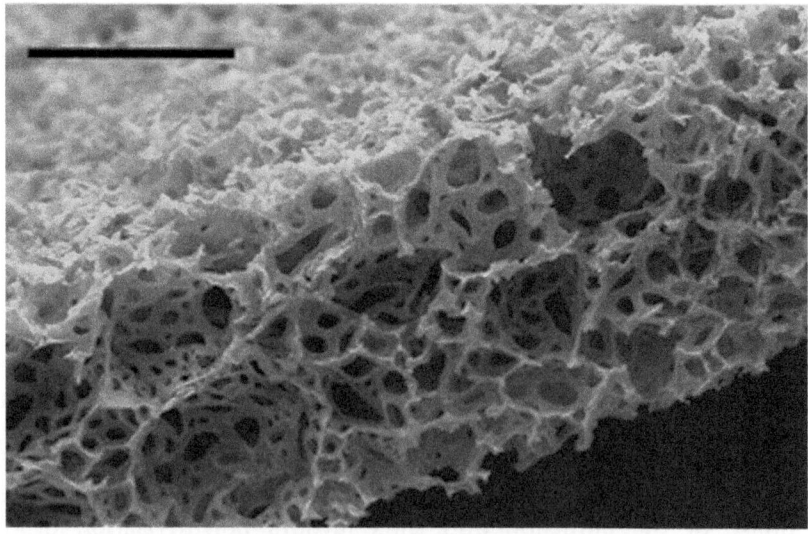

Figure 6.5 Structure of the Alvetex® scaffold by scanning electron microscopy. The material is supplied as thin 200 µm membranes that fit conveniently into the bottom of traditional 2D welled plates.
Reinnervate is acknowledged for supplying the image and allowing reprint in this chapter. Scale bar = 100 µm.

The main disadvantage of the material is surface chemistry. Whilst polystyrene is the conventional substrate for *in vitro* cell growth, it is not an optimum surface for cell growth. To rectify this, Alvetex® can be easily coated with ECM proteins such as collagen or fibronectin to improve cell adhesion and migration. Furthermore, recent studies have also shown that the material can be chemically functionalised with ECM mimics, such as pendent galactose residues, to enhance cell adhesion.[15]

6.3 Methods to Examine Cell Behaviour in Three Dimensions: Alvetex® Case Study

Materials that offer a 3D interface for cell growth must still allow researchers to obtain information about the biological status of the cells during and after the growth period. There is little point culturing in three dimensions if a researcher cannot 'talk' or 'listen' to cells in that environment. Being able to image cells, examine their cell structure, assess gene expression and understand protein activity are just some of the crucial needs of researchers working in the field of 3D cell culture, regenerative medicine and tissue engineering. The remainder of this chapter will therefore focus on some examples of examining cell behaviour in three dimensions, using the polymeric Alvetex® scaffold as an example to illustrate basic methods. It is important to note that these methods are merely intended to highlight a few of the general procedures applicable for examining cells at the 3D biointerface, and that more detailed protocols of each technique can be obtained from the Reinnervate website (http://reinnervate.com/).

6.3.1 Imaging of Live Cells

Constantly monitoring cell status under a bright-field light microscope is standard practice for 2D cultures; however, this is much more difficult at the 3D interface. Three-dimensional cultures approximate the complexity and structure of native tissue, creating issues with light scattering and thus preventing clear, crisp images from being obtained. Furthermore, most 3D substrates are not transparent, which again limits the use of conventional bright-field microscopes for direct imaging. In order to image live cells in three dimensions, either the use of a chemical stain or an advanced confocal microscope is needed.

For Alvetex®, it was found that applying a common non-toxic stain (neutral red; Sigma) was a very useful way of gaining an indication of cell attachment and confluency with a standard bright field microscope. The stain, which only needs to be exposed to the cells for a matter of minutes, quickly penetrates through cell membranes and accumulates in the lysosomes. Imaging with a standard bright field microscope then offers a simple red visualisation of the cells (Figure 6.6). The stain can also be subsequently washed away, allowing the culture to proceed as normal, although it is

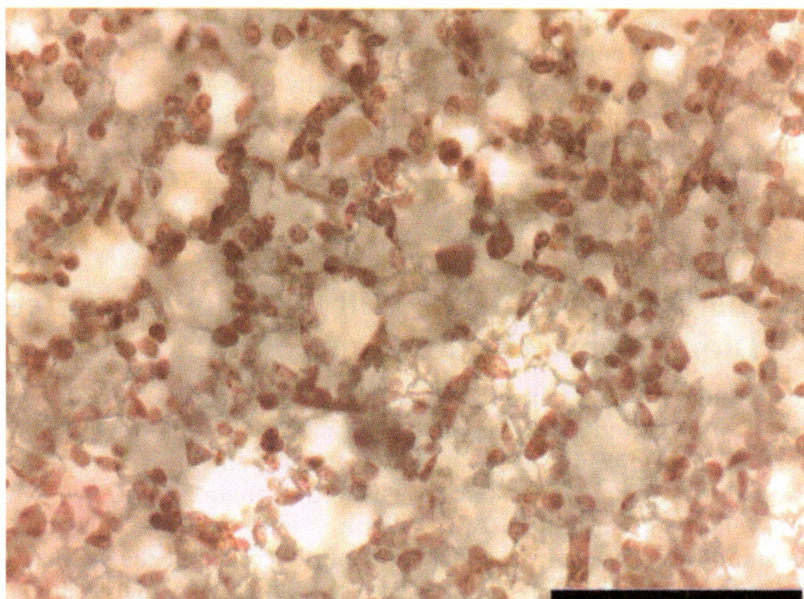

Figure 6.6   Neutral red staining of CHO-K1 cells cultured in Alvetex® for 2 days and imaged using a standard bright-field microscope. Note the clear visualisation of the cells even though light is scattered around the 3D interface. Reinnervate is acknowledged for supplying the image and allowing reprint in this chapter. Scale bar = 200 μm.

generally recommended to use sacrificial cultures where possible as any residual stain may impact the experiment.

> **TIP:** *For 3D imaging with neutral red staining, dye exposure will likely need to be optimised for cell type, scaffold and the microscope being used. However, a useful starting point is the neutral red protocol for Alvetex®, described on the Reinnervate website (http://reinnervate.com/using-alvetex/protocols/).*

Using confocal microscopy in conjunction with a fluorescent marker (such as CellTracker™ CM-DiI; Invitrogen) is another tool to visualise live cells in three dimensions. With this method, optical sections are used to construct impressive 3D depth profiles (Z-stacks) of the entire cell mass in real-time. This allows for important structural changes to be monitored during cell differentiation, division or migration in three dimensions. Furthermore, depending on the thickness of the sample and the capacity of the microscope, almost entire 3D tissue blocks (typically 100 μm) can be examined, offering useful insights into the organisation of cells at the 3D interface.

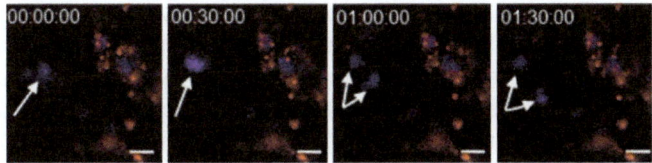

Figure 6.7   Live 3D imaging of dividing CHO-K1 cells on Alvetex®. Cells were stained with CellTracker™ CM-DiI (1/1000 dilution) immediately prior to seeding onto the material. After 5 days, cultures were imaged on a ZEISS LSM 510 confocal microscope. Cells were stained with Hoechst 33342 the evening prior to imaging. White arrows illustrate the dividing cells.
Reinnervate is acknowledged for supplying the image and allowing reprint in this chapter. Scale bars = 20 μm.

> **TIP:** *For 3D imaging using fluorescent markers and confocal microscopes, it is crucial that the fluorescent marker employed produces a good signal-to-noise ratio. This often requires extensive optimisation of the marker-staining process. A poor signal-to-noise ratio masks positive fluorescent staining and thus gives low-resolution images of the cells during confocal microscopy.*

There are some potential drawbacks to be aware of for such live imaging techniques. First, operating confocal microscopes for live 3D imaging requires specialist training, is time-consuming and needs the appropriate equipment (*e.g.*, stage inserts are required to maintain 37 °C and 5% CO_2 for most animal cell experiments). Second, repeated and/or long-term exposure to dyes and fluorescent light can sometimes lead to photobleaching and/or light toxicity, jeopardising experimental results. Under certain working conditions there may be the issue of auto-fluorescence by the 3D substrate although this can be reduced by optimisation of the technique. Nonetheless, the tool is highly powerful and probably the most applicable way of visualising live cells at the 3D interface. An example of live cell imaging on Alvetex® is shown in Figure 6.7, using CellTracker™ CM-DiI and Hoechst counter-staining to monitor dividing cells in three dimensions.

6.3.2   Histology and Immunocytochemistry

Histological and immunocytochemical analysis of tissue samples is a well-known method of analysing cell structure and status at a particular time point. Hospitals frequently use these techniques to assess biopsy tissue samples from patients as a tool to diagnose certain diseases, including cancers. These techniques can easily be applied to cell cultures at the 3D interface, providing that the 3D substrate is sufficiently permeable and chemically stable to withstand the processing procedures.

A broad range of histological procedures is available, using different fixatives, chemical stains and antibodies. As the purpose of this chapter is to merely highlight some of the general techniques to examine cells at the 3D interface, only a brief example of a common processing procedure is described. The first step in the process is to fix the cells on the 3D substrate with the appropriate fixative. Examples include Bouin's fixative (preferred for haematoxylin and eosin staining procedures) and paraformaldehyde (usually preferred for immunocytochemistry applications). After fixation the samples are then dehydrated through a series of ethanol gradients and then placed into a clearing agent (*e.g.* Histo-Clear; National Diagnostics). Samples are then embedded into paraffin wax and then sectioned into sections of 5–10 μm thickness before being placed onto a glass slide. Once the sample is on the glass slide, a variety of different staining procedures can then be performed to image cells, proteins and other biomolecules. Simple histology stains like haematoxylin and eosin can be used to obtain valuable cross-sections of cells in the 3D environments, such as the example from Alvetex® in Figure 6.8. Alternatively, immunocytochemical procedures can be employed, where antigen specific antibodies coupled to a fluorophore or colorimetric stain (directly or indirectly) can be used to detect almost any cell marker.

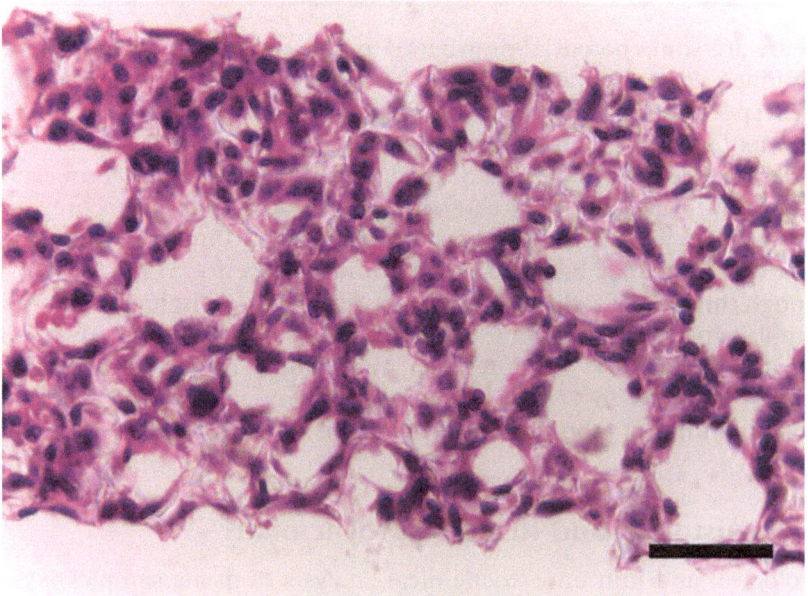

Figure 6.8 Histological analysis of HepG2 cells cultured on Alvetex®, using haematoxylin and eosin staining. In this example cell nuclei have been stained dark blue (haematoxylin) and cell cytoplasms have been stained pink (eosin); however, a range of alternative stains are available.
Reinnervate is acknowledged for supplying the image and allowing reprint in this chapter. Scale bar = 50 μm.

6.3.3 Electron Microscopy

Electron microscopy is a powerful tool to probe cell structure and ultra-structure and works perfectly well in three dimensions, again providing that the substrate can survive the chemical processing procedure. Cells in three dimensions can be imaged using both scanning electron microscopy (for cell surface assessments) as well as transmission electron microscopy (for cell ultrastructure assessments). The first step in the procedure is fixing the cells, usually with a combination of paraformaldehyde and glutaraldehyde fixatives followed by osmium tetroxide. For SEM experiments, samples are then dehydrated through a series of ethanol gradients and then dried using a critical point dryer. For TEM experiments, samples are embedded into a hard resin after dehydration and then sectioned into ultra-thin sections (typically 70 nm) before being stained and imaged.

> **TIP:** *The 3D scaffolds intended for analysis via electron microscopy should be chemically and mechanically stable. Scaffolds that are too soft and unstable will not survive the extensive electron microscopy processing procedure; this will likely destroy the 3D organisation of the cells before imaging.*

Figure 6.9 shows a TEM image of a hepatocyte cell cultured on Alvetex®. The morphology and ultrastructure of the hepatocyte is 3D, with healthy signs of microvilli on the surface of the cell. The scaffold can also be seen in the image.

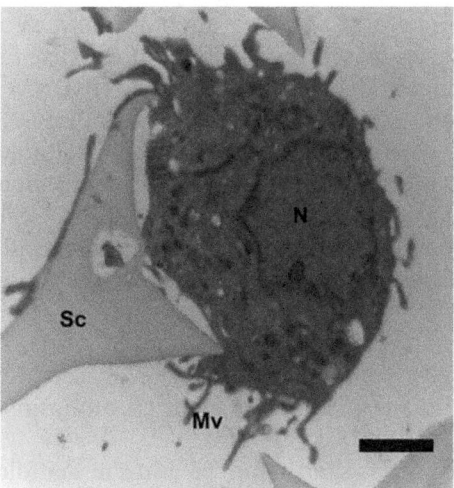

Figure 6.9 Ultrastructural analysis of HepG2 cells cultured on Alvetex® using transmission electron microscopy. Sc, Alvetex® scaffold; N, nucleus; Mv, microvilli.
Reinnervate is acknowledged for supplying the image and allowing reprint in this chapter. Scale bar = 2 μm.

6.3.4   Metabolic Assays and Other Functional Experiments

Probing metabolic and functional activity in three dimensions is generally less challenging than imaging cells in three dimensions. Many commercial assays rely on analysing the culture media, where the detection of secreted metabolites, proteins, hormones and other biomolecules can simply be performed by taking samples of the culture media without disrupting the culture. However, for substrates such as hydrogels this may be more challenging due to slower media diffusion through the gel (see Section 6.2.5). Similarly, simulating cell function in two and three dimensions is much the same, as most stimulants can be added to the culture media (*e.g.*, dissolving a drug in the culture media would be the same for both 2D and 3D cultures). Other metabolic assays, such as the MTT [3-(4,5-dimethylthiazol-2-yl)-2,5-diphenyltetrazolium bromide] assay, can also be easily applied to a 3D culture to gain preliminary insights into the culture's viability over the growth period.

Often with metabolic and functional data it is important to employ some form of normalisation to cell number. For example, it might be important to know how many cells have contributed to a total measured protein level obtained from a media sample. This then requires measuring cell number, which is usually a challenge in three dimensions. For conventional 2D cultures, normalisation is fairly straightforward as cells can either be counted directly on the plastic, or subsequently extracted and counted with tools such as a haemocytometer or a flow cytometric assay. However, retrieving and counting cells from 3D cultures can be more difficult (see Section 6.3.6). Cells in three dimensions will form tight junctions with their neighbours, as well as secrete their own ECM which anchors them to one another and the 3D interface. To bypass this, it is possible to normalise functional data to the level of double-stranded deoxyribonucleic acid (dsDNA) using a commercial assay kit (PicoGreen®; Invitrogen). Double-stranded DNA is much easier to extract than entire cells as it can be obtained by simply lysing cells, rather than having to disrupt the strong protein complexes anchoring the cell to the substrate. Once the DNA is collected, a PicoGreen® calibration curve with known cell numbers can be used to extrapolate the number of cells in the 3D population. A brief protocol/example of PicoGreen® analysis of cells cultured on Alvetex® is given below:

1. Hepatocyte cells were first cultured on Alvetex® for 5 days.
2. Culture medium was analysed for total albumin secretion, representative of the entire 3D population.
3. Culture medium was then removed and the scaffolds were then treated with a standard lysis solution. The mixture was then vortexed and homogenised to extract all DNA from the scaffold culture into the lysis solution.
4. The lysis solution was then incubated with the PicoGreen® reagent and the absorbance measured. In parallel, known cell numbers from 2D

cultures (*e.g.*, cells pre-counted on a haemocytometer) were also lysed and incubated with PicoGreen®. Absorbances of the known 2D cultures were used to correlate 3D absorbance with cell number.

5. The total albumin level obtained from the initial media sample was then divided by the 3D cell number to provide a reading of albumin secretion per cell. This then allowed for a direct comparison with albumin secretion of 2D cell cultures.

6.3.5 Genomics and Proteomics

The study of genes and proteins is an important aspect of cell biology, especially in cancer biology. Hence being able to successfully extract mRNA and total protein from 3D cultures is crucial. Encouragingly, the process is straightforward for most 3D substrates. Usually all that is needed is the application of a lysis buffer to burst open the cell membranes and release the contents, including the cellular proteins and genetic material. Once the lysate is collected, typical protein or RNA isolation and purification procedures can be applied as normal. For Alvetex®, mRNA and total protein are easily extracted using IGEPAL®CA-630 (Sigma), although a slightly more aggressive homogenisation process compared to 2D cultures is normally needed to ensure all genetic material is obtained from a 3D sample.

6.3.6 Cell Retrieval

One of the main issues with growing cells at the 3D interface is retrieving cells at the end of the culture period. Cells that are densely packed into a 3D structure will secrete their own ECM to tightly anchor themselves to one another and to the 3D substrate. In addition, most 3D substrates cannot be easily opened or disrupted, especially chemically cross-linked hydrogels, electrospun fibres and polymeric scaffolds. As a result, the short incubation times with proteases such as trysin–EDTA that were suitable for retrieving cells from 2D cultures are much less effective in three dimensions.

Whilst full cell retrieval may not be an issue for tissue engineering or regenerative medicine applications, where the biodegradable material is intended to be implanted into the body along with the cells, it does pose a potential issue for *in vitro* cell culture applications. Not being able to continually passage proliferative cells in three dimensions is a major challenge, in that cells stocks need to be pre-cultured in two dimensions before being applied to 3D experiments. Similarly, certain *in vitro* biological techniques require cells to be extracted from the substrate, including flow cytometry or using transfection kits.

Some 3D materials do allow for partial cell retrieval. For example, some physically cross-linked hydrogels or polymeric scaffolds can be de-gelled or broken down by certain external stimuli but, again, care needs to be taken that this process does not harm the cells. For Alvetex®, which is a chemically cross-linked material, partial cell retrieval (*ca.* 30–50%) is possible with several trypsin–EDTA incubations, interspersed with culture media rests so as not to expose the cells to the protease for prolonged periods. Extracted cells *via* this method appeared healthy and performed normal cell functions. They were also subsequently seeded onto new scaffolds and continued to grow successfully in three dimensions, indicating the potential for 3D passaging opportunities.

TIP: *For 3D cell extraction from Alvetex® using trypsin–EDTA solutions, typically 3×5 min trypsin–EDTA incubations interspersed with 2×3 min culture media incubations is sufficient to extract some cells from Alvetex®.*

6.4 Future Direction of Biointerfaces in Three Dimensions

Growing cells at the 3D interface is rapidly becoming standard practice for many research laboratories. This new benchmark enables scientists to approximate the structural and physical aspects of the native environment that were not possible with conventional 2D interfaces. Future work has now focused on further developing and optimising these 3D models to include additional aspects of the *in vivo* environment, such as specific biochemical or mechanical components. For example, 3D co-culture is now receiving a lot of attention as a means of improving the biochemical environment. Many organs and tissues are composed of multiple cell types that provide various biochemical stimuli for one another, such as secreted growth factors or specific cell surface ligands or receptors. Mimicking this scenario in three dimensions would therefore be advantageous, especially if the different cell types could also undergo discrete structural organisations representative of the native tissue. Another development area for 3D culture models is replicating blood flow using perfusion or bioreactor systems. Not only does this help to replicate some of the mechanical stresses introduced from blood flow, it also helps to distribute media throughout the 3D cell mass, offering opportunities to create nutrient gradients between cells. Hepatocytes, for example, constantly experience blood flow coming from the digestive system. Depending on where a particular hepatocyte is located in the 3D cord, it will experience either a nutrient-rich or nutrient-poor blood supply. This heterogeneity is actually very important for hepatocyte function, thus mimicking this with 3D perfusion models could also enhance functional behaviour in the laboratory. Finally, a significant proportion of research has now turned to improving the surface chemistry of synthetic 3D substrates. Consequently the field of synthetic polymer biomaterials is now thriving with synthetic 3D materials that have been functionalised with peptides, carbohydrates, proteins and other biomolecules.

In summary, 3D biointerfaces are now an integral part of cell biology and biomedicine. A range of natural and synthetic materials can be used to create such interfaces, with the choice of material depending heavily on the desired application. There are also several techniques available to examine cells at the 3D interface, including methods to image live cells and probe function. Future work is now progressing towards synthetic 3D models that employ co-cultures, media perfusion and targeted surface chemistries so that native cell function can be more closely replicated in the laboratory.

References

1. A. Abbott, Cell culture: Biology's new dimension, *Nature*, 2003, **424**, 870–872.
2. F. Pampaloni, E. G. Reynaud and E. H. K. Stelzer, The third dimension bridges the gap between cell culture and live tissue, *Nat. Rev. Mol. Cell Biol.*, 2007, **8**, 839–845.
3. D. J. Maltman and S. A. Przyborski, Developments in three-dimensional cell culture technology aimed at improving the accuracy of in vitro analyses, *Biochem. Soc. Trans.*, 2010, **38**, 1072–1075.
4. K. M. Yamada and E. Cukierman, Modeling Tissue Morphogenesis and Cancer in 3D, *Cell*, 2007, **130**, 601–610.
5. L. C. Kimlin, G. Casagrande and V. M. Virador, In vitro three-dimensional (3D) models in cancer research: An update, *Mol. Carcinog.*, 2013, **52**, 167–182.
6. M. P. Lutolf and J. A. Hubbell, Synthetic biomaterials as instructive extracellular microenvironments for morphogenesis in tissue engineering, *Nat. Biotechnol.*, 2005, **23**, 47–55.
7. E. T. Pashuck and M. M. Stevens, Designing Regenerative Biomaterial Therapies for the Clinic, *Sci. Transl. Med.*, 2012, **4**, 160–164.
8. J. C. Y. Dunn, M. L. Yarmush, H. G. Koebe and R. G. Tompkins, Hepatocyte function and extracellular-matrix geometry – long-term culture in a sandwich configuration, *FASEB J.*, 1989, **3**, 174–177.
9. B. Swift, N. D. Pfeifer and K. L. R. Brouwer, Sandwich-cultured hepatocytes: an in vitro model to evaluate hepatobiliary transporter-based drug interactions and hepatotoxicity, *Drug Metab. Rev.*, 2010, **42**, 446–471.
10. J. L. Drury and D. J. Mooney, Hydrogels for tissue engineering: scaffold design variables and applications, *Biomaterials*, 2003, **24**, 4337–4351.
11. T. J. Sill and H. A. von Recum, Electro spinning: applications in drug delivery and tissue engineering, *Biomaterials*, 2008, **29**, 1989–2006.
12. H. Seitz, W. Rieder, S. Irsen, B. Leukers and C. Tille, Three-dimensional printing of porous ceramic scaffolds for bone tissue engineering, *J. Biomed. Mater. Res., Part B*, 2005, **74**, 782–788.
13. R. J. Carnachan, M. Bokhari, S. A. Przyborski and N. R. Cameron, Tailoring the morphology of emulsion-templated porous polymers, *Soft Matter*, 2006, **2**, 608–616.
14. I. Pulko and P. Krajnc, High internal phase emulsion templating - a path to hierarchically porous functional polymers, *Macromol. Rapid Commun.*, 2012, **33**, 1731–1746.

15. A. S. Hayward, *et al.*, Galactose-Functionalized PolyHIPE Scaffolds for Use in Routine Three Dimensional Culture of Mammalian Hepatocytes, *Biomacromolecules*, 2013, **14**, 4271–4277.
16. A. S. Hayward, N. Sano, S. A. Przyborski and N. R. Cameron, Acrylic-Acid-Functionalized PolyHIPE Scaffolds for Use in 3D Cell Culture, *Macromol. Rapid Commun.*, 2013, **34**, 1844–1849.
17. E. Knight, B. Murray, R. Carnachan and S. Przyborski, Alvetex (R): Polystyrene Scaffold Technology for Routine Three Dimensional Cell Culture, in *3D Cell Culture: Methods and Protocols*, ed. J. W. Haycock, 2011, vol. 695, pp. 323–340.
18. N. R. Cameron and A. Barbetta, The influence of porogen type on the porosity, surface area and morphology of poly(divinylbenzene) PolyHIPE foams, *J. Mater. Chem.*, 2000, **10**, 2466–2472.
19. M. Bokhari, R. J. Carnachan, N. R. Cameron and S. A. Przyborski, Culture of HepG2 liver cells on three dimensional polystyrene scaffolds enhances cell structure and function during toxicological challenge, *J. Anat.*, 2007, **211**, 567–576.
20. M. W. Hayman, K. H. Smith, N. R. Cameron and S. A. Przyborski, Growth of human stem cell-derived neurons on solid three-dimensional polymers, *J. Biochem. Biophys. Methods*, 2005, **62**, 231–240.
21. M. Schutte, *et al.*, Rat Primary Hepatocytes Show Enhanced Performance and Sensitivity to Acetaminophen During Three-Dimensional Culture on a Polystyrene Scaffold Designed for Routine Use, *Assay Drug Dev. Technol.*, 2011, **9**, 475–486.

Section C
Multi-functional Biointerfaces

CHAPTER 7

Interfaces in Composite Materials

ENSANYA A. ABOU NEEL,*[a,b,c] WOJCIECH CHRZANOWSKI[d]
AND ANNE M. YOUNG[c]

[a] Operative and Aesthetic Dentistry Department, Biomaterials Division, King Abdulaziz University, P.O. Box 80200, Jeddah 21589, Saudi Arabia; [b] Faculty of Dentistry, Biomaterials Department, Tanta University, Tanta, Egypt; [c] Biomaterials and Tissue Engineering Division, UCL, Eastman Dental Institute, 256 Gray's Inn Road, London WC1X 8LD, UK; [d] The Faculty of Pharmacy, The University of Sydney, Sydney, NSW 2006, Australia
*Email: eabouneel@kau.edu.sa; e.abouneel@ucl.ac.uk

7.1 Introduction

On a macroscopic scale, the word 'composite' refers to combination of two or more components; the aim is to synergistically combine positive features of these components and provide improved functionality of the material (composite). These components may differ in composition, morphology and physical properties.[1] In this chapter, the word 'composite' refers to combinations of either different polymeric matrices or polymer matrix and filler phase. The region between the different components of the composite (*e.g.*, the matrix and filler) can be a surface (*i.e.*, interface) or a third phase (*i.e.*, interphase). The properties of a composite are therefore dependent on the composition, properties and the interfacial interaction of its components. Composites can be designed to provide materials with tailored chemical, physical and mechanical properties to fulfill specific applications. Most

RSC Smart Materials No. 10
Biointerfaces: Where Material Meets Biology
Edited by Dietmar Hutmacher and Wojciech Chrzanowski
Published by the Royal Society of Chemistry, www.rsc.org

human tissues, *e.g.*, bone and teeth, are composites. The amount, morphology, distribution and characteristics of the components determine the overall behavior of the resulting tissue.[1]

Over the last 50 years, the use of composites has progressively increased. Applications include aeronautical, automotive, naval and, more recently, biomedical. Many biomedical composites are now commercially available, but many more are under investigation. New applications include drug delivery devices and tissue engineering scaffolds. For biomedical applications, biocompatibility and biomimicity are important.

7.1.1 Biomedical Applications of Composites

Synthetic composites can be designed to reproduce the form, mechanical characteristic and function of a damaged tissue.[1] This section focuses on orthopedic and dental applications of composites.

7.1.1.1 *Orthopedic*

Composites are regarded as third-generation biomaterials for orthopedic applications.[2] They are commonly employed as bone fixation resorbable plates,[3] hip joint replacement implants,[4] bone cements,[5] and bone grafts.[6] Commercially available composite bone plate examples include laminated continuous carbon fiber reinforced poly(L-lactide)[3] and continuous poly(L-lactide) fiber reinforced poly(L-lactide) composites.[7] These are commonly used for fixing comminuted zygomatic arch[8] and distal femur fractures.[9] Furthermore, they are employed in craniofacial surgery[10] and to correct the form of soft tissues as in cleft lips.[11] Other composites developed as potential bone plates include phosphate glass fiber reinforced polylactide[12] and cross-linked polypropylene fumarate–poly(lactide-*co*-glycolide)–hydroxyapatite.[13] Furthermore, polyether ether ketones (PEEK), which have become a polymer of choice for fixation of hard tissue, primarily in spinal applications, has been combined with hydroxyapatite, carbon fillers and fibers to enhance the material integration within the body.[14–18] The use of some fillers (*e.g.*, carbon fibers) reinforces the material, thus allowing the fabrication of even stronger devices for load bearing applications.[19]

7.1.1.2 *Dental*

Non-degradable dental resin composites are extensively used in restorative dentistry. These consist of a polymerizable resin matrix, typically high molecular weight hydrophobic dimethacrylate monomers, silane-treated silica-based inorganic filler particles, and an initiator–activator system. They set *via* free radical polymerization upon exposure to a blue light source. The polymerized composites are cross-linked three-dimensional networks with suitable esthetic, physical and mechanical properties for dental applications.[20] Dental composites can be bonded to tooth structure using resin adhesives. The adhesive is often chemically similar to the composite but with reduced filler content. Composite uses include filling materials,[21,22] cements,[23] adhesives,[24] direct laminate veneers,[25,26] and fixed partial dentures.[27]

Dental composites shrink upon set by 0.9 to 3.5 vol%.[28] The percent of shrinkage is reduced by raising the volume percentage of filler or increasing the monomer molecular weight. This shrinkage creates stresses[29] as high as 10–25 MPa within the composite and at its interface with the tooth structure.[30] The shrinkage stresses severely strain the interfacial bond creating micron dimension gaps at the composite-tooth interface. Subsequently, the invasion of fluid and bacteria can lead to post-operative sensitivity, pain, tooth discoloration, caries and, finally, restoration failure. The elastic properties of the polymer matrix and the ability of the polymer chains to rearrange and relieve stresses have been shown to influence the final polymerization stresses.[30] Water sorption (typically ~1.5%)[31] and the subsequent expansion can relieve some polymerization stresses. Water sorption, however, can be a slow process. These two processes of shrinkage and expansion can make gaining a strong, resilient bond between dental composites and tooth structure, dentin in particular, difficult.[29]

Several strategies have been attempted to reduce shrinkage stresses. These include modifying the composition (*e.g.*, increasing filler content, using high molecular weight monomer, or low shrink monomer), placement technique (curing the composite in thin consecutive layers) or the curing methods (*e.g.*, soft start and pulse delay).[29]

7.1.1.3 The Future

Composites may be employed in the future as scaffolds for hard (*e.g.*, poly(ε-caprolactone)–bioactive glass composite[32]) or soft tissue regeneration (*e.g.*, poly(2-hydroxyethylmethacrylate)–carbon nanotubes hydrogel composites[33]). Other future potential applications of composites includes oral (*e.g.*, polylactide-*co*-glycolide–bioactive glass composite[34]) and ophthalmic implants (*e.g.*, silicon polycaprolactone composites[35]). Moreover, composites may be used as wound dressing (*e.g.*, silk fibroin–hydroxyapatite composites[36]) and biosensors/electronic devices (*e.g.*, polysaccharide–polyaniline composites[37]). Furthermore, composites can be used as delivery vehicles for drugs (*e.g.*, polycarbonate–silica xerogel nano-composites for the release of rifampicine[38]), proteins (*e.g.*, poly(ethylene glycol) fumarate and gelatin microparticles for the release of bone morphogenetic protein-2[39]) or cells.[40]

7.2 Composite Interfaces

The boundary between any two layers or phases of different chemistry and/or microstructure can be described as an 'interface'. Unlike the interface, interphase describes a gradual but not abrupt change in properties from one phase to another and is three-dimensional in nature.[41] Generally, either the interface or interphase plays a key role in defining the overall properties of a composite.[42] For simplification, the term interface will be used throughout this chapter to describe both terminologies.

The following discussion involves the interfaces within (*e.g.*, filler–matrix) and around the composite (composite-surrounding environment). These interfaces significantly influence the composite mechanics (*e.g.*, strength, modulus and toughness) and behavior (its interaction with surrounding tissues, circulating cells, macromolecules and biologics) respectively. At these interfaces, dynamic processes such as ion diffusion/precipitation and cell attachment are usually occurring. Appreciation of the dynamics of these processes helps in tailoring material interfaces with desired properties.

7.2.1   Filler–Matrix

The filler can be bonded primarily through ionic or covalent bonds or secondarily through van der Waals bonds to the resin matrix. The level of filler–matrix interaction affects fillers loading/dispersibility, stress-transfer across the filler–matrix interface and hence the composite performance.[43] With low filler–matrix interaction, macroscopic phase separation,[43] stress build-up, cracks, delamination or debonding can occur at the interface.[44,45] This condition would become even worse in composites continually subjected to fluctuating thermal and mechanical stresses. Improving interfacial properties is therefore important for enhancing the durability of these composites. This could be achieved by tuning the morphology of the filler,[46] *e.g.*, with mesoporous fillers, the polymer chains act as a necklace passing through the nano-sized channels in the filler.[47] Functionalization[48–50] and nano-structuring[51] of the interface has also been attempted to improve the interfacial properties.

Functionalization of the interface involves surface treatment of fillers to achieve covalent interactions with the matrix.[48–50] It improves the distribution of filler particles within the matrix[52] and subsequently immobilization of the matrix in the vicinity of the filler.[43] Surface treatments of fillers include silanization using coupling agents[53–55] [*e.g.*, 3-aminopropyltriethoxysilane (AMPTES), methyltriethoxysilane (MTES) and 3-mercaptopropyltrimethoxy-silane (MPTMS),[56] γ-methacryloyloxypropyltrimethoxy-silane (γ-MPS)[53]], electrolytic or gas phase modification[41] or grafting polymer chains.[43,57,58] The coupling agents have both hydrolysable (*e.g.*, trialkyloxysilane) and non-hydrolysable organofunctional (*e.g.*, amino, methacrylate or vinyl) groups. These groups form strong interactions with filler particles (by forming covalent and hydrogen bonds with silanols or similar groups on the filler surface)[59] and polymer matrix,[60] respectively. For effective grafting of polymer chains on the surface of fillers, and hence a strong filler–matrix interaction, appropriate selection of the side chains of the grafted polymer is a key factor.[61]

Nano-structuring the interface is another option to improve the interfacial interactions between the filler and resin matrix. In nano-composites, the high surface area/volume ratio contributes to much greater effects of the interface on the composite properties. The structure of this interface is influenced by the number of anchoring points the polymer matrix chains make with the filler[51] (Figure 7.1, panel I). An effective adsorption of the polymer chains onto the surface of the filler particles produces a compact interface. A diffuse interface with long polymer loops and tails, however, would result from poor adsorption of the polymer chains on the surface of filler particles.[62] The strength of interaction of polymer chains with the filler particles controls both the conformations of the polymer chains at the interface and the entanglement distribution in a larger region surrounding the filler particles.[51] This will then be reflected on the overall elastic moduli and behavior (*e.g.*, necking and drawing after yielding) of nano-composites.[50]

7.2.1.1 Methodology of Examination

The filler–matrix interface has been investigated using many different techniques. The structure and density of the interface, for example, has been examined using thermal gravimetric analysis (TGA), transmission electron microscopy (TEM), Fourier transform infrared spectroscopy (FTIR)[51] and molecular dynamic simulation (MDS).[50] Analysis of surface chemistry by X-ray photoelectron spectroscopy (XPS), Auger electron spectroscopy (AES) and time of flight secondary ion mass spectrometry (ToF-SIMS) has also been attempted. Thermodynamics of the interface has been investigated using dynamic contact angle analysis (DCAA) and inverse gas chromatography (IGC).[41] The gap between the filler and matrix has been monitored using optical microscopy, light attenuation and scanning electron microscopy.[63] Dispersion and spatial distribution of the filler within the matrix has been quantified using coarse-grained molecular dynamics simulation[43] and scanning electron microscopy.[64] The reinforcing effect of the filler has been assessed through mechanical testing[64] and stress-transfer between phases by X-ray diffraction.[61] Information on the interaction of the coupling agent with the filler and matrix has also been gained using thermogravimetric analysis (TGA) and Fourier transform infrared spectroscopy (FTIR).[55]

7.2.2 Composite-surrounding Environment

The traditional concept of biocompatibility relied on the use of inert materials that do not instigate toxic effects. More recently, this concept has shifted to the use of reactive, stimuli responsive materials that interact with the local environment for full tissue integration. Studying the interface between a biomaterial and surrounding environment provides new insights on biophysical phenomena and molecular mechanisms underlying complex biological processes. At this interface, the circulating cells can attach, spread, proliferate and differentiate or die. The circulating biomolecules can also be adsorbed, bound to – and/or released from – a biomaterial surface.

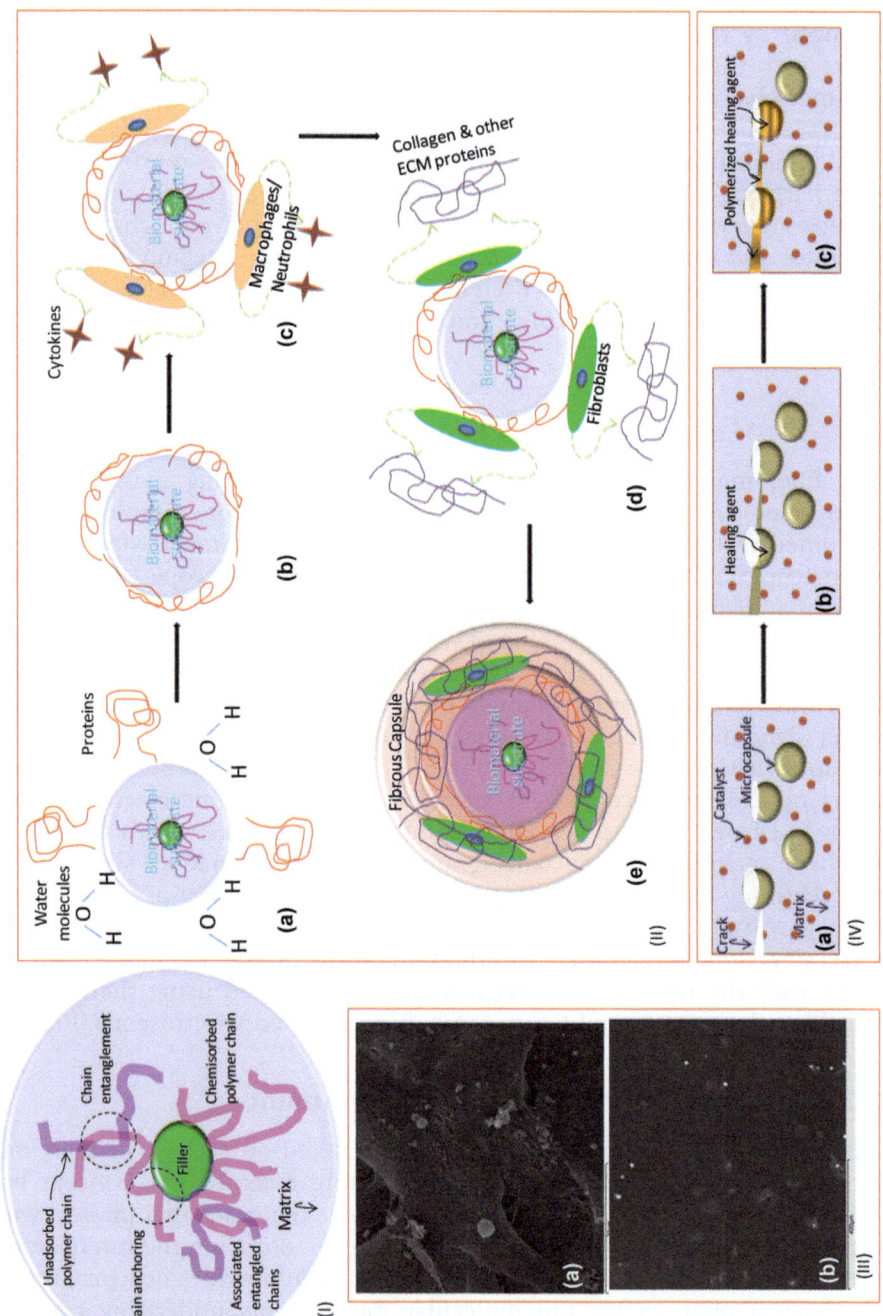

The behavior of these circulating cells and biomolecules can be modulated by varying the interfacial physical and chemical properties (Figure 7.1 panels II and III).

When implanted into the body, biomaterials can be recognized by an intricate recognition system. This includes scavenger (SR), toll-like (TLR) and integrin (IR) receptors and other surface proteins of host cells, *e.g.*, immune and endothelial cells. Hydrophilic materials such as hydrogels tend to be recognized by SR and TLR of both macrophages and endothelial cells. Hydrophobic materials, however, tend to be recognized by IR of neutrophils, monocytes and macrophages. The magnitude and type of the body response varies from one material to another.

The hydrophobicity/hydrophilicity of the interface can affect the behavior of anchorage-dependent cells, *e.g.*, osteoblasts. With hydrophobic interfaces, osteoblasts may be unable to generate sufficient adhesion signals necessary for their adhesion which is a prerequisite for other cellular activities.[65] The interface could also affect the recruitment of leukocytes and the inflammatory response.[66] This is particularly important for implantable

---

Figure 7.1 (I) Diagrammatic representation of the interface between a filler particle and a polymer matrix in a composite; the interface composed of adsorbed polymer chain that has point contact (anchor or trains) with the filler particle, unadsorbed polymer segment (*e.g.*, loops and tails) that are entangled with other chains adsorbed on filler surface. Modified from Ciprari *et al.*[51] (II) Diagrammatic representation of cell–biomaterial surface interaction: once the biomaterial is implanted into the body, blood contact occurs; water molecules and proteins from the blood become adsorbed onto the surface of the biomaterial (a), then a specific and complex protein layer forms over the surface providing binding sites for cell surface integrins (b), immune cells *e.g.*, macrophages and neutrophils reach the implanted site to phagocytose the implanted material otherwise they fuse together forming foreign-body giant cells that release cytokines to recruit fibroblasts (c); fibroblasts infiltrate the surface and secrete collagen and other proteins of the extracellular matrix (ECM) (d) that subsequently form a fibrous capsules around the implanted biomaterial (e). Modified from Harvey *et al.*[73] (III) Scanning electron microscopy images showing the behavior of human mesenchymal stem cells (HMSCs) attached to composite (calcium phosphate filled polylactic-*co*-glycolic acid dimethacrylate) surface: HMSCs extend filopodia to sense the surface, when they encounter an optimal surface condition (a protein layer, which is vital to specific-cell interaction, adsorbed on the surface in such case), they showed extended like-morphology (a) but rounded morphology when the surface is not an optimal (lack of adsorbed protein in such case) (b). (IV) Diagrammatic representation of self-healing composites: once the crack, initiated within the composite, reaches the healing microcapsules (a), the healing agent incorporated into microcapsules will be released to fill the micro-crack (b), the polymerization of the released healing agent will then occur upon its contact with the catalyst incorporated into the polymeric matrix resulting in complete crack closure (c). Adopted from White *et al.*[97]

glucose biosensors designed for 'real-time' monitoring of glucose level in diabetic patients.[67]

The products released from materials can also affect the behavior of surrounding cells and the function of the immune system. With incomplete polymerization of dental composite, the release of triethylene glycol dimethacrylate (TEGDMA) or 2-hydroxyethyl methacrylate (HEMA) monomers can cause cell death and DNA damage due to the formation of reactive oxygen species at the interface. They can also delay cell cycles; inhibit the function of odontoblasts and pulp-derived stem cells. For development of more effective dental therapies and smart composites used in contact with different oral tissues, thorough understanding these events is required.[68]

7.2.2.1   *Methodology of Examination*

The interaction of cells with the composite surface can be quantified by investigating cell attachment, growth, proliferation, gene expression, extracellular matrix production and mineralization (in the case of bone).[69] Dynamics of cell–matrix interaction including cell–matrix anchorage, cell–cell communication and real-time shear stress-induced adhesion dynamics, have been examined using surface plasmon resonance imaging ellipsometry (SPRIE).[70] Dynamics of the cytoskeletal polymers, *e.g.*, microtubules and actin filaments of migrating or dividing cells have been investigated by total internal fluorescent speckle microscopy (TIR-FSM) and wide field epi-fluorescence FSM.[71]

7.3   State-of-the-Art 'Talking' to Living Organisms

Surface features, including chemistry,[72] topography (*e.g.*, grooves and ridges or pits and pillars),[73] roughness,[74] wettability, free energy, and compliance,[73] of a biomaterial significantly affect its interaction with cells and/or biomolecules. Some modern materials are designed to sense changes in the surrounding environment and then respond to such changes. In many respects, these materials behave like living organisms and therefore could be described as 'smart'. These smart materials should be able to adapt to a dynamically evolving environment and maintain harmony with it.[75] These smart multi-functional composites represent the future generation platforms for designing three-dimensional scaffolds for both drug delivery and tissue engineering applications.[76] In this chapter, composites having biomimetic or on-demand responsive matrices or fillers that enable flexible design of macroscopic and microscopic architectures are also considered smart.

The next section covers biomimetic and smart materials as well as the most up-to-date knowledge and bio-inspired approaches to control living organisms (*e.g.*, cells, viruses, microorganisms, or biomolecules) or tissue responses to synthetic biomaterials or implantable devices.

7.3.1 Smart Composites Based on Biomimetic Matrices

During development, the shape of tissues is highly influenced by the mutual interaction between cells and the surrounding extracellular matrix (ECM). Biomimetic materials, that may be a protein of ECM, *e.g.*, collagen, are widely proposed for biomedical use.[77] Protein-based biomaterials are derived from many biological sources that vary in size from microorganisms to mammals and this would affect the cost of manufacture and processing. Those derived from plants, *e.g.*, wheat, are relatively cheap and abundant but often require chemical modification to improve their properties.[78] Some protein-based biomaterials with similar structural composition to those obtained from mammals can also be derived from prokaryotic bacteria offering an alternative route for biomaterials production.[79] Silk-based proteins are produced by a wide variety of larvae of insects including silkworms, mites and flies.[80] Commercially available collagen is typically derived from bovine or porcine sources.

Protein-based biomaterials have been exploited for a number of biomedical applications. Their fragile mechanical properties, associated risk of infection, antigenicity and possible deterioration after long-term implantation, however, restrict their widespread use as a pure biomaterial. Some of these problems may be overcome by combining different proteins to make composites, *e.g.*, fibrinogen/factor XIII/fibronectin and thrombin.[81,82] Due to their biocompatibility and availability, they have considerable potentials as matrices for drug delivery and tissue engineering applications. Fibrin as a form of fibrous protein has been used as a hemostatic agent, sealant or glue,[81,82] in cosmetic surgery[83] and ophthalmology.[84] With recent advances in biotechnology, protein-based biomaterials derived from living organisms could be designed with specific chemical, mechanical or structural properties[85] and used in a form of hydrogels, cross-linked hydrogels or decellularized native tissues.[86]

7.3.2 On-demand Responsive Composites

On-demand responsive composites have been prepared by the combination of hydrogel matrices with other polymers or fillers. The hydrogels convey the on-demand responsive property to the composite. Hydrogel benefits include tunable hydration and softness which make them extracellular matrix-like. Virtually all water-soluble polymers can be assembled into hydrogels that can be tailored to respond controllably to different external or internal stimuli (*e.g.*, temperature,[87,88] light, pH, ionic concentration, chemical reaction[89] and electric current[90]). A key feature of the smartness of these hydrogels involves their ability to return to their original state after stimulus removal. To be stimulus-responsive, a hydrogel (produced from natural[91] or synthetic polymers[88]) may be coupled with a specific functional group that is dependent upon the intended functionality. The hydrogel becomes functionally active when treated with the correct stimulus. Due to their excellent biocompatibility and tunable chemical, physical and mechanical properties, stimuli-responsive hydrogels may be used in a diverse range of biomedical applications. Examples include smart biomimetic three-dimensional scaffolds[92] for minimally invasive, *e.g.*, cartilage repair[89] and stem cell therapy[87,88] and drug delivery applications.[93] They could also be used to engineer organs or tissues with spatial complexity in composition and organization, *e.g.*, zonal articular cartilage repair.[94] Example composites for a range of applications, prepared by combining hydrogels with a variety of filler phases or other polymers, will be described below.

7.3.2.1 Thermo-responsive Composites

Reversible thermo-responsive composites, produced by incorporating a thermo-responsive gel layer between two plastic or glass sheets[75] or relying on temperature-dependent match/mismatch of the refractive indices of the nanofiller and matrix,[52] have been used as smart optical switches/lenses for sensors and eye protection.[52,75] They exhibit an abrupt transparent–opaque transition when the temperature exceeds a critical value. When the temperature goes down, they become transparent. The thermo-responsive composites can also be used for targeting drug delivery applications. Controlling the *in situ* gelling of the hydrogel phase and the way by which this phase aggregates controls the drug release kinetics and the magnitude of burst and overall drug release.[95]

7.3.2.2 Electro-responsive Composites

Smart composites based on electrically conducting polymers, *e.g.*, polypyrrole, can be used for controlled release of a diversity of biologically active molecules (*e.g.*, proteins or polysaccharides or heparin) or cellular signal factors (*e.g.*, growth factors). These composites have been shown to support endothelia cell growth.[96]

7.3.2.3 Thermo- and Mechanico-responsive (Self-healing) Composites

The integrity of a structural composite can be jeopardized by thermal and mechanical fatigue induced micro-cracks. Smart composites, with the ability to repair cracks, have been produced by incorporation of healing agent-loaded microcapsules within the matrix. Once intruded by a crack, these microcapsules release the healing agent that polymerizes upon contact with a catalyst embedded within the polymeric matrix. This self-healing process yielded ~75% recovery in the composite toughness[97] (Figure 7.1, panel IV). Inclusion of microcapsules, however, can reduce the initial mechanical properties of composites. Use of a thermally mendable matrix, *e.g.*, mendomer 401, has been an alternative approach.[98]

Biologically inspired, self-healing composites, based on incorporation of reactive fillers (*e.g.*, monocalcium phosphate monohydrate and β-tricalcium phosphate) within a degradable polymer matrix, have been developed as bone adhesives/cements. Upon water sorption, these fillers undergo chemical reaction to form a new phase (*e.g.*, brushite). The resultant phase occupies greater volume than the starting phases and potentially fills the cracks/defects produced within the composite.[99–103] Step-by-step preparation of this example of self-healing composites will be described in detail later in this chapter.

7.3.3   Composites Based on Smart Fillers

Inorganic nanoparticles with diversified functionalities and self-assembling ability could be a material for future biomedical use. Self-assembly of nanoparticles enables flexible design of materials with different microscopic and macroscopic architectures. Composites containing Raman-active nanoparticles, for example, could be used as ultrasensitive molecular detection and imaging platform *e.g.*, rhodamine 6G (R6G).[104] Composites with diverse set of on-demand properties, *e.g.*, regions of variable expansion, conductivity and strength properties, have also been produced by incorporating various fillers into a single flexible polymer matrix. Applications include bioelectronics and sensors. Inclusion of microfluidic channels, that can expand on demand, within the flexible matrix extends the application of these composites for live analysis of captured rare cells, *e.g.*, circulating tumor cells.[105] Composites with anti-bacterial properties have also been produced by plasma grafting of anti-bacterial nanoparticles onto a polymer matrix.[106] Various smart, anti-cariogenic, pH-responsive, dental nanocomposites have been prepared by incorporating different calcium phosphate fillers (*e.g.*, amorphous calcium phosphate, nano-tetracalcium phosphate, mixture of mono-calcium phosphate monohydrate and β-tricalcium phosphate). Under acidic conditions, these composites release Ca and PO_4. The released species have the ability to neutralize the acidity produced by cariogenic bacteria, stop caries, and re-mineralize dental lesions.[107–110]

7.3.4   Functionalized Composites

Designing new functional materials with tailored physical and chemical properties as well as versatile and reliable processing capabilities represents a great challenge in modern materials science. Cell adhesion is a prerequisite for other cellular activities.[65] Introducing nano-scale surface topography on biomaterials[74] can therefore mediate adsorption of cell adhesive proteins, *e.g.*, fibronectin,[111] which are necessary for cell adhesion. Hierarchically organized nano-structured composites can be fabricated *via* nano-imprint lithography,[112] photo-imprinting[72] and atomic layer deposition.[113]

Mimicking the structure of biological materials is another interesting direction for developing composites with biomimetic properties, *e.g.*, electrospun composite nanofibers. Due to their similarity to the extracellular matrix and ease of incorporation of drugs (*e.g.*, chemotherapeutic drugs,[114] antibiotics[115,116]), growth factors, natural materials,[117] and inorganic nanofillers,[118] electrospun composite nanofibers have been proposed for drug delivery[117,119,120] and tissue engineering applications.[121–124] Shape memory electrospun composite nanofibers have emerged as a class of active biomaterials that are capable of changing shape in a predefined way under controlled external stimulus such as heat and alternating magnetic field. This class of electrospun composite nanofibers has potential as multifunctional tissue engineering scaffolds.[125]

To convey the multi-functional character of biologic modulators, surface modification of synthetic composites with RGD (integrins binding domains) has been attempted. Growth factors, morphogenetic proteins, specific oligopeptide cleaving sequences specific for matrix metalloproteinases (MMPs), anti-thrombin sequences and plasmin degradation sites have also been attempted.[126] For example, adhesion ligands[127] and recombinant osteopontin (rOPN17-169)[65] have been observed to improve the adhesion of anchorage-dependent cells, *e.g.*, osteoblasts to a hydrophobic surface. Both adhesion ligands and rOPN17-169 have the capacity to bind to various integrins; rOPN17-169 bind to $\alpha_v\beta_3$,[128] $\alpha_v\beta_1$,[129] $\alpha_v\beta_5$,[129] or $\alpha_a\beta_1$[130] integrins.[65]

Coating a composite or implantable device with a smart coat is an alternative way of developing functionalized composites. Drug-releasing engineered surface coatings are of paramount importance for various biomedical applications from implantable devices to tissue engineering. For example, lipogel-coated–liposome-functionalized poly(vinyl alcohol) composite hydrogels have been made cell adhesive by coating the lipogel with poly(dopamine).[131] While enhancing the *in vivo* functionality of the implantable device, it can also be important to reduce the foreign body reaction. Coating glucose biosensors with poly(lactic-*co*-glycolic) acid (PLGA) microspheres dispersed in poly(vinyl alcohol) (PVA) hydrogels controlled both inflammation and fibrous tissue formation while allowing for rapid glucose diffusion to the sensor.[67,93]

7.4 Interfaces of Clinical Importance

This next section focuses on interfaces having clinical importance and the recent advances in composites used to restore them. These interfaces include osteochondral, neuromuscular-prosthesis and tooth-restoration.

7.4.1 The Osteochondral Interface

Understanding the structural and morphological changes occurring at the osteochondral interface is of prime importance in treating many diseases affecting this transitional zone. The osteochondral junction encompasses the region between the deep layers of articular cartilage and the underlying subchondral bone. In healthy joints, the osteochondral junction has a complex structure; it comprises of deep layer of non-calcified cartilage, tidemark, calcified cartilage, cement line and subchondral bone (Figure 7.2).[132] This junction is essential for providing both functional and structural integration of cartilage into the underlying bone.

Osteoarthritis is a major cause of physical disability. It is a degenerative joint disease characterized by structural, chemical and mechanical changes at the osteochondral junction. These changes have been identified as loss of integrity and plasticity at the osteochondral junction. With the disruption of the integrity of osteochondral junction, its barrier action will be compromised. The subchondral bone becomes exposed to the endothelial growth factors (VEGF) produced by articular chondrocytes and bone turnover increases. Reciprocally, invasion of the articular cartilage by vascular channels exposes the chondrocytes to differentiation factors produced from

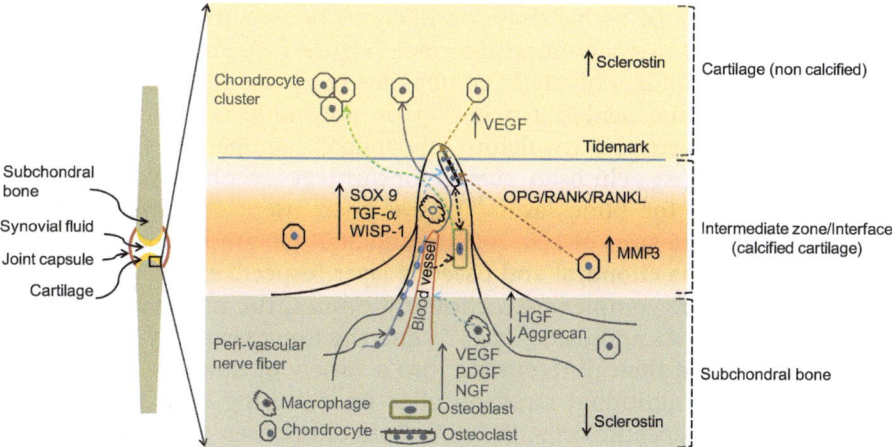

Figure 7.2   Molecular cross-talk (as indicated by dashed arrows; the direction of each arrow shows the direction of cross-talk) between subchondral bone and cartilage at the osteochondral interface. A vascular channel passing from subchondral bone to non-calcified cartilage through the osteochondral interface (calcified cartilage). This channel contains a blood vessel, osteoblasts, osteoclasts and macrophages. These cells interact with each other through cytokines, growth factors and other signals; local interactions lead to stimulation of angiogenesis, nerve growth, and osteoblastic activity and new bone formation. In the case of osteoarthritis, these interactions are enhanced by signals across the osteochondral interface. With the disruption of tidemark, subchondral bone becomes exposed to the endothelial growth factors (VEGFs) produced by articular chondrocytes nearer the interface. Reciprocally, invasion of the articular cartilage by vascular channels exposes the chondrocytes to differentiation factors produced from subchondral bone, and both chondrocytes hypertrophy and endochondral ossification occur.
Modified from Suri and Walsh.[132]

subchondral bone, and both chondrocytes hypertrophy and endochondral ossification occur (Figure 7.2).[132] These changes in both subchondral bone and cartilage are associated with joint pain, tenderness, swelling and stiffness.

The prevalence of osteoarthritis increases with increasing age and obesity of the population; it affects approximately 8 million people in UK but nearly 27 million in USA.[133] Treatment guidelines generally first recommend a non-operative approach (weight loss, exercise and anti-inflammatory drugs) before surgical options such as joint replacement. Unlike bone, cartilage regeneration is limited.[134] Currently there is no widely accepted treatment to repair osteochondral defects[135] although tissue engineering is emerging as a promising future option. Simultaneous engineering of articular cartilage and subchondral bone with a stable interface, that separates these two histologically, mechanically and functionally different tissues, is very challenging, however. Designing stratified, *i.e.*, bilayered[136–141] or multilayered,[142–147] scaffolds with spatially controlled, continuous gradients of materials composition/properties and growth factors[148] to mimic the complex elegance of native tissue is likely to be essential for effective regeneration of the osteochondral interface (Figure 7.3). For the construction of stratified scaffolds, polymeric composites (*e.g.*, silk–fibroin sponge[142] or collagen–hyaluronic acid–fibrin gel[134]) or polymeric–ceramic composites (*e.g.*, polycaprolactone–β-tricalcium phosphate nanoparticles)[149] or combination of both (*e.g.*, chitosan–gelatin for cartilage layer but hydroxyapatite-chitosan–gelatin for bone layer[136]) have been used. For effective tissue regeneration, pore sizes of 100–200 μm and 300–450 μm has been shown to be optimal for the chondral and osseous layer respectively.[150]

To engineer the complex osteochondral tissues, the native cross-talk between chondrocytes and osteoblasts[132] and the innate capacity of stem cells to proliferate and then differentiate into a variety of lineages[151] have been employed. The traditional strategy relies on culturing each phase of the scaffold with its specific cells; the differentiated phases can then be joined (*e.g.*, using fibrin glue[136] or self-assembling peptides[142,152]) and co-cultured to form the osteochondral construct. Co-culturing allows the cross-talk between different cell populations and formation of a more complete osteochondral construct with an interface of hypertrophic chondrocytes, type X collagen and MMP-13.[142,152] *In vivo* implantation of bilayered integrated, spatially controlled composites supported the osteochondral regeneration in a lapine knee defect.[136,153]

Growth factors, *e.g.*, transforming growth factor-β1 (TGF-β1) and bone morphogenetic protein-2 (BMP-2) have also been incorporated within the scaffold to enhance stem cell differentiation.[154–156] Yet, the dose and spatial distribution of growth factors within the scaffolds is difficult to control. Furthermore, the dual release of growth factors [*i.e.*, both TGF-β3 and insulin growth factor-1 (IGF-1)] does not necessarily synergistically enhance the quality of engineered osteochondral tissues over the single release (*e.g.*, IGF-1).[156] More importantly, the growth factor proteins are unstable and

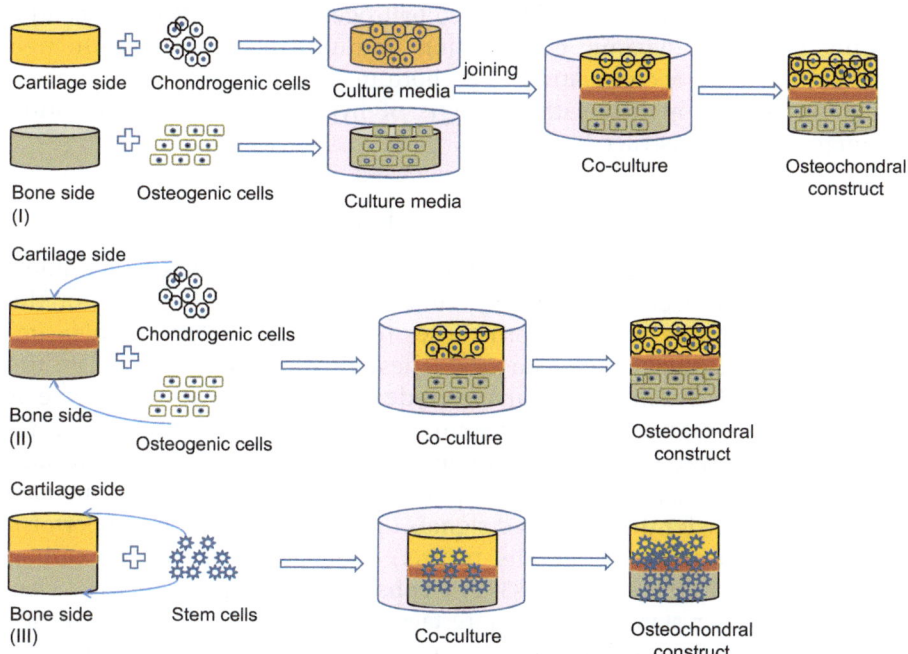

Figure 7.3 Diagrammatic representation of various strategies of bilayered constructs used for osteochondral regeneration. The scaffold consists of an upper chondral phase, required to reproduce the lubricating and shock-absorbing function of articular cartilage and the lower osseous phase required to provide the mechanical support for the subchondral bone. These two phases were then joined by suturing or using a material that simulates the extracellular matrix (I) scaffolds where each phase was separately cultured with the specific cells; the differentiated phases were then joined and co-cultured to form the osteochondral construct, (II) single but heterogeneous scaffolds, where each portion is seeded with the proper cell population (III) single and homogenous scaffolds where the whole scaffold was seeded with stem cells that would be differentiated into both chondrogenic and osteogenic lineage according to the inducing conditions.
Modified from Nooeaid *et al.*[222]

have short half-lives *in vivo*.[154,155] To overcome this problem, platelet-rich plasma functionalized scaffolds or gene-activated scaffolds have been utilized.[157,158]

Not only the osteochondral interface has a complex structure, but its cartilaginous component has three separate structural zones with depth-dependent composition and organization.[159] To engineer such spatial complexity of zonal gradients of articular cartilage, recapitulating the gradients in regulatory signals, required to drive the spatial changes in stem cell differentiation during development and maturation, is essential. This was obtained by modulating the mechanical environment and oxygen tension thorough the depth of stem cells-seeded hydrogels. Combination of a spatial confinement and dynamic compression produced a construct with high glycoaminoglycan accumulation in its bottom half but high collagen accumulation in its top half with no signs of hypertrophy and calcification throughout the construct.[94]

7.4.2 Neuromuscular Prosthesis–Tissue Interface

The electrical properties of the neural and cardiac prosthesis–tissue interface have an impact on the safety, function and hence longevity of the prosthesis. Examples of these prostheses (electrodes) include cochlea implants[160] and artificial pacemakers.[161] These medical devices translate the electrical impulses delivered by an electrode contacting the acoustic nerve or cardiac muscle into perception of sound or regulation of the heart beat respectively. The stability of the neural and cardiac electrode–tissue interface determines the sensitivity of the electrode to small electrical signals and also its ability to transfer the processed signals to the adjacent tissues to perform the required function.[162] In this respect, composite multi-walled carbon nanotube–polyelectrolytes (MWNT–PE) substantially outperformed as electrodes than other state of the art interface materials [*e.g.*, electrochemically deposited iridium oxide (IrOx) and poly(3,4-ethylenedioxythiophene) (PEDOT)].[162] Composites such as polypyrrole–single-walled carbon nanotube (PPy–SWCNT) were also used as a thin film coating on the traditional platinum microelectrodes. This composite coating improved the microelectrode electrochemical interaction (by reducing its impedance while improving the signal-to-noise ratio)[163] and neurite outgrowth pheochromocytoma cells when implanted in the cortex of rats.[164]

Understanding brain tissue responses represents the forefront of neural engineering. Recent trends in advanced biomaterials and neural interfaces[165,166] involve bio-inspired exploration of new prototypes with improved biocompatibility and lifetime while maintaining neuronal viability. Recent advances in technology, *e.g.*, brain activity analysis and computational algorithms as well as the brain–machine interface (BMI)/brain–computer interface (BCI) have gained importance as a strategy for improving impaired neuromuscular systems.[166] Developing long lasting, highly selective, implantable neural interface devices that can record the neural activity from small, targeted groups of neurons for years with high fidelity and reliability has been attempted. These devices are based on composite electrodes consisting of a carbon-fiber core, a thin film coating based on poly(*p*-xylylene) as a dielectric barrier and a recording pad based on poly(thiophene). These devices may be used as brain-controlled prosthetic devices. They were ultrasmall (*i.e.*, one order of magnitude smaller) but more mechanically complaint with the brain tissue than the traditional recording electrodes. They also elicited much reduced chronic reactive tissue reactions and enabled single-neuron recording capability in acute and early chronic experiments in rats.[167]

7.4.3 Composite–Dentin Interface

The long-term success and clinical safety of resin composites dental restorations depends, among other factors, on the integrity and durability of tooth/restoration interface. Compromised interface integrity leads to invasion of fluid and bacteria, postoperative sensitivity, pain, tooth discoloration, caries and finally failure of restorations.

For a durable composite restoration–tooth interface, several adhesive systems, that depend on separate or combined applications of etching, priming and bonding, are evolved.[168] Etching with ∼35% phosphoric acid selectively removes hydroxyapatite crystals, that act as a 'natural shelter' around collagen fibrils,[169,170] from dental tissues (Figure 7.4). Accordingly demineralized collagen fibrils, of dentin in particular,[171,172] collapse upon subsequent air drying. The application of a hydrophilic primer then expands the collapsed fibrils to enable subsequent penetration of the bonding monomer. Polymerization and entanglement of the bonding monomer within the demineralized collagen fibrils network creates a hybrid layer or inter-diffusion zone of resin-reinforced collagen. With some primers (self-etch primers that combine etching and priming steps), a very thin (∼1 μm) acid–base resistant layer adjacent to the hybrid layer was observed. This layer has been considered as a part of the hybrid layer and can prevent secondary caries development.[173] The thickness of the hybrid layer affects the durability of the resin–dentin interface.[45] Insufficient resin infiltration/encapsulation of collagen fibrils makes the hybrid layer vulnerable to enzymatic degradation.[174] The stability of the hybrid layer is therefore dependant mainly on the effective coupling of the adhesive monomer with the infiltrated substrate. In addition to the mechanical retention in the

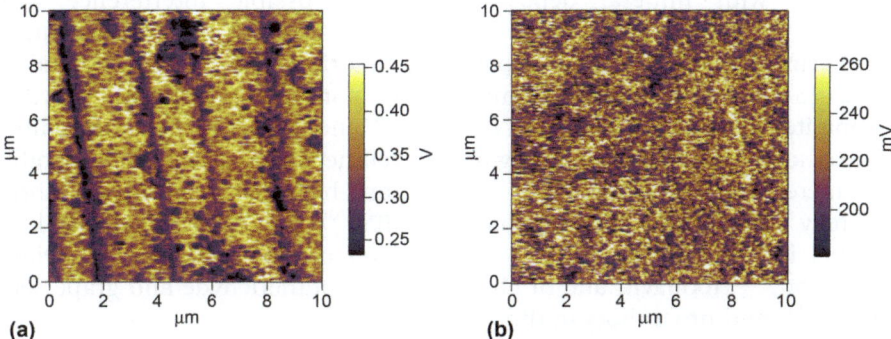

Figure 7.4 Atomic force microscopy images showing dentin before (a) and after (b) acid etching. Acid etching removes the hydroxyapatite crystals from dentin leaving only the collagen network. Collapse of collagen network could happen upon excessive air drying preventing further penetration of adhesive; therefore maintaining wet collagen is essential for proper penetration of adhesive resin into the collagen network to form the hybrid layer of resin-reinforced collagen.

hybrid layer, an ionic interaction between carboxylate or phosphate groups on the bonding monomer and remaining hydroxyapatite crystals can also occur.[169]

The oral environment with associated chemical and mechanical stresses continually challenges the integrity of the tooth-restoration interfaces.[175] For example, the durability of this interface is highly vulnerable to fatigue failure by progressive localized damage under cyclic loading.[176] The durability of this interface could also be jeopardized by the endogenous matrix metallo-proteinases (MMPs), *e.g.*, MMP-2, MMP-8 and MMP-9[177,178] and cysteine cathepsins (B, K and L)[179,180] that attack the demineralized collagen network left after acid-etching. The activation of MMPs could occur upon exposure to acid produced by bacteria in biofilms[181,182] or by acid-etch reagents.[183] Preserving the integrity of the collagen matrix *via* controlling the proteo-lytic enzyme activity is therefore a key issue in optimizing the durability of the tooth/restoration interface.[179] Several attempts include application of MMPs inhibitor as a pre-conditioning step of dentin or by its inclusion into dentin adhesives (etchant or primer or bonding). Among these inhibitors are: benzalkonium chloride,[174] galardin,[184] BB94 and GM6001,[185] chlor-hexidine digluconate[186,187] and ZnO particles.[188] Pre-conditioning of dentin with ethylenediaminetetraacetic acid (EDTA),[189] sodium oxalate,[190] titanium tetrafluoride and caffeic acid phenethyl ester,[191] a phosphoric acid–sodium hypochlorite (NaOCl) mixture[192] can inhibit MMPs activity and accordingly hybrid-layer degradation nano-leakage at the interface without compromising the bond strength of the adhesives.

Protecting the exposed collagen network from the action of MMPs is another strategy used to improve the durability of the resin-bond interface. Incorporation of nanofillers or nanogels could reinforce the hybrid layer and maintain its stability through their protective and/or therapeutic effects associated with mineral deposition and/or possible interference with MMPs.[193] Examples of these fillers are: zirconia (20–50 nm),[194] biomimetic or thermally produced hydroxyapatite nanoparticles (20–70 nm),[195] colloidal silica (5–40 nm) or barium aluminosilicate nanofillers (400 nm),[196] hydro-xyapatite nanorods,[197] reactive nanogel[198] and bioactive calcium/sodium phosphosilicate.[193] Cross-linking of collagen networks enhances its stability by increasing the stiffness[199,200] and nano-hardness of collagen fibrils thereby reducing its degradation by MMPs.[200] Examples of cross-linking agents include: tannic acid,[201] combination of riboflavin and ultraviolet radiation,[199] riboflavin and/or chitosane,[202] glutaraldehyde and grape seed extract[200] and proanthocyanidins.[203]

7.4.3.1   Methodology of Resin–Dentin Interface Examination

The interaction between treated dentin surface and resin can be measured through morphology, chemical changes, bond strength or nanoleakage across the interface. This section deals with different methodological techniques used to study this interface.

The effectiveness of the interface as measured by the interfacial micro-morphology,[204] continuity, quality[205] and thickness of hybrid layer, resin tags length,[206] nano-leakage[207] and micropermeability[208] across the interface has been carried out using various different techniques including scanning elec-tron microscopy,[209] confocal laser scanning microscopy,[193] transmission electron microscopy (TEM),[207] optical coherence tomography,[210] micro-tomography, two-photon laser microscopy,[211] field emission scanning elec-tron microscopy/energy dispersive spectroscopy (FE-SEM/EDS)[212] and focused/broad ion beam (FIB and BIB).[213]

The strength of resin–dentin interface was measured using micro-tensile,[207] shear[214] and micropush-out[215] bond strength test. Micro or nanomechanical characterization, including modulus of elasticity and nano-hardness, across the interface has been detected using atomic force micro-scopy,[216] a customized Triboindenter[217] and finite element analysis.[218] Furthermore, microstrains developed across the interface has been meas-ured with microscopic Moiré interferometry.[219]

The chemical and structural changes of the adhesive dentin interface was studied using X-ray diffraction (XRD), energy-dispersive X-ray spectroscopy (EDS)[169] and micro-Raman spectroscopy.[202] Enzymatic degradation of hy-brid layer was performed by fluorometric assay and zymography.[185,220]

7.5 Step-by-Step Preparation of an Example of Smart Composites

This section describes step-by-step the preparation of a self-healing flexible polymer-modified brushite composite for fibular osteotomy repair.[99–103] The flexibility of the polymeric matrix and its complete degradation enabled complete healing of fractured bone.[99,103] The reactivity of the filler phase (monocalcium phosphate monohydrate and β-tricalcium phosphate), how-ever, is responsible for the bulk and self-healing ability of this smart com-posite. The filler phase has the ability to form brushite within (to correct any internal defect) or on the surface (to provide better cell attachment) of the composite.

The preparation includes the synthesis/purification of the monomeric component that upon polymerization will be the matrix phase of the com-posite. It also includes the formation of the composite by simple mixing of the monomer and fillers.

The preparation steps are given in Sections 7.5.1 to 7.5.3.

7.5.1 Synthesis of Poly(lactide-*co*-propylene glycol-*co*-lactide) co-oligomer

The co-oligomer is prepared by ring opening polymerization reaction of cyclic DL-lactide ($C_6H_8O_4$, Aldrich) using polypropylene glycol (PPG with molecular weight of 425 g mol^{-1}, Aldrich) at a 1 : 4 molar ratio, and stannous octoate [0.05% (w/w) of PPG] as a catalyst. (By mass, polypropylene glycol (PPG) : polylactide (L) : stannous octate is 100 : 138 : 0.05.) The reaction is carried out under vacuum and a nitrogen atmosphere at 150 °C for 6 h. After cooling, the products are dissolved in isopropanol and isolated from insoluble un-reacted lactide by filtration. Isopropanol is removed *via* rotary evaporation and chemical product purity and molecular weight determined by NMR.

7.5.2   Methacrylation of Poly(lactide-*co*-propylene glycol-*co*-lactide) co-oligomer

Reaction products from the first step (Section 7.5.1) are cooled in a reaction flask using an ice bucket. Dichloromethane is added to obtain a solution containing 0.02 mol of PPG–100 mL dichloromethane solvent. Triethylamine (TEA) and methacryloyl chloride (MAC) (1 : 1 molar ratio) are slowly added at 1 : 4 PPG : MAC molar ratio and then left to react at room temperature overnight. After filtration, solvent is removed by evaporation. The resultant product is further purified by repeated dissolution in acetone, filtering and rotary evaporation until a clear, product is obtained. Further purification is achieved by product redisolution in dichloromethane and addition of an equal volume of 1% aqueous HCl. After repeated washing and separation of organic and aqueous phases the organic phase is further washed with first followed by distilled water. After final rotary evaporation of solvent, the product composition can be confirmed using nuclear magnetic resonance (NMR).

7.5.3   Preparation of the Composite

The resultant purified dimethacrylate is combined with 10 wt% hydroxyethyl-methacrylate (HEMA) (Sigma-Aldrich) containing 10 wt% camphorquinone (CQ) initiator or 10 wt% *N,N*-dimethyl-*p*-toluidine (DMPT) activator (Sigma-Aldrich). An equal mass of initiator and activator containing monomers are then combined with equimolar monocalcium phosphate monohydrate (MCPM) and β-tricalcium phosphate (β-TCP) filler at levels up to 80 wt%. The resultant paste can be cured using blue light (*e.g.*, a Dentsply Trubyte Triad® 2000TM light box or dental curing gun). Typically 20–60 s light exposure is required with the dental gun to cure 1 mm thick specimens.

7.6   Advances and Challenges

The future of biomaterials is to deliver composite interfaces that instigate specific cellular functions and direct cell–cell interactions. These smart interfaces should be capable of sensing and accordingly interacting with the dynamic surrounding environment. The smart multi-functional composites represent the future generation platforms for designing three-dimensional scaffolds for various biomedical applications.[76] Recent trends in advanced biomaterials and neural interfaces as an example represent the forefront of neural engineering for improving impaired neuromuscular systems. For instance ultra-small brain-controlled prosthetic devices/chips/electrodes with single- or multiple-neuron recording capabilities can be employed to control the brain activity in depressed or epileptic patients. This chapter provided an update on recent advances in the developments of smart multi-functional composites with the focus on four main areas: biomimetic matrices, on-demand responsive hydrogels, smart fillers, and functionalized composites.

From a clinical and regulatory perspective, the use of smart composites poses the intriguing challenge of coaching *in situ* tissue regeneration.[221] Designing and engineering biomaterials capable of controlling biological processes such as cell fate and tissue formation without losing their synthetic characteristics is another challenge that requires tremendous effort in the near future.[221] Despite the substantial advances in biomedical materials used today, the body still often recognizes them as foreign entities. The magnitude of this response varies from one material to another. This is a further challenge that requires more investigation. Controlling the degradation of composites to enable full tissue repair is another challenge.

References

1. E. Salernitano and C. Migliaresi, Composite materials for biomedical applications: a review, *J. Appl. Biomater. Biomech.*, 2003, **1**(1), 3–18.
2. H. Liu and T. J. Webster, Mechanical properties of dispersed ceramic nanoparticles in polymer composites for orthopedic applications, *Int. J. Nanomed.*, 2010, **5**, 299–313.
3. M. A. Slivka, D. B. Spenciner, H. B. Seim 3rd, W. C. Welch, H. A. Serhan and A. S. Turner, High rate of fusion in sheep cervical spines following anterior interbody surgery with absorbable and nonabsorbable implant devices, *Spine*, 2006, **31**(24), 2772–2777.
4. K. S. Katti, Biomaterials in total joint replacement, *Colloids Surf., B*, 2004, **39**(3), 133–142.
5. K. T. Chu, Y. Oshida, E. B. Hancock, M. J. Kowolik, T. Barco and S. L. Zunt, Hydroxyapatite/PMMA composites as bone cements, *Biomed. Mater. Eng.*, 2004, **14**(1), 87–105.
6. A. Sobczak-Kupiec, D. Malina, M. Piatkowski, K. Krupa-Zuczek, Z. Wzorek and B. Tyliszczak, Physicochemical and biological properties of hydrogel/gelatin/hydroxyapatite PAA/G/HAp/AgNPs composites modified with silver nanoparticles, *J. Nanosci. Nanotechnol.*, 2012, **12**(12), 9302–9311.
7. A. Saikku-Backstrom, J. E. Raiha, T. Valimaa and R. M. Tulamo, Repair of radial fractures in toy breed dogs with self-reinforced biodegradable bone plates, metal screws, and light-weight external coaptation, *Vet. Surg.*, 2005, **34**(1), 11–17.

8. P. Chen, B. Liu, H. Z. Zhang and J. Q. Bu, A modified preauricular-temporal approach for fixing comminuted and redisplaced zygomatic arch fractures with the resorbable bone plate, *Chin. J. Traumatol.*, 2012, **15**(5), 288–290.

9. D. Lin, K. Lian, J. Hong, Z. Ding and W. Zhai, Pediatric physeal slide-traction plate fixation for comminuted distal femur fractures in children, *J. Pediatr. Orthop.*, 2012, **32**(7), 682–686, DOI: 10.1097/BPO0b013e3182694e21.

10. J. Y. Choi, J. W. Kim, C. K. Yoo, P. Y. Yun, S. H. Baek and Y. K. Kim, Evaluation of post-surgical relapse in maxillary surgery using resorbable plate, *J. Craniomaxillofac. Surg.*, 2011, **39**(8), 578–582, DOI: 10.1016/jjcms201012003, Epub 2011 Feb 5.

11. D. Varghai, H. Chim and A. K. Gosain, Mechanical analysis of resorbable plates for long-term soft-tissue molding, *J. Craniofac. Surg.*, 2011, **22**(2), 543–545, DOI: 10.1097/SCS0b013e318208ba49.

12. L. T. Harper, I. Ahmed, R. M. Felfel and C. Qian, Finite element modelling of the flexural performance of resorbable phosphate glass fibre reinforced PLA composite bone plates, *J. Mech. Behav. Biomed. Mater.*, 2012, **15**, 13–23, DOI: 10.1016/jjmbbm201207002, Epub 2012 Jul 11.

13. V. Hasirci, K. U. Lewandrowski, S. P. Bondre, J. D. Gresser, D. J. Trantolo and D. L. Wise, High strength bioresorbable bone plates: preparation, mechanical properties and in vitro analysis, *Biomed. Mater. Eng.*, 2000, **10**(1), 19–29.

14. D. M. Devine, J. Hahn, R. G. Richards, H. Gruner, R. Wieling and S. G. Pearce, Coating of carbon fiber-reinforced polyetheretherketone implants with titanium to improve bone apposition, *J. Biomed. Mater. Res., Part B*, 2013, **101**(4), 591–598.

15. M. Kern and F. Lehmann, Influence of surface conditioning on bonding to polyetheretherketon (PEEK), *Dent. Mater.*, 2012, **28**(12), 1280–1283.

16. B. Stawarczyk, F. Beuer, T. Wimmer, D. Jahn, B. Sener, M. Roos, *et al.*, Polyetheretherketone-A suitable material for fixed dental prostheses?, *J. Biomed. Mater. Res., Part B*, 2013, **6**(10), 32932.

17. E. L. Steinberg, E. Rath, A. Shlaifer, O. Chechik, E. Maman and M. Salai, Carbon fiber reinforced PEEK Optima–a composite material biomechanical properties and wear/debris characteristics of CF-PEEK composites for orthopedic trauma implants, *J. Mech. Behav. Biomed. Mater.*, 2013, **17**, 221–228.

18. B. Wang, Y. Zhu, Y. Jiao, F. Wang, X. Liu, H. Zhu, *et al.*, A New Anterior-Posterior Surgical Approach for the Treatment of Cervical Facet Dislocations, *J. Spinal Disord. Tech.*, 2013, **3**, 3.

19. K. Fujihara, Z. M. Huang, S. Ramakrishna, K. Satknanantham and H. Hamada, Feasibility of knitted carbon/PEEK composites for orthopedic bone plates, *Biomaterials*, 2004, **25**(17), 3877–3885.

20. B. Zimmerli, M. Strub, F. Jeger, O. Stadler and A. Lussi, Composite materials: composition, properties and clinical applications.

A literature review, *Schweiz. Monatsschr. Zahnmed.*, 2010, **120**(11), 972–986.

21. P. S. Stein, J. Sullivan, J. E. Haubenreich and P. B. Osborne, Composite resin in medicine and dentistry, *J. Long-Term Eff. Med. Implants*, 2005, **15**(6), 641–654.

22. E. S. Duke, Packable composites for posterior clinical applications, *Compend. Contin. Educ. Dent.*, 2000, **21**(7), 604–605.

23. Y. Wang, J. J. Lee, I. K. Lloyd, O. C. Wilson Jr., M. Rosenblum and V. Thompson, High modulus nanopowder reinforced dimethacrylate matrix composites for dental cement applications, *J. Biomed. Mater. Res., Part A*, 2007, **82**(3), 651–657.

24. P. Senawongse, A. Srihanon, A. Muangmingsuk and C. Harnirattisai, Effect of dentine smear layer on the performance of self-etching adhesive systems: A micro-tensile bond strength study, *J. Biomed. Mater. Res., Part B*, 2010, **94**(1), 212–221, DOI: 10.1002/jbmb31643.

25. Y. O. Zorba, Y. Z. Bayindir and C. Barutcugil, Direct laminate veneers with resin composites: two case reports with five-year follow-ups, *J. Contemp. Dent. Pract.*, 2010, **11**(4), E056–E062.

26. S. Tahmasbi, F. Heravi and S. M. Moazzami, Fracture characteristics of fibre reinforced composite bars used to provide rigid orthodontic dental segments, *Aust. Orthod. J.*, 2007, **23**(2), 104–108.

27. S. Garoushi and P. Vallittu, Fiber-reinforced composites in fixed partial dentures, *Libyan J. Med.*, 2006, **1**(1), 73–82, DOI: 10.4176/060802.

28. W. Weinmann, C. Thalacker and R. Guggenberger, Siloranes in dental composites, *Dent. Mater.*, 2005, **21**, 68–74.

29. L. F. Schneider, L. M. Cavalcante and N. Silikas, Shrinkage Stresses Generated during Resin-Composite Applications: A Review, *J. Dent. Biomech.*, 2010, 1–14.

30. J. C. Kleverlaan and A. J. Feilzer, Polymerization shrinkage and contraction stress of dental resin composites, *Dent. Mater.*, 2005, **21**, 1150–1157.

31. N. Martin and N. Jedynakiewicz, Measurement of water sorption in dental composites, *Biomaterials*, 1998, **19**(1–3), 77–83.

32. B. Lei, K. H. Shin, D. Y. Noh, I. H. Jo, Y. H. Koh, H. E. Kim, *et al.*, Sol-gel derived nanoscale bioactive glass (NBG) particles reinforced poly (epsilon-caprolactone) composites for bone tissue engineering, *Mater. Sci. Eng. C Mater. Biol. Appl.*, 2013, **33**(3), 1102–1108, DOI: 10.1016/jmsec201211039, Epub 2012 Dec 8.

33. X. Meng, D. A. Stout, L. Sun, R. L. Beingessner, H. Fenniri and T. J. Webster, Novel injectable biomimetic hydrogels with carbon nanofibers and self assembled rosette nanotubes for myocardial applications, *J. Biomed. Mater. Res., Part A*, 2013, **101**(4), 1095–1102, DOI: 10.1002/jbma34400, Epub 2012 Sep 24.

34. N. Hild, P. N. Tawakoli, J. G. Halter, B. Sauer, W. Buchalla, W. J. Stark, *et al.*, pH-dependent antibacterial effects on oral microorganisms through pure PLGA implants and composites with nano-sized bioactive glass, *Acta Biomater.*, 2013, DOI: 10.1016/jactbio201306030.

35. S. Kashanian, F. Harding, Y. Irani, S. Klebe, K. Marshall, A. Loni, *et al.*, Evaluation of mesoporous silicon/polycaprolactone composites as ophthalmic implants, *Acta Biomater.*, 2010, **6**(9), 3566–3572, DOI: 10.1016/jactbio201003031, Epub 2010 Mar 27.

36. R. Okabayashi, M. Nakamura, T. Okabayashi, Y. Tanaka, A. Nagai and K. Yamashita, Efficacy of polarized hydroxyapatite and silk fibroin composite dressing gel on epidermal recovery from full-thickness skin wounds, *J. Biomed. Mater. Res., Part B*, 2009, **90**(2), 641–646, DOI: 10.1002/jbmb31329.

37. A. S. Leppanen, C. Xu, J. Liu, X. Wang, M. Pesonen and S. Willfor, Anionic polysaccharides as templates for the synthesis of conducting polyaniline and as structural matrix for conducting biocomposites, *Macromol. Rapid Commun.*, 2013, **34**(13), 1056–1061, DOI: 10.1002/marc201300275, Epub 2013 May 17.

38. M. C. Costache, A. D. Vaughan, H. Qu, P. Ducheyne and D. I. Devore, Tyrosine-derived polycarbonate-silica xerogel nanocomposites for controlled drug delivery, *Acta Biomater.*, 2013, **9**(5), 6544–6552, DOI: 10.1016/jactbio201301034, Epub 2013 Feb 5.

39. L. A. Kinard, C. Y. Chu, Y. Tabata, F. K. Kasper and A. G. Mikos, Bone Morphogenetic Protein-2 Release from Composite Hydrogels of Oligo(poly(ethylene glycol) fumarate) and Gelatin, *Pharm. Res.*, 2013, **30**(9), 2332–2343.

40. F. Wang, Z. Li, M. Khan, K. Tamama, P. Kuppusamy, W. R. Wagner, *et al.*, Injectable, rapid gelling and highly flexible hydrogel composites as growth factor and cell carriers, *Acta Biomater.*, 2010, **6**(6), 1978–1991, DOI: 10.1016/jactbio200912011, Epub 2009 Dec 23.

41. D. A. Jesson and J. F. Watts, The interface and interphase in polymer matrix composites: Effect on mechanical properties and methods for identification, *Polym. Rev.*, 2012, **52**, 321–354.

42. Md. Abdulla Al Masud and A. K. M. Masud, Effect of interphase characteristic and property on axial modulus of carbon nanotube based composites, *J. Mech. Eng.*, 2010, **ME 41**(1), 15–24.

43. J. Liu, Y. Gao, D. Cao, L. Zhang and Z. Guo, Nanoparticle Dispersion and Aggregation in Polymer Nanocomposites: Insights from Molecular Dynamics Simulation, *Langmuir*, 2013, **27**(12), 7926–7933.

44. Y. X. Gan, Effect of Interface Structure on Mechanical Properties of Advanced Composite Materials, *Int. J. Mol. Sci.*, 2009, **10**(12), 5115–5134.

45. K. L. Van Landuyt, J. De Munck, A. Mine, M. V. Cardoso, M. Peumans and B. Van Meerbeek, Filler debonding & subhybrid-layer failures in self-etch adhesives, *J. Dent. Res.*, 2010, **89**(10), 1045–1050, DOI: 10.1177/0022034510375285, Epub 2010 Jul 14.

46. Y. Zhang and K. E. Tanner, Effect of filler surface morphology on the impact behaviour of hydroxyapatite reinforced high density polyethylene composites, *J. Mater. Sci.: Mater. Med.*, 2008, **19**(2), 761–766, Epub 2007 Jul 10.

47. S. P. Samuel, S. Li, I. Mukherjee, Y. Guo, A. C. Patel, G. Baran, *et al.*, Mechanical properties of experimental dental composites containing a combination of mesoporous and nonporous spherical silica as fillers, *Dent. Mater.*, 2009, **25**(3), 296–301, DOI: 10.1016/jdental200807012, Epub 2008 Sep 19.

48. J. M. Gonzalez-Dominguez, Y. Martinez-Rubi, A. M. Diez-Pascual, A. Anson-Casaos, M. Gomez-Fatou, B. Simard, *et al.*, Reactive fillers based on SWCNTs functionalized with matrix-based moieties for the production of epoxy composites with superior and tunable properties, *Nanotechnology*, 2012, **23**(28), 285702, DOI: 10.1088/0957-4484/23/28/285702, Epub 2012 Jun 21.

49. T. Huang, R. Lu, C. Su, H. Wang, Z. Guo, P. Liu, *et al.*, Chemically modified graphene/polyimide composite films based on utilization of covalent bonding and oriented distribution, *ACS Appl. Mater. Interfaces*, 2012, **4**(5), 2699–2708, DOI: 10.1021/am3003439, Epub 2012 Apr 19.

50. S. Yang, J. Choi and M. Cho, Elastic stiffness and filler size effect of covalently grafted nanosilica polyimide composites: molecular dynamics study, *ACS Appl. Mater. Interfaces*, 2012, **4**(9), 4792–4799, Epub 2012 Sep 7.

51. D. Ciprari, K. Jacob and R. Tannenbaum, Characterization of polymer nanocomposite interphase and Its Impact on mechanical properties, *Macromolecules*, 2006, **39**, 6565–6573.

52. D. F. Cheng and A. Hozumi, Highly Loaded Silicone Nanocomposite Exhibiting Quick Thermoresponsive Optical Behavior, *ACS Appl. Mater. Interfaces*, 2013, **3**(7), 2219–2223.

53. H. Zhang and M. Zhang, Effect of surface treatment of hydroxyapatite whiskers on the mechanical properties of bis-GMA-based composites, *Biomed. Mater.*, 2010, **5**(5), 054106, DOI: 10.1088/1748-6041/5/5/054106, Epub 2010 Sep 28.

54. Y. Yoshida, K. Shirai, Y. Nakayama, M. Itoh, M. Okazaki, H. Shintani, *et al.*, Improved filler-matrix coupling in resin composites, *J. Dent. Res.*, 2002, **81**(4), 270–273.

55. C. M. Vaz, R. L. Reis and A. M. Cunha, Use of coupling agents to enhance the interfacial interactions in starch-EVOH/hydroxylapatite composites, *Biomaterials*, 2002, **23**(2), 629–635.

56. D. Metin, F. Tihminlioglu, D. Balkose and S. Uiki, The effect of interfacial interactions on the mechanical properties of polypropylene/natural zeolite composites, *Composites, Part A*, 2004, **35**(1), 23–32.

57. W. Yuan, W. Li, Y. Mu and M. B. Chan-Park, Effect of side-chain structure of rigid polyimide dispersant on mechanical properties of single-walled carbon nanotube/cyanate ester composite, *ACS Appl. Mater. Interfaces*, 2011, **3**(5), 1702–1712, DOI: 10.1021/am2002229, Epub 2011 May 9.

58. W. Noohom, K. S. Jack, D. Martin and M. Trau, Understanding the roles of nanoparticle dispersion and polymer crystallinity in controlling the mechanical properties of HA/PHBV nanocomposites, *Biomed. Mater.*,

2009, **4**(1), 015003, DOI: 10.1088/1748-6041/25/1/015003, Epub 2008 Nov 4.

59. N. B. Cramer, J. W. Stansbury and C. N. Bowman, Recent advances and developments in composite dental restorative materials, *J. Dent. Res.*, 2011, **90**(4), 402–416.

60. H. Demir and D. S. Balkse, Influence of surface modification of fillers and polymer on flammability and tensile behaviour of polypropylene-composites, *Polym. Degrad. Stab.*, 2006, **91**(5), 1079–1085.

61. A. Jalal Uddin, J. Araki and Y. Gotoh, Toward "strong" green nano-composites: polyvinyl alcohol reinforced with extremely oriented cellulose whiskers, *Biomacromolecules*, 2011, **12**(3), 617–624, DOI: 10.1021/bm101280f, Epub 2011 Feb 4.

62. R. Tannenbaum, M. Zubris, K. David, K. I. Jacob, I. Jasiuk and N. Dan, Characterization of metal-polymer interfaces in nanocomposites and the implications to mechanical properties, *J. Phys. Chem. B*, 2006, **110**(5), 2227–2232.

63. L. Feng, B. I. Suh and A. C. Shortall, Formation of gaps at the filler-resin interface induced by polymerization contraction stress: Gaps at the interface, *Dent. Mater.*, 2010, **26**(8), 719–729.

64. F. K. Moreira, D. C. Pedro, G. M. Glenn, J. M. Marconcini and L. H. Mattoso, Brucite nanoplates reinforced starch bionanocomposites, *Carbohydr. Polym.*, 2013, **92**(2), 1743–1751, DOI: 101016/jcarbpol201211019, Epub 2012 Nov 14.

65. Y. J. Lee, S. J. Park, W. K. Lee, J. S. Ko and H. M. Kim, MG63 osteoblastic cell adhesion to the hydrophobic surface precoated with recombinant osteopontin fragments, *Biomaterials*, 2003, **24**(6), 1059–1066.

66. C. T. Herman, G. K. Potts, M. C. Michael, N. V. Tolan and R. C. Bailey, Probing dynamic cell-substrate interactions using photochemically generated surface-immobilized gradients: application to selectin-mediated leukocyte rolling, *Integr. Biol.*, 2011, **3**(7), 779–791, DOI: 10.1039/c0ib00151a, Epub 2011 May 26.

67. Y. Wang, F. Papadimitrakopoulos and D. J. Burgess, Polymeric "smart" coatings to prevent foreign body response to implantable biosensors, *J. Controlled Release*, 2013, **169**(3), 341–347.

68. S. Krifka, G. Spagnuolo, G. Schmalz and H. Schweikl, A review of adaptive mechanisms in cell responses towards oxidative stress caused by dental resin monomers, *Biomaterials*, 2013, **34**(19), 4555–4563, DOI: 10.1016/jbiomaterials201303019, Epub 2013 Mar 27.

69. A. Gupta, M. D. Woods, K. D. Illingworth, R. Niemeier, I. Schafer, C. Cady, *et al.*, Single walled carbon nanotube composites for bone tissue engineering, *J. Orthop. Res.*, 2013, **31**(9), 1374–1381.

70. S. H. Kim, W. Chegal, J. Doh, H. M. Cho and D. W. Moon, Study of cell-matrix adhesion dynamics using surface plasmon resonance imaging ellipsometry, *Biophys. J.*, 2011, **100**(7), 1819–1828, DOI: 10.1016/jbpj201101033.

71. M. C. Adams, A. Matov, D. Yarar, S. L. Gupton, G. Danuser and C. M. Waterman-Storer, Signal analysis of total internal reflection fluorescent speckle microscopy (TIR-FSM) and wide-field epi-fluorescence FSM of the actin cytoskeleton and focal adhesions in living cells, *J. Microsc.*, 2004, **216**(Pt 2), 138–152.

72. J. Sun, Y. Ding, N. J. Lin, J. Zhou, H. Ro, C. L. Soles, *et al.*, Exploring Cellular Contact Guidance Using Gradient Nanogratings, *Biomacromolecules*, 2010, **11**(11), 3067–3072.

73. A. G. Harvey, E. W. Hill and A. Bayat, Designing implant surface topography for improved biocompatibility, *Expert Rev. Med. Devices*, 2013, **10**(2), 257–267, DOI: 10.1586/erd1282.

74. G. Wang, X. Liu, H. Zreiqat and C. Ding, Enhanced effects of nano-scale topography on the bioactivity and osteoblast behaviors of micron rough ZrO2 coatings, *Colloids Surf., B*, 2011, **86**(2), 267–274, DOI: 10.1016/jcolsurfb201104006, Epub 2011 Apr 17.

75. Z. Miklos, S. Andras, F. Genoveva, F. Jozsef, S. Jozsef and M. Gabor, Smart gel-glass based on the responsive properties of polymer gels, *Polym. Adv. Technol.*, 2001, **12**, 501–505.

76. R. A. Perez, J. E. Won, J. C. Knowles and H. W. Kim, Naturally and synthetic smart composite biomaterials for tissue regeneration, *Adv. Drug Delivery Rev.*, 2013, **65**(4), 471–496, DOI: 10.1016/jaddr201203009, Epub 2012 Mar 20.

77. S. J. Lee and A. Atala, Scaffold technologies for controlling cell behavior in tissue engineering, *Biomed. Mater.*, 2013, **8**(1), 010201, DOI: 10.1088/1748-6041/8/1/010201, Epub 2013 Jan 25.

78. L. Kurniawan, G. G. Qiao and X. Zhang, Formation of wheat-protein-based biomaterials through polymer grafting and crosslinking reactions to introduce new functional properties, *Macromol. Biosci.*, 2009, **9**(1), 93–101.

79. R. Han, A. Zwiefka, C. C. Caswell, Y. Xu, D. R. Keene, E. Lukomska, *et al.*, Assessment of prokaryotic collagen-like sequences derived from streptococcal Scl1 and Scl2 proteins as a source of recombinant GXY polymers, *Appl. Microbiol. Biotechnol.*, 2006, **72**(1), 109–115.

80. D. G. Harkin, K. A. George, P. W. Madden, I. R. Schwab, D. W. Hutmacher and T. V. Chirila, Silk fibroin in ocular tissue reconstruction, *Biomaterials*, 2011, **32**(10), 2445–2458.

81. D. M. Albala and J. H. Lawson, Recent clinical and investigational applications of fibrin sealant in selected surgical specialties, *J. Am. Coll. Surg.*, 2006, **202**, 685–697.

82. P. A. Carless, D. A. Henry and D. M. Anthony, Fibrin sealant use for minimising peri-operative allogeneic blood transfusion, *Cochrane Database Syst. Rev.*, 2003, **2**, CD004171.

83. A. Vogel, K. O'Grady and D. M. Toriumi, Surgical tissue adhesives in facial plastic and reconstructive surgery, *Facial Plast. Surg.*, 1993, **9**(1), 49–57.

84. F. M. Lagoutte, L. Gauthier and P. R. Comte, A fibrin sealant for perforated and preperforated corneal ulcers, *Br. J. Ophthalmol.*, 1989, **73**(9), 757–761.

85. S. Basu, Y. Gerchman, C. H. Collins, F. H. Arnold and R. Weiss, A synthetic multicellular system for programmed pattern formation, *Nature*, 2005, **434**, 1130–1134.

86. J. P. Stegemann, S. N. Kaszuba and S. L. Rowe, Review: advances in vascular tissue engineering using protein-based biomaterials, *Tissue Eng.*, 2007, **13**(11), 2601–2613.

87. F. Cellesi, Thermoresponsive hydrogels for cellular delivery, *Ther. Delivery*, 2012, **3**(12), 1395–1407, DOI: 10.4155/tde12114.

88. S. Garty, N. Kimelman-Bleich, Z. Hayouka, D. Cohn, A. Friedler, G. Pelled, *et al.*, Peptide-modified "smart" hydrogels and genetically engineered stem cells for skeletal tissue engineering, *Biomacromolecules*, 2010, **11**(6), 1516–1526, DOI: 10.1021/bm100157s.

89. A. A. Amini and L. S. Nair, Injectable hydrogels for bone and cartilage repair, *Biomed. Mater.*, 2012, **7**(2), 024105, DOI: 10.1088/1748-6041/7/2/024105, Epub 2012 Mar 29.

90. H. Cui, J. Shao, Y. Wang, P. Zhang, X. Chen and Y. Wei, PLA-PEG-PLA and its Electroactive Tetraaniline Copolymer as Multi-interactive Injectable Hydrogels for Tissue Engineering, *Biomacromolecules*, 2013, **14**(6), 1904–1912.

91. T. D. Johnson and K. L. Christman, Injectable hydrogel therapies and their delivery strategies for treating myocardial infarction, *Expert Opin. Drug Delivery*, 2013, **10**(1), 59–72, DOI: 10.1517/174252472013739156, Epub 2012 Nov 9.

92. C. H. Lee, J. Shin, J. S. Lee, E. Byun, J. H. Ryu, S. H. Um, *et al.*, Bioinspired, Calcium-free Alginate Hydrogels with Tunable Physical and Mechanical Properties and Improved Biocompatibility, *Biomacromolecules*, 2013, **14**(6), 2004–2013.

93. Y. Wang, S. Vaddiraju, L. Qiang, X. Xu, F. Papadimitrakopoulos and D. J. Burgess, Effect of dexamethasone-loaded poly(lactic-co-glycolic acid) microsphere/poly(vinyl alcohol) hydrogel composite coatings on the basic characteristics of implantable glucose sensors, *J. Diabetes Sci. Technol.*, 2012, **6**(6), 1445–1453.

94. S. D. Thorpe, T. Nagel, S. F. Carroll and D. J. Kelly, Modulating Gradients in Regulatory Signals within Mesenchymal Stem Cell Seeded Hydrogels: A Novel Strategy to Engineer Zonal Articular Cartilage, *PLoS One*, 2013, **8**(4), e60764, DOI: 10.1371/journalpone0060764, Print 2013.

95. D. Sivakumaran, D. Maitland, T. Oszustowicz and T. Hoare, Tuning drug release from smart microgel-hydrogel composites via cross-linking, *J. Colloid Interface Sci.*, 2013, **392**, 422–430.

96. B. Garner, A. J. Hodgson, G. G. Wallace and P. A. Underwood, Human endothelial cell attachment to and growth on polypyrrole-heparin is vitronectin dependent, *J. Mater. Sci.: Mater. Med.*, 1999, **10**(1), 19–27.

97. S. R. White, N. R. Sottos, P. H. Geubelle, J. S. Moore, M. R. Kessler, S. R. Sriram, *et al.*, Autonomic healing of polymer composites, *Nature*, 2001, **409**(15), 794–817.

98. J. S. Park, H. S. Kim and H. Thomas Hahn, Healing behavior of a matrix crack on a carbon fiber/mendomer composite, *Compos. Sci. Technol.*, 2009, **69**(7–8), 1082–1087.

99. E. A. Abou Neel, V. Salih, P. A. Revell and A. M. Young, Brushite and self-healing flexible polymer-modified brushite bone adhesives for fibular osteotomy repair, *Adv. Eng. Mater.*, 2014, **16**(2), 218–230.

100. E. A. Abou Neel, G. Palmer, J. C. Knowles, V. Salih and A. M. Young, Chemical, modulus and cell attachment studies of reactive calcium phosphate filler-containing fast photo-curing, surface-degrading, polymeric bone adhesives, *Acta Biomater.*, 2010, **6**(7), 2695–2703.

101. E. A. Abou Neel, V. Salih, P. A. Revell and A. M. Young, Viscoelastic and biological performance of low-modulus, reactive calcium phosphate-filled, degradable, polymeric bone adhesives, *Acta Biomater.*, 2012, **8**(1), 313–320, DOI: 10.1016/jactbio201108008, Epub 2011 Aug 17.

102. A. M. Young, S. Man Ho, E. A. Abou Neel, I. Ahmed, J. E. Barralet, J. C. Knowles, *et al.*, Chemical characterization of a degradable polymeric bone adhesive containing hydrolysable fillers and interpretation of anomalous mechanical properties, *Acta Biomater.*, 2009, **5**(6), 2072–2083, DOI: 10.1016/jactbio200902022, Epub 2009 Feb 20.

103. P. A. Revell, E. A. Abou Neel, V. Salih and M. A. Young, A novel calcium phosphate-containing, fast photo-curing, degradable, polymeric bone adhesive promotes osteotomy healing, *Eur. Conf. Biomater.*, *24th*, 2011, 121–127.

104. J. S. Wi, E. S. Barnard, R. J. Wilson, M. Zhang, M. Tang, M. L. Brongersma, *et al.*, Sombrero-shaped plasmonic nanoparticles with molecular-level sensitivity and multifunctionality, *ACS Nano*, 2011, **5**(8), 6449–6457, DOI: 10.1021/nn201649n, Epub 2011 Jul 12.

105. J. Loomis, P. Xu and B. Panchapakesan, Stimuli-responsive transformation in carbon nanotube/expanding microsphere-polymer composites, *Nanotechnology*, 2013, **24**(18), 185703, DOI: 10.1088/0957-4484/24/18/185703, Epub 2013 Apr 10.

106. Y. de Rancourt, B. Couturaud, A. Mas and J. J. Robin, Synthesis of antibacterial surfaces by plasma grafting of zinc oxide based nanocomposites onto polypropylene, *J. Colloid Interface Sci.*, 2013, **402**, 320–326.

107. L. Cheng, M. D. Weir, P. Limkangwalmongkol, G. D. Hack, H. H. Xu, Q. Chen, *et al.*, Tetracalcium phosphate composite containing quaternary ammonium dimethacrylate with antibacterial properties, *J. Biomed. Mater. Res., Part B*, 2012, **100**(3), 726–734, DOI: 10.1002/jbmb32505, Epub 2011 Dec 21.

108. T. Uysal, M. Amasyali, A. E. Koyuturk, S. Ozcan and D. Sagdic, Amorphous calcium phosphate-containing orthodontic composites. Do they prevent demineralisation around orthodontic brackets?, *Aust. Orthod. J.*, 2010, **26**(1), 10–15.

109. T. Uysal, M. Amasyali, S. Ozcan, A. E. Koyuturk, M. Akyol and D. Sagdic, In vivo effects of amorphous calcium phosphate-containing

orthodontic composite on enamel demineralization around ortho-dontic brackets, *Aust. Dent. J.*, 2010, **55**(3), 285–291, DOI: 10.1111/j1834-7819201001236x.

110. I. Mehdawi, E. A. Neel, S. P. Valappil, G. Palmer, V. Salih, J. Pratten, *et al.*, Development of remineralizing, antibacterial dental materials, *Acta Biomater.*, 2009, **5**(7), 2525–2539, DOI: 10.1016/jactbio200903030, Epub 2009 Mar 31.

111. D. Khang, S. Y. Kim, P. Liu-Snyder, G. T. R. Palmore, S. M. Durbin and T. J. Webster, Enhanced fibronectin adsorption on carbon nanotube/poly(carbonate) urethane: Independent role of surface nano-roughness and associated surface energy, *Biomaterials*, 2007, **28**(32), 4756–4768.

112. M. Tamborra, M. Striccoli, M. L. Curri, J. A. Alducin, D. Mecerreyes, J. A. Pomposo, *et al.*, Nanocrystal-based luminescent composites for nanoimprinting lithography, *Small*, 2007, **3**(5), 822–828.

113. I. M. Szilagyi, G. Teucher, E. Harkonen, E. Farm, T. Hatanpaa, T. Nikitin, *et al.*, Programming nanostructured soft biological surfaces by atomic layer deposition, *Nanotechnology*, 2013, **24**(24), 245701, DOI: 10.1088/0957-4484/24/24/245701, Epub 2013 May 16.

114. T. Amna, M. S. Hassan, K. T. Nam, Y. Y. Bing, N. A. Barakat, M. S. Khil, *et al.*, Preparation, characterization, and cytotoxicity of CPT/Fe(2)O(3)-embedded PLGA ultrafine composite fibers: a synergistic approach to develop promising anticancer material, *Int. J. Nanomed.*, 2012, **7**, 1659–1670, DOI: 10.2147/IJNS24467, Epub 2012 Mar 27.

115. R. Qi, R. Guo, F. Zheng, H. Liu, J. Yu and X. Shi, Controlled release and antibacterial activity of antibiotic-loaded electrospun halloysite/poly (lactic-co-glycolic acid) composite nanofibers, *Colloids Surf., B*, 2013, **110**, 148–155, DOI: 10.1016/jcolsurfb201304036.

116. X. Sun, L. Zhang, Z. Cao, Y. Deng, L. Liu, H. Fong, *et al.*, Electrospun composite nanofiber fabrics containing uniformly dispersed anti-microbial agents as an innovative type of polymeric materials with superior antimicrobial efficacy, *ACS Appl. Mater. Interfaces*, 2010, **2**(4), 952–956, DOI: 10.1021/am100018k.

117. X. Y. Dai, W. Nie, Y. C. Wang, Y. Shen, Y. Li and S. J. Gan, Electrospun emodin polyvinylpyrrolidone blended nanofibrous membrane: a novel medicated biomaterial for drug delivery and accelerated wound heal-ing, *J. Mater. Sci.: Mater. Med.*, 2012, **23**(11), 2709–2716, DOI: 10.1007/s10856-012-4728-x, Epub 2012 Aug 9.

118. W. Ji, F. Yang, H. Seyednejad, Z. Chen, W. E. Hennink, J. M. Anderson, *et al.*, Biocompatibility and degradation characteristics of PLGA-based electrospun nanofibrous scaffolds with nanoapatite incorporation, *Biomaterials*, 2012, **33**(28), 6604–6614, DOI: 10.1016/jbiomater-ials201206018, Epub 2012 Jul 5.

119. S. Wang, Y. Zhao, M. Shen and X. Shi, Electrospun hybrid nanofibers doped with nanoparticles or nanotubes for biomedical applications, *Ther. Delivery*, 2012, **3**(10), 1155–1169.

120. Z. X. Meng, X. X. Xu, W. Zheng, H. M. Zhou, L. Li, Y. F. Zheng, *et al.*, Preparation and characterization of electrospun PLGA/gelatin nano-fibers as a potential drug delivery system, *Colloids Surf., B*, 2011, **84**(1), 97–102, DOI: 10.1016/jcolsurfb201012022, Epub 2010 Dec 21.

121. R. Sridhar, S. Sundarrajan, J. R. Venugopal, R. Ravichandran and S. Ramakrishna, Electrospun inorganic and polymer composite nano-fibers for biomedical applications, *J. Biomater. Sci., Polym. Ed.*, 2013, **24**(4), 365–385, DOI: 10.1080/092050632012690711, Epub 2012 Aug 13.

122. J. B. Lee, H. N. Park, W. K. Ko, M. S. Bae, D. N. Heo, D. H. Yang, *et al.*, Poly(ʟ-lactic acid)/hydroxyapatite nanocylinders as nanofibrous structure for bone tissue engineering scaffolds, *J. Biomed. Nanotechnol.*, 2013, **9**(3), 424–429.

123. Y. Kishimoto, F. Ito, H. Usami, E. Togawa, M. Tsukada, H. Morikawa, *et al.*, Nanocomposite of silk fibroin nanofiber and montmorillonite: fabrication and morphology, *Int. J. Biol. Macromol.*, 2013, **57**, 124–128, DOI: 10.1016/jijbiomac201303016, Epub 2013 Mar 13.

124. M. E. Frohbergh, A. Katsman, G. P. Botta, P. Lazarovici, C. L. Schauer, U. G. K. Wegst, *et al.*, Electrospun hydroxyapatite-containing chitosan nanofibers crosslinked with genipin for bone tissue engineering, *Biomaterials*, 2012, **33**(36), 9167–9178.

125. T. Gong, W. Li, H. Chen, L. Wang, S. Shao and S. Zhou, Remotely actuated shape memory effect of electrospun composite nanofibers, *Acta Biomater.*, 2012, **8**(3), 1248–1259.

126. F. Rosso, G. Marino, A. Giordano, M. Barbarisi, D. Parmeggiani and A. Barbarisi, Smart materials as scaffolds for tissue engineering, *J. Cell. Physiol.*, 2005, **203**, 465–470.

127. J. D. Humphries, A. Byron and M. J. Humphries, Integrin ligands at a glance, *J. Cell Sci.*, 2006, **119**, 3901–3903.

128. C. Gabler, D. A. Chapman and G. J. Killian, Expression and presence of osteopontin and integrins in the bovine oviduct during the oestrous cycle, *Reproduction*, 2003, **126**(6), 721–729.

129. D. D. Hu, E. C. K. Lin, N. L. Kovachi, J. R. Hoyer and J. W. Smith, A Biochemical Characterization of the Binding of Osteopontin to Integrins avb1 and avb5, *J. Biol. Chem.*, 1995, **270**(44), 26232–26238.

130. K. J. Bayless, G. A. Meininger, J. M. Scholtz and G. E. Davis, Osteopontin is a ligand for the a4b1 integrin, *J. Cell Sci.*, 1998, **111**, 1165–1174.

131. B. E. Jensen, L. Hosta-Rigau, P. R. Spycher, E. Reimhult, B. Stadler and A. N. Zelikin, Lipogels: surface-adherent composite hydrogels assembled from poly(vinyl alcohol) and liposomes, *Nanoscale*, 2013, **5**(15), 6758–6766.

132. S. Suri and D. A. Walsh, Osteochondral alterations in osteoarthritis, *Bone*, 2012, **51**(2), 204–211.

133. M. D. Van Manen, J. Nace and M. A. Mont, Management of primary knee osteoarthritis and indications for total knee arthroplasty for general practitioners, *J. Am. Osteopath. Assoc.*, 2012, **112**(11), 709–715.

134. M. Rampichova, M. Buzgo, B. Krizkova, E. Prosecka, M. Pouzar and L. Strajtova, Injectable hydrogel functionalised with thrombocyte-rich solution and microparticles for accelerated cartilage regeneration, *Acta Chir. Orthop. Traumatol. Cech.*, 2013, **80**(1), 82–88.

135. W. Swieszkowski, B. H. S. Tuan, K. J. Kurzydlowski and D. W. Hutmacher, Repair and regeneration of osteochondral defects in the articular joints, *Biomol. Eng.*, 2007, **24**(5), 489–495.

136. J. Chen, H. Chen, P. Li, H. Diao, S. Zhu, L. Dong, *et al.*, Simultaneous regeneration of articular cartilage and subchondral bone in vivo using MSCs induced by a spatially controlled gene delivery system in bilayered integrated scaffolds, *Biomaterials*, 2011, **32**(21), 4793–4805, DOI: 10.1016/jbiomaterials201103041, Epub 2011 Apr 13.

137. G. I. Im, J. H. Ahn, S. Y. Kim, B. S. Choi and S. W. Lee, A hyaluronate-atelocollagen/beta-tricalcium phosphate-hydroxyapatite biphasic scaffold for the repair of osteochondral defects: a porcine study, *Tissue Eng., Part A*, 2010, **16**(4), 1189–1200, DOI: 10.1089/tenTEA20090540.

138. M. Liu, Z. Xiang, F. Pei, F. Huang, S. Cen, G. Zhong, *et al.*, Repairing defects of rabbit articular cartilage and subchondral bone with biphasic scaffold combined bone marrow stromal stem cells, *Zhongguo Xiu Fu Chong Jian Wai Ke Za Zhi*, 2010, **24**(1), 87–93.

139. J. H. Ahn, T. H. Lee, J. S. Oh, S. Y. Kim, H. J. Kim, I. K. Park, *et al.*, Novel hyaluronate-atelocollagen/beta-TCP-hydroxyapatite biphasic scaffold for the repair of osteochondral defects in rabbits, *Tissue Eng., Part A*, 2009, **15**(9), 2595–2604, DOI: 10.1089/tenTEA20080511.

140. H. Zhang, W. Wang, D. Chu, Y. Liu and J. Guan, Preliminary study on chitosan/HAP bilayered scaffold, *Zhongguo Xiu Fu Chong Jian Wai Ke Za Zhi*, 2008, **22**(11), 1358–1363.

141. X. Z. Zhou, V. Y. Leung, Q. R. Dong, K. M. Cheung, D. Chan and W. W. Lu, Mesenchymal stem cell-based repair of articular cartilage with polyglycolic acid-hydroxyapatite biphasic scaffold, *Int. J. Artif. Organs*, 2008, **31**(6), 480–489.

142. K. Chen, P. Shi, T. K. Teh, S. L. Toh and J. C. Goh, In vitro generation of a multilayered osteochondral construct with an osteochondral interface using rabbit bone marrow stromal cells and a silk peptide-based scaffold, *J. Tissue Eng. Regener. Med.*, 2013, DOI: 10.1002/term1708.

143. E. Kon, M. Delcogliano, G. Filardo, M. Busacca, A. Di Martino and M. Marcacci, Novel nano-composite multilayered biomaterial for osteochondral regeneration: a pilot clinical trial, *Am. J. Sports Med.*, 2011, **39**(6), 1180–1190, DOI: 10.1177/0363546510392711, Epub 2011 Feb 10.

144. E. Kon, M. Delcogliano, G. Filardo, M. Fini, G. Giavaresi, S. Francioli, *et al.*, Orderly osteochondral regeneration in a sheep model using a novel nano-composite multilayered biomaterial, *J. Orthop. Res.*, 2010, **28**(1), 116–124, DOI: 10.1002/jor20958.

145. E. Kon, M. Delcogliano, G. Filardo, D. Pressato, M. Busacca, B. Grigolo, *et al.*, A novel nano-composite multi-layered biomaterial for treatment

of osteochondral lesions: technique note and an early stability pilot clinical trial, *Injury*, 2010, **41**(7), 693–701, DOI: 10.1016/jinjury200911014, Epub 2009 Dec 24.

146. A. Heymer, G. Bradica, J. Eulert and U. Noth, Multiphasic collagen fibre-PLA composites seeded with human mesenchymal stem cells for osteochondral defect repair: an in vitro study, *J. Tissue Eng. Regener. Med.*, 2009, **3**(5), 389–397, DOI: 10.1002/term175.

147. D. Schumann, A. K. Ekaputra, C. X. Lam and D. W. Hutmacher, Biomaterials/scaffolds. Design of bioactive, multiphasic PCL/collagen type I and type II-PCL-TCP/collagen composite scaffolds for functional tissue engineering of osteochondral repair tissue by using electrospinning and FDM techniques, *Methods Mol. Med.*, 2007, **140**, 101–124.

148. N. Mohan, N. H. Dormer, K. L. Caldwell, V. H. Key, C. J. Berkland and M. S. Detamore, Continuous gradients of material composition and growth factors for effective regeneration of the osteochondral interface, *Tissue Eng., Part A*, 2011, **17**(21–22), 2845–2855, DOI: 10.1089/tentea20110135, Epub 2011 Aug 4.

149. C. Erisken, D. M. Kalyon and H. Wang, Viscoelastic and biomechanical properties of osteochondral tissue constructs generated from graded polycaprolactone and beta-tricalcium phosphate composites, *J. Biomech. Eng.*, 2010, **132**(9), 091013, DOI: 10.1115/14001884.

150. P. Duan, Z. Pan, L. Cao, Y. He, H. Wang, Z. Qu, *et al.*, The effects of pore size in bilayered poly(lactide-co-glycolide) scaffolds on restoring osteochondral defects in rabbits, *J. Biomed. Mater. Res., Part A*, 2014, **102**(1), 180–192.

151. M. T. Rodrigues, M. E. Gomes and R. L. Reis, Current strategies for osteochondral regeneration: from stem cells to pre-clinical approaches, *Curr. Opin. Biotechnol.*, 2011, **22**(5), 726–733, DOI: 10.1016/jcopbio201104006, Epub 2011 May 6.

152. K. Chen, K. S. Ng, S. Ravi, J. C. Goh and S. L. Toh, In vitro generation of whole osteochondral constructs using rabbit bone marrow stromal cells, employing a two-chambered co-culture well design, *J. Tissue Eng. Regener. Med.*, 2013, DOI: 10.1002/term1716.

153. J. Xie, Z. Han, M. Naito, A. Maeyama, S. H. Kim, Y. H. Kim, *et al.*, Articular cartilage tissue engineering based on a mechano-active scaffold made of poly(L-lactide-co-epsilon-caprolactone): In vivo performance in adult rabbits, *J. Biomed. Mater. Res., Part B*, 2010, **94**(1), 80–88, DOI: 10.1002/jbmb31627.

154. X. Guo, H. Park, G. Liu, W. Liu, Y. Cao, Y. Tabata, *et al.*, In vitro generation of an osteochondral construct using injectable hydrogel composites encapsulating rabbit marrow mesenchymal stem cells, *Biomaterials*, 2009, **30**(14), 2741–2752, DOI: 10.1016/jbiomaterials200901048, Epub 2009 Feb 20.

155. X. Wang, E. Wenk, X. Zhang, L. Meinel, G. Vunjak-Novakovic and D. L. Kaplan, Growth factor gradients via microsphere delivery in biopolymer scaffolds for osteochondral tissue engineering, *J. Controlled*

Release, 2009, **134**(2), 81–90, DOI: 10.1016/jjconrel200810021, Epub 2008 Nov 17.

156. K. Kim, J. Lam, S. Lu, P. P. Spicer, A. Lueckgen, Y. Tabata, *et al.*, Osteochondral tissue regeneration using a bilayered composite hydrogel with modulating dual growth factor release kinetics in a rabbit model, *J. Controlled Release*, 2013, **168**(2), 166–178.

157. H. R. Lee, K. M. Park, Y. K. Joung, K. D. Park and S. H. Do, Platelet-rich plasma loaded hydrogel scaffold enhances chondrogenic differentiation and maturation with up-regulation of CB1 and CB2, *J. Controlled Release*, 2012, **159**(3), 332–337, DOI: 10.1016/jjconrel201202008, Epub 2012 Feb 15.

158. A. Marmotti, M. Bruzzone, D. E. Bonasia, F. Castoldi, R. Rossi, L. Piras, *et al.*, One-step osteochondral repair with cartilage fragments in a composite scaffold, *Knee Surg. Sports Traumatol. Arthrosc.*, 2012, **20**(12), 2590–2601, DOI: 10.1007/s00167-012-1920-y, Epub 2012 Feb 21.

159. C.-B. James and T. L. Uh, A review of articular cartilage pathology and the use of glucosamine sulfate, *J. Athl. Train.*, 2001, **36**(4), 413–419.

160. G. M. Clark, The multiple-channel cochlear implant: the interface between sound and the central nervous system for hearing, speech, and language in deaf people-a personal perspective, *Philos. Trans. R. Soc. London, Ser. B*, 2006, **361**(1469), 791–810.

161. M. Schaldach, New aspects in electrostimulation of the heart, *Med. Prog. Technol.*, 1995, **21**(1), 1–16.

162. E. Jan, J. L. Hendricks, V. Husaini, S. M. Richardson-Burns, A. Sereno, D. C. Martin, *et al.*, Layered carbon nanotube-polyelectrolyte electrodes outperform traditional neural interface materials, *Nano Lett.*, 2009, **9**(12), 4012–4018, DOI: 10.1021/nl902187z.

163. G. Baranauskas, E. Maggiolini, E. Castagnola, A. Ansaldo, A. Mazzoni, G. N. Angotzi, *et al.*, Carbon nanotube composite coating of neural microelectrodes preferentially improves the multiunit signal-to-noise ratio, *J. Neural Eng.*, 2011, **8**(6), 066013, DOI: 10.1088/1741-2560/8/6/066013, Epub 2011 Nov 8.

164. Y. Lu, T. Li, X. Zhao, M. Li, Y. Cao, H. Yang, *et al.*, Electrodeposited polypyrrole/carbon nanotubes composite films electrodes for neural interfaces, *Biomaterials*, 2010, **31**(19), 5169–5181, DOI: 10.1016/jbiomaterials201003022, Epub 2010 Apr 10.

165. F. Zhang, M. Aghagolzadeh and K. Oweiss, A Fully Implantable, Programmable and Multimodal Neuroprocessor for Wireless, Cortically Controlled Brain-Machine Interface Applications, *J. Signal. Process. Syst.*, 2012, **69**(3), 351–361, Epub 2011 Jun 15.

166. J. Ushiba, Brain-machine interface–current status and future prospects, *Brain Nerve*, 2010, **62**(2), 101–111.

167. T. D. Kozai, N. B. Langhals, P. R. Patel, X. Deng, H. Zhang, K. L. Smith, *et al.*, Ultrasmall implantable composite microelectrodes with bioactive surfaces for chronic neural interfaces, *Nat. Mater.*, 2012, **11**(12), 1065–1073, DOI: 10.1038/nmat3468, Epub 2012 Nov 11.

168. F. Ozer and M. B. Blatz, Self-etch and etch-and-rinse adhesive systems in clinical dentistry, *Compend. Contin. Educ. Dent.*, 2013, **34**(1), 12–14, 16, 18; quiz 20, 30.

169. Y. Yoshida, K. Yoshihara, N. Nagaoka, S. Hayakawa, Y. Torii, T. Ogawa, *et al.*, Self-assembled Nano-layering at the Adhesive interface, *J. Dent. Res.*, 2012, **91**(4), 376–381, DOI: 10.1177/0022034512437375, Epub 2012 Feb 1.

170. L. Solhi, M. Atai, A. Nodehi, M. Imani, A. Ghaemi and K. Khosravi, Poly(acrylic acid) grafted montmorillonite as novel fillers for dental adhesives: Synthesis, characterization and properties of the adhesive, *Dent. Mater.*, 2012, **28**(4), 369–377.

171. G. Leloup, W. D'Hoore, D. Bouter, M. Degrange and J. Vreven, Meta-analytical review of factors involved in dentin adherence, *J. Dent. Res.*, 2001, **80**, 1605–1614.

172. A. Poitevin, J. De Munck, K. Van Landuyt, E. Coutinho, M. Peumans, P. Lambrechts, *et al.*, Critical analysis of the influence of different parameters on the microtensile bond strength of adhesives to dentin, *J. Adhes. Dent.*, 2008, **10**, 7–16.

173. G. Inoue, T. Nikaido, A. Sadr and J. Tagami, Morphological categorization of acid-base resistant zones with self-etching primer adhesive systems, *Dent. Mater. J.*, 2012, **31**(2), 232–238, Epub 2012 Mar 23.

174. D. H. Pashley, F. R. Tay, L. Breschi, L. Tjaderhane, R. M. Carvalho, M. Carrilho, *et al.*, State of the art etch-and-rinse adhesives, *Dent. Mater.*, 2011, **27**(1), 1–16, DOI: 10.1016/jdental201010016, Epub 2010 Nov 27.

175. J. L. Drummond, Degradation, fatigue, and failure of resin dental composite materials, *J. Dent. Res.*, 2008, **87**(8), 710–719.

176. R. Belli, L. N. Baratieri, M. Braemb, A. Petschelt and U. Lohbauer, Tensile and bending fatigue of the adhesive interface to dentin, *Dent. Mater.*, 2010, **26**, 1157–1165.

177. L. N. Niu, L. Zhang, K. Jiao, F. Li, Y. X. Ding, D. Y. Wang, *et al.*, Localization of MMP-2, MMP-9, TIMP-1, and TIMP-2 in human coronal dentine, *J. Dent.*, 2011, **39**(8), 536–542, DOI: 10.1016/jjdent201105004, Epub 2011 May 27.

178. Y. Shono, M. Terashita, J. Shimada, Y. Kozono, R. M. Carvalho, C. M. Russell, *et al.*, Durability of resin-dentin bonds, *J. Adhes. Dent.*, 1999, **1**(3), 211–218.

179. L. Tjaderhane, F. D. Nascimento, L. Breschi, A. Mazzoni, I. L. S. Tersariol, S. Geraldeli, *et al.*, Optimizing dentin bond durability: Control of collagen degradation by matrix metalloproteinases and cysteine cathepsins, *Dent. Mater.*, 2013, **29**(1), 116–135.

180. P. M. Scaffa, C. M. Vidal, N. Barros, T. F. Gesteira, A. K. Carmona, L. Breschi, *et al.*, Chlorhexidine inhibits the activity of dental cysteine cathepsins, *J. Dent. Res.*, 2012, **91**(4), 420–425, DOI: 10.1177/0022034511435329, Epub 2012 Jan 19.

181. D. H. Pashley, F. R. Tay and S. Imazato, How to increase the durability of resin-dentin bonds, *Compend. Contin. Educ. Dent.*, 2011, **32**(7), 60–64, 66.

182. M. M. Mutluay, K. Zhang, H. Ryou, M. Yahyazadehfar, H. Majd, H. H. Xu, *et al.*, On the fatigue behavior of resin-dentin bonds after degradation by biofilm, *J. Mech. Behav. Biomed. Mater.*, 2013, **18**, 219–231, DOI: 10.1016/jjmbbm201210019, Epub 2012 Nov 17.

183. N. Lehmann, R. Debret, A. Romeas, H. Magloire, M. Degrange, F. Bleicher, *et al.*, Self-etching increases matrix metalloproteinase expression in the dentin-pulp complex, *J. Dent. Res.*, 2009, **88**(1), 77–82, DOI: 10.1177/0022034508327925.

184. L. Breschi, P. Martin, A. Mazzoni, F. Nato, M. Carrilho, L. Tjäderhane, *et al.*, Use of a specific MMP-inhibitor (galardin) for preservation of hybrid layer, *Dent. Mater.*, 2010, **26**(6), 571–578.

185. A. Almahdy, G. Koller, S. Sauro, J. W. Bartsch, M. Sherriff, T. F. Watson, *et al.*, Effects of MMP inhibitors incorporated within dental adhesives, *J. Dent. Res.*, 2012, **91**(6), 605–611, DOI: 10.1177/0022034512446339, Epub 2012 Apr 18.

186. C. Sabatini, Effect of a Chlorhexidine-containing Adhesive on Dentin Bond Strength Stability, *Oper. Dent.*, 2013, **38**(6), 609–617.

187. R. Osorio, M. Yamauti, E. Osorio, M. E. Ruiz-Requena, D. Pashley, F. Tay, *et al.*, Effect of dentin etching and chlorhexidine application on metalloproteinase-mediated collagen degradation, *Eur. J. Oral Sci.*, 2011, **119**(1), 79–85, DOI: 10.1111/j1600-0722201000789x.

188. M. Toledano, M. Yamauti, M. E. Ruiz-Requena and R. Osorio, A ZnO-doped adhesive reduced collagen degradation favouring dentine remineralization, *J. Dent.*, 2012, **40**(9), 756–765, DOI: 10.1016/jjdent201205007, Epub 2012 Jun 1.

189. J. M. Thompson, K. Agee, S. J. Sidow, K. McNally, K. Lindsey, J. Borke, *et al.*, Inhibition of endogenous dentin matrix metalloproteinases by ethylenediaminetetraacetic acid, *J. Endod.*, 2012, **38**(1), 62–65, DOI: 10.1016/jjoen201109005, Epub 2011 Nov 3.

190. I. C. De Moraes Porto, A. K. De Andrade, L. C. Alves and R. Braz, Effect of dentin pretreatment with potassium oxalate: analysis of microtensile bond strengths and morphologic aspects, *Microsc. Res. Tech.*, 2012, **75**(2), 239–244, DOI: 10.1002/jemt21048, Epub 2011 Aug 1.

191. M. Dundar, M. Ozcan, M. E. Comlekoglu and B. H. Sen, Nanoleakage inhibition within hybrid layer using new protective chemicals and their effect on adhesion, *J. Dent. Res.*, 2011, **90**(1), 93–98, DOI: 10.1177/0022034510382547, Epub 2010 Oct 12.

192. S. Sauro, F. Mannocci, M. Toledano, R. Osorio, D. H. Pashley and T. F. Watson, EDTA or H3PO4/NaOCl dentine treatments may increase hybrid layers' resistance to degradation: a microtensile bond strength and confocal-micropermeability study, *J. Dent.*, 2009, **37**(4), 279–288, DOI: 10.1016/jjdent200812002, Epub 2009 Jan 19.

193. A. C. Profeta, F. Mannocci, R. M. Foxton, I. Thompson, T. F. Watson and S. Sauro, Bioactive effects of a calcium/sodium phosphosilicate on the resin-dentine interface: a microtensile bond strength, scanning electron microscopy, and confocal microscopy study, *Eur. J.*

Oral Sci., 2012, **120**(4), 353–36210.1111/j1600-0722201200974x, Epub 2012 Jul 3.

194. U. Lohbauer, A. Wagner, R. Belli, C. Stoetzel, A. Hilpert, H. D. Kurland, *et al.*, Zirconia nanoparticles prepared by laser vaporization as fillers for dental adhesives, *Acta Biomater.*, 2010, **6**(12), 4539–4546, DOI: 10.1016/jactbio201007002, Epub 2010 Aug 3.

195. A. Wagner, R. Belli, C. Stotzel, A. Hilpert, F. A. Muller and U. Lohbauer, Biomimetically- and Hydrothermally-grown HAp Nanoparticles as Reinforcing Fillers for Dental Adhesives, *J. Adhes. Dent.*, 2013, **15**(5), 413–422.

196. V. Di Hipolito, A. F. Reis, S. B. Mitra and M. F. de Goes, Interaction morphology and bond strength of nanofilled simplified-step adhesives to acid etched dentin, *Eur. J. Dent.*, 2012, **6**(4), 349–360.

197. M. Sadat-Shojai, M. Atai, A. Nodehi and L. N. Khanlar, Hydroxyapatite nanorods as novel fillers for improving the properties of dental adhesives: Synthesis and application, *Dent. Mater.*, 2010, **26**(5), 471–482.

198. R. R. Moraes, J. W. Garcia, N. D. Wilson, S. H. Lewis, M. D. Barros, B. Yang, *et al.*, Improved dental adhesive formulations based on reactive nanogel additives, *J. Dent. Res.*, 2012, **91**(2), 179–184, DOI: 10.1177/0022034511426573, Epub 2011 Oct 21.

199. Y.-S. Chiang, Y.-L. Chen, S.-F. Chuang, C.-M. Wu, P.-J. Wei, C.-F. Han, *et al.*, Riboflavin-ultraviolet-A-induced collagen cross-linking treatments in improving dentin bonding, *Dent. Mater.*, 2013, **29**(6), 682–692.

200. P. H. dos Santos, S. Karol and A. K. Bedran-Russo, Nanomechanical properties of biochemically modified dentin bonded interfaces, *J. Oral Rehabil.*, 2011, **38**(7), 541–546, DOI: 10.1111/j1365-2842201002175x, Epub 2010 Nov 9.

201. A. K. Bedran-Russo, K. J. Yoo, K. C. Ema and D. H. Pashley, Mechanical properties of tannic-acid-treated dentin matrix, *J. Dent. Res.*, 2009, **88**(9), 807–811, DOI: 10.1177/0022034509342556.

202. U. Daood, K. Iqbal, L. I. Nitisusanta and A. S. Fawzy, Effect of chitosan/riboflavin modification on resin/dentin interface: Spectroscopic and microscopic investigations, *J. Biomed. Mater. Res., Part A*, 2013, **101**(7), 1846–1856.

203. R. R. Liu, M. Fang, S. J. Zhao, F. Li, L. J. Shen and J. H. Chen, The potential effect of proanthocyanidins on the stability of resin-dentin bonds against thermal cycling, *Zhonghua Kou Qiang Yi Xue Za Zhi*, 2012, **47**(5), 268–272.

204. M. O. Barceleiro, J. B. de Mello, C. L. Porto, K. R. Dias and M. S. de Miranda, Hybrid layer thickness and morphology: Influence of cavity preparation with air abrasion, *Gen. Dent.*, 2011, **59**(6), e242–e247.

205. T. D. Azevedo, A. C. Bezerra, J. Faber and O. A. de Toledo, Evaluation of chlorhexidine on the quality of the hybrid layer in noncarious primary teeth: an in vitro study, *J. Dent. Child.*, 2010, **77**(1), 25–31.

206. R. H. Sundfeld, T. A. Valentino, R. S. de Alexandre, A. L. Briso and M. L. Sundefeld, Hybrid layer thickness and resin tag length of a self-etching adhesive bonded to sound dentin, *J. Dent.*, 2005, **33**(8), 675–681, Epub 2005 Apr 1.

207. M. T. de Oliveira, A. F. Reis, C. A. Arrais, A. N. Cavalcanti, A. C. Aranha, C. de Paula Eduardo, *et al.*, Analysis of the interfacial micromorphology and bond strength of adhesive systems to Er:YAG laser-irradiated dentin, *Lasers Med. Sci.*, 2013, **28**(4), 1069–1076.

208. D. P. Chen, D. D. Pei, Y. K. Wang, C. Huang, Adl ti and S. Y. Liu, Study on the micropermeability of resin-dentin bonding interfaces with ethanol-wet bonding technique, *Zhonghua Kou Qiang Yi Xue Za Zhi*, 2011, **46**(12), 755–758.

209. L. R. Abou-Id, L. F. Morgan, G. A. Silva, L. T. Poletto, L. D. Lanza and C. Albuquerque Rde, Ultrastructural evaluation of the hybrid layer after cementation of fiber posts using adhesive systems with different curing modes, *Braz. Dent. J.*, 2012, **23**(2), 116–121.

210. A. K. Braz, R. E. de Araujo, T. Y. Ohulchanskyy, S. Shukla, E. J. Bergey, A. S. Gomes, *et al.*, In situ gold nanoparticles formation: contrast agent for dental optical coherence tomography, *J. Biomed. Opt.*, 2012, **17**(6), 066003, DOI: 10.1117/1JBO176066003.

211. P. H. D'Alpino, J. C. Pereira, N. R. Svizero, F. A. Rueggeberg, R. M. Carvalho and D. H. Pashley, A new technique for assessing hybrid layer interfacial micromorphology and integrity: two-photon laser microscopy, *J. Adhes. Dent.*, 2006, **8**(5), 279–284.

212. Y. Yuan, Y. Shimada, S. Ichinose and J. Tagami, Qualitative analysis of adhesive interface nanoleakage using FE-SEM/EDS, *Dent. Mater.*, 2007, **23**(5), 561–569, Epub 2006 Jun 12..

213. E. Coutinho, T. Jarmar, F. Svahn, A. A. Neves, B. Verlinden, B. Van Meerbeek, *et al.*, Ultrastructural characterization of tooth-biomaterial interfaces prepared with broad and focused ion beams, *Dent. Mater.*, 2009, **25**(11), 1325–1337, DOI: 10.1016/jdental200906002, Epub 2009 Jul 10..

214. R. Freeman, S. Varanasi, I. A. Meyers and A. L. Symons, Effect of air abrasion and thermocycling on resin adaptation and shear bond strength to dentin for an etch-and-rinse and self-etch resin adhesive, *Dent. Mater. J.*, 2012, **31**(2), 180–188, Epub 2012 Mar 23.

215. K. Bitter, S. Paris, C. Pfuertner, K. Neumann and A. M. Kielbassa, Morphological and bond strength evaluation of different resin cements to root dentin, *Eur. J. Oral Sci.*, 2009, **117**(3), 326–333, DOI: 10.1111/j1600-0722200900623x.

216. S. Sauro, R. Osorio, T. F. Watson and M. Toledano, Assessment of the quality of resin-dentin bonded interfaces: an AFM nano-indentation, muTBS and confocal ultramorphology study, *Dent. Mater.*, 2012, **28**(6), 622–631, DOI: 10.1016/jdental201202005, Epub 2012 Mar 17.

217. P. H. Dos Santos, S. Karol and A. K. Bedran-Russo, Long-term nanomechanical properties of biomodified dentin-resin interface components,

J. Biomech., 2011, **44**(9), 1691–1694, DOI: 10.1016/jjbiomech201103030, Epub 2011 Apr 30.

218. A. Misra, P. Spencer, O. Marangos, Y. Wang and J. L. Katz, Micromechanical analysis of dentin/adhesive interface by the finite element method, *J. Biomed. Mater. Res., Part B*, 2004, **70**(1), 56–65.

219. J. D. Wood, P. Sobolewski, V. Thakur, D. Arola, A. Nazari, F. R. Tay, *et al.*, Measurement of microstrains across loaded resin-dentin interfaces using microscopic moire interferometry, *Dent. Mater.*, 2008, **24**(7), 859–866, Epub 2007 Nov 28.

220. A. Mazzoni, F. D. Nascimento, M. Carrilho, I. Tersariol, V. Papa, L. Tjaderhane, *et al.*, MMP activity in the hybrid layer detected with in situ zymography, *J. Dent. Res.*, 2012, **91**(5), 467–472, DOI: 10.1177/0022034512439210, Epub 2012 Feb 21.

221. L. Moroni, A. Nandakumar, F. B. de Groot, C. A. van Blitterswijk and P. Habibovic, Plug and play: combining materials and technologies to improve bone regenerative strategies, *J. Tissue Eng. Regener. Med.*, 2013, DOI: 10.1002/term1762.

222. P. Nooeaid, V. Salih, J. P. Beier and A. R. Boccaccini, Osteochondral tissue engineering: scaffolds, stem cells and applications, *J. Cell. Mol. Med.*, 2012, **16**(10), 2247–2270, DOI: 10.1111/j1582-4934201201571x.

CHAPTER 8

Bioactive Conducting Polymers for Optimising the Neural Interface

JOSEF GODING, RYLIE GREEN, PENNY MARTENS AND
LAURA POOLE-WARREN*

Graduate School of Biomedical Engineering, University of New South
Wales, Australia
*Email: l.poolewarren@unsw.edu.au

8.1   Introduction

Since the seminal research of MacDiarmid, Heegerand and Shirakawa in the 1970s, interest in conducting polymers has largely focussed on improving the stability and electrical properties of this extraordinary class of polymers. More recently, interest in their application for biomedical uses has increased, with particular focus on surface electrodes, drug delivery devices, biosensors and implantable electrodes interfacing with the nervous system. While forms of conducting polymers that exhibit high stability and good electrical properties have been identified, limitations associated with their mechanical properties and biological performance restrict their widespread application to medical devices. However, with the increasing electrode complexity and the drive to reduce electrode size in many neural applications, metal electrodes are reaching their limits. There is clearly a need for electrode materials that support safe and sustainable delivery of the appropriate stimulation to excitable tissues.

RSC Smart Materials No. 10
Biointerfaces: Where Material Meets Biology
Edited by Dietmar Hutmacher and Wojciech Chrzanowski
© The Royal Society of Chemistry 2015
Published by the Royal Society of Chemistry, www.rsc.org

With their characteristic conjugated backbone structure, intrinsically electrically conductive polymers offer superior electrical properties for charge transfer in biological environments when compared to bare metal electrodes. They can also be incorporated within currently existing device designs with little to no modification due to the simplicity of electro-chemical fabrication methods. However, their greatest potential as electrode materials and a significant advantage they have over metal electrodes comes from their ability to accommodate biofunctionality through the incorporation of bioactive molecules. This chapter will review the range of systems that have been developed with the common aim of producing high quality interfaces between neuroprosthetic electrodes and the surrounding neural tissue. Furthermore it will examine the current challenges and limitations surrounding the use of conducting polymers and some of the proposed solutions for optimising the neural interface.

8.2 Conducting Polymers

8.2.1 Mechanism of Conduction

Electrical conduction within conducting polymers originates from the conjugated backbone structure.[1] This conjugated structure is characterised by alternating single and double bonds between sequential carbon atoms as seen in the chemical structure of polyacetylene in Figure 8.1.

In addition to the strong sigma (σ) bond present between each carbon in the polymer backbone, the alternating double bonds also contain a delocalised pi (π) bond. Repeated conjugation along the backbone gives rise to a conduction valence band within the conducting polymer. The incorporation of an ionic dopant molecule within the conducting polymer will oxidise (p-dope) or reduce (n-dope) the polymer at the site of the π-bond by removing or donating an electron.[2] This creates a free radical which pairs with the dopant to form a polaron, which in turn can be further oxidised or reduced to form bipolarons as illustrated in Figure 8.2.

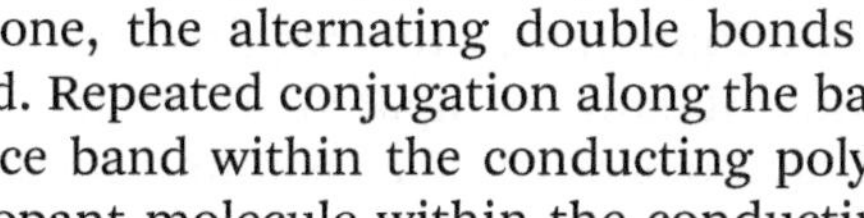

Figure 8.1   Chemical structure of polyacetylene.

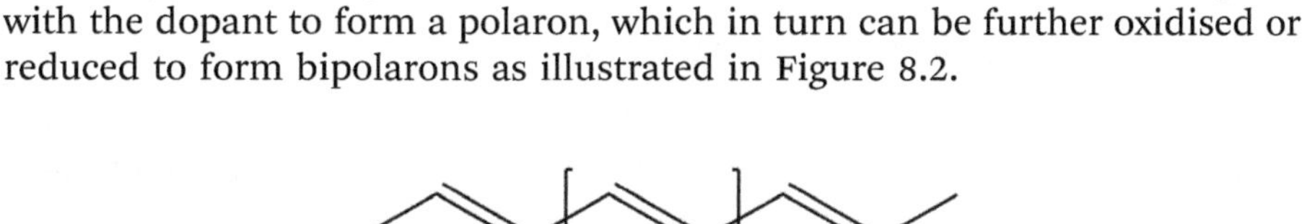

Figure 8.2   Bipolaron conduction along polypyrrole.

It is these bipolaron sites which facilitate electrical conduction by providing a path for the movement of charge carriers along the polymer backbone when the material is exposed to an electrical potential. Electrical conduction not only occurs along the polymer backbone but also between neighbouring polymer strands.[3] Although not as well understood, conduction between polymer strands is critical to the observed macroscopic conduction within conducting polymers due to their highly disordered structure. The choice of dopant molecule plays a critical role in determining the properties of the resultant conducting polymer film. Conventional dopants, including perchlorates and sulfonated aromatic compounds such as *p*-toluenesulfonate (pTS) and poly(styrenesulfonate) (PSS), are used due to their efficiency of oxidising the conducting polymer backbone.

8.2.2 Fabrication

8.2.2.1 Chemical Synthesis

Early research on conducting polymers primarily utilised chemical methods for polymerisation. In chemical fabrication routes, the monomer is dissolved in an organic solvent and is added to a solution of mild Lewis-acid catalysts.[3] The catalyst enables oxidative coupling of monomer units to form polymer strands. Conducting polymer films can then be formed using solvent casting techniques. While this fabrication route gives a high degree of control and allows for high-volume fabrication of highly conductive polymers on both conductive and non-conductive substrates, it is rarely used for biomedical applications due to the required use of toxic solvents and oxidative species. Additionally, compared to electrochemical methods of synthesis, chemical methods are time consuming and typically require post-fabrication processing. As such, the bulk of research concerning biomedical conducting polymers utilises electrochemical deposition.

8.2.2.2 Electrochemical Deposition

Due to its relative speed, ease, reliability and degree of control electrodeposition is heavily favoured for fabricating biomedical conducting polymers.[4,5] Electrodeposition is generally carried out in a three-electrode cell as seen in Figure 8.3.

The monomer unit is dissolved in the presence of a charged dopant species, typically an anionic dopant. During electrodeposition an oxidising potential, either galvanostatic or potentiostatic, is applied through the precursor solution.

TIP: *The potentiostatic mode allows for control of oxidation/over-oxidation whereas the galvanostatic mode allows for control over the kinetics of deposition (growth rate and film thickness). It is the authors' opinion that galvanostatic deposition produces more reliable samples.*

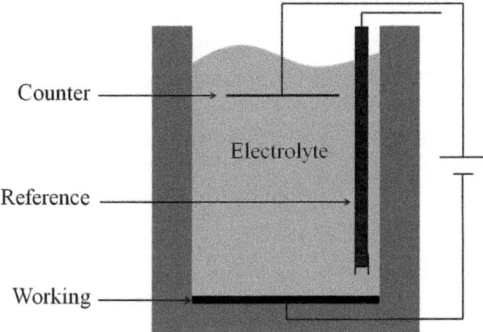

Figure 8.3 Three-electrode cell for electrochemical deposition.

This oxidising potential causes the monomer unit to deprotonate and form a radical ion intermediate capable of initiating polymerisation. As the polymerisation progresses the growing oligomers become insoluble and precipitate onto the working electrode, continued growth and cross-linking of conducting polymer chains eventually forms a thin polymer film. The thickness of the electrodeposited film is controlled by the total charge passed through the electrode cell. Tallman *et al.* describe a model for the thickness of deposited films, δ, based upon charge passed, as seen in eqn (8.1):

$$\delta = \frac{j \cdot t}{F} \times \frac{m_{EW}}{\rho} \qquad (8.1)$$

where j is current density (A cm^{-2}), t is the deposition time (s), F is the Faraday constant, ρ is polymer density and m_{EW} is the equivalent weight of polymer (molar mass of monomer plus associated dopant, divided by the number electrons transferred per monomer unit during polymerisation).[6]

Increasing the deposition charge, by increasing the current density or deposition time will result in thicker, more textured films. As the electrodeposition process continues the film develops a characteristic nodular surface morphology, greatly increasing the films surface area as seen in Figure 8.4.

TIP: *Films will initially deposit in a two-dimensional manner, but will begin to shift to a three-dimensional nodular growth as film thickness increases. Smooth two-dimensional growth can be encouraged through the use of low current densities (i.e., slower deposition) and by limiting film thickness. The use of poly(styrene sulfonate) as a dopant will also result in smoother films compared to films doped with p-toluenesulfonate.*

This nodular structure and high surface area are important to the electrical superiority of conducting polymers over traditional metal electrodes as this characteristic morphology is largely responsible for the high charge storage capacities and low impedances observed in conducting polymers.

Figure 8.4 SEM image of surface morphology of PEDOT/pTS at ×2500 magnification.

8.2.3 Characterisation

There are several key design criteria of conducting polymers for biomedical applications. These include surface morphology, electrochemical and mechanical stability, charge transfer characteristics, and biocompatibility. A summary of common analytical techniques used to characterise these and other properties is presented below.

8.2.3.1 Surface Properties

The critical surface properties typically evaluated include surface chemistry and topography. There is a range of techniques used to measure these properties and these are outlined briefly in this section. It is particularly important to understand how these properties may change as a result of incorporation of biological molecules and different dopants since they all directly impact on the cellular and tissue responses.

A critical component of the characterisation of conducting polymer electrodes is its chemical composition. This is of particular importance when working with novel dopants (such as bioactive dopants) or when incorporating non-conventional components into conducting polymer films. X-ray photoelectron spectroscopy (XPS) is the most valuable methodology for probing the elemental composition, as well as the chemical and electrical state of conducting polymers. In XPS the sample is irradiated with X-rays causing electrons to be ejected from the top 10 nm of the material. Measurement of the number and kinetic energy (binding energy) of ejected electrons provides a characteristic elemental spectrum. By analysing the relevant elemental spectra it is possible to determine the doping ratio (how many dopant molecules per monomer unit) as well as the redox state (the extent of doping) of the conducting polymer through deconvolution of characteristic elemental peaks. Combining XPS with time-of-flight secondary ion mass spectrometry provides a powerful combination of analytical techniques that allows determination of the presence of biological species on or within the top 1–2 nm of the materials surface.

> **TIP:** *Before chemical examination using XPS, samples should be cleaned with a suitable media (e.g., alcohol, water) to remove contamination that remain on the surface after synthesis, handling or are adsorbed from air. A typical contaminant adsorbed from air is carbon ($E_B = 285$ eV) and its high concentration skews the results obtained for other elements. Binding energies for double bonded carbon that is present in the polymer structure is $E_B = 285.7$ eV and deconvolution of the carbon peak allows for determination of contributions of each component to the total amount of detected carbon. A typical procedure to remove carbon contamination from polymers is sonication in alcohol for 5 min (or other adequate solvent) following by sonication in ultrapure water for 5 min and drying/purging with nitrogen (providing that both alcohol and water do not interact with or dissolve the polymer).*

Another useful technique for analysing elemental composition, although providing less detailed information than XPS and also probing to a greater depth, is energy-dispersive X-ray spectroscopy (EDS) imaging offered by scanning and transmission electron microscopes. EDS has the advantage of spatially mapping chemical composition, however this method is limited in its resolution and accuracy (detection limit of 0.1–0.5 wt%) compared to XPS.

Since charge transfer is dominantly a surface based phenomenon, it is also important to characterise the surface morphology of conducting polymers. The surface morphology of a conducting polymer film is dependent upon choice of dopant (and non-dopant inclusions) as well as the electrodeposition parameters.[7] Analysis of surface topography is most commonly conducted using scanning electron microscopy, which can provide a qualitative analysis of surface morphology and roughness at the nanoscale.

Techniques such as optical profilometry and atomic force microscopy (AFM) can be used to give more detailed, quantitative analyses of surface properties. Optical profilometry uses light-based probes to scan the surface and determine its roughness with resolution in the micro- to nanoscale. AFM characterises surfaces properties by scanning a material surface with a nanoprobe and measuring the probe–surface interactions. In addition to topography and roughness, AFM is capable of spatial mapping of elastic modulus and electrical conductivity at the nanoscale.

8.2.3.2 Electrochemical Properties

There are several key electrical properties which require characterisation in order to evaluate the electrical performance of a conducting polymer.

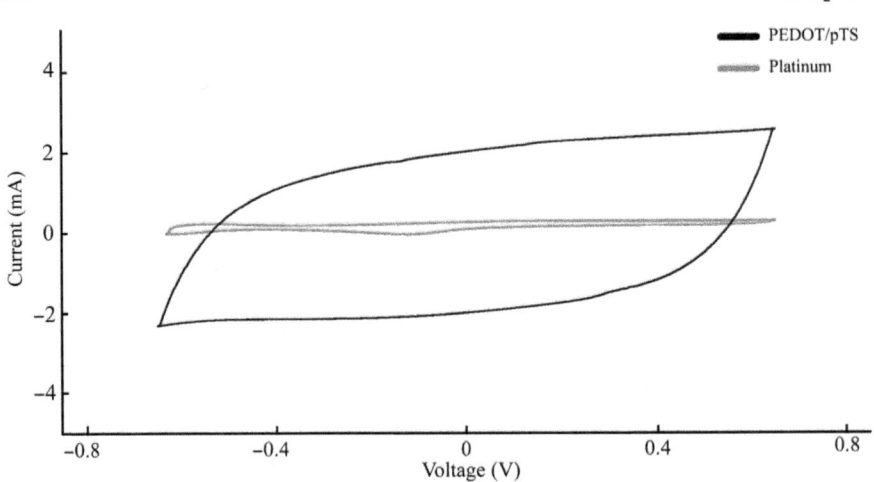

Figure 8.5 Cyclic voltammetry of platinum and PEDOT/pTS.

The two most common methods are cyclic voltammetry (CV) and electro-chemical impedance spectroscopy (EIS). CV is used to evaluate the electrical activity and stability of electroactive materials.[8] In CV the material is scanned between two set voltage potentials. Electrochemical activity within the material is observed as redox peaks in the voltage *versus* current hysteresis loop as shown in Figure 8.5.

> **TIP:** *CV utilises a three-electrode cell. The counter-electrode should be inserted at the top of the cell (maximise working distance) and the reference should be placed in close proximity to the working electrode (the sample of interest) at the bottom of the cell. CV should be conducted in a relevant electrolyte solution such as saline, DPBS or serum-containing media. Typical scanning range is from −0.6 to 0.8 V with scan rates of 50 to 150 mV s^{-1}.*

Integrating the area within each loop determines the charge storage capacity (amount of charge transduced per unit of geometric surface area). Comparing changes in charge storage capacity with increasing cycle number allows assessment of the electrochemical stability of the conducting polymer.

EIS is used to evaluate the frequency dependent charge transfer characteristics of an electrode.[8] The electrode is probed with a small amplitude sinusoidal current or voltage. This measurement is made over a wide range of stimulating frequencies, typically from 1 MHz down to below 1 Hz. The current–voltage response is measured and compared to the input signal to yield an impedance magnitude and phase lag which can be represented in a Bode plot, as seen in Figure 8.6, allowing examination of the faradaic and non-faradaic charge transfer characteristics of the electrode.

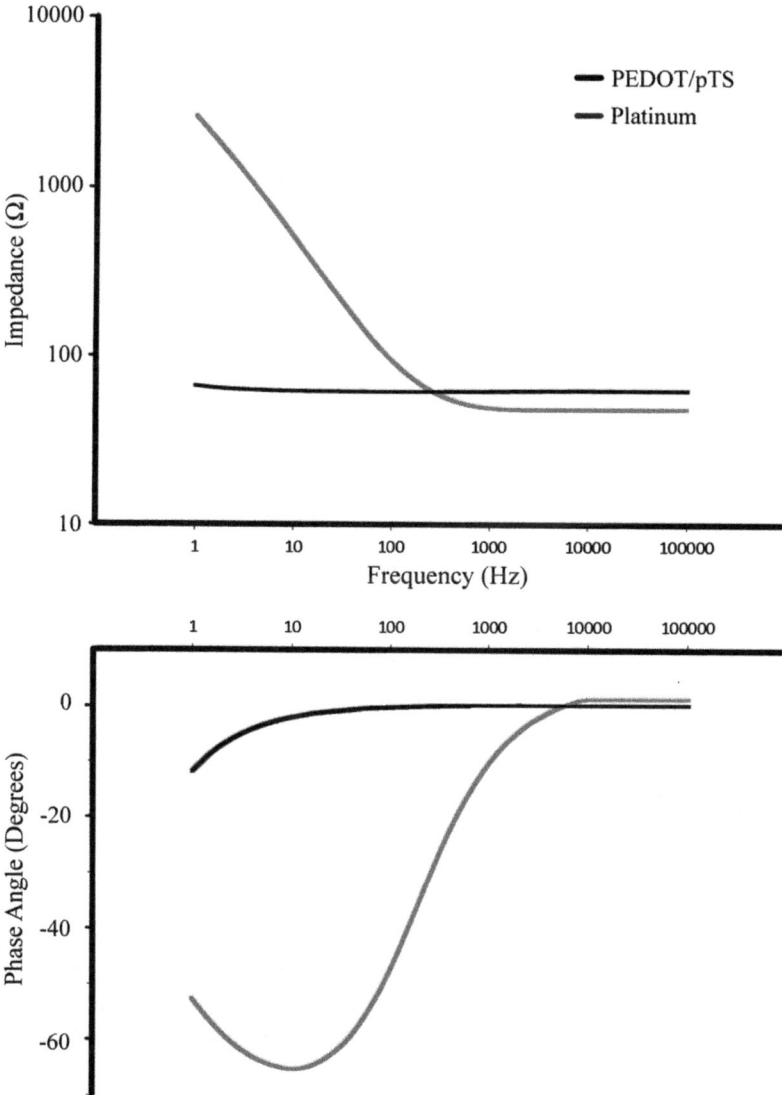

Figure 8.6   Bode plot of platinum and PEDOT/pTS.

TIP: *During EIS sufficient stimulation amplitude (~ 100 mV AC) is required to elicit a measurable response. It may be necessary to use a Faraday cage to electrically isolate the sample from environmental interference.*

8.2.3.3   Mechanical Properties

Very little has been reported on the mechanical properties of conducting polymers. Due to the thin film, substrate-dependent nature of conducting polymers many conventional techniques, such as tensile testing, do not apply.[3] Yang and Martin used nanoindentation to evaluate the mechanical properties of PEDOT/LiClO$_4$. They found that films with higher surface area due to increased roughness, and hence lower impedances, were the most mechanically compliant films.[9] Baek *et al.* used nanoindentation to examine the effect of dopant properties on the mechanical properties of PEDOT and observed that increasing dopant size correlated with increased stiffness. Specialised AFM techniques can also be used to mechanically characterise polymer coatings. Green *et al.* mechanically characterised conducting polymer, hydrogel, and conducting hydrogel coatings using peak-force quantitative nanomechanical mapping atomic force microscopy.[10]

Film adhesion and delamination can be tested using standard test methods for pull-off strength and adhesion of coatings (ASTM D-4541-95 and D3359). Examination of adhesion/delamination and friability (lack of cohesion) are of particular importance when working with bioactive dopants as the resultant polymer films tend to be more prone to mechanical failure *via* delamination and brittle fracture than conventionally doped films.[11]

TIP: *The authors use a modified ASTM tape adhesion test (D3359-02) in order to evaluate coating adhesion and friability. An X-cut is scored into the polymer coating with a scalpel before application of standard 3M masking tape. The tape is removed after 5 min and images of the coating before and after delamination are analysed to determine the extent of delamination. Examining delamination images with images of polymer retained on the tapes surface allows for a relative measure of polymer friability.*

8.2.3.4   Biological Properties

Techniques which characterise the biological properties of conducting polymers fall under one of two categories: *in vitro* or *in vivo* studies. *In vitro* studies allow for examination of specific biological interactions; however they can only serve as a model to give an indication of the potential biological response in the *in vivo* environment.

In vitro biological characterisation of conducting polymers utilises a range of cellular models. The first step is typically determining the cytotoxic potential of the material using cellular growth inhibition and viability assays. Cytotoxicity can be investigated using a range of cells including standard clonal cells (such as L929 fibroblasts), clonal neural cells (such as PC12 neural-like pheochromocytoma cells) as well as primary cells such as neuroglia and neural ganglion cells and can be quantified through the use of live/dead staining (viability) cell counting (inhibition) or through biochemical techniques such as MTT assays (cytostatic activity).[50] The toxicity of conducting polymers is largely determined by the choice of dopant molecule.[7,12,13] When in an electrochemically reduced state, the charge interaction between the polymer backbone and the anionic dopant is lost, allowing the dopant to diffuse out into the surrounding environment. Therefore it is important to consider the toxicity and mobility of dopant ions within biomedical applications. Biomedical conducting polymers such as polypyrrole (PPy) and poly(3,4-ethylenedioxythiophene) (PEDOT) have generally been found to be non-cytotoxic materials.[14]

Other *in vitro* assays will examine more application specific cellular responses. Perhaps the most common is culturing PC12 neural-like cells and analysing cell and neurite density.[14] Other models used include primary neural tissue explants, typically neurons and glial cells.[15-17] Studies have shown that a range of factors affect *in vitro* neural cell response to conducting polymers, including synthesis technique, film roughness and morphology as well as dopant chemistry and mobility.[3]

TIP: *PC12 is a non-adherent cell line which will differentiate and grow neuronal processes when cultured in low-serum media and exposed to nerve growth factor. Samples will require coating with either laminin or poly-L-lysine to facilitate cellular attachment. When plating, PC12s are moved from high-serum media (RPMI + 10% horse serum + 5% foetal bovine serum) to a low serum media (RPMI + 1% horse serum) containing 50 ng mL^{-1} nerve growth factor, plating at 20 000 cells cm^{-2}. Media is refreshed at 48 h and samples are imaged at 96 h. Neurite density can be determined by tracing neurites in ImageJ software (National Institutes of Health) using the NeuronJ plugin.*

In vivo characterisation conventionally involves implantation studies assessing electrical recording and stimulating capabilities as well as histological examinations of the tissue–electrode interface within the central nervous system to characterise the biological response to implantation.[18-20] For a more in-depth review of *in vivo* studies of conducting polymers the reader is referred to Ateh *et al.*[21]

8.2.4 Biomedical Applications of Conducting Polymers

Conducting polymers are being investigated for a variety of applications within the field of biomedical engineering including stimulating-recording electrodes, electrically controlled drug-release systems, biosensors and nerve grafts and conduits. Due to the high demand for the chemical and electrical stability for these applications, heterocyclic compounds such as PPy and polythiophenes, whose chemical structure are illustrated in Figure 8.7, are favoured for biomedical applications.[4] A range of molecules are used as conventional dopants for biomedical applications, including *p*-toluenesulfonate (pTS), polystyrenesulfonate (PSS) and perchlorates.[14]

The focus of research concerning biomedical conducting polymers falls heavily on their use as neuroprosthetic electrode coatings; however, conducting polymers have also shown great promise in applications such as controlled drug delivery systems, nerve grafts/conduits and biosensors.

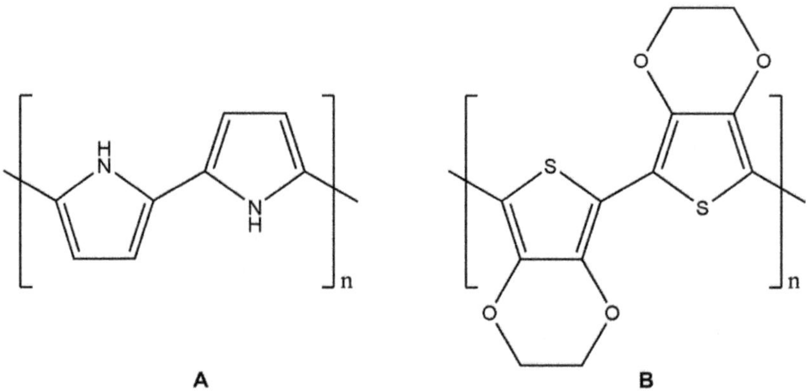

Figure 8.7 Chemical structure of (A) polypyrrole and (B) poly(3,4-ethylene dioxythiophene).

8.2.4.1   *Neuroprosthetic Electrodes*

Bioelectrodes are used in a wide variety of neuroprosthetic devices to stimulate or record neuronal excitation, such as cochlear implants, deep brain stimulators, and motor prosthetics. Current electrode designs typically involve metal-to-tissue interfaces, resulting in poor long-term electrical stability and efficacy.[8,22] This poor stability greatly impacts upon a neuroprosthetic devices ability to chronically stimulate or record neural activity. The root of this poor device performance is the inflammatory response at the tissue–electrode interface. The inflammatory response is characterised by fibrotic encapsulation of the device which results in an increase in interfacial impedance. Furthermore, the persistent presence of the device can result in chronic inflammation at the interface, leading to local tissue damage and neuronal death. The effect of the fibrotic encapsulation and chronic inflammation is an increase in separation distance between the electrode surface and the target cells for stimulation.[4,14,23] For stimulating electrodes the combination of increased impedance and distance to target cell population acts to decrease the quality of signal transmission, requiring more charge to be injected (coupled with increased power requirements) in order to successfully stimulate neuronal cells. However, metal electrodes are severely limited in the amount of charge they can safely inject without causing undesirable irreversible reactions at the interface, known as the charge injection limit. As modern neuroprosthetic electrodes shrink in size, the charge injection limit of metal electrodes is quickly becoming a key limitation in the safe and efficacious operation of neuroprosthetics. Conducting polymers have charge injection limits two orders of magnitude larger than metal electrodes, with PEDOT having an injection limit of up to 15 mC cm^{-2} compared to just 50–250 μC cm^{-2} for bare platinum.[8] This means that for the same stimulation threshold a conducting polymer electrode can be significantly smaller than a traditional bare metal electrode.

Because conducting polymers are softer than metal electrodes they experience less strain mismatch at the interface.[10,22] Strain mismatch is a contributing factor for chronic inflammation so it is desirable to use electrode materials with mechanical properties closer to that of neural tissue. The highly textured, nodular surface morphology not only helps to provide conducting polymers with superior electrical properties, but it may also provide a preferable surface for soft tissue integration. Several studies have shown that neural cells will preferentially adhere to coated electrodes.[3,24–26]

8.2.4.2 Drug Delivery Systems

Depending on the size and charge of an incorporated bioactive molecule it is possible for conducting polymer electrodes to also act as controlled drug delivery devices.[5,27] If the bioactive molecule is mobile within the conducting polymer it can be released in a controlled fashion *via* electrical stimulation. This is discussed in greater detail in Section 8.3.3.

8.2.4.3 Nerve Grafts/Conduits

Nerve grafts are devices designed to repair or bridge a defect in neural tissue. The most common application is nerve conduits for spinal injury, however, nerve grafts can also be used to repair damage in the peripheral nervous system. The formation of an ordered, effective neural conduction path is the key goal of nerve grafts. Conducting polymers are a promising material platform for nerve grafts because their electroactivity can be used to not only guide the directional growth of neural cells but to also encourage neurite outgrowth. Schmidt *et al.* have demonstrated the use of PPy in the fabrication of a nerve guidance channel which showed favourable interactions with PC12 cells and chicken sciatic nerve explants.[28] Furthermore, the PPy guidance channels were able to promote a significant increase in neurite outgrowth by stimulating the cells with a 100 mV potential for 2 h. Many of the goals and approaches discussed in this chapter hold true for conducting polymers in nerve graft applications.

8.2.4.4 Biosensors

Biosensors are analytical electrodes designed to detect specific molecules of interest in applications ranging from health care to environmental monitoring.[29] In these biosensors the conducting polymer is used as a matrix for the entrapment of biomolecules which act as a sensing element, chosen due to their biorecognition of the analyte (the molecule to be detected). Interaction between the sensing element and the analyte results in a chemical change that is detected by the conducting polymer matrix. The conducting polymer acts as a transducer converting the chemical signal into an electrical signal from which the concentration of analyte can be determined.

Within the field of biomedical engineering these biosensors have been used in the detection of blood glucose levels, drug compounds, DNA, cholesterol, antibodies, viruses and neurotransmitters.

8.2.4.5 Application-specific Characterisation

In addition to the materials based characterisation techniques discussed above, there is a sub-set of application specific characterisations commonly reported in the literature. Important application specific considerations include safe and efficacious delivery of stimulation charge, long-term stability and incorporation into biomedical device manufacturing processes including sterilisation.

For all electrode materials there exists a stimulation potential limit past which harmful, irreversible reactions will occur at the electrode interface. Typically these limits are defined by the potentials at which electrolysis of water will occur; however, other reactions such as dissolution of platinum electrodes may occur at lower levels of charge injection. The charge injection limits of an electrode material can be determined through analysis of voltage transients arising from biphasic stimulation as seen in Figure 8.8.

The charge injection limit is determined to occur when the maximum residual potential, E_{MC}, reaches the reduction potential for water $(E_{MC} = -0.6 \text{ V})$.[8,30] E_{MC} can be determined by subtracting the access voltage of the electrolyte, V_a, from the maximum negative voltage, V_t.

The maximum voltage polarisations of the electrode are determined over a range of stimulation-relevant current densities and pulse widths and

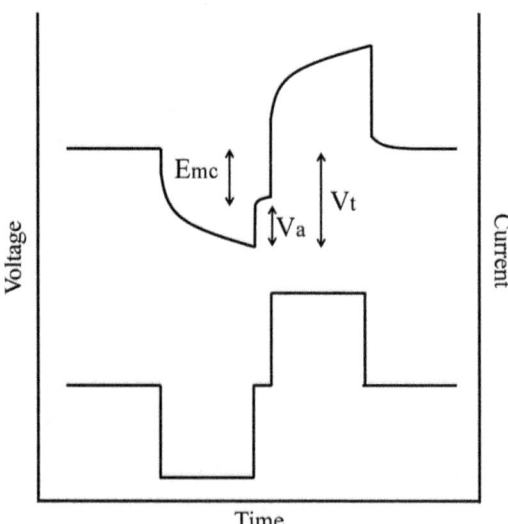

Figure 8.8   Voltage transient in biphasic stimulation.

compared to known maximum safe potentials (*i.e.*, the voltage window outside of which electrolysis occurs).

> **TIP:** *It is important to stimulate between electrodes fabricated from the same material to prevent bias of the measured waveform from dissimilar materials. Additionally, the electrolyte should be an adequate representation of the fluid in which the device is intended to operate following implantation. The authors use artificial perilymph or cell culture media. Due to its lack of protein content, saline is not considered to be an appropriate electrolyte for simulating implantation performance.*

If conducting polymers are to be used in biomedical devices, then they need to meet the design criteria for the specific device. Some devices, such as the cochlear implant are required to have operating lifetimes in excess of 70 years. As such, it is important to investigate the long-term stability and function of conducting polymers. Boretius *et al.* investigated the long-term stability of PEDOT coatings under chronic, pulsed electrical stimulation such as experienced in neuroprosthetic applications.[31] Stimulation consisted of biphasic pulses each injecting 20 nC of charge; periodic EIS was used to assess coating stability. A mean-time-to-failure of 127 million pulses was calculated (suitable for sub-chronic implantation), after which time the electrodes still operated within safe parameters for the platinum electrode substrate.

Similar studies by Green *et al.* tested the stability of PEDOT/pTS under 2.8 billion stimulation cycles, each injecting 100 nC of charge, in 15% serum supplemented cell culture medium.[32] A 16% reduction in charge storage capacity of PEDOT/pTS electrodes was observed over 2.8 billion cycles, compared to a 15% reduction in non-stimulated electrodes. It was determined that continual stimulation of PEDOT had negligible effects on electrode performance. Further work demonstrated that roughening of the underlying platinum substrate acts to increase the long-term stability of PEDOT coatings.[30] In the same studies, the effect of ethylene oxide (ETO) and steam sterilisation on electrical properties and surface morphology of PEDOT were examined. ETO sterilisation is favoured for neuroprosthetic devices due to the presence of electrical components which may be damaged during steam sterilisation. No change in biphasic stimulation was observed post-ETO sterilisation; however, a decrease in charge storage capacity was found and was thought to be related to formation of microcracks caused by variation in moisture levels throughout the treatment process. Decreases of 20–40% of original charge storage capacity were observed post steam sterilisation; however, the bulk electroactivity remained intact.

> **TIP:** *The authors use ethanol and/or UV irradiation to sterilise samples for use in biological characterisation experiments. These methods are less destructive than steam and ETO sterilisation.*

8.2.5   Limitations of Conducting Polymers in Neural Interfaces

Conducting polymers have been shown to have clear advantages over metal electrodes in the *in vitro* setting. However, in the *in vivo* environment this benefit is reduced, largely due to the effects of the inflammatory response on the neural interface. This was exemplified in a study by Abidian *et al.* in which gold electrodes coated in conducting polymer nanotubes were implanted in the barrel cortex of rats.[19] Prior to implantation, coating the electrodes resulted in the impedance decreasing by two orders of magnitude and the charge transfer capacity increasing by three orders of magnitude. Upon implantation the measured impedance of the electrodes rapidly increased from 841 ± 7 to 908 ± 5 kΩ for uncoated electrodes and from 17 ± 4 to 87 ± 8 kΩ for coated electrodes (measured at 1 kHz). By day 8 of implantation, impedance had risen to 1250 ± 3 kΩ for uncoated and 546 ± 30 kΩ for coated electrodes, before finally settling out to 980 ± 15 kΩ for coated and 521 ± 18 kΩ for coated electrodes by day 49 of implantation. Overall, the benefit gained from conducting polymer coatings decreased from nearly three orders of magnitude to less than a two-fold increase. This demonstrates that the biological response and the quality of the neural interface established are key to the functional performance of the conducting polymer electrode. One approach to overcome this limitation is the biofunctionalisation of conducting polymers *via* the incorporation of bioactive molecules within conducting polymer films.

8.3   Biofunctionalisation of Conducting Polymers

The main goal of incorporating bioactive molecules is to elicit a desirable cellular response which will result in the establishment of high quality neural interfaces.[4,14] Two common approaches to this include attenuating the inflammatory response using anti-inflammatory drugs and encouraging neuronal survival and intimacy at the interface through the use of growth factors and cell adhesion molecules.

8.3.1   Incorporation of Bioactive Molecules

There are two main routes through which bioactive molecules can be incorporated within conducting polymer films. These are electrostatic incorporation of charged molecules as bioactive dopants, or through physical entrapment of neutral or weakly charged molecules as non-dopant inclusions.[3,27] Less commonly used methods of incorporation include physical adsorption and covalent attachment to the surface of conducting polymer films. The method of incorporation is typically determined by the properties of the bioactive molecule in question as well as the desired means of biological presentation.

8.3.1.1   Incorporation of Charged Biomolecules as Dopants

As stated previously, the presence of a charged dopant molecule is required during electrochemical deposition in order to form a functional conducting polymer. If the biomolecule of interest has one or multiple sites of ionic charges (dopant groups) then it can be electrostatically incorporated within the conducting polymer as a dopant. There is typically one dopant group associated with every three to four monomer units, and as such the bioactive dopant needs to be supplied at a high enough concentration within the electrolyte precursor solution to ensure successful polymerisation.[5] Large, bulky, biological dopants may experience mass transport limitations during polymerisation resulting in a concentration deficit at the deposition interface, interfering with the electrodeposition process.[3,33]

A critical aspect of creating bioactive conducting polymers is the retention of the dopants biofunctionality. In particular, biofunctionality may be compromised during electrodeposition due to chemical or electrical denaturation as a result of the organic solvents and electrical charge used during electrodeposition.

TIP: *Both PPy and PEDOT can be fabricated from aqueous solutions in the absence of organic solvents. Unlike PPy, PEDOT has very poor solubility in water (the authors typically use a 1 : 1 water–acetonitrile solution for PEDOT when organic solvents are not a concern). Due to the low concentration of PEDOT in aqueous solutions, low current densities should be used during deposition to avoid mass-transport limitations.*

Also of critical importance is the effect of dopant choice on the properties of the resultant conducting polymer film.[7] The three main factors determining conducting polymer properties are choice of monomer (conducting polymer chemistry), electrodeposition parameters (solvent, pH, current density, time) and choice of dopant. As such, when using bioactive dopants there is typically a trade-off between biofunctionality and dopant efficacy.

8.3.1.2   Incorporation of Neutral and Weakly Charged Biomolecules as Non-dopant Inclusions

There are three pathways through which bioactive molecules can be incorporated within conducting polymers as non-dopant inclusions; physical entrapment within the polymer matrix or attachment to the polymer surface through physical adsorption or covalent tethering.[3,26,34]

Non-dopant bioactive molecules can be incorporated within conducting polymer matrices through their inclusion in the electrolyte solution used for electrodeposition. As the conducting polymer deposits on the electrode the bioactive molecules present in the electrolyte become physically entrapped within the polymer matrix as it forms. Incorporation of bioactive molecules using this method can disrupt the formation of the conducting polymer, resulting in a degradation of film morphology as well as electrical and mechanical properties.[3,11]

Physical adsorption and covalent tethering of molecules to the surface of a conducting polymer film have the advantage of leaving the bulk properties of the conducting polymer intact.[26,34] However, this comes at the expense of limited capacity for incorporation. Adsorption to the surface may better retain biofunctionality in the short term, due to the absence of solvents or electrochemical stimulation which may act to denature these molecules. However, the biomolecules are only held in place by weak physical interactions and will likely be displaced by molecules with stronger affinities upon implantation, greatly limiting the duration of the coatings biofunctionality.

Covalent tethering of biomolecules to the conducting polymer surface provides a more permanent means of attachment although may require the use of solvents increasing the risk of biomolecular denaturation. Covalent tethering can also give control over the spatial presentation of bioactive molecules. Both physical adsorption and covalent tethering require post-fabrication processing that may not be practical for application within neuroprosthetic devices; this is reflected in the literature with the majority of biomolecular incorporation utilising either bioactive dopants or physical entrapment.

Depending upon the method of incorporation and the specific molecule in question, the presentation of biologically active molecules will fall into one of two categories. In the case of covalent tethering and/or the use of large, entrapped molecules the presentation will be surface only; alternatively, conducting polymers can act as drug delivery systems when utilising smaller, more mobile, biologically active molecules.

8.3.2   Conducting Polymers as Bioactive Surfaces

Under certain circumstances incorporated bioactive molecules may be immobilised within or on the surface of the conducting polymer matrix, thus limiting their biological interaction to presentation of molecules on the electrode surface. Immobilisation of bioactive molecules is dependent upon either the method of incorporation or on the properties of the molecule in question. Covalent tethering and physical adsorption of biomolecules will obviously lead to a surface-only presentation, however, bioactive dopants and entrapped molecules may also become immobile within the polymer matrix due to their size or specific polymer interactions. Because the volume of interaction is limited to only the surface, bioactive molecules incorporated in this fashion tend to have bioactivity focused on neural cell attachment or presentation of bioactive molecules that are effective at very low concentrations.

Several studies have examined incorporating laminin peptide fragments as dopants within conducting polymer films. Stauffer and Cui individually cultured primary cortical rat neurons and cortical murine astrocytes on PPy doped with two laminin fragments from cell binding domains.[15] Laminin doped coatings were found to support higher levels of neuron density and neurite length while having significantly less astrocyte adhesion compared to gold electrodes. Green *et al.* found the co-incorporation of laminin peptides and nerve growth factor created PEDOT films capable of supporting PC-12 neural-like cells in the absence of externally supplied growth factor.[11]

George *et al.* developed a novel system for presentation of biomolecules using the biotin–streptavidin binding complex to tether biomolecules to the surface.[35] PPy was co-doped with sodium dodecylbenzenesulfonate (NaDBS) and biotin. The PPy film was then incubated with streptavidin which will bind to any biotin dopant present on the films surface. The film can then be incubated with any biotinylated molecule for surface only incorporation. This system was used to bind NGF to the surface of PPy/NaDBS–biotin coatings. This technique has the benefit of achieving biofunctionality without compromising the bulk properties of the conducting polymer.

Collazos-Castro *et al.* covalently attached two cell-adhesion molecules, *N*-cadherin and L1, to the surface of PEDOT/PSS-*co*-(maleic acid) *via* the covalent immobilisation of anti-IgG anti-bodies.[26] The electrical properties of the film were not degraded by the covalent attachment and the attachment of cell adhesion molecules resulted in improved axonal elongation and branching when cultured with embryonic rat cerebral cortex neurons.

Bax *et al.* utilised a plasma surface treatment to covalently attach the cell adhesive protein tropoelastin to PPy surfaces.[34] The plasma treatment (plasma immersion ion implantation) allows for covalent attachment of molecules, without employing chemical linking molecules *via* generation of free radicals in the treated polymer surface. Covalent attachment of tropoelastin resulted in improved adhesion and spreading of human dermal fibroblasts; masking during the plasma treatment allowed for selective patterning of cells on the PPy surface.

8.3.3 Conducting Polymers as Drug Delivery Devices

When a polymer electrode is in an electrochemically reduced state the attractive forces associating dopant and polymer backbone are neutralised and the dopant may become mobile. Electrical stimulation also causes volumetric changes within the coating; continually cycling a conducting polymer between redox states results in repeated volumetric expansion and contraction. This actuation allows the coating to be used as a reservoir from which mobile molecules (both dopants and non-dopant inclusions) can be pumped into the surrounding environment. Drug release can be controlled through material design (composition and structure of the conducting polymer coating) and through electrical stimulation. By controlling the nature and intensity of electrical stimulation it is possible to control the timing, dose size and release curves of the bioactive molecule.

Dexamethasone phosphate, an anionic form of a powerful anti-inflammatory drug, has been incorporated in several conducting polymer systems as a dopant. The presence of the phosphate group allows the molecule to function adequately as a dopant, although a decrease in electrical properties is observed compared to films produced using a conventional dopant. Wadhwa *et al.* fabricated PPy/dexamethasone films capable of releasing 16 μg cm^{-2} over 30 stimulation cycles, resulting in an estimated concentration of 1 μM within 500 μm of the electrode surface.[36] This level of release was successfully able to lower the number of reactive astrocytes and inhibit microglial growth when cultured with primary murine cerebellum glia.

Thompson *et al.* physically entrapped neurotrophin-3 (NT3) within PPy/pTS films.[37] It was found that 26 μm thick films were capable of releasing approximately 5 ng cm^{-2} and 9 ng cm^{-2}, respectively, for unstimulated and stimulated electrodes. Further studies demonstrated that when cultured with primary rat auditory neuron explants, stimulated PPy/pTS-NT3 films promote neuron survival and neurite outgrowth.[20,38]

8.3.4   Limitations of Biofunctionalised Conducting Polymers

While it has been demonstrated that incorporation of bioactive molecules within conducting polymer films shows significant promise in altering the biological response to neuroprosthetic electrodes, there remain several challenges and limitations to the use of bioactive conducting polymers. The primary concern with incorporation of biomolecules is the negative impact on the mechanical and electrical properties of the conducting polymer that may occur. As stated previously, biofunctionality of a dopant molecule typically comes at the cost of its efficacy as a dopant. Similarly, non-dopant inclusions can have a significant impact upon the formation of the polymer matrix. Green *et al.* reported a significant decrease in the nodularity and surface area of PEDOT when doped with laminin peptide fragments compared to PEDOT/pTS.[24] Doping PEDOT with laminin peptides resulted in a four-fold decrease in charge storage capacity (200 mC cm^{-2} compared to 50 mC cm^{-2}) and was found to produce films more prone to delamination compared to PEDOT/pTS. Generally, it has been found that incorporation of larger molecules within conducting polymer matrices results in softer, less adherent coatings.

Another consideration when incorporating larger molecules for drug release is that of mobility. Molecular weight and structure, as well as electrostatic interactions can affect the mobility of molecules within conducting polymer matrices and this may limit the rate and total volume of release such that a sustained therapeutic effect cannot be achieved. It is important to note that size is not the only factor determining mobility release. Kontturi *et al.* found nicotinic acid, a small anionic dopant, to be immobile within PPy due to the nitrogen atom in the nicotinic acid extracting a proton from the nitrogen in the pyrrole rings.[39]

There are also application specific limitations to consider such as electrode size and bioactive loading capacity. Modern neuroprosthetic devices are tending towards the use of ever decreasing electrode sizes, with electrodes as small as 5–50 μm in diameter being reported. At this scale, electrode size will become the limiting factor in the total amount of bioactive molecule that can be incorporated and presented/released to the surrounding biological environment. Studies examining the release of entrapped NT3 from within 1.66 mm^2 PPy/pTS coated electrodes found a passive rate of release of 0.03 ng per day and a stimulated rate of release of 0.1 ng per day.[20] When cultured with spiral ganglion neurons it was found that only stimulated release of NT3 was capable of promoting neural survival. However, the electrodes were depleted of NT3 after 21 days of stimulated release after which no therapeutic effect was observed. Achieving long-term therapeutic effect from microelectrodes will be an increasing challenge as electrode size and hence volume for incorporation continues to decrease. Despite the promise shown by bioactive conducting polymers, the limitations discussed above demonstrate several critical shortcomings for their use in neural interfaces.

8.4 Modified Biofunctional Conducting Polymers for Neural Interfaces

Several strategies have been developed to address the limitations associated with bioactive conducting polymers. These approaches focus on either modifying existing material platforms through the creation of conducting polymer layered constructs and modification of conducting polymer structures or through creating new conducting polymer based composite materials.

8.4.1 Structured Conducting Polymers

8.4.1.1 Bilayered Conducting Polymers

One simple method of addressing the mechanical shortcomings of bioactive conducting polymers is the use of bilayered structures. This approach combines two layers of conducting polymer; typically one will be a bioactive layer providing biofunctionality and the other a conventional conducting polymer layer providing electrical and mechanical stability. Utilising a pre-layer of PPy/PSS underneath a bioactive layer of PPy/hyaluronic acid was found to produce a coating electrically and mechanically superior compared to a single component PPy/hyaluronic acid coating.[18] Alternatively, coating a bioactive layer with a conventional conducting polymer can not only improve electrical properties but can also be used to give a greater degree of control over the release of biological factors. Massoumi *et al.* minimised the amount of passive release of dexamethasone from within a PPy film by depositing a layer of poly(*N*-methylpyrrole)/PSS on top of the PPy/dexamethasone layer.[40] However, the bilayered structure was also found to reduce the total release volume of dexamethasone.

8.4.1.2 *Nanostructured Conducting Polymers*

Several studies have examined the effect of structuring conducting polymer coatings at the nanoscale. This typically involves the use of a nanotemplate during electrodeposition to either impart nanoporosity or to create nanowires or nanotubes. Kang *et al.* created nanoporous PPy/PSS coatings loaded with nerve growth factor (NGF) by electrodepositing onto electrodes coated with polystyrene beads.[41] Electrodes were soaked in tetrahydrofuran after deposition in order to remove the polystyrene template. The nanoporous PPy/PSS coatings had improved electrical properties due to increased surface area. The nanoporous electrodes were also capable of releasing greater volumes of NGF as a result of increased surface area and the continuous porosity acting as a low-resistance diffusion pathway. Luo *et al.* utilised a similar system to create dual-release electrode coatings.[42] Fluorescein, a model drug dopant, was incorporated within nanoporous PPy as a dopant. The coating was then subjected to an ethanol soaking process to make the film hydrophobic. The electrodes were then incubated with a solution containing dexamethasone phosphate, allowing the drug to absorb to the nanoporous structure. This method of incorporation significantly increased the loading of dexamethasone and allowed for the electrically controlled release of two different molecules.

A major limitation to the use of template systems is the required use of solvents to remove the template. Kim *et al.* used alumina templates to create conducting polymer nanowires.[43] The use of hydrofluoric acid or sodium hydroxide to remove the template resulted in a decrease in conductivity from 30 S cm^{-1} to 7 S cm^{-1} and 0.01 S cm^{-1}, respectively.[43,44]

Abidian *et al.* developed a system for creating conducting polymer nanotubes without the need to use organic solvents to remove the nanotemplate.[45] The nanotemplate was formed by electrospinning dexamethasone-loaded poly(lactide-*co*-glycolide) (PLGA) nanofibres onto the electrode surface. Electrodeposition of PEDOT around the PLGA fibres resulted in the formation of PEDOT nanotubes ranging from 100 to 600 nm with wall thickness varying from 50 to 100 nm. Delivery of dexamethasone is accomplished through both the passive degradation of the PLGA fibres and through the active electrical stimulation of the PEDOT nanotubes. It was found that under passive conditions less than 25% of incorporated dexamethasone was released over a 54 day period. Periodic electrical stimulation was found to result in a cumulative release of approximately 80% of the incorporated dexamethasone. No indication was given as to the biofunctionality or therapeutic effect of released dexamethasone, however this level of release is expected to be therapeutically significant.[46]

8.4.2 Conducting Polymer Composites

Another approach to overcoming the limitations of bioactive conducting polymers is the creation of conducting polymer composites. The aim of creating these composites is to combine the electrical functionality of the conducting polymer with the benefits of the other composite component. Two such systems, conducting polymer–carbon nanotube composites and conducting hydrogels are discussed below. Another class of composites attracting increasing focus is that of degradable conducting polymer composites. These are systems in which conducting polymers or oligomers are used to form either composites or co-polymer blends for use in tissue engineering and drug delivery. Readers are referred to Guo *et al.* for a more detailed discussion of degradable systems.[47]

8.4.2.1 Conducting Polymer–Carbon Nanotube Composites

Carbon nanotubes (CNTs) can be incorporated within conducting polymer electrodes to increase conductivity and electrostability while decreasing electrical impedance.[48] Conducting polymer–CNT composites are typically fabricated *via* electrodeposition using a CNT-loaded electrolyte solution. In addition to improved electrical properties, CNT composites can also be used for controlled drug delivery.

Luo *et al.* deposited PPy around CNT that had been preloaded with dexamethasone.[49] The PPy matrix encapsulated the CNTs and prevented the passive release of dexamethasone from within. When simulated, the PPy–CNT composites released larger and more sustained doses of dexamethasone compared to conventional PPy/dexamethasone films. Released dexamethasone was found to retain its bioactivity as demonstrated by its *in vitro* reduction of microglial (HAPI cell line) activation.

Concerns regarding the cytotoxicity of CNTs have been raised in the past; however, studies have shown that incorporation of CNTs within conducting polymer coatings does not result in cellular inhibition.[48] This is likely due to the fact that the CNTs are immobilised within the conducting polymer matrix and as such have only very limited physical interaction (uptake) with nearby cells.

8.4.2.2 Conducting Hydrogels

Hydrogels are cross-linked polymer networks that swell in aqueous environments due to their highly hydrophobic nature. Conducting hydrogels are composites of conducting polymers and hydrogels integrated at the molecular level. The goal of conducting hydrogels is to combine the electrical functionality of conducting polymers with the mechanical and biological properties of hydrogels by creating interpenetrating polymer networks.[50] Hydrogels have mechanical properties akin to those of neural tissue, reducing strain mismatch at the neural interface. Furthermore, the use of conducting polymer–hydrogel composites may avoid the degradation of mechanical properties upon incorporation of bioactive molecules such as that experienced by conventional conducting polymers. Due to the open, swollen hydrogel network, conducting hydrogels can accommodate larger

Table 8.1 Review of conducting hydrogel systems.

Hydrogel	Conducting polymer	Dopant	Ref.
Polyacrylamide	PPy	pTS, NaNo$_3$	Kim[51]
Polyacrylamide	PPy	Potassium persulfate	Barthus[52]
Polyacrylate	PPy	FeCl$_3$	Lin[53]
Poly(acrylic acid)	PEDOT	PSS	Dai[54]
Poly(vinyl alcohol)/poly(acrylic acid)	PEDOT	PSS	Lu[55]
Poly(vinyl alcohol)/heparin	PEDOT	Heparin (gel)	Green[10]
Poly(hydroxyethylmethacrylate-*co*-3-sulfopropylmethacrylate)	PPy	Sulfopropylmethacrylate (gel)	Justin[56]

volumes of non-dopant biomolecules than conventional conducting polymer coatings. Several composite systems have been developed, a summary of which is provided in Table 8.1.

Barthus *et al.* developed and characterised a PPy/sodium nitrate–polyacrylamide conducting hydrogel for controlled drug delivery.[52] The distribution of PPy/sodium nitrate within the polyacrylamide was found to be heterogeneous (semi-interpenetrating networks), tending towards a more homogenous distribution with greater hydrogel mesh sizes. The conducting hydrogel coating was capable of controlled release of safranin, a model drug compound.

Despite the number of systems reported, there has been minimal confirmation of totally interpenetrating networks with many composites having poor network integration resulting in distinct polymer phases. The degree of polymer network integration is dependent upon the fabrication technique, namely the manner in which the conducting polymer component is doped.[10]

The conventional fabrication route for creating conducting hydrogels is the electrochemical deposition of conducting polymer within a preformed hydrogel matrix. The hydrogel component is physically or covalently cross-linked from an aqueous precursor solution of oligomer or macromer onto an electrode substrate. The hydrogel network is then swelled in conducting polymer precursor solution before deposition of a conducting polymer matrix within the hydrogel network. Dopants are commonly incorporated as mobile molecules in solution within the mesh network of the hydrogel. This causes electrodeposition to be localised to the electrode surface where the oxidation potential is highest, resulting in minimal polymer integration.

TIP: *The authors deposit a thin pre-layer of conducting polymer onto the working electrode before forming the hydrogel layer. This pre-layer allows the hydrogel layer to mechanically 'grip' onto the electrode. To encourage integration of the two polymer networks when depositing conducting polymer through the hydrogel the authors utilise dopant that is covalently associated with the gel network, either through direct modification of hydrogel macromer or by covalently cross-linking dopant containing groups into the gel network.*

Green *et al.* fabricated conducting hydrogels with substantially inter-penetrating networks using a two-component biosynthetic hydrogel with co-valently incorporated dopant.[10] The gel component consisted of 18 wt% poly(vinyl alcohol) (PVA) and 2 wt% heparin (highly anionic glycosaminogly-can), cross-linked together using methacrylate functional groups and photo-initiated polymerisation. Covalent incorporation of the dopant throughout the hydrogel guided the nucleation and deposition of PEDOT within the gel during electrodeposition. The resulting conducting hydrogel had a slight in-crease in impedance and approximately half the CSC of conventional PEDOT/pTS electrodes, but retained a considerable electrical improvement compared to bare platinum electrodes. The conducting hydrogels were found to be capable of supporting PC12 cell attachment and neurite growth.

The electrical properties of conducting hydrogels can be improved through nanostructuring of the polymer hybrid. Baek *et al.* fabricated nanostructured thin film conducting hydrogels by depositing PEDOT/pTS around poly(2-hydroxyethyl methacrylate) brushes which had been bound to the surface of a gold electrode *via* surface initiated atom-transfer radical-polymerisation.[57] The hybrid material was found to have increased CSC and neural cell growth (PC12 cell line) compared to conventional PEDOT/pTS.

Research on conducting hydrogels is still in its infancy; moving forward key areas of research involves investigating the role of the dopant and interaction between the two polymer components during polymerisation and the affect these have on the resultant composite structures and prop-erties as well as the role of biofunctionality within conducting hydrogels.

8.5 Conclusion

Conducting polymers are a promising alternative to bare metal electrodes for use in neuroprosthetic devices and are an enabling technology for biosensors and nerve guides. They have been proven to be beneficial in functional *in vivo* tests. Imparting biofunctionality to conducting polymer coatings through incorporation of biologically active molecules has the potential of creating high quality neural interfaces for electrical recording and stimulation of neural tissue. However, major challenges exist in the development of bioac-tive conducting polymers such as balancing biofunctionality with electrical and mechanical stability. Several approaches to overcoming these limitations have been developed. Creating conducting polymer composites, such as conducting hydrogels allows for the development of electrode coatings with all the benefits of bioactive conducting polymers while maintaining electrical and mechanical stability, providing a material platform for the creation of high quality neural interfaces for next-generation neuroprosthetic devices.

A Appendix: Methodologies and Practical Advice

The following section details methods commonly used by the authors in the fabrication and characterisation of conducting polymers with a focus on providing practical advice for these methods.

A.1   Fabrication: Electrochemical Deposition

- During deposition it is important to keep the applied potential within a certain window, dictated by the specific conducting polymer in use. The minimum value is the potential at which the monomer unit is oxidised and can begin to form oligomers and deposit onto the working electrode. The maximum value is the potential at which the electrochemical stability of the conducting polymer is exceeded, resulting in an irreversible inactivation of the electrochemical properties of the polymer. PEDOT is much more electrochemically stable than PPy (over-oxidation at 1.6 V and 0.6 V, respectively).
- Electrochemical deposition can be carried out in either galvanostatic or potentiostatic modes. The potentiostatic mode allows for control of oxidation/over-oxidation whereas the galvanostatic mode allows for control over the kinetics of deposition (growth rate and film thickness). It is the authors' opinion that galvanostatic deposition produces more reliable samples.
- Depositing films will initially grow in a 2D manner, but will begin to shift to a 3D nodular growth as film thickness increases. Smooth 2D growth can be encouraged through the use of low current densities (*i.e.*, slower deposition). The use of poly(styrene sulfonate) as a dopant will also result in smoother films compared to films doped with *p*-toluenesulfonate.
- Smaller working electrodes with counter-electrodes of equivalent size and lower current densities will help to produce more uniform coatings.
- The type of electrode substrate chosen for deposition will considerably impact upon the quality of the electrode coating. Platinum and gold electrodes have been found to produce mechanically and electrically superior coatings compared to other substrate types.
- The authors typically use a precursor solution containing 0.1 M EDOT monomer and 0.05 M dopant in a 1:1 water–acetonitrile solution and deposit galvanostatically at current densities in the range of 0.5–3 mA cm^{-2}. Deionised water should always be used to minimise contamination with ionic species. PPy has better aqueous solubility than PEDOT and can be deposited from entirely aqueous solutions; however, it is less stable in aqueous environments and typically requires distillation and purging.

A.2   Characterisation

A.2.1   Cyclic Voltammetry

A three-electrode set-up is utilised for CV. The coated electrode (sample) is the working electrode placed at the base of the cell. A length of platinum wire is used as the counter electrode and is inserted into the top of the cell (to maximise the distance between counter and working electrodes). A Ag/AgCl reference electrode is inserted into the cell and placed down near (but not in contact with) the surface of the working electrode. The cell is then filled with an electrolyte solution (physiological saline, DPBS or serum containing media). The voltage limits for the scan must remain inside the water window, a typical range is −0.6 to 0.8 V (*versus* Ag/AgCl). Voltage scan rates of 50–150 mV s^{-1} are used, with higher scan rates producing larger, more distinct redox peaks.

A.2.2 Electrochemical Impedance Spectroscopy

The physical setup for EIS is the same as for CV; however, the stimulation technique differs. A sufficient stimulation amplitude (∼100 mV AC) is required to elicit a measurable response. It may be necessary to use a Faraday cage to electrically isolate the sample from environmental interference.

A.2.3 Charge Injection Limit/Biphasic Stimulation

It is necessary to assess not only the material properties of conducting polymers, using electrochemistry characterisation techniques, but also appropriate application specific properties. Most devices use biphasic stimulation to activate tissue and as such the authors adopt a current-controlled charge balanced biphasic stimulus to examine the voltage transients which result from active stimulation. These are square waveforms (refer to Figure 8.8) which can be applied between a pair of stimulation sites, usually comprised of microelectrodes within an array. It is important to stimulate between electrodes fabricated from the same material to prevent bias of the measured waveform from dissimilar materials. Additionally, the electrolyte should be an adequate representation of the fluid in which the device is intended to operate following implantation. The authors use artificial perilymph or cell culture media. Saline is not suitable as it does not contain appropriate protein content to adequately reflect the neural interface.

A.2.4 Film Delamination

The authors use a modified ASTM tape adhesion test (D3359-02) in order to evaluate coating adhesion and friability. This method involves using a scalpel to make an X-cut in the surface of the coating to reveal the underlying substrate. Standard 3M 'Low-medium adhesion masking tape' is then applied to the coated surface and allowed to sit for 5 min. The tape is slowly pulled off the electrode surface. An optical microscope is used to image the electrode coating before the application of the tape and after its removal. Analysis of images involves converting to greyscale and thresholding to isolate the coating from the substrate. A pixel count allows for the determination of the percentage of coating that was entirely removed as a measure of delamination. Optical imaging of the coating left on the surface of the tape accounts for partial removal of the coating and gives an indication of the friability of the coatings.

A.2.5　PC12 Neurite Outgrowth Assay

PC12 is a cell line derived from a pheochromocytoma of a rat adrenal medulla. It is a non-adherent cell line which when exposed to NGF will differentiate and grow neuronal processes. The cells are cultured in RPMI media containing 10% horse serum and 5% foetal bovine serum. Although being non-adherent, the cells will clump together and therefore require aspiration before counting and plating. The cells can be aspirated through repeated ($\sim$70 times) drawing and expulsion through a 1000 µL pipette. Samples will require coating in laminin in order to encourage cellular attachment; however, this may be omitted if the sample contains its own biological molecules for cellular attachment. For the outgrowth assay cells are plated at 20 000 cells cm^{-2} in RPMI containing 1% horse serum and 50 ng mL^{-1} NGF. A two-thirds media refresh is carried out after 48 h and the samples are analysed at 96 h using a live/dead stain. Neurite density can be determined by tracing neurites in ImageJ using the NeuronJ plug-in.

References

1. N. K. Guimard, N. Gomez and C. E. Schmidt, *Prog. Polym. Sci.*, 2007, **32**, 876–921.
2. T. F. Otero, J. G. Martinez and J. Arias-Pardilla, *Electrochim. Acta*, 2012, **84**, 112–128.
3. R. A. Green, S. Baek, N. H. Lovell and L. A. Poole-Warren, *Nanostructured Conductive Polymers*, John Wiley & Sons, Ltd, 2010, pp. 707–736.
4. L. Poole-Warren, N. Lovell, S. Baek and R. Green, *Expert Rev. Med. Devices*, 2010, **7**, 35–49.
5. D. Svirskis, J. Travas-Sejdic, A. Rodgers and S. Garg, *J. Controlled Release*, 2010, **146**, 6–15.
6. D. E. Tallman, C. Vang, G. G. Wallace and G. P. Bierwagen, *J. Electrochem. Soc.*, 2002, **149**, C173.
7. S. Baek, R. A. Green and L. A. Poole-Warren, *J. Biomed. Mater. Res., Part A*, 2013, 2743–2754.
8. S. F. Cogan, *Annu. Rev. Biomed. Eng.*, 2008, **10**, 275–309.
9. J. Yang and D. C. Martin, *J. Mater. Res.*, 2006, 5(21), 1124–1132.
10. R. A. Green, R. T. Hassarati, J. A. Goding, S. Baek, N. H. Lovell, P. J. Martens and L. A. Poole-Warren, *Macromol. Biosci.*, 2012, **12**, 494–501.
11. R. A. Green, N. H. Lovell and L. A. Poole-Warren, *Acta Biomater.*, 2010, **6**, 63–71.
12. P. M. George, A. W. Lyckman, D. A. LaVan, A. Hegde, Y. Leung, R. Avasare, C. Testa, P. M. Alexander, R. Langer and M. Sur, *Biomaterials*, 2005, **26**, 3511–3519.
13. J. M. Fonner, L. Forciniti, H. Nguyen, J. D. Byrne, Y.-F. Kou, J. Syeda-Nawaz and C. E. Schmidt, *Biomed. Mater.*, 2008, **3**, 034124.
14. R. A. Green, N. H. Lovell, G. G. Wallace and L. A. Poole-Warren, *Biomaterials*, 2008, **29**, 3393–3399.

15. W. R. Stauffer and X. T. Cui, *Biomaterials*, 2006, **27**, 2405–2413.
16. S. Y. Kim, K.-M. Kim, D. Hoffman-Kim, H.-K. Song and G. T. R. Palmore, *ACS Appl. Mater. Interfaces*, 2011, **3**, 16–21.
17. B. C. Thompson, S. E. Moulton, R. T. Richardson and G. G. Wallace, *Biomaterials*, 2011, **32**, 3822–3831.
18. J. H. Collier, J. P. Camp, T. W. Hudson and C. E. Schmidt, *J. Biomed. Mater. Res.*, 2000, **50**, 574–584.
19. M. R. Abidian, K. A. Ludwig, T. C. Marzullo, D. C. Martin and D. R. Kipke, *Adv. Mater.*, 2009, **21**, 3764–3770.
20. R. T. Richardson, A. K. Wise, B. C. Thompson, B. O. Flynn, P. J. Atkinson, N. J. Fretwell, J. B. Fallon, G. G. Wallace, R. K. Shepherd, G. M. Clark and S. J. O'Leary, *Biomaterials*, 2009, **30**, 2614–2624.
21. D. D. Ateh, H. A. Navsaria and P. Vadgama, *J. R. Soc., Interface*, 2006, **3**, 741–752.
22. V. S. Polikov, P. A. Tresco and W. M. Reichert, *J. Neurosci. Methods*, 2005, **148**, 1–18.
23. A. Mercanzini, P. Colin, J.-C. Bensadoun, A. Bertsch and P. Renaud, *IEEE Trans. Biomed. Eng.*, 2009, **56**, 1909–1918.
24. R. A. Green, N. H. Lovell and L. A. Poole-Warren, *Biomaterials*, 2009, **30**, 3637–3644.
25. X. Liu, Z. Yue, M. J. Higgins and G. G. Wallace, *Biomaterials*, 2011, **32**, 7309–7317.
26. J. E. Collazos-Castro, G. R. Hernández-Labrado, J. L. Polo and C. García-Rama, *Biomaterials*, 2013, **34**, 3603–3617.
27. J. Goding, L. Poole-warren, R. Green and P. Martens, *Ther. Delivery*, 2012, **3**, 1–7.
28. C. E. Schmidt, V. R. Shastri, J. P. Vacanti and R. Langer, *Proc. Natl. Acad. Sci. U. S. A.*, 1997, **94**, 8948–8953.
29. M. Gerard, A. Chaubey and B. D. Malhotra, *Biosens. Bioelectron.*, 2002, **17**, 345–359.
30. R. A. Green, R. T. Hassarati, L. Bouchinet, C. S. Lee, G. L. M. Cheong, J. F. Yu, C. W. Dodds, G. J. Suaning, L. A. Poole-Warren and N. H. Lovell, *Biomaterials*, 2012, **33**, 5875–5886.
31. T. Boretius, M. Schuettler and T. Stieglitz, *Artif. Organs*, 2011, **35**, 245–248.
32. R. A. Green, P. B. Matteucci, R. T. Hassarati, B. Giraud, C. W. D. Dodds, S. Chen, P. J. Byrnes-Preston, G. J. Suaning, L. A. Poole-Warren and N. H. Lovell, *J. Neural Eng.*, 2013, **10**, 016009.
33. W. Schuhmann, C. Kranz, H. Wohlschläger and J. Strohmeier, *Biosens. Bioelectron.*, 1997, **12**, 1157–1167.
34. D. V. Bax, R. S. Tipa, A. Kondyurin, M. J. Higgins, K. Tsoutas, A. Gelmi, G. G. Wallace, D. R. McKenzie, A. S. Weiss and M. M. M. Bilek, *Acta Biomater.*, 2012, **8**, 2538–2548.
35. P. M. George, D. A. LaVan, J. A. Burdick, C.-Y. Chen, E. Liang and R. Langer, *Adv. Mater.*, 2006, **18**, 577–581.
36. R. Wadhwa, C. F. Lagenaur and X. T. Cui, *J. Controlled Release*, 2006, **110**, 531–541.

37. B. C. Thompson, S. E. Moulton, J. Ding, R. Richardson, A. Cameron, S. O'Leary, G. G. Wallace and G. M. Clark, *J. Controlled Release*, 2006, **116**, 285–294.
38. R. T. Richardson, B. Thompson, S. Moulton, C. Newbold, M. G. Lum, A. Cameron, G. Wallace, R. Kapsa, G. Clark and S. O'Leary, *Biomaterials*, 2007, **28**, 513–523.
39. K. Kontturi, P. Pentti and G. Sundholm, *J. Electroanal. Chem.*, 1998, **453**, 231–238.
40. B. Massoumi and A. Entezami, *J. Bioact. Compat. Polym.*, 2002, **17**, 51–62.
41. G. Kang, R. Ben Borgens and Y. Cho, *Langmuir*, 2011, **27**, 6179–6184.
42. X. Luo and X. T. Cui, *Electrochem. Commun.*, 2009, **11**, 1956.
43. B. H. Kim, D. H. Park, J. Joo, S. G. Yu and S. H. Lee, *Synth. Met.*, 2005, **150**, 279–284.
44. Y. Li and R. Qian, *Synth. Met.*, 1988, **26**, 139–151.
45. M. R. Abidian, D.-H. Kim and D. C. Martin, *Adv. Mater.*, 2006, **18**, 405–409.
46. W. Shain, L. Spataro, J. Dilgen, K. Haverstick, S. Retterer, M. Isaacson, M. Saltzman and J. N. Turner, *IEEE Trans. Neural Syst. Rehabil. Eng.*, 2003, **11**, 186–188.
47. B. Guo, L. Glavas and A.-C. Albertsson, *Prog. Polym. Sci.*, 2013, **38**, 1263–1286.
48. R. A. Green, C. M. Williams, N. H. Lovell and L. A. Poole-Warren, *J. Mater. Sci.: Mater. Med.*, 2008, **19**, 1625–1629.
49. X. Luo, C. Matranga, S. Tan, N. Alba and X. T. Cui, *Biomaterials*, 2011, **32**, 6316–6323.
50. R. A. Green, S. Baek, L. A. Poole-Warren and P. J. Martens, *Sci. Technol. Adv. Mater.*, 2010, **11**, 014107.
51. B. C. Kim, G. M. Spinks, G. G. Wallace and R. John, *Polymer*, 2000, **41**, 1783–1790.
52. R. C. Barthus, L. M. Lira and S. I. C. De Torresi, 2008, **19**, 630–636.
53. J. Lin, Q. Tang, J. Wu and Q. Li, 2010, **116**, 1376–1383.
54. T. Dai, X. Qing, Y. Lu and Y. Xia, *Polymer*, 2009, **50**, 5236–5241.
55. Y. Lu, Y. Li, J. Pan, P. Wei, N. Liu, B. Wu, J. Cheng, C. Lu and L. Wang, *Biomaterials*, 2012, **33**, 378–394.
56. G. Justin and A. Guiseppi-Elie, *Biomacromolecules*, 2009, **10**, 2539–2549.
57. S. Baek, R. Green, A. Granville, P. Martens and L. Poole-Warren, *J. Mater. Chem. B*, 2013, **1**, 3803.

CHAPTER 9

Polycaprolactone-based Scaffolds Fabricated Using Fused Deposition Modelling or Melt Extrusion Techniques for Bone Tissue Engineering

PATRINA S. P. POH,[a] MICHAL BARTNIKOWSKI,[a,b]
TRAVIS J. KLEIN,[b] GILES T. S. KIRBY[a] AND
MARIA A. WOODRUFF*[a]

[a] Biomaterials and Tissue Morphology Group, Institute of Health and Biomedical Innovation, Queensland University of Technology, Australia; [b] Cartilage Regeneration Laboratory, Institute of Health and Biomedical Innovation, Queensland University of Technology, Australia
*Email: mia.woodruff@qut.edu.au

9.1 Introduction

9.1.1 Bone Tissue Engineering

Modern tissue engineering is a convergence of different disciplines. Broadly, tissue engineering is a combination of successes from clinical medicine, engineering and science. The current driving force for developments in tissue engineering is a lack of transplants and a shortage of efficacious implants for human organ or tissue replacements. Tissue engineering has

RSC Smart Materials No. 10
Biointerfaces: Where Material Meets Biology
Edited by Dietmar Hutmacher and Wojciech Chrzanowski
© The Royal Society of Chemistry 2015
Published by the Royal Society of Chemistry, www.rsc.org

become synonymous with regenerative medicine as the boundaries between replacement and regeneration become blurred.

Bone is a connective tissue with a number of important roles. Bone supports and protects organs, transfers loads, produces red and white blood cells, stores minerals, stores growth factors and fats and even has a role in sound transduction. Millions of years of evolution have led to mechanisms allowing bone to accomplish these tasks. Bone is capable of considerable healing and remodelling. Bone is constantly remodelling so, eventually, sites of repair become indistinguishable from native bone. Unfortunately, healing is not always spontaneous; cases such as non-union fractures, cancer explant sites or joint replacement surgeries may require a greater degree of intervention.[1] When additional bone is required to structurally and biologically support a defect, the clinical gold standard is to use autologous bone graft as filler. This contains an effective mixture of minerals, growth factors and cells which facilitate osteoconduction (bone growth on the surface of a material), osteoinduction (the stimulation of undifferentiated or pluripotent cells into bone-forming cells) and osteogenesis (the process of osteoblasts laying down new bone).[2] Unfortunately, supply of this valuable material is very limited. The most effective grafts are autologous and this requires removal of bone from a donor site, often leading to complications such as donor site morbidity. What is required is a filler material with equivalent (or greater) efficacy as autologous graft. This is where bone tissue engineering cones in. It is hypothesised that we can generate biodegradable synthetic scaffolds with the required mechanical and biological properties to support the target site and induce effective bone regeneration. The recent resurgence of poly-caprolactone (PCL) research[3] has coincided with the tissue engineering additive manufacturing (AM) revolution. PCL has excellent biomechanical properties, suited to bone repair and has a melting point of just 59–64 °C, lower that other medical grade polyesters, meaning scaffolds can be formed using melt techniques.

9.1.2 Polymers Used in Bone Tissue Engineering

Natural and synthetic polymers have been investigated extensively for orthopaedic repair. Natural polymers are generally animal derived, potentially leading to batch-to-batch variation and a risk of pathogenic contamination. Synthetic polymer synthesis can be easily manipulated and the material purities can be more easily controlled. Consequently, the mechanical and physical properties of synthetic polymers can be accurately predicted and reproduced. For these reasons, a large proportion of scaffold research has diverged from natural polymers towards the use of synthetic polymers. Synthetic polymers widely used as scaffold materials in tissue engineering are detailed in Table 9.1.

PCL belongs to the family of aliphatic polyesters, as do poly(lactic acid) (PLA) and poly(glycolic acid) (PGA) which are known for their biomedical applications. PCL exhibits superior rheological and viscoelastic properties

Table 9.1   Selected synthetic polymers with their respective melting points, glass transition temperatures, degradation time and degradation products.

Polymer	Melting point (°C)	Glass transition (°C)	Degradation time (months)	Degradation products
Poly(lactide)	173 to 178	60 to 65	6 to 12	L-Lactic acid
Poly(glycolide)	225 to 230	35 to 40	>24	Glycolic acid
Poly(caprolactone)	59 to 64	−65 to −60	>24	Caproic acid
Poly(DL-lactide-*co*-glycolide	Amorphous	45 to 55	5 to 6	DL-Lactic acid and glycolic acid
Poly(L-lactide-*co*-DL-lactide)	Amorphous	55 to 60	12 to 16	Lactic acid
Polydioxanone	—	−10 to 0	6 to 12	Glyoxylic acid
Poly(glycolide-*co*-γ-caprolactone)	Amorphous	—	1 to 2	Lysine, glycolic and caproic acids
Poly(DL-lactide-*co*-caprolactone)	Amorphous	—	>24	Lactic acid and caproic acid

over most other aliphatic polyesters, exhibiting a degradation temperature (T_d) of 350 °C,[4] an extremely low glass transition temperature (T_g) of −60 °C, a low melting temperature (T_m) of 59–64 °C[3,4] and an *in vivo* degradation time in excess of 24 months.[5] It is these properties that make PCL such a promising candidate for long term biodegradable implants.[6,7] These properties also allow PCL to be easily manipulated into variety of shapes with relative ease. PCL has already been approved by the US Food and Drug Administration (FDA) for use in several biomedical devices. This gives PCL added value as a candidate in tissue engineering research because further PCL medical devices should have a much easier route-to-market with fewer regulatory boundaries and a mapped quality control framework. An extensive review on PCL properties, applications and resurgence in the biomedical field has been published by Woodruff and Hutmacher,[3] thus further detail will not be presented here but rather this brief overview will focus on specific additive manufacturing techniques used to produce three-dimensional PCL scaffolds for intended use in bone tissue engineering.

9.2   Scaffold Fabrication Techniques

Fabrication of scaffolds may be achieved through conventional or rapid prototyping (RP) techniques. Conventional fabrication techniques include particulate leaching, phase separation, fibre meshing or bonding, melt moulding, gas foaming, membrane lamination, hydrocarbon templating, freeze drying, solution casting and emulsion freeze drying.[8] These conventional techniques have a number of limitations. They usually lack the interconnected channel microstructure required for good tissue infiltration and vascular ingrowth and organic solvents are often required for processing (harmful organic solvents present a potential regulatory issue). These conventional methods are generally manual techniques providing inconsistent and inflexible processing procedures meaning the processes are not easily scalable.[8]

Rapid prototyping is a general term that covers a number of techniques that convert a digital blueprint of an item into a three-dimensional (3D) model. These techniques are either *additive* (where material is added to build a model) or *subtractive* (where material is removed to reveal a model). Solid freeform fabrication and 3D printing are both terms used to describe AM technologies. AM is the official industry standard term (ASTM F2792) for all applications of the technology, defined by Wohlers Associates as the process of joining materials to make objects from 3D model data, usually layer upon layer.[9] The two methods of AM explored in this chapter will be fused deposition modelling (FDM) and melt extrusion (ME); these are the most well investigated additive techniques for bone tissue engineering.

Over the past two decades, AM technologies have been developed and commercialised with focus on the rapid manufacturing of prototypes for non-biomedical applications. As the field of tissue engineering has evolved, AM technologies have been adopted. These techniques offer precise control over the matrix architecture (size, shape, interconnectivity, branching, geometry and orientation), yielding biomimetic structures varying in design and even material composition. Through this the ability to control mechanical properties, biological integration and degradation kinetics of the scaffolds has been advanced. AM techniques are easily automated and integrated with imaging techniques with the potential to generate patient and/or application-specific scaffolds.[10] A review by Melchels *et al.*[11] has extensively discussed AM for tissue and organ printing.

9.2.1 Principles of Additive Manufacturing: Fused Deposition Modelling and Melt Extrusion

Typical AM processes use a number of methods to generate layer parameters that are translated into tool paths. The most common involve digitally slicing a computer aided design (CAD) model, digitally slicing computerised tomography (CT) data or direct generation of tool paths using mathematical algorithms. Using a CAD method provides the most hands-on approach as the final product can be visualised on a computer screen and easily manipulated. This also allows the generation of complex shapes with ease. Using CT data has the advantage of being the most physiologically accurate when generating implants. However, both these methods rely on pre-defined generic algorithms to translate the data into layer parameters and ultimately tool paths. While powerful and enabling, these algorithms have the limitation that they lack precise control of the scaffold geometry and architecture. If this is an important requirement then direct mathematical generation of tool paths is the most flexible and tailorable.

Various types of code can be used to communicate tool paths. Most commercial systems rely on confidential approaches through which CAD models are sliced and transferred to instructional parameters. Most research laboratories rely on a script known as G-code. In all cases, the slices are formed on the *x, y* plane, with the *z* axis presenting a variable layer spacing or thickness. An AM machine follows tool path directions in the *x, y* axes to generate a deposition layer. Once the layer is complete, the *z* axis is incremented by one slice thickness and the process is repeated until the three dimensional model is generated. Deposited microfilaments are deposited as a raster (a grid of filaments) and sometimes a contour is used to better define the edge of the structure. Laying down of patterns of filaments can be controlled and can be specified anywhere between 0° and 180° (with respect to the *x*-axis). This angle of deposition is known as the raster angle (RA).

FDM is a particular type of AM process, which uses spools of polymer filament (Figure 9.1a). Many spool systems contain two spools for the build and support materials. The support material is used to maintain the scaffold

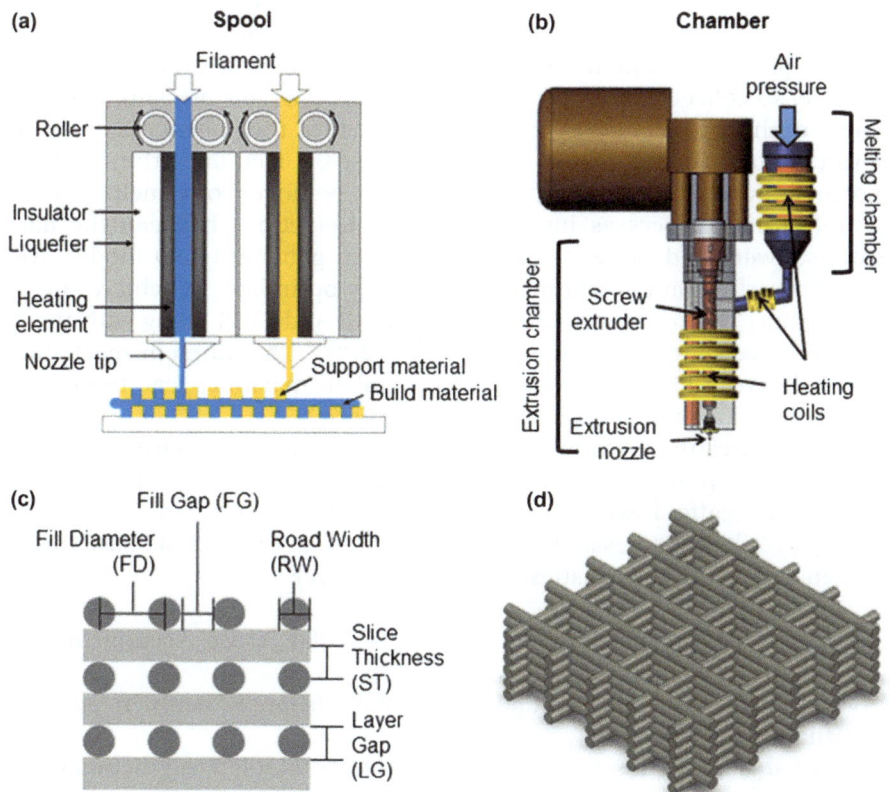

Figure 9.1 Schematic of the extrusion elements of spool (a) and chamber (b) FDM machines. Cross-section of a typical FDM fabricated scaffold (c). Three-dimensional image of an FDM fabricated scaffold (d).
Adapted from Zein *et al.*[13] and Domingos *et al.*[12]

structure and shape during the fabrication process. Once a polymer has been heated and deposited, it remains soft until it cools and the generation of complex overhanging shapes can lead to sagging and ultimate deformation of the final structure. This is where a support material is crucial. The support material is often water soluble facilitating easy removal. Driven by rollers, the polymer filaments are fed through the extrusion head and into the liquefier. The liquefier consists of heating elements which can be accurately controlled. This temperature is very important as it affects polymer flow characteristics; specific layer parameters are tailored to certain flow characteristics. As the filament passes through the liquefier, it is melted and extruded through a nozzle onto the collecting platform (Figure 9.1a).

Another variation of AM process, known as ME, uses a melting chamber (rather than spools) (Figure 9.1b) allows the use of different and exotic polymers that are unavailable in spools.[12] Medical grade polymers ordered in small custom batches can be fabricated into scaffolds using this method. The temperature of melting and extrusion chambers can be accurately controlled. Air pressure is applied to the inlet of the melting chamber to push molten polymer against the inlet of the extrusion chamber. This positive pressure gradient then allows a rotating screw within the extrusion chamber to collect molten polymer and exert a constant pressure along the length of the screw. Once the nozzle tip is reached the polymer is deposited onto the collecting platform. This rotating screw facilitates the mixing of molten polymer thereby producing a more homogeneous melt. Additive manufacturing processes involve complex interactions between the hardware, software and material properties. When generating 3D models with ME, processing parameters must be carefully optimised to achieve the desired result. The processing parameters of ME and FDM along with the respective effects are summarised in Table 9.2.[13,14] The different mechanical set-ups between FDM (spool) and ME (chamber) machines lead to different terminology in some cases as shown in Table 9.2.

When fabricating a scaffold for bone tissue engineering purposes, it is of the utmost importance to optimise the parameters listed in Table 9.2. to generate a scaffold with favourable porosity and mechanical strength. Changes to these processing parameters can affect the scaffold structure by influencing the characteristics below (Figure 9.1c):

- *Road width*: the diameter of the circular cross section of laid microfilament
- *Fill gap*: the edge-to-edge horizontal distance between adjacent filaments
- *Fill diameter*: the centre-to-centre horizontal distance between two consecutive filaments in the same layer
- *Slice thickness*: the vertical distance between the filament centre of adjacent layers
- *Layer gap*: the edge-to-edge vertical distance between layers of the same microfilament alignment

Table 9.2 Processing parameters using FDM and ME with their respective effects.[13,14]

FDM (spool)	ME (chamber)	Effects
Roller speed	Air pressure and rotating screw speed	Determines the material feed rate into the liquefier (FDM)/extrusion chamber (ME) and the molten material out of the nozzle. Changes in roller speed/air pressure/rotating screw speed will affect the flow rate of molten material out of the nozzle and the road width (RW) of the laid microfilament.
Liquefier temperature	Temperature of the heating coils	Changes in temperature affect the viscosity of the material. Raising the temperature beyond the material melting temperature can reduce material viscosity. Excess temperature may damage the material, or induce degradation of molecular weight over time.
Speed of the extrusion head	Speed of the collecting platform	Determines the output of machine. Increased extrusion had/collecting platform speed = increased machine output.
Direction of deposition		Determines the lay-down pattern/raster angle (RA) of scaffold.
Nozzle size		Determines the diameter/RW of the extruded microfilament.
Nozzle translational speed		Ensures consistent and uniform extrusion of the microfilament. Can be regulated to allow for stretching of the microfilament. This parameter works in conjunction with the flow rate.
Fill gap (FG)		Determines the distance between laid microfilaments. This parameter determines the stability and porosity of the engineered construct.
Flow rate		Defined as the speed of extrusion of molten material. Determined by the viscosity of molten material, roller speed

9.3 Polycapralactone Scaffolds

9.3.1 Physical Characteristics of Polycaprolactone Scaffolds

The requirements of a scaffold for bone TE are very demanding. As well as meeting the requirements of a 3D scaffold to support cell attachment and survival, these scaffolds also need to provide suitable mechanical support post-implantation and throughout the duration of healing. A number of candidate polymers possess suitable properties and this chapter will focus on PCL and composites thereof. A number of structures have been explored using PCL for bone tissue engineering and several are detailed in Figure 9.2. The recent resurgence of PCL research has coincided with the tissue engineering AM revolution. Initial studies focused on two main areas: (1) the porosity/strength compromise of FDM/ME PCL scaffolds, and (2) *in vivo* and *in vitro* biocompatibility of these scaffolds.

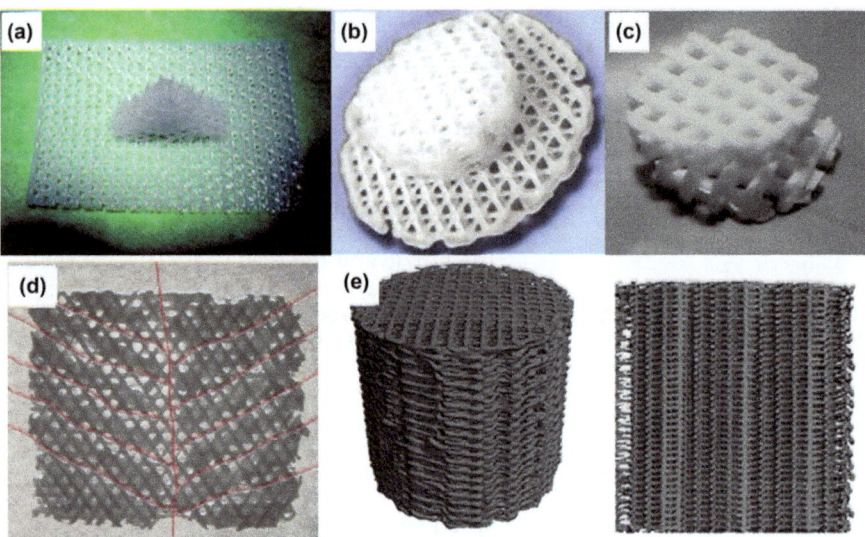

Figure 9.2   A selection of PCL-based scaffolds fabricated using AM techniques. (a) A customised PCL scaffold used for mandibular defect reconstructions.[81] (b) An Osteoplug™ implant designed to cover burr-hole skull defects.[80] (c) A bobbin-shaped scaffold to maximise empty space and increase vascularisation.[75] (d) A scaffolds with incorporated vascular channels (indicated in red) in an arborising pattern following an open channel design.[74] (e) A customised scaffold used to repair a segmental defect in a long bone.[21]

In 2002, Zein *et al.*[13] explored the effects of different fabrication parameters for FDM PCL porous scaffolds and how these affect porosity and, in turn, mechanical strength. Different raster angles (0/90° and 0/60/120°) and FG lengths (0.508, 0.610 or 0.711 mm) were explored along with two different nozzle tip sizes (254 or 406 μm).[13] All fabricated scaffolds exhibited a regular honeycomb-like pattern and fully interconnected pore channels. As would be expected, smaller tip sizes produced scaffolds with larger channel diameters [as long as the FG (Figure 9.1c) was kept constant]. This decrease in tip size increased porosity from 48–61% to 71–77%.[13] It was observed, however, that with smaller tip sizes, de-lamination was more likely to occur between layers. The fabricated scaffolds were mechanically tested and it was observed that increased porosity led to decreased compressive strength, yield strength and yield strain, regardless of lay-down pattern, FG and RW. This inverse relationship between porosity and scaffold mechanical strength means that there will likely be a compromise when fabricating a scaffold for bone tissue engineering applications. This is because the scaffold obviously needs to be strong but it also requires suitable porosity for tissue in growth and integration. The porosity allows the movement of cells through the scaffold as well as the diffusion of nutrients/waste. Ideally, this porous structure will support and assist vascular integration into the scaffold. A similar relationship between scaffold strength and porosity was reported by Hoque *et al.*[15] Increased deposition speeds (240–360 mm min^{-1}) resulted in reduced RW and consequently higher porosity (45 $\pm$ 1.80 to 75 $\pm$ 3.0%) for PCL scaffolds.[15] This lead to a proportional decrease in yield strength (4.58 $\pm$ 0.18 to 2.12 $\pm$ 0.08 MPa).[15]

It should be noted that Zein *et al.*,[13] carried out mechanical testing following the ASTM F451 standard for testing acrylic bone cement. This standard has no requirement for mimicking *in vivo* conditions such as moisture during testing so it could be argued that the mechanical strength values obtained are potentially not physiologically relevant.[16] It has in fact been shown that similarly fabricated PCL scaffolds behave very differently when mechanically tested in saline at 37 °C.[17] PCL scaffolds, fabricated with two different raster angles (0/60/120° and 0/72/144/36/108°), each with 61% porosity, were all shown to be weaker when tested in saline at 37 °C for both stiffness and yield strength.[17] Stiffness dropped from 41.9 $\pm$ 3.5 to 29.4 $\pm$ 4.0 MPa and yield strength dropped from 3.1 $\pm$ 0.1 to 2.3 $\pm$ 0.2 MPa.[17] Another factor that few consider is the sterilisation method used prior to cell culture/implantation. There is some evidence to suggest that ethanol sterilisation has a dramatic effect on the mechanical properties of long-chain polyesters.[18,19] If PCL constructs are to be applied to load-bearing situations based solely on mechanical testing data then it is very important that this data applies to the actual *in vivo* situation and it is recommended to always test samples as close to the physiologically relevant scenario as possible.

By nature, PCL will never have the same mechanical properties as bone but it is important to consider this during the initial fabrication/characterisation stages. The exact requirements of scaffolds for bone tissue engineering remain largely undefined; there are no target design tolerances for these materials but for load bearing applications the scaffold must behave as predicted *in vivo*. It appears that mechanical testing under physiological conditions is a necessity for analysis of additive manufactured PCL based scaffolds. PCL based scaffolds for load bearing applications are often used with some type of internal or external fixator.[20,21] Although these scaffolds are designed to provide structural support, they are not load-bearing.

9.3.2 Polycaprolactone Scaffold *In Vitro* Analysis

We know that the raster pattern affects mechanical properties, but how do these differences on the macro-scale influence cells? Hoque *et al.*[15] demonstrated that different raster patterns (0/30°, 0/60° and 0/90°) had no significant influence on the attachment or proliferation of smooth muscle cells. This is unsurprising considering the macro-scale of the scaffold struts, although it should be noted that a traditional static seeding method was used. The differences in attachment between the scaffolds may have varied *in vivo* or in a perfusion culture due to different flow characteristics imposed by the different scaffold architectures. This type of analysis was beyond the scope of this research paper and is an area of current scientific interest.

It is widely accepted that PCL, alone and without surface modification, is a poor substrate for cells to attach to and modification techniques are widely reported.[22-24] It should be noted, however, that cells can and have been cultured successfully on untreated PCL. It would appear that initially, cells attach and retain a spherical shape but after 2–3 days, matrix deposition and cell migration lead to cell–scaffold interactions typically seen in treated scaffolds.[15,17]

It also appears that the raster angle of PCL scaffolds has little effect on cell attachment and proliferation within parameters and under methods typically assessed. Hoque *et al.*[15] demonstrated that with three different RAs (0/30°, 0.60° and 0/90°) there was no significant influence on attachment or proliferation of rat smooth muscle cells.

Studies of untreated PCL scaffolds were taken a step further by Schantz *et al.*[25,26] when they assessed porous PCL constructs *in vivo*. Porous PCL scaffolds were generated with a triangular structure (0/60/120°) and an etching process was first reported whereby sodium hydroxide was used to roughen the PCL surface with the aim of increasing cell attachment by way of increasing the surface hydrophilicity of the PCL. Cells were seeded onto the scaffolds and fibrin glue was utilised to assist cell seeding and attachment. Attachment, proliferation and extracellular matrix (ECM) synthesis were all observed within the construct.[25]

9.4   Polycaprolactone-based Composite Scaffolds

Initial studies indicated that PCL scaffolds are a viable option for bone scaffolds, owing to their superior formability, relative low cost, high mechanical strength, biocompatibility, potential for high purity and ease of characterisation. The next step was to increase the bioactivity of these constructs with the aim of producing a bioactive scaffold able to compete with the clinical gold standard of autograft. PCL is bioinert; this is an advantage in that it does not initiate a response or interact with surrounding biological tissue in a negative way; unfortunately, it is also unable to induce bone formation. To overcome this drawback, various PCL-based composite scaffolds have been developed for tissue engineering applications. For bone regeneration, a bioactive material is capable of creating an osteogenic environment; promoting the development of a mineralised interface as a natural bonding junction between living tissues and non-living materials (*i.e.*, the scaffold itself) and subsequently encouraging regeneration of the defect sites into which the scaffold has been placed.[27] It is thus an imperative to emulate this through creation of an artificial 3D bioactive scaffold.

9.4.1   Polycaprolactone with Hydroxyapatite

Bioactive ceramics are biocompatible compounds that can either be related to the body's own material or more durable metal oxides. Bioceramics have a long history of use in the dental and orthopaedic fields. Examples include hydroxyapatite [HA; $Ca_{10}(PO_4)_6(OH)_2$] and tricalcium phosphate [TCP; $Ca_3(PO_4)_2$]. These have both been combined with PCL with the aim of improving the bioactivity of the scaffolds. Schantz *et al.*[28] performed *in vitro* studies on PCL–HA (75 : 25 wt%) scaffolds, using fibrin glue as the cell delivery matrix, to control and improve the cell seeding process and assist attachment and proliferation. Fibrin glue is a mixture of the plasma glycoprotein fibrinogen and the enzyme thrombin (both in solution); thrombin converts the soluble fibrinogen into insoluble fibrin strands. This is how a blood clot forms and it is a useful way to immobilise cells within a hydrogel that is initially injectable. PCL–HA composite scaffolds degraded three to four times faster than PCL-only scaffolds.[28] Histological and biochemical analyses have shown that both PCL-only and PCL–HA scaffolds seeded with bone marrow-derived mesenchymal stromal cells (MSCs) led to proliferation within the scaffold pores and differentiation of cells down the osteogenic lineage. The PCL-based scaffolds provided a physically stable framework for bone formation, while the fibrin glue acted as an effective cell delivery vehicle suitable for cell proliferation and differentiation.[28]

Shor *et al.*[29] indicated that both PCL-only and PCL–HA (75 : 25 wt%) scaffolds cultured for 21 days with primary foetal bovine osteoblasts supported cell growth and proliferation. Between the two groups, PCL–HA scaffolds produced more mineralised matrix and improved osteoblast differentiation compared to PCL-only scaffolds, which was confirmed both by SEM images and alkaline phosphatase activity assays.[29]

In a contrasting study, Chim *et al.*[30] demonstrated no significant difference between PCL only and PCL–HA (85 : 15 wt%) composite scaffolds when seeded with human calvarial osteoblasts. *In vitro* studies spanning 3 weeks showed that PCL and PCL–HA scaffolds had a constant rate of cell proliferation and increasing tissue growth throughout the depth of the implants, confirmed through light and confocal imaging microscopy.[30] Subsequent *in vivo* studies involved seeding the scaffolds with human calvarial osteoblasts and 21 days of culture days prior to subcutaneous implantation into balb C nude mice.[30] At 14 weeks, explanted PCL and PCL–HA constructs showed excellent mechanical strength and evidence of mineralisation identified using X-ray imaging. There were no significant differences detected between PCL and PCL–HA scaffolds. The ratio of HA to PCL in the composites (15 wt%) was relatively low, which may be the reason no significant differences were seen observed.[30]

9.4.2 Polycaprolactone with Tricalcium Phosphate

PCL–TCP and PCL–HA composites have demonstrated favourable mechanical and biochemical properties as well as favourable degradation and resorption kinetics in comparison to PCL alone. Lam *et al.*[31] have undertaken extensive studies on the degradation of FDM fabricated PCL and PCL–TCP scaffolds. Studies have shown that the addition of TCP particles within the

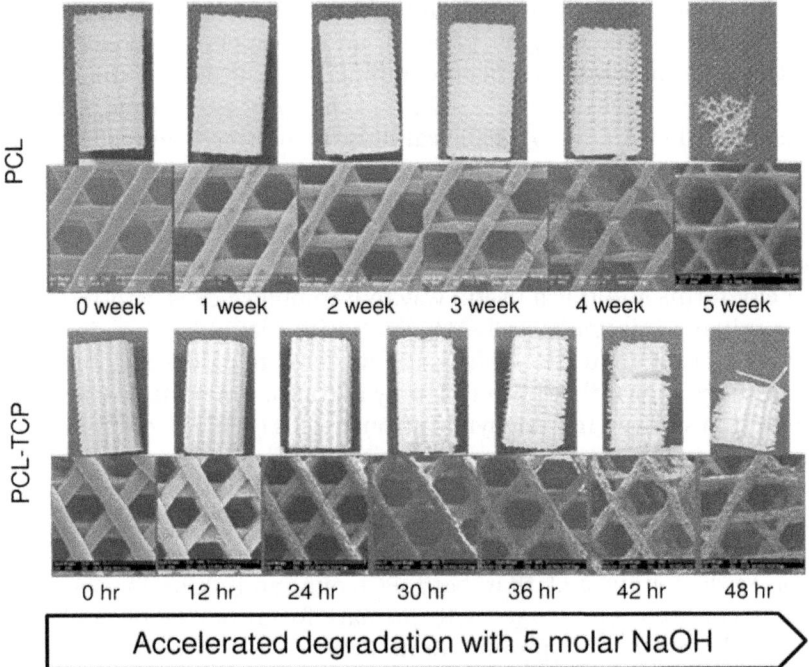

Figure 9.3   A macroscopic and microscopic overview of two scaffolds degrading over time in a 5 M NaOH solution. The scaffolds are fabricated from PCL alone and a mixture of PCL with TCP (80:20 wt%). Note the different time scales.
Adapted from Lam *et al.*[31]

PCL bulk increases the hydrophilicity of the bulk polymer.[31] This greatly increased the degradation rate of PCL–TCP scaffolds compared to PCL only scaffold.[5,31–33] Figure 9.3 shows the macroscopic and microscopic topography of PCL and PCL–TCP (80:20 wt%) when immersed in a 5 M sodium hydroxide solution. It was determined that PCL–TCP scaffolds had a faster degradation rate and almost completely degraded after just 48 h. In contrast, the PCL-only scaffolds gradually degraded over a 5 week period.[31] It has long been known that adding hydrophilic components to degradable polyesters is a method to control degradation rates. This is important because an ideal scaffold should degrade at a rate whereby natural tissues can regenerate and take over. Inclusion of different amounts of TCP is potentially a way to modulate degradation rates of scaffolds destined for different physiological regions.

This concept was taken a step further by Kim and Kim[34] They explored the concept of functional grading on PCL–TCP composites. The TCP concentration in each layer was varied (5–40 wt%). Two control groups of plain PCL–TCP (80:20 and 90:10 wt%) without functional grading were used. Figure 9.4 illustrates two scaffold compositions, one homogeneous mixture and one functionally graded.[34] The structures were analysed *in vitro* using

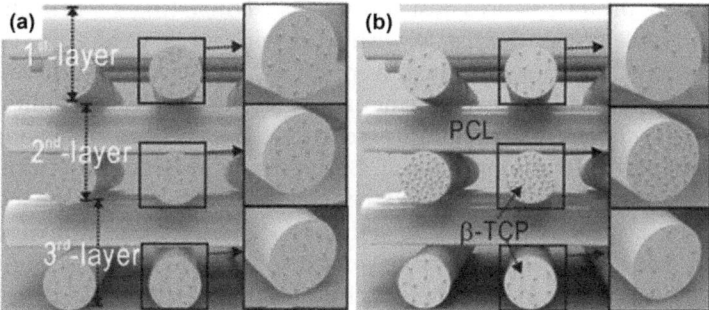

Figure 9.4 (a) Homogeneous PCL–TCP scaffold. (b) Functionally graded PCL–TCP scaffold. Note the different amounts of TCP in each layer.[34]

osteoblast-like cells (MG63). It was found that a 30% increase in water absorption was induced by functional grading, although the mechanical properties of this design were lower than that of the homogeneous control; this was thought to be caused by non-uniform stress distribution. Furthermore, an increase in cell attachment, proliferation and calcium deposition were recorded in the functionally graded groups after 14 days. However, cell viability in the functionally graded groups was lower than in the control groups.[34] Functional grading is a promising and cutting edge technique sure to be the focus of further research.

Zhou *et al.*[35] undertook a study comparing the osteogenic potential of FDM fabricated PCL with PCL–TCP (80 : 20 wt%) scaffolds (RA: 0/60/120°). The study confirmed that PCL–TCP scaffolds had improved hydrophilicity and mechanical properties over the control group.[35] Similarly, *in vitro* studies conducted using human alveolar osteoblasts also indicated that cells on PCL–TCP scaffolds demonstrate better proliferation, formation of denser multilayered cell sheets and earlier expression of bone matrix-related proteins than on PCL-only scaffolds.[35] It was found, however, that the bone specific proteins were mostly found at the border region of the constructs, with only a few in the interior. This uneven distribution of the matrix proteins was attributed to the potentially limited supply of nutrients available in static culture, as well as the formation of mineralised tissue initially occurring on the outer surface of the constructs, thereby possibly hindering the exchange of nutrients and waste to the interior part of the scaffolds. Authors suggested that the uneven tissue formation may be prevented by the use of a perfusion system which could assist the exchange of nutrients or *via* incorporation of osteoinductive factors into the central part of the scaffolds to promote tissue mineralisation.[35]

The same research group next combined FDM engineered PCL–TCP scaffolds with cell sheets (cells that had been grown to confluency and detached as a single sheet containing cells and extracellular matrix).[36] Studies *in vitro* revealed cells within the constructs had undergone osteogenic differentiation as indicated by elevated alkaline phosphatase levels and bone-related protein expression. Constructs consisting of cell sheets and scaffolds were assessed *in vivo* with and without 8 weeks pre-culture in osteogenic media. The constructs were then subcutaneously implanted into nude rats, which allowed assessment of the osteogenic capabilities of the scaffold–cell sheet constructs by assessing the ability to form bone in an ectopic site, away from bone tissue. Both groups indicated the formation of neo-cortical and well-vascularised cancellous bone within the constructs with up to a 40% bone volume. Histological and immunohistochemical analyses revealed that new bone formation followed predominantly an endochondral pathway, whereby chondrocytes proliferated, underwent hypertrophy and apoptosis whilst the ECM space was invaded by blood vessels, osteoclasts, bone marrow cells and osteoblasts. The osteoblasts deposited bone matrix into the ECM space and gradually the woven bone matrix matured into fully mineralised compact bone, exhibiting the histological markers of native bone.[36]

Abbah *et al.*[37] employed the same approach as Zhou *et al.*[36] whereby differentiated autogenous porcine bone marrow stromal cell sheets were wrapped around PCL–TCP (80 : 20 wt%) scaffolds. These scaffolds were then implanted into a pig model to assess spinal interbody fusion. Scaffolds without bone marrow stromal cell sheets served as a control. New bone formation was evident at 3 months in the cell group. At 6 months, bone fusion was observed in five out of eight scaffolds in the cell groups but no bone as observed in the control groups. Biomechanical analysis revealed that segmental stability was significantly increased in the cell group compared to the control group at 6 months post-implantation.[37] These two studies have shown that combining PCL–TCP scaffolds with cell sheet technology can contribute towards extensive *in vivo* bone formation thus have great potential in the field of bone tissue engineering.

It can be deduced from the aforementioned studies that effective cell delivery and distribution within scaffolds is essential in ensuring cell proliferation, differentiation and relevant extracellular matrix production; important prerequisites to enable new bone formation. Fibrin glue has been shown to improve the stability and cell attachment properties of scaffolds and hence enhance cell seeding efficiency as well as subsequent cell proliferation.[38] Investigations into the potential of fibrin glue and lyophilised collagen as cell-delivery matrices within PCL scaffolds were undertaken by Leong *et al.*[39] Adipose tissue-derived stem cells were seeded onto PCL–TCP scaffolds with either fibrin glue or lyophilised collagen. Results from *in vitro* studies showed that both fibrin and collagen were effective matrices for cell delivery into scaffolds, with adipose tissue-derived stem cells showing good adherence to the PCL–TCP scaffold surfaces and formation of bridging and overlapping cell sheets, as well as substantial ECM deposition within both types of scaffold matrix systems.[39]

Collagen was also used by Lee and Kim[40] to improve cell attachment onto PCL–TCP (60 : 40 wt%) scaffolds. It was shown that collagen coated PCL–TCP scaffolds exhibited improved surface hydrophilicity, enhanced initial cell attachment (especially at the central core of scaffold) and improved water-absorption ability compared to PCL–TCP-only scaffolds.[40] The same group also showed that electrospun PCL fibres embedded within PCL–TCP (60 : 40 wt%) scaffolds and subsequently coated with a mixture of collagen and hydroxyapatite improved cell viability and accelerated osteoblastic differentiation and mineralisation compared to uncoated scaffolds.[41] Long-term studies are still required to validate the findings.

Fibrin glue is still a popular choice for cell delivery. Rai *et al.*[42] fabricated scaffolds from PCL–TCP (80 : 20 wt%) (raster angle: 0/60/120°) and used these scaffolds as a cell transportation vehicle. Human mesenchymal stem cells with fibrin as the cell delivery matrix were assessed *in vivo* in critical-sized femoral defects in nude rats. The scaffolds were able to support cell proliferation and osteogenic differentiation. Fluorescently labelled human mesenchymal stem cells survived in the defect site up to 3 weeks after implantation. However, only 50% of the implants showed significant new bone formation.[42]

Much the same as standard PCL, PCL–TCP composite scaffolds can be easily fabricated in any shape or size using FDM techniques. The production of anatomically relevant constructs was explored by Lim *et al.*[43] PCL–TCP scaffold wedges were fabricated and used in a pig osteotomy model, wherein empty defects served as a control. Computerised tomography scans and bone mineral density analysis revealed that after 6 months, scaffold groups had a higher BMD compared to the control group. However, augmentation with a PCL–TCP scaffold wedge had no additional beneficial effects on bone consolidation when compared to the control group (empty defect).[43] This was an unexpected result, most likely due to specific nuances of the model such as defect location, scaffold composition or scaffold structure.

The results from all these studies have indicated that PCL or PCL–TCP scaffolds can act as transportation vehicles for cells into defect sites, and the use of fibrin glue or collagen as cell delivery matrices can effectively enhance the cell attachment and subsequent cell proliferation, differentiation and ECM secretion required to form appropriate bone microstructure. Furthermore, Zhang *et al.*[44,45] have shown that the use of a bioreactor to culture cell–scaffold constructs may also increase cell viability, proliferation, differentiation and secretion of mineralised bone matrix. This approach also significantly increases new bone formation and neovascularisation, when implanted into ectopic or femoral defects in a rat model.

In the absence of cells, PCL or PCL–TCP scaffolds alone may not be sufficiently bioactive in nature to promote bone regeneration, as indicated by Reichert *et al.*[21] Cylindrical PCL–TCP and PLDLLA–TCP–PCL scaffolds were implanted into a 2 cm tibial segmental defect sheep model. Groups with empty defects or autologous bone grafts (which are classed as the *gold standard* of clinical treatment) served as negative and positive controls. At 3 months, both scaffolds showed good biocompatibility and sufficient mechanical strength. The extent of healing was not comparable, however, to the positive control group which showed the highest amount of new bone formation (47%).[21]

Similar results were also obtained in a long term study (12 months) carried out by Cipitria *et al.*[46] They implanted PCL–TCP scaffolds into a 3 cm tibial segmental defect sheep model. At 12 months, bony bridging across the defect site and greater bone formation around the scaffold were observed.[46] The amount of bone formation was insufficient for clinically relevant bridging of the defect. Thus, to stimulate sufficient bone formation within defect sites (especially when treating large segmental bone defects) it is advisable to combine FDM/ME-fabricated 3D scaffolds with either osteogenic cells or bone growth stimulating agents.

More recently, Berner *et al.*[47] investigated the effect of lay-down pattern (raster angle) of PCL–TCP scaffolds on bone regeneration. PCL–TCP scaffolds with two different raster angles (0/90° and 0/60/120°) were fabricated and implanted into rat critically sized cranial defects for 12 weeks (Figure 9.5).[47] Micro-CT, micro-compression and histological analysis all revealed that the 0/60/120° scaffolds showed lower bone formation in comparison to the 0/90° scaffolds.[47] This was the first time (to the authors' knowledge) that differences in *in vivo* bone regeneration have been reported from differences in architecture of scaffold struts. It was speculated that differences in porosity, surface area may have affected nutrient diffusion and attachment of new matrix onto the scaffold. This study highlights the importance of scaffold macro-architecture. It is often easy to forget this important aspect of scaffold design when looking at the micro-architectural features such as scaffold surface roughness and functionalisation.

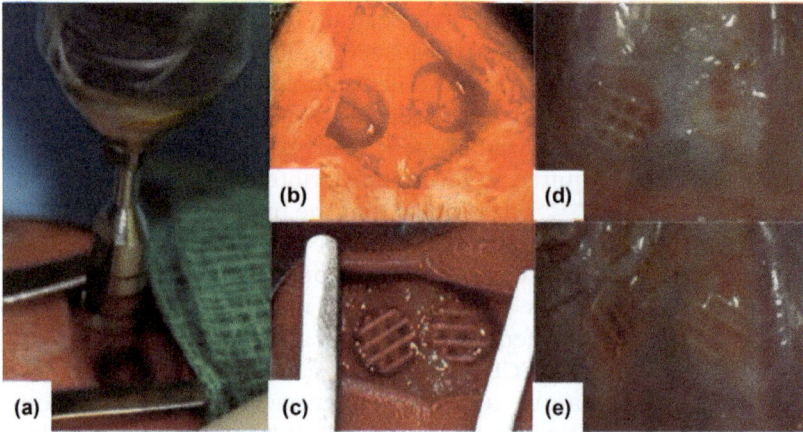

Figure 9.5   (a) Cranial defects were drilled with a 5 mm dental drill, (b) the bone and the surrounding periosteum were carefully removed without damaging the dura, (c) the scaffolds were press fitted with good contact between the host bone and the scaffold. This was evident from the explanted specimens which showed excellent scaffold integration into the host bone with no fibrous capsule evident at 12 weeks in either the 0/90° laydown pattern group (d) and the 0/60/120° laydown pattern group (e).[47]

9.5 Surface Functionalisation of Polycaprolactone-based Scaffolds

Surface coatings have been employed to improve the functionality of PCL while retaining the properties of the material. Surface coating can be achieved through chemical reactions by altering the compounds or molecules on the existing surface or coating the existing surface with a different material such as bone marrow aspirate, heparin and collagen. The surface functionality of scaffolds is important in tissue engineering because the surface of scaffolds is what interfaces with cells. These surface properties have the ability to influence cell proliferation, differentiation and migration. Most surface modification techniques also have no effect on the bulk properties of the scaffold, so mechanical properties are not affected.

9.5.1 Polycaprolactone with Bone Marrow Aspirate

Rohner *et al.*[48] coated PCL scaffolds with bone marrow aspirate prior to *in vivo* implantation. The scaffolds were implanted into surgically created orbital defects in a porcine model. Histomorphometric analysis, 3 months post-implantation, revealed that the bone marrow-coated PCL scaffolds showed 14.1% new bone formation compared to just 4.5% in non-coated PCL scaffolds. The formation of new bone within the bone marrow coated-PCL scaffold was postulated as being due to actions of the mesenchymal cells which were present within the implanted bone marrow. Although some foreign-body giant cells were detected around PCL implants, there were no other signs of clinical inflammation.[48] The authors suggested that long-term studies would be required to clarify and verify the amount of foreign-body reaction, bone formation and degradation of bone-marrow-coated PCL scaffolds in comparison to non-coated scaffolds.[48]

9.5.2 Polycaprolactone with Heparin

Heparin is a highly-sulfated glycosaminoglycan, localised within the secretory granules of mast cells. Clinically, heparin has been widely used as an anti-coagulant,[49] but long-term treatment with heparin is often associated with drug-induced osteoporosis.[50] Although studies *in vitro* have shown that high concentrations of heparin can inhibit matrix deposition and mineralisation, low concentration of heparin (5–500 ng mL^{-1}) can actually promote ECM deposition and mineralisation in osteoblast-like Saos-2 cells.[51] With the potential beneficial effects of heparin in mind, Chum *et al.*[52] incorporated heparin onto collagen coated PCL–TCP scaffolds. The effects of these scaffolds on the bone-forming potential of porcine mesenchymal progenitor cells was investigated. It was found that there were no significant differences detected in the amount of osteogenesis between the PCL–TCP–collagen-heparin and PCL–TCP–collagen scaffold groups. The authors of this study concluded that delivering of tissue-relevant heparan sulfate instead of heparin might prove to be more stimulatory for osteogenesis as heparan sulfates are known to control stem/progenitor cells by virtue of their effects on developmental signalling cascades.[52]

9.5.3 Collagen-mimetic Peptide: The Peptide Sequence GFOGER

The effect of biological substances on cell attachment is generally a result of specific polypeptide integrins interacting with cells. Rather than using an undefined ECM mixture it may be advantageous for efficacy and regulatory reasons or cost to consider pure forms of the integrins of interest. Fibronectin is a common protein used as an attachment factor, of which sometimes simply the RGD sequence motif is used. In the same way, the attachment integrins for collagen are DGEA and GFOGER. Wojtowicz *et al.*[53] coated PCL scaffolds with a triple helical peptide containing the hexapeptide sequence GFOGER (PCL-GFOGER) and subsequently used the scaffold in a critical-sized rat femoral defect model. The GFOGER was passively adsorbed onto the PCL scaffolds and the effectiveness of PCL-GFOGER scaffolds in promoting bone repair was examined and compared to uncoated PCL scaffolds and empty defects. Histological, μ-CT, and X-ray analysis on implants retrieved at 4, 8 and 12 weeks revealed that PCL-GFOGER scaffolds promoted significantly faster and greater bone formation, mineralisation and defect bridging compared to PCL scaffolds without the peptide addition.[53] Furthermore, the extent of new bone formation was found to be dependent on the PCL-GFOGER scaffold surface area available for host tissue interaction, wherein scaffolds with greater surface area to volume ratio showed higher bone volume compared to scaffolds with lower surface area to volume ratio.[53]

9.5.4 Plasma Modification and Fibronectin Coating

Plasma modification allows the creation of micro-scale roughness and/or formation of controllable chemical functional groups on the material surfaces homogeneously, without changing the bulk properties of the biosubstrate; thereby enhancing cell–material interaction.[54] Fibronectin is an adhesive glycoprotein present in ECM that has been widely incorporated as a bioactive ligand onto scaffold surfaces to improve the binding strength of cells. A study carried out by Yildirim *et al.*[55] investigated subjecting PCL scaffolds to oxygen-plasma modification, fibronectin coating or a combination of both. Atomic force microscopy analysis revealed that plasma modification significantly increased the scaffold surface roughness compared to unmodified and fibronectin coated PCL scaffolds. Scaffolds in this study were also seeded with 7F2 mouse osteoblasts cells and cultured for 21 days. Among all groups (unmodified, plasma modified, fibronectin coated or plasma modified and fibronectin coated), scaffolds with plasma modification and fibronectin exhibited the highest proliferation rates and significantly higher osteoblast differentiation rates as indicated by alkaline phosphatase and osteocalcin expression.[55] Higher osteoblast differentiation means that a greater number of bone cells are being produced. Alkaline phosphatase is often expressed by osteoblasts during/after differentiation and osteocalcin is secreted solely by osteoblasts. In a similar study, Berneel *et al.*[56] pre-activated PCL surfaces by subjecting the scaffolds to 30 s of argon-plasma treatment, coating the scaffolds with gelatin type B and finally dip-coating the scaffolds with fibronectin. SEM, X-ray photoelectron spectroscopy and confocal laser scanning microscopy indicated homogenous gelatin type B and fibronectin coating on the entire scaffolds. Scaffolds were also seeded with human foreskin fibroblasts and cultured for up to 21 days. Histological sections of cell–gelatin type B–fibronectin–PCL scaffold constructs revealed that at day 7, cells migrated towards the centre of the scaffolds and at day 21, the entire scaffold was colonised.[56] This behaviour was not observed in PCL-only scaffolds. However, this study only presented qualitative results; further validation of the findings through quantitative means would be necessary to draw significant conclusions.

9.5.5 Hydroxyl Functionalisation

The intrinsic hydrophobicity, coupled with an absence of bioactive functional groups makes PCL unfavourable for cell adhesion and proliferation. Seyednejad *et al.*[57,58] developed a hydroxyl-functionalised co-polymer based on PCL, known as poly(hydroxylmethylglycolide-*co*-ε-caprolactone) (pHMGCL). In their studies, 3D pHMGCL scaffolds were fabricated using an FDM technique with PCL scaffolds as controls. Assessments *in vitro* indicated that pHMGCL scaffolds supported superior cell adhesion and proliferation rates compared to PCL scaffolds. Cells were filling most of the pores at day 7 of culture. Additionally, alkaline phosphatase activity was significantly higher in pHMGCL scaffolds compared to PCL scaffolds. This indicated improved osteogenic differentiation.[58] Following on from this, pHMGCL and PCL scaffolds were implanted subcutaneously into balb/C mice. Both pHMGCL and PCL scaffolds indicated good biocompatibility *in vivo*. Tissue infiltration was observed on both pHMGCL and PCL scaffolds 1 month post-implantation. However, it was shown that pHMGCL scaffolds induced a more intense tissue–biomaterial interaction, with the presence of mononuclear inflammatory cells within the infiltrated tissue (1 month post-implantation), progressing to a mild chronic inflammatory response as indicated by presence of giant cells and progressing fibrosis (seen 3 months post-implantation). Notably, pHMGCL scaffolds induced significantly higher vascularisation within the infiltrated tissue than PCL scaffolds, indicating better neotissue–scaffold integration.[57]

9.5.6   Carbonated Hydroxyapatite–Gelatin Composite

Low interfacial bonding and ineffective stress transfer between the ceramic filler and polymer matrix imposes limitations on the mechanical properties of composite scaffolds. In an attempt to circumvent this problem, Arafat *et al.*[59] modified TCP using 3-glycidoxypropyl trimethoxysilane (GPTMS), covalently-bonded the modified TCP with PCL by annealing at 120 °C for 2 h and fabricated 3D composite scaffolds utilising FDM technology. The scaffolds were coated with CHA–gelatin with the aim of improving their osteoconductive properties. It was shown that the covalently bonded PCL–TCP scaffolds had significantly higher compressive moduli and compressive yield compared to non-covalently bonded PCL–TCP scaffolds. *In vitro* studies indicated that CHA–gelatin composite coating on PCL–TCP scaffolds improved the proliferation and osteoconductive capability of the scaffolds significantly.[59]

9.5.7   Phlorotannin Conjugations

Further investigations into PCL–TCP scaffolds have examined the *in vitro* response to the addition of phlorotannin conjugations (at 1, 3 and 5 wt%) into 80 : 20 wt% PCL–TCP scaffolds.[60] The subsequent effects on physical characteristics such as surface roughness, water absorption ability and mechanical properties were analysed. The viability and differentiation of osteoblast-like cells (MG63) was also investigated. A greater than four-fold increase in cell attachment was observed in the phlorotannin treated groups. Additionally, cell viability was also increased dramatically in the phlorotannin groups. Energy dispersive spectroscopy was used to examine the amounts of elemental calcium and phosphorus. A marked increase was shown in the phlorotannin 5 wt% group compared with the plain PCL–TCP controls.[60]

9.6   Polycaprolactone-based Scaffolds as Drug Carriers

9.6.1   Antibiotics

Teo *et al.*[61] explored the suitability of FDM fabricated PCL–TCP (80 : 20 wt%) scaffolds as carriers for gentamicin sulfate (GS) to treat infected wounds. It was demonstrated through *in vivo* studies that a PCL–TCP–GS scaffolds with 5 wt% of GS had inadequate bacterial-elimination ability while scaffolds with 25 wt% of GS were shown to decrease cell viability due to the levels being possibly toxic. An effective middle ground appeared to be scaffolds with 15 wt% of GS, which were shown to efficiently eliminate both Gram-positive and Gram-negative bacteria within 2 h and had good biocompatibility with dermal fibroblasts.[61] Murine full thickness wounds treated with scaffolds containing 15 wt% GS showed greater healing rates than the control group (wound treated with sterile gauze). It was also proven that the PCL–TCP–GS scaffolds could be a highly effective local antibiotic delivery vehicle as shown by their ability to eliminate bacterial infection within 7 days.[61]

9.6.2 Adeno-associated Virus Encoding Bone Morphogenetic Protein-2

Bone morphogenetic proteins (BMPs) are one of the most potent growth factors regulating the formation of mineralised tissue. One current and serious limitation to widespread clinical use is that safe delivery of controlled doses have not yet been fully optimised. Extreme cases of undesirable results include ectopic bone formation and swelling in the neck and throat, leading to compression of the airway and nearby neurological structures.[62] Studies have been conducted into AM scaffolds as delivery vehicles for BMPs. These constructs aim to provide a safe delivery strategy and controlled release into the target site, promoting bone healing whilst circumventing adverse reactions associated with potential toxic doses. Delivery of bone morphogenetic protein 2 (BMP-2) signals through gene therapy vectors can be used as an alternative to the delivery of large doses of this growth factor which has been associated with side effects such as inflammation and ectopic bone formation. Dupont *et al.*[63] coated PCL scaffolds with either self-complimentary adeno-associated virus (scAAV), a transcapsidated vector expressing the human BMP-2 gene (scAAV-BMP2; experimental group) or AAV-luciferase (AAV-Luc; control group). Acellular scaffolds or hMSCs seeded scaffolds were then implanted into immuno-compromised athymic nude rats with bilateral 8 mm full thickness diaphyseal segmental defects. After 12 weeks, mineral formations were restricted to the immediate vicinity of the defect site across all groups (accellular-scAAV-BMP2, accellular-scAAV-Luc, hMSCs-scAAV-BMP2 and hMSCs-scAAV-Luc). It was also shown that all experimental groups had significantly higher mineral volumes, maximum torque, and torsional stiffness compared to the control group.[63] It should be noted that accellular-scAAV-BMP2 scaffolds led to more bony bridging (5/10 defects bridged) at the defect site than hMSCs-scAAV-BMP2 scaffolds (3/10 defects bridged). Gene therapy hence was shown to potentially be highly beneficial in promoting bone regeneration. This was, however, a small sample size so the results here should be thought of as an indication. This technology bears limitations, such as undesired host immune responses, insertional mutagenesis (for integrating viruses), viral immunogenicity and possible carcinogenesis.[64] Clinical use of gene therapy for bone regeneration will require the highest possible safety standards due to the chronic and non-fatal nature of non-union or large bone fractures. Careful considerations have to be made for the combination of gene therapy with 3D bone scaffolds to ensure that the clinical advantages outweigh the potential disadvantages of gene therapy for bone regeneration.

9.6.3   Bone Morphogenetic Protein

Rai *et al.* used fibrin as a carrier matrix for recombinant human (rh) BMP-2 in FDM/ME fabricated PCL-based scaffolds.[65,66] The effects of varied concentrations of rhBMP-2 (0, 10, 100, 1000 ng mL^{-1}) on osteogenic signals from canine osteoblasts seeded onto PCL–TCP scaffolds were investigated. Results indicated that rhBMP-2 enhanced the differentiation function of canine osteoblasts in a non-dose dependent manner, resulting in accelerated mineralisation, followed by apoptosis of osteoblasts as they underwent terminal differentiation.[65] The release kinetics of rh-BMP2 from PCL–TCP scaffolds were investigated next. Either 10 or 20 μg mL^{-1} of rhBMP-2 was loaded onto the PCL–TCP scaffold using fibrin glue as a carrier matrix. Analysis revealed that rhBMP-2 was well dispersed and uniformly distributed across the PCL–TCP scaffolds but formed large clumps which were sparsely distributed across the PCL only scaffold. The release profile of rhBMP-2 within the PCL–TCP–fibrin construct was dose dependent; 20 μg mL^{-1} rhBMP-2 had a triphasic release profile and 10 μg mL^{-1} rhBMP-2 had a diphasic release profile.[66] In the case of PCL–fibrin constructs, rhBMP-2 showed a diphasic release profile at both rhBMP-2 concentrations. The addition of TCP appears to delay rhBMP-2 release and thus PCL–TCP scaffolds have the potential to be used as vehicles for the controlled release of rhBMP-2.

Sawyer *et al.*[67] utilised collagen-I as carrier matrix for rhBMP-2 within FDM-fabricated PCL–TCP (80 : 20 wt%) scaffolds. For the experiments, 5 μg of rhBMP-2 was loaded onto the PCL–TCP–collagen scaffolds prior to either release assays or the implantation into critical-sized rat calvarial defects (Figure 9.6a). Release assays over 48 h indicated that rhBMP-2 underwent initial burst release, reaching 100% release within the first 6 h. Post-implantation, functional, histological and mechanical analyses clearly demonstrated that bone healing efficacy of PCL–TCP–collagen–rhBMP-2 implants was greater than PCL–TCP–collagen implants without the inclusion of rhBMP-2.[67] At 15 weeks, PCL–TCP–collagen–rhBMP-2 implants stimulated significant bone repair, with most defects reaching closure with full bridging from the calvarial host bone edges, as indicated by the μ-CT images and the stained histological sections (Figure 9.7b and c). For the PCL–TCP–collagen implants without rhBMP-2, new bone formation was minimal, growing only from the periphery of the defect.[67]

In a similar fashion, Abbah *et al.*[68] loaded rhBMP-2 (0.6 mg per scaffold) onto collagen-coated PCL–TCP (80 : 20 wt%) scaffolds and subsequently analysed these in an anterior-lumbar spinal pig model (Figure 9.7a) to evaluate their function as a biological interbody cage device for bone regeneration and spinal fusion. Histological and μ-CT analyses indicated that after 3 months the rhBMP-2 loaded scaffolds induced notable new bone formation with complete defect bridging (Figure 9.7c and d). At 6 months, advanced bone maturation was indicated with increased amounts of horizontal trabeculae interlacing with the columns of vertical trabeculae.[68] Graft

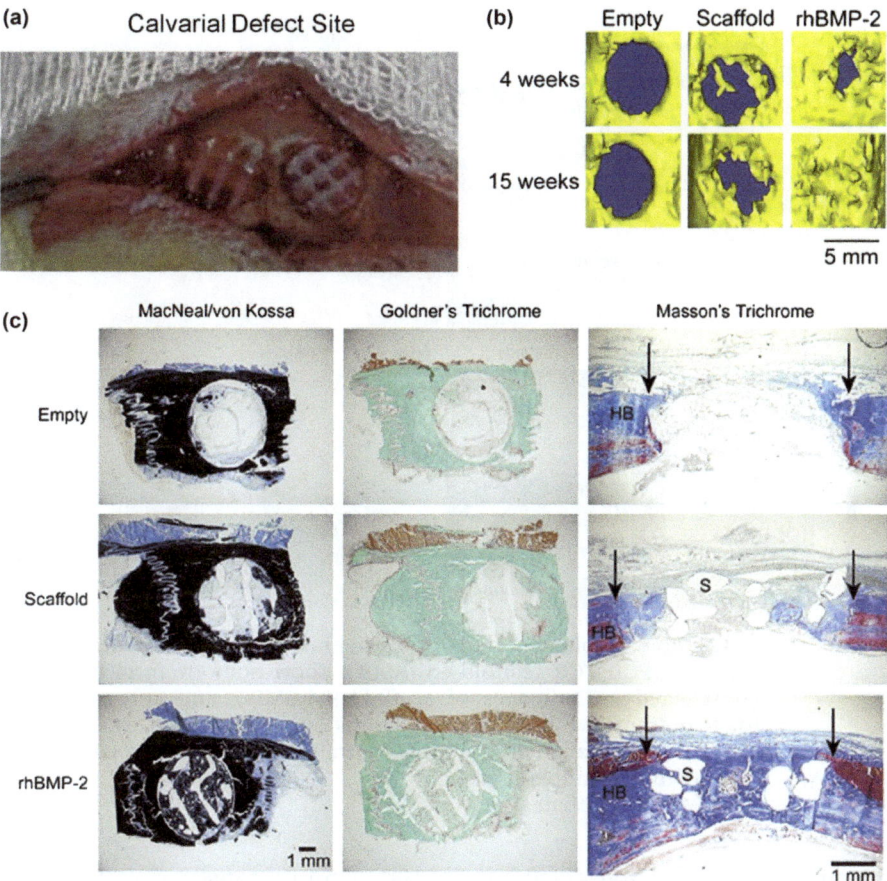

Figure 9.6 (a) PCL–TCP–collagen scaffold was implanted into a rat calvarial defect site. (b) Representative μCT analysis shows full closure in BMP-2 treated sites at 15 weeks. (c) Bone specific staining of histological sections (left to right); Von Kossa, Goldner's trichrome and Masson's trichrome. Extensive bone healing is shown in scaffolds treated with BMP-2.
Adapted from Sawyer *et al.*[67]

fracture and pseudoarthrosis, which were observed in the control group (autograft bone), were absent at both time points in the experimental rhBMP-2 loaded PCL–TCP–collagen scaffold implants.[68] This short-term study demonstrated that a porous PCL–TCP–collagen scaffold designed for bone healing at a load-bearing site, when supplemented with low doses of rhBMP-2, can promote bone regeneration and bone fusion in a large animal model of anterior lumbar interbody fusion.

In a more recent study, Kang *et al.*[69] compared loading PCL scaffolds with cells and BMP-2. PCL scaffolds were loaded with BMSCs and/or BMP-2 using fibrin glue as a carrier matrix. These scaffolds were implanted into

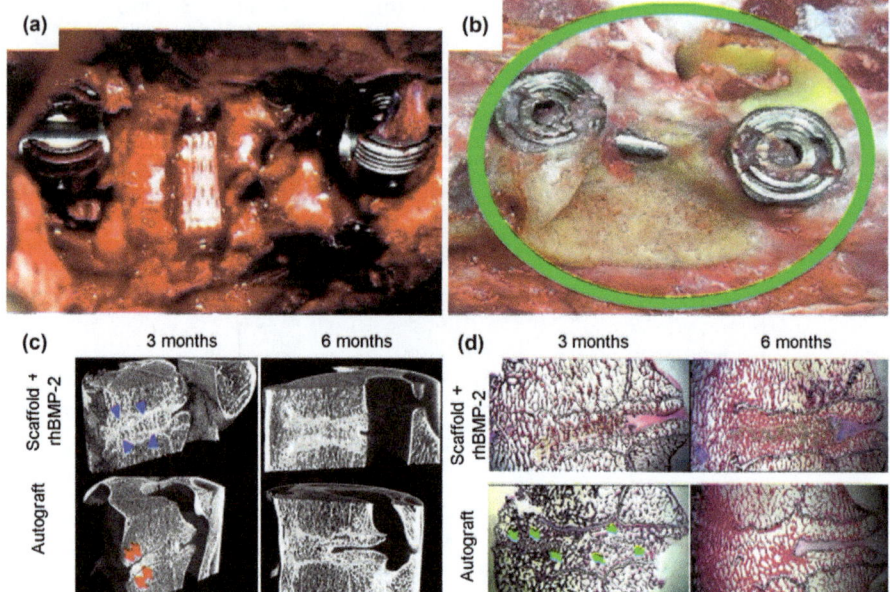

Figure 9.7   (a) Scaffold implanted into intervertebral disc. (b) Harvested motion
segment; PCL–TCP scaffold + rhBMP-2 show bone overgrowth. (c) Re-
constructed µ-CT images of scaffold + rhBMP-2 and autograft bone.
(d) Representative histological sections (sagittal view) of explant. Note
the fairly well aligned columns of trabecular bone formed in scaffold + rhBMP-2 implant group.
Adapted from Abbah *et al.*[68]

20 mm ulnar segmental defects in rabbits. Groups consisted of PCL scaffolds, PCL–fibrin scaffolds, and animals with empty defects serving as
controls. At 1, 2 and 3 months post-implantation, the explanted tissue–
scaffold constructs were analysed with X-ray radiograph imaging, µ-CT
imaging and histological H&E staining.[69] Three months post-implantation,
all control groups showed minimal bone formation, whilst in PCL-BMSCs-
BMP2 treatment groups defect bridging was already apparent 2 months
post-implantation, with complete defect bridging achieved at 3 months
post-implantation. PCL-BMP2 and PCL-BMSCs treatment groups indicated
defect bridging only after 3 months post-implantation. This study indicated
that combinations of BMSCs and BMP-2 with PCL scaffolds can enhance
bone healing efficacy.[69] This supports the general opinion that a combination of scaffold, bioactive molecules and cells will provide the most therapeutically efficacious constructs.

Reichert *et al.*[70] explored a combination of MSCs or rhBMP-7 with PCL–
TCP (80 : 20) scaffolds in comparison to autologous cancellous bone grafts
(ABG). MSCs and rhBMP-7 were loaded onto the scaffolds using platelet rich
plasma (PRP) and bovine collagen type I as carrier matrices, prior to implantation into 3 cm mid-diaphysial tibial segmental defects using an ovine

model. Three months post-implantation, bridging was observed for both ABG and rhBMP-7 treatment groups. Twelve months post-implantation, rhBMP-7 treatment group show significantly greater bone formation and mechanical strength compared to the ABG treatment group as illustrated by μ-CT imaging (Figure 9.8) and biomechanical analyses. At all time points, the level of bone formation in the cell-free scaffold group and MSC treatment group were not comparable to those of ABG and rhBMP-7 treatment groups.[70]

Low levels of BMP combinations to better mimic physiological conditions were explored by Yilgor *et al.*[71–73] PCL scaffolds were coated with PGLA-BMP2 and PHBV-BMP7 nanocapsules using alginate as a carrier. The nanocapsules were stabilised onto the PCL scaffold surfaces by cross-linking of alginate with 5% calcium chloride. An earlier study had shown that a sequential release of BMP-2 followed by BMP-7 from their respective nanocapsules could be achieved.[71] Nanocapsules containing BMP-2 and BMP-7 were loaded onto PCL scaffolds with varying architecture. It was shown that co-administration of BMP-2 and BMP-7 in a sequential manner could promote *in vitro* osteogenic activity regardless of the scaffold architecture.[72] Following these positive outcomes, PLGA-BMP2 and/or PHBV-BMP7 were loaded into PCL scaffolds with varying pore geometry and sizes.[73] The BMP loaded scaffolds were implanted into rats with bilateral 5 mm burr-hole pelvis

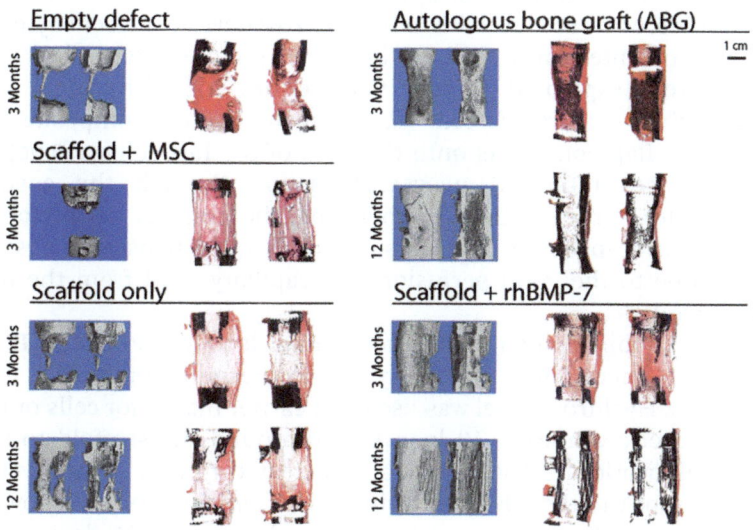

Figure 9.8 Cylindrical PCL–TCP scaffolds produced *via* FDM with an outer diameter of 20 mm, inner diameter of 8 mm and height of 30 mm implanted into an ovine segmental defect. Histological and imaging analyses of bone defects after 3 and 12 months. μ-CT reconstructions are shown next to safranin orange/von Kossa-stained histological sections (orientation: top, proximal; left, medial).
Adapted from Reichert *et al.*[70]

defects. Results have shown that interconnected pores and large pore surface area enhanced bone regeneration regardless of pore geometry. Furthermore, sequential deliveries of BMP-2 followed by BMP-7 further enhanced *in vivo* bone regeneration.[73]

It is clear that FDM/ME PCL scaffolds can be an effective, localised carrier for BMPs into defect sites promoting bone regeneration whilst preventing ectopic bone formation which may result from high burst release and concomitant high physiological doses of BMPs.

9.7 Advanced Scaffold Architecture

9.7.1 Scaffolds with Vascular Channels

Neovascularisation plays a major role in bone regeneration. During a normal bone repair process, the formation of granulation tissue is very important. This granulation tissue is a collection of small blood vessels and fibrous tissue providing structure and the delivery of nutrients. This plays a key role in the reparative phase where progenitor cells migrate into the injury site and differentiate. The design of scaffolds for bone TE is driven towards the generation of bioactive constructs which can stimulate vascularisation and subsequent guidance of vasculature. In a recent study, Muller *et al.*[74] fabricated a PCL–HA (90 : 10 wt%) scaffold with a patient-specific vessel network (Figure 9.9a) by combining a RP approach with CT data and CAD technology. The scaffolds were seeded with bone marrow-derived mesenchymal stem cells and implanted into a rat groin model (Figure 9.9c). After 3 weeks, scaffolds were explanted (Figure 9.9d) either for histological analysis (Figure 9.9f–h) or transferred *via* microsurgery as composite tissue–polymer-free flap constructs onto the neck of the rats (Figure 9.9e). Histological analyses indicated neovascularisation through the customised channels within the scaffold.[74] Following the transfer of prefabricated composite tissue–polymer-free flaps to the neck, the flaps were observed to be viable due to the good perfusion and capillary refill from the induced vascular network.

A novel bobbin-shaped poly(L-lactide-*co*-DL-lactide) (PLDLLA)–PCL–TCP scaffold was mentioned previously as an example of a PCL scaffold (Figure 9.2c). Hyaluronan gel was used as a carrier matrix for cells or rhBMP-2 and an arterio-venous (A-V) loop placed around the scaffold to support axial vascularisation.[75] The A-V loop was created from a 20 mm vein graft harvested from the right femoral vein of a rat. *In vitro* studies showed that rhBMP-2 was gradually released from the gel for up to 35 days and some rhBMP-2 was retained inside the gel. Osteoblasts indicated good viability within the gel.[75] PLDLLA–PCL–TCP scaffolds loaded with hyaluronan hydrogels containing rhBMP-2 (either 500 ng mL^{-1} or 2.5 µg mL^{-1}), or 3 million osteoblasts were implanted into Lewis rats for 8 weeks.[75] Scaffolds with a plain hydrogel served as controls. The gradual degradation of hyaluronan gel and thus the gradual released of rhBMP-2 induced bone-related

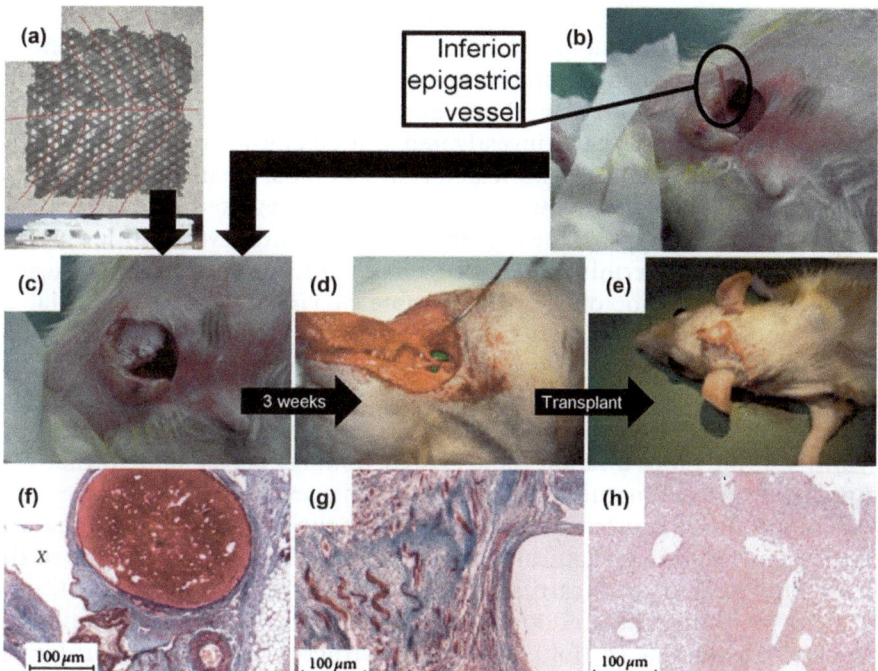

Figure 9.9   (a) Top and side view of fabricated PCL–HA (9 : 1 wt%) scaffolds with incorporated vascular channels. (b) Isolated and ligated inferior epigastric vessel inserted into the vascular channel of the scaffold. (c) Explanted scaffolds were transferred as composite tissue–polymer-free flaps (d) to the scalp of rats (e) or stained with haematoxylin and eosin (H&E) (f–h). (F) Evidence of blood vessel formation (red) with surrounding connective tissue (blue); these areas together represent a vascular channel that was prefabricated within the PCL–HA scaffold. (g) Multiple capillaries (red) are also seen in areas of connective tissue, indicating extensive neovascularisation. (h) Control scaffolds only show granulated tissue without notable evidence of neovascularisation. Adapted from Muller *et al.*[74]

gene expression, but formation of mineralised bone was not detected histologically. Micro-CT analysis of explants revealed all composite scaffolds showed successful vascularisation with osteoblast transplantation having a considerable effect on increasing the blood vessel outgrowth from the A-V loop.[75] This demonstrated that vein grafts may be used to induce vascularisation and eventually allow the generation of axially vascularised tissues with minimal donor site morbidity.

9.7.2   Hybrid Osteochondral Scaffolds

The clinical applications of FDM fabricated PCL scaffolds is not limited to bone regeneration. By altering the processing parameters, PCL scaffolds of different sizes, shapes and mechanical strengths can be generated to suit specific applications. The potential of FDM fabricated PCL-based scaffolds for regeneration of osteochondral defects (localised areas of joint damage involving both the underlying bone and the overlying cartilage) has been explored. Cao *et al.*[76] have investigated the co-culture of osteogenic and chondrogenic cells on FDM-fabricated PCL scaffolds. Results *in vitro* showed that both cell types proliferated, migrated and integrated at the interface, but the final phenotype of the osteoblasts and chondrocytes within the PCL scaffold were undetermined. This research indicates that FDM-fabricated PCL scaffolds are not only biocompatible *in vitro* but have potential as osteochondral constructs in tissue engineering. Hutmacher's group[77,78] has investigated the potential of two types of hybrid scaffolds *in vivo*. In the first hybrid design, fibrin served as a cartilage component whilst the FDM-fabricated PCL scaffold served as the bone component.[77] Implantation of the hybrid scaffolds into a rabbit model osteochondral defect showed that PCL supports growth of mature trabecular bone; however, cartilage regeneration was compromised due to rapid degradation of the fibrin glue matrix and the consequent loss of mechanical support, required for the sustenance of the newly formed immature cartilage.[77] The authors concluded that PCL scaffolds are a promising matrix for osteochondral bone regeneration, but fibrin glue is not a suitable scaffold for the reconstruction of articular cartilage at load-bearing sites.[77]

Following the observations that fibrin glue is a poor cartilage substrate, a second hybrid scaffold was created. FDM-fabricated PCL scaffolds served as the cartilage component while PCL–TCP scaffolds comprised the bone phase.[78] The hybrid scaffolds were implanted into osteochondral defects within a rabbit model along with BMSCs and without BMSCs. Analyses of tissues at 3 and 6 months after implantation indicated that BMSC groups showed superior repair results compared to control groups.[78] With BMSC implantation, the hybrid scaffolds provided sufficient support for new osteochondral tissue formation. The bone regeneration was consistently good from 3 to 6 months, with firm integration to the host tissue. The cartilage regeneration was more complicated; at 3 months, samples showed formation of cartilage tissue around the PCL scaffold filaments.[78] However, at 6 months, several samples revealed some degree of cartilage degradation whilst others maintained a desirable appearance. Biomechanical tests showed that the Young's modulus of the experimental group cartilage was 0.72 MPa. This was deemed noteworthy when compared with the 0.81 MPa value of healthy cartilage.[78] *In vivo* viability of implanted cells was demonstrated by the retention for 6 weeks in the scaffolds. This study demonstrated that PCL-based hybrid scaffolds in combination with BMSCs could be an alternative treatment for osteochondral defects in loading sites.[78] Both of these studies by the Hutmacher group have demonstrated that FDM fabricated PCL and PCL–TCP scaffolds have the potential to promote bone healing. In the context of cartilage tissue engineering, PCL scaffolds performed better than fibrin as the degradation rate of PCL is much lower and thus it is able to provide sufficient mechanical support for cellular development and subsequent ECM excretion necessary for successful cartilage regeneration.

9.8 Clinical Studies

Medical-grade PCL has been approved by the FDA for use in several medical devices. These include monofilament sutures, subdermal contraceptive implants and polymer-based root canal filling material. Recently, FDM engineered 3D PCL scaffolds have received FDA approval (patent number: US 7968026 B1), and are sold under the trade name Osteoplug or Osteomesh (Osteopore International Ltd, Singapore). Clinical studies have been undertaken by Schantz *et al.*[79] and Low *et al.*[80] using FDM-fabricated PCL scaffolds as burr-hole plugs for cranioplasty on patients diagnosed with chronic subdural haematoma. For both studies, follow-up reviews with all patients showed good osteointegration of implants with the surrounding calvarial bone, and no case of infection or adverse systemic reaction was reported.[79,80] Schuckert *et al.*[81] utilised custom-made FDM fabricated PCL scaffolds (Figure 9.2a) in combination with platelet-rich plasma (PRP) and human recombinant bone morphogenetic protein-2 (rhBMP-2) to treat a critical-sized defect in the anterior mandible of a 71-year-old female patient, which had been caused by bacterial infection in her previous two dental implants. An examination at 1 week post-operative, indicated that her wounds healed without any complications. Six months post-operative, *de novo*-grown bone was observed in the anterior mandible. Histological examination of bone biopsies revealed that newly formed bone was identical to bordering local bone and presented vital laminar bone.[81]

9.9 Conclusions

The fabrication techniques described here provide high flexibility in design and allow the generation of 3D porous PCL-based scaffolds of almost any size and shape such as those found in complex anatomical sites. Through software manipulation, researchers are able to take patient-specific data from medical imaging technologies and generate patient-specific scaffolds for bone defects. A current limitation to this approach is the ability of software to generate anatomically accurate constructs while ensuring the tool-paths required produce a scaffold with favourable biological and mechanical properties.

FDM/ME PCL scaffolds have demonstrated tremendous potential in bone tissue engineering research; their applications are not limited to this field. The relevance of these scaffolds has been demonstrated in the cancer research field[82–84] and these more physiologically relevant tissue culture/engineering constructs will no doubt prove useful in other areas of biological research. As cell and material engineering disciplines develop and the knowledge behind growth factor delivery strategies progresses, the development of efficacious customised scaffolds for bone tissue engineering remains a fruitful endeavour and moves ever closer to the clinic. Real-world off-the-shelf alternatives to the current and very limited therapeutic approaches to treat large bone defects are not far away. Ongoing advancements in biofabrication means we can expect these additive technologies to become more commonplace for osteochondral applications.

A Appendix: Methods Section

A.1 Mixing Polycaprolactone with Bioactive Glass Particles by Fast Precipitation into Excess 100% Ethanol[85]

This method of PCL–BG composite mixing is inexpensive and can be easily adapted in any laboratory with basic set-up. This method works on the principle of displacing the chloroform into 100% ethanol, precipitating solid PCL–BG composite. When mixing composites using this method, it is recommended that the total volume of PCL–BG mixture does not exceed 100 mL. This method can also be adapted for mixing other types of particles (*e.g.*, carbon nanotubes, tricalcium phosphate particles, or hydroxyapatite particles) into PCL bulk.

1. Dissolve PCL pellets in chloroform to make 10% (w/v) PCL solution at room temperature.
2. Add the desired wt% of BG particles relative to the PCL mass into the PCL solution. Ensure that the PCL is completely dissolved before the addition of BG particles.

> **NOTE:** *To avoid large agglomeration of BG particles, it is recommended to sieve the BG while adding it to the PCL solution. This will also help in standardising the stirring time for Step 3.*

3. Stir the PCL–bioactive glass mixture using a magnetic stirrer until homogenous. Stirring time should be standardised across all batches.
4. *Critical step*: with a slow and steady rate, pour the homogeneous PCL–bioactive glass mixture into 100% ethanol under constant stirring. Ensure the amount of 100% ethanol used is at least five-fold excess of PCL–bioactive glass solution.

> **NOTE:** *The PCL–BG mixture should form a precipitate upon contact with excess 100% ethanol. The precipitate can appear gluey if the amount of ethanol used is insufficient to effectively separate the solvent from the PCL–bioactive glass mixture.*

5. Isolate the solid PCL–bioactive glass composite by Büchner filtration.

6. Air dry the solid-PCL–bioactive glass composite until constant mass is observed for two consecutive days.
7. Store the dried solid composite in a dry cabinet/desiccator.

A.2 Mathematical Derivation of Melt Extruded Scaffold Microfilament Diameter from Translational Velocity

Please refer to Figure 9.1 when dealing with this derivation. The theoretical scaffold filament diameter can be predicted using eqn (A.1):

$$d_o = d_i \times \sqrt{\frac{v_i}{v_o}} \tag{A.1}$$

where d_o is output filament diameter, d_i is input filament diameter, v_i is input velocity and v_o is output velocity (extruder head speed). Eqn (A.1) is derived with the assumption that input material volume (V_i) equals to output material volume (V_o) and thermal expansion of polymer is not taken into account:

$$V_i = V_o \tag{A.2}$$

As the extruded microfilament is cylindrical shaped, the volume of a cylinder is assumed as shown in eqn (A.3):

$$\text{volume} = \pi \times \text{radius}^2 \times \text{length} \tag{A.3}$$

So,

$$V_i = \left(\frac{d_i}{2}\right)^2 \times \pi \times L_i \tag{A.4}$$

$$V_o = \left(\frac{d_o}{2}\right)^2 \times \pi \times L_o \tag{A.5}$$

where L_i is input filament length and L_o is output filament length.
By substituting eqn (A.4) and (A.5) into eqn (A.2):

$$\left(\frac{d_i}{2}\right)^2 \times \pi \times L_i = \left(\frac{d_o}{2}\right)^2 \times \pi \times L_o \tag{A.6}$$

After simplification of eqn (A.6):

$$d_o = d_i \times \sqrt{\frac{L_i}{L_o}} \tag{A.7}$$

As length of microfilament is equivalent to distance travelled:

$$L_i = v_i \times t \tag{A.8}$$

$$L_o = v_o \times t \tag{A.9}$$

where t is time.

By substituting eqn (A.8) and (A.9) into eqn (A.7):

$$d_o = d_i \times \sqrt{\frac{v_i t}{v_o t}} \tag{A.10}$$

After simplification of eqn (A.10), we derived eqn (A.1).

> **NOTE:** v_o, *which represents the translational velocity, can be set in the fabrication parameter. On the other hand, v_i is variable across machine/extruder type, which has to be calculated separately.*

Acknowledgements

We would like to acknowledge support from the ARC linkage funds LP110200082 and LP100200084.

References

1. M. A. Woodruff, C. Lange, J. Reichert, A. Berner, F. Chen, P. Fratzl, J.-T. Schantz and D. W. Hutmacher, *Mater. Today*, 2012, **15**, 430–435.
2. T. Albrektsson and C. Johansson, *Eur. Spine J.*, 2001, **10**, S96–S101.
3. M. A. Woodruff and D. W. Hutmacher, *Prog. Polym. Sci.*, 2010, **35**, 1217–1256.
4. L. Suggs, S. Moore and A. Mikos, *Synthetic Biodegradable Polymers for Medical Applications*, ed. J. E. Mark, Springer, New York, 2007, pp. 939–950.
5. C. X. F. Lam, D. W. Hutmacher, J.-T. Schantz, M. A. Woodruff and S. H. Teoh, *J. Biomed. Mater. Res., Part A*, 2009, **90**, 906–919.
6. A. Peister, M. A. Woodruff, J. J. Prince, D. P. Gray, D. W. Hutmacher and R. E. Guldberg, *Stem Cell Res.*, 2011, 7, 17–27.
7. N. Bock, M. A. Woodruff, D. W. Hutmacher and T. R. Dargaville, *Polymers*, 2011, **3**, 131–149.
8. E. Sachlos, J. T. Czernuszka, S. Gogolewski and M. Dalby, *Eur. Cells Mater.*, 2003, **5**, 29–40.
9. W. A. Inc., Report of Additive manufacturing and 3D printing, http://wholersassociates.com/press56.htm.
10. D. W. Hutmacher, M. Sittinger and M. V. Risbud, *Trends Biotechnol.*, 2004, **22**, 354–362.
11. F. P. W. Melchels, M. A. N. Domingos, T. J. Klein, J. Malda, P. J. Bartolo and D. W. Hutmacher, *Prog. Polym. Sci.*, 2012, 1079–1104.
12. M. Domingos, D. Dinucci, S. Cometa, M. Alderighi, P. J. Bartolo and F. Chiellini, *Int. J. Biomater.*, 2009, Article ID: 239643.
13. I. Zein, D. W. Hutmacher, K. C. Tan and S. H. Teoh, *Biomaterials*, 2002, **23**, 1169–1185.

14. D. Schumann, A. K. Ekaputra, C. X. F. Lam and D. W. Hutmacher, *Biomaterials/scaffolds. Design of bioactive, multiphasic PCL/collagen type I and type II-PCL-TCP/collagen composite scaffolds for functional tissue engineering of osteochondral repair tissue by using electrospinning and FDM techniques*, ed. H. R. Hauser and M. Fussenegger, Humana Press, 2007, vol. 140, pp. 101–124.

15. M. E. Hoque, W. Y. San, F. Wei, S. Li, M. H. Huang, M. Vert and D. W. Hutmacher, *Tissue Eng., Part A*, 2009, **15**, 3013–3024.

16. L. Wu, J. Zhang, D. Jing and J. Ding, *J. Biomed. Mater. Res., Part A*, 2005, **76**, 264–271.

17. D. W. Hutmacher, T. Schantz, I. Zein, K. W. Ng, S. H. Teoh and K. C. Tan, *J. Biomed. Mater. Res.*, 2001, **55**, 203–216.

18. C. E. Holy, C. Cheng, J. E. Davies and M. S. Shoichet, *Biomaterials*, 2000, **22**, 25–31.

19. M. Selim, A. J. Bullock, K. A. Blackwood, C. R. Chapple and S. MacNeil, *BJU Int.*, 2011, **107**, 296–302.

20. A. Berner, J. C. Reichert, M. A. Woodruff, S. a. A. J. Saifzadeh, D. R. Epari, M. Nerlich, M. A. Schuetz and D. W. Hutmacher, *Acta Biomater.*, 2013, **9**, 7874–7884.

21. J. C. Reichert, M. E. Wullschleger, A. Cipitria, J. Lienau, T. K. Cheng, M. A. Schütz, G. N. Duda, U. Nöth, J. Eulert and D. W. Hutmacher, *Int. Orthop.*, 2011, **35**, 1229–1236.

22. A. Karakecili, C. Satriano, M. Gumusderelioglu and G. Marletta, *J. Mater. Sci.: Mater. Med.*, 2007, **18**, 317–319.

23. H. Zhang and S. Hollister, *J. Biomater. Sci., Polym. Ed.*, 2009, **20**, 1975–1993.

24. G. Marletta, G. Ciapetti, C. Satriano, S. Pagani and N. Baldini, *Biomaterials*, 2005, **26**, 4793–4804.

25. J.-T. Schantz, S. H. Teoh, T. C. Lim, M. Endres, C. X. F. Lam and D. W. Hutmacher, *Tissue Eng.*, 2003, **9**, 113–126.

26. J.-T. Schantz, D. W. Hutmacher, C. X. F. Lam, M. Brinkmann, K. M. Wong, T. C. Lim, N. Chou, R. E. Guldberg and S. H. Teoh, *Tissue Eng.*, 2003, **9**, 127–139.

27. L. L. Hench, R. J. Splinter, W. C. Allen and T. K. Greenlee, *J. Biomed. Mater. Res.*, 1972, **5**, 117–141.

28. J.-T. Schantz, A. Brandwood, D. Hutmacher, H. Khor and K. Bittner, *J. Mater. Sci.: Mater. Med.*, 2005, **16**, 807–819.

29. L. Shor, S. U. Geri, X. Wen, M. Gandhi and W. Sun, *Biomaterials*, 2007, **28**, 5291–5297.

30. H. Chim, D. W. Hutmacher, A. M. Chou, A. L. Oliveira, R. L. Reis, T. C. Lim and J. T. Schantz, *Int. J. Oral Max. Surg.*, 2006, **35**, 928–934.

31. C. X. F. Lam, S. H. Teoh and D. W. Hutmacher, *Polym. Int.*, 2007, **56**, 718–728.

32. C. X. F. Lam, M. M. Savalani, S. H. Teoh and D. W. Hutmacher, *Biomed. Mater.*, 2008, **3**, Article ID: 034108.

33. J. C. Reichert, M. A. Woodruff, T. Friis, V. Quent, S. Gronthos, G. N. Duda, M. A. Schütz and D. W. Hutmacher, *J. Tissue Eng. Regener. Med.*, 2010, **4**, 565–576.

34. Y. B. Kim and G. Kim, *Appl. Phys. A: Mater. Sci. Process.*, 2012, **108**, 949–959.

35. Y. Zhou, D. W. Hutmacher, S. L. Varawan and T. M. Lim, *Polym. Int.*, 2007, **56**, 333–342.

36. Y. Zhou, F. Chen, S. T. Ho, M. A. Woodruff, T. M. Lim and D. W. Hutmacher, *Biomaterials*, 2007, **28**, 814–824.

37. S. A. Abbah, C. X. F. Lam, K. A. Ramruttun, J. C. H. Goh and H.-K. Wong, *Tissue Eng., Part A*, 2009, **17**, 809–817.

38. J. J. Thorn, H. Sorensen, U. Weis-Fogh and M. Andersen, *Int. J. Oral Max. Surg.*, 2004, **33**, 95–100.

39. D. T. Leong, K. N. Wee, A. Gupta, D. W. Hutmacher and M. A. Woodruff, *Curr. Drug Discovery Technol.*, 2008, **5**, 319–327.

40. H. Lee and G. Kim, *J. Mater. Chem.*, 2011, **21**, 6305–6312.

41. M. G. Yeo and G. H. Kim, *Chem. Mater.*, 2011, 903–913.

42. B. Rai, J. L. Lin, Z. X. H. Lim, R. E. Guldberg, D. W. Hutmacher and S. M. Cool, *Biomaterials*, 2010, **31**, 7960–7970.

43. H. C. Lim, J. H. Bae, H. R. Song, S. H. Teoh, H. K. Kim and D. H. Kum, *J. Bone Joint Surg., Br. Vol.*, 2011, **93**, 120–125.

44. Z.-Y. Zhang, S. H. Teoh, W.-S. Chong, T.-T. Foo, Y.-C. Chng, M. Choolani and J. Chan, *Biomaterials*, 2009, **30**, 2694–2704.

45. Z. Y. Zhang, S. H. Teoh, M. S. Chong, E. S. Lee, L. G. Tan, C. N. Mattar, N. M. Fisk, M. Choolani and J. Chan, *Biomaterials*, 2010, **31**, 608–620.

46. A. Cipitria, C. Lange, H. Schell, W. Wagermaier, J. C. Reichert, D. W. Hutmacher, P. Fratzl and G. N. Duda, *J. Bone Miner. Res.*, 2012, **27**, 1275–1288.

47. A. Berner, M. A. Woodruff, C. X. F. Lam, M. T. Arafat, S. Saifzadeh, R. Steck, J. Ren, M. Nerlich, A. K. Ekaputra, I. Gibson and D. W. Hutmacher, *Int. J. Oral Max. Surg.*, 2014, **43**(4), 506–513.

48. D. Rohner, D. W. Hutmacher, T. K. Cheng, M. Oberholzer and B. Hammer, *J. Biomed. Mater. Res., Part B*, 2003, **66**, 574–580.

49. D. L. Rabenstein, *Nat. Prod. Rep.*, 2002, **19**, 312–331.

50. E. Lefkou, M. Khamashta, G. Hampson and B. J. Hunt, *Lupus*, 2010, **19**, 3–12.

51. H. J. Hausser and R. E. Brenner, *J. Cell. Biochem.*, 2004, **91**, 1062–1073.

52. Z. Z. Chum, M. A. Woodruff, S. M. Cool and D. W. Hutmacher, *Acta Biomater.*, 2009, **5**, 3305–3315.

53. A. M. Wojtowicz, A. Shekaran, M. E. Oest, K. M. Dupont, K. L. Templeman, D. W. Hutmacher, R. E. Guldberg and A. J. García, *Biomaterials*, 2010, **31**, 2574–2582.

54. Y.-P. Jiao and F.-Z. Cui, *Biomed. Mater.*, 2007, **2**, R24.

55. E. D. Yildirim, R. Besunder, D. Pappas, F. Allen, S. Güçeri and W. Sun, *Biofabrication*, 2010, **2**, 013109.

56. E. Berneel, T. Desmet, H. Declercq, P. Dubruel and M. Cornelissen, *J. Biomed. Mater. Res., Part A*, 2012, **100**, 1783–1791.
57. H. Seyednejad, D. Gawlitta, R. V. Kuiper, A. De Bruin, C. F. Van Nostrum, T. Vermonden, W. J. A. Dhert and W. E. Hennink, *Biomaterials*, 2012, **33**, 4309–4318.
58. H. Seyednejad, D. Gawlitta, W. J. A. Dhert, C. F. van Nostrum, T. Vermonden and W. E. Hennink, *Acta Biomater.*, 2011, 7, 1999–2006.
59. M. T. Arafat, C. X. F. Lam, A. K. Ekaputra, S. Y. Wong, X. Li and I. Gibson, *Acta Biomater.*, 2011, 7, 809–820.
60. M. Yeo, W.-K. Jung and G. Kim, *J. Mater. Chem.*, 2012, **22**, 3568–3577.
61. E. Y. Teo, S.-Y. Ong, M. S. Khoon Chong, Z. Zhang, J. Lu, S. Moochhala, B. Ho and S.-H. Teoh, *Biomaterials*, 2011, **32**, 279–287.
62. D. Benglis, M. Y. Wang and A. D. Levi, *Neurosurgery*, 2008, **62**, ONS423-431; discussion ONS431.
63. K. M. Dupont, J. D. Boerckel, H. Y. Stevens, T. Diab, Y. M. Kolambkar, M. Takahata, E. M. Schwarz and R. E. Guldberg, *Cell Tissue Res.*, 2012, **347**, 575–588.
64. D. J. Gould and P. Favorov, *Gene Ther.*, 2003, **10**, 912–927.
65. B. Rai, S. H. Teoh, K. H. Ho, D. W. Hutmacher, T. Cao, F. Chen and K. Yacob, *Biomaterials*, 2004, **25**, 5499–5506.
66. B. Rai, S. H. Teoh, D. W. Hutmacher, T. Cao and K. H. Ho, *Biomaterials*, 2005, **26**, 3739–3748.
67. A. A. Sawyer, S. J. Song, E. Susanto, P. Chuan, C. X. F. Lam, M. A. Woodruff, D. W. Hutmacher and S. M. Cool, *Biomaterials*, 2009, **30**, 2479–2488.
68. S. A. Abbah, C. X. L. Lam, D. W. Hutmacher, J. C. H. Goh and H.-K. Wong, *Biomaterials*, 2009, **30**, 5086–5093.
69. S.-W. Kang, J.-H. Bae, S.-A. Park, W.-D. Kim, M.-S. Park, Y.-J. Ko, H.-S. Jang and J.-H. Park, *Biotechnol. Lett.*, 2012, **34**, 1375–1384.
70. J. C. Reichert, A. Cipitria, D. R. Epari, S. Saifzadeh, P. Krishnakanth, A. Berner, M. A. Woodruff, H. Schell, M. Mehta, M. A. Schuetz, G. N. Duda and D. W. Hutmacher, *Sci. Transl. Med.*, 2012, **4**, 141ra193.
71. P. Yilgor, N. Hasirci and V. Hasirci, *J. Biomed. Mater. Res., Part A*, 2010, **93**, 528–536.
72. P. Yilgor, R. Sousa, R. Reis, N. Hasirci and V. Hasirci, *J. Mater. Sci.: Mater. Med.*, 2010, **21**, 2999–3008.
73. P. Yilgor, G. Yilmaz, M. B. Onal, I. Solmaz, S. Gundogdu, S. Keskil, R. A. Sousa, R. L. Reis, N. Hasirci and V. Hasirci, *J. Tissue Eng. Regener. Med.*, 2013, 687–696.
74. D. Muller, H. Chim, A. Bader, M. Whiteman and J.-T. Schantz, *Stem Cells Int.*, 2011, Article ID 547247.
75. S. N. Rath, G. Pryymachuk, O. A. Bleiziffer, C. X. F. Lam, A. Arkudas, S. T. B. Ho, J. P. Beier, R. E. Horch, D. W. Hutmacher and U. Kneser, *J. Mater. Sci.: Mater. Med.*, 2011, 1–13.
76. T. Cao, K.-H. Ho and S.-H. Teoh, *Tissue Eng.*, 2003, **9**, 103–112.

77. X. X. Shao, D. W. Hutmacher, S. T. Ho, J. C. H. Goh and E. H. Lee, *Biomaterials*, 2006, **27**, 1071–1080.

78. X. Shao, J. C. H. Goh, D. W. Hutmacher, E. H. Lee and G. Zigang, *Tissue Eng.*, 2006, **12**, 1539–1551.

79. J.-T. Schantz, T.-C. Lim, C. Ning, S. H. Teoh, K. C. Tan, S. C. Wang and D. W. Hutmacher, *Neurosurgery*, 2006, **58**, ONS-E176.

80. S. W. Low, Y. J. Ng, T. T. Yeo and N. Chou, *Singapore Med. J.*, 2009, **50**, 777–780.

81. K.-H. Schuckert, S. Jopp and S.-H. Teoh, *Tissue Eng., Part A*, 2008, **15**, 493–499.

82. S. Sieh, A. A. Lubik, J. A. Clements, C. C. Nelson and D. W. Hutmacher, *Organogenesis*, 2010, **6**, 181–188.

83. D. W. Hutmacher, D. Loessner, S. Rizzi, D. L. Kaplan, D. J. Mooney and J. A. Clements, *Trends Biotechnol.*, 2010, **28**, 125–133.

84. D. W. Hutmacher, *Nat. Mater.*, 2010, **9**, 90–93.

85. P. S. P. Poh, D. W. Hutmacher, M. M. Stevens and M. A. Woodruff, *Biofabrication*, 2013, Article ID: 045005.

Section D
Chemo-structural Biointerfaces

CHAPTER 10

High Throughput Techniques for the Investigation of Cell–Material Interactions

LAUREN R. CLEMENTS,[a] HELMUT THISSEN[b] AND NICOLAS H. VOELCKER*[c]

[a] School of Chemical and Physical Sciences, Flinders University, Bedford Park, 5042 SA, Australia; [b] CSIRO Materials Science and Engineering, Bayview Avenue, Clayton, 3168 VIC, Australia; [c] Mawson Institute, University of South Australia, Mawson Lakes, 5095 SA, Australia
*Email: nico.voelcker@unisa.edu.au

10.1 Introduction

Understanding cell–material interactions is of fundamental importance for many biomedical applications. Cells and living tissue often come into direct contact with non-natural materials and the nature of the surface has been shown to have a direct influence on the cellular response.[1–3] Many surface properties including topography,[4,5] elasticity,[6] surface chemistry,[7] wettability[8,9] and the presence of chemical functional groups or biological signals[10,11] have been demonstrated to influence the cellular response.[1,2] Considerable attention has been directed towards the surface modification of substrates to either enhance or minimise cellular attachment as well as improving cell metabolism and function.[12,13] In many instances, cells do not react to the substrates themselves, but rather to adsorbed or surface bound biomolecules. The use of proteins, such as fibronectin (Fn), collagen, and

RSC Smart Materials No. 10
Biointerfaces: Where Material Meets Biology
Edited by Dietmar Hutmacher and Wojciech Chrzanowski
© The Royal Society of Chemistry 2015
Published by the Royal Society of Chemistry, www.rsc.org

peptides have also been shown to promote cellular attachment[10,14,15] while on the other hand poly(ethylene glycol) (PEG) and certain other polymers are effective in discouraging protein fouling on the surface.[16-18]

The interactions of cells with a given surface are by no means generic. For a start, different cell types favour different surface conditions. Screening for a favourable cellular response can be a tedious and time-consuming process. For example, identification of biological factors or factor combinations on biomaterial surfaces requires the screening of a large combinatorial space. Using conventional methods, throughput is limited to cell culture well formats and sample size as well as large cell culture volumes. For this reason, several high throughput techniques have emerged to screen for optimal surface conditions giving the optimum cellular response.[11,19,20] Cell microarrays and gradient surfaces are gaining increasing popularity in biomaterials research due to the large number of surface parameters that can be examined in a high-throughput manner.[21]

Cell microarrays allow for the simultaneous analysis of several properties at once, by generating a library of discreetly different surface properties. The advantage of the microarray format is that many different combinations of surface properties and chemistries can be screened simultaneously. Gradient surfaces on the other hand display a gradual variation of one or two surface properties along the x or y axis, allowing the optimisation of a limited set of variables. These techniques are complementary and can both be used effectively for the screening of cell–surface interactions and optimising material properties. The advantage of gradient surfaces is that there is virtually an unlimited number of data points along the gradient which allows the determination of limits through which each surface characteristic is effective, allowing specific tailoring of a surface to deliver an optimum cell response.[22] The use of a continuously graded surface can be both time and material efficient and also offers advantages in regard to the minimisation of systematic errors normally associated with discreet, individual experiments. Thus, with the use of gradients, it is possible to effectively reduce sample sizes and analysis time of optimisation experiments.

A further step in this context is the development of two-dimensional (2D) or orthogonal gradients, whereby two different gradient surfaces are deposited perpendicular to each other. By using two orthogonal gradients on the same substrate, a large surface modification space can be screened simultaneously[23] which allows for even faster screening of the optimum surface conditions for cell attachment and growth. Importantly, the orthogonal gradient experiments take into account the interplay between parameters and their combined influence on the cellular response.

10.1.1 Influence of Material Surface Chemistry and Topography on Cell Behaviour

The way cells interact with their surrounding environment is of significant interest to many emerging medical applications including biomaterials, biosensors and tissue engineering, just to name a few.[24] Distinct types of cell have different demands in that the surface characteristics required for successful cellular response will vary between cell types. For example Kunzler *et al.*[25] found that fibroblasts showed preferential attachment towards smooth surfaces whereas osteoblasts favoured rougher surfaces.

The properties of a given surface have long been known to affect responses such as the attachment and proliferation of cells. Surface characteristics including wettability, topography, chemical functionalities, biomolecule density and elasticity have been identified as critical parameters (Figure 10.1).[5,9,11] In many instances, it is the adsorbed biomolecules on the surface, usually originating from serum, which have the greatest influence on cellular response.[26] Therefore, the composition of this protein layer requires considerable attention when aiming at tailoring the surface for an optimum cellular response. For this reason, researchers have invested significant research into methods of either preventing protein adsorption or conversely the immobilisation of proteins (or parts thereof) to a surface to help direct a cellular response.[1,11,27]

The extracellular matrix (ECM) is the extracellular part of tissues that provides structural support to cells. Besides structural support, the ECM performs many other functions including assistance with anchorage of cells, segregation of tissues from one another and regulating intercellular communication. The ECM is composed of many different proteins, the most commonly known include collagen, vitronectin, Fn, elastin and laminin. All of these proteins are involved in the attachment of cells through interacting with integrin receptors present on the cell surface.[28,29] Thus, the ability to control the presentation of ECM proteins on a substrate is a promising approach towards the tailoring of implant devices and other biomaterials. For this reason, considerable attention has been directed towards cell–surface interactions on ECM modified substrates.[30–34] Subunits of ECM proteins

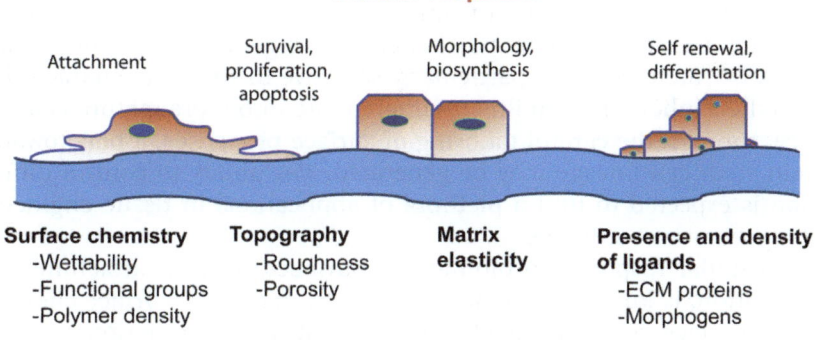

Figure 10.1   Schematic demonstrating the wide variety of material surface characteristics that influence cellular response with regards to attachment, proliferation, self-renewal or differentiation.

have been identified that interact with integrins such as the arginine–glycine–aspartic acid (RGD) peptide. This particular peptide represents a cell attachment-mediating sequence present in Fn, and is known to promote cell attachment when immobilised on surfaces.[31,33,35] Through careful tailoring of biomaterial surface properties for the cell type of choice, the desired degree of attachment, spreading or proliferation can be achieved. Conversely, polymers such as polyethylene oxide (PEO), also known as PEG, have been shown to minimise cell and biomolecule adsorption on a surface.[16,36–39] PEO is particularly effective in regard to the reduction and prevention of protein adsorption and, therefore, cell attachment, and this effectiveness depends on the density, structure and molecular weight of surface immobilised PEO.[40,41]

In addition, it has been shown previously that the cellular response can be directly influenced by surface topography alone.[42–45] Relevant topographical characteristics including porosity,[5,46–48] grooved surfaces[49,50] and surface roughness[25,51,52] have been demonstrated to influence the cellular response. For example, ridges (2 mm wide and 3–5 mm high) have been formed on surfaces and 95% of cells were observed to grow along the axis of these ridges, regardless of any additional chemistry applied to the surface.[42] Commonly termed 'contact guidance,' this phenomenon has been used to direct cell growth. Similarly, Teixeira *et al.*[43] found that keratocytes and fibroblast cell lines aligned themselves along pre-formed grooves on the substrate. Human neuroblastoma cells were found to selectively attach to the coated regions. In another study, patterned poly-L-lactide (PLLA) films were prepared to investigate the differentiation of multipotent mouse bone marrow stromal precursors.[53] It was found that the presence of evenly spaced 3 mm groves on the surface led to faster cell attachment and differentiation after 10 days compared to flat PLLA films.

10.1.2 Stem Cell–Material Surface Interactions

Researchers are increasingly turning their attention towards stem cell research for its obvious potential benefits in regenerative medicine applications.[54,55] Stem cells are particularly sensitive to substrate characteristics due to their inherent capability to differentiate into more mature cell types. Hence, through the careful tailoring of surface properties, a bias towards a certain stem cell lineage may be generated. The ability to control differentiation is expected to find a plethora of applications in tissue engineering and regenerative medicine.

The cellular response to biomaterial surfaces, such as attachment and proliferation, is known to be mediated by the material's surface chemistry. Indeed, surface chemistry is also able to instruct cell function and direct cell differentiation.[56–63] However, it has proven difficult to predict cell fate outcomes based on the molecular composition of the polymer substrate.[58,64] Recent work on carboxylic acid based plasma polymer (octadiene–acrylic acid) gradient surfaces demonstrated a relationship between retention of 'stemness' in mouse embryonic stem cells (mESCs) and the degree of cell attachment.[65] Increased cell spreading and the absence of three-dimensional (3D) colony organisation were observed to coincide with loss of stem cell marker expression. However, others have suggested that a comparatively larger colony size is optimal for maintaining cell pluripotency.[66]

Preliminary studies have demonstrated that tailoring a surface's mechanical properties can direct stem cell differentiation to a particular cell type.[2,67–70] For example, mesenchymal stem cells (MSCs) were shown to be very sensitive to substrate elasticity, allowing for elasticity-directed cell differentiation (Figure 10.2A).[2] Polyacrylamide-based gradient gels were prepared where the matrix ranged from soft to relatively rigid depending on the extent of cross-linking. The soft substrates, mimicking the elasticity of brain tissue, resulted in the stem cells displaying a neuronal phenotype. On the rigid surfaces whereby the surface had comparable elasticity to bone, osteoblast-like characteristics were observed from the cells. Similarly, on the intermediate elasticity surfaces, muscle cells were evident. Expression of transcriptional markers associated with neuronal, muscle and osteogenic cell lines were consistent with morphology observations.

Curran *et al.*[62,71] investigated the effect of different surface functional groups on the differentiation of MSCs (Figure 10.2B). A variety of silanes with different terminal groups including $-NH_2$, $-CH_3$, $-SH$, $-OH$ and $-COOH$ were deposited onto glass slides. Real time polymerase chain reactions were used to observe the expression of several differentiation markers including: β-actin (increase in expression), orinithine decarboxylase (indication of proliferation), collagen II (chondrocytes), CBFA1 (bone transcription factor), collagen I (MSC marker), TGF-β3 (ECM production)). MSCs cultures maintained their stem-cell phenotype on $-CH_3$ modified surfaces but $-COOH$, $-OH$, $-NH_2$ and $-SH$ surfaces promote osteogenesis. McBeath *et al.*[72] demonstrated that the cell shape, namely the spreading area, influences hMSC differentiation through growing cells on 'spots' of Fn proteins of different sizes (Figure 10.2C). hMSCs that were allowed to spread out (*i.e.*, larger spots) were shown to differentiate into osteoblasts, whilst hMSCs with restricted spreading underwent adipogenesis.

Through the use of microarrays displaying 576 combinations of 25 polymers (derivatives of acrylate, diacrylate, dimethacrylate and triacrylate monomers), Anderson *et al.*[56] observed that the majority of such combinations were able to influence the differentiation of human embryonic stem cells into epithelial-like cells. This technique is promising for tissue engineering applications to create specific tissue constructs based on the composition of the underlying polymer surface. Similarly, Derda *et al.*[73] have utilised microarrays of peptides immobilised onto self-assembled monolayers (SAMs) to determine peptidic surfaces that support embryonic stem cell growth and self-renewal. They investigated 18 different laminin-containing peptides to identify which properties supported proliferation and self-renewal. It was found that five of the 18 peptides supported proliferation

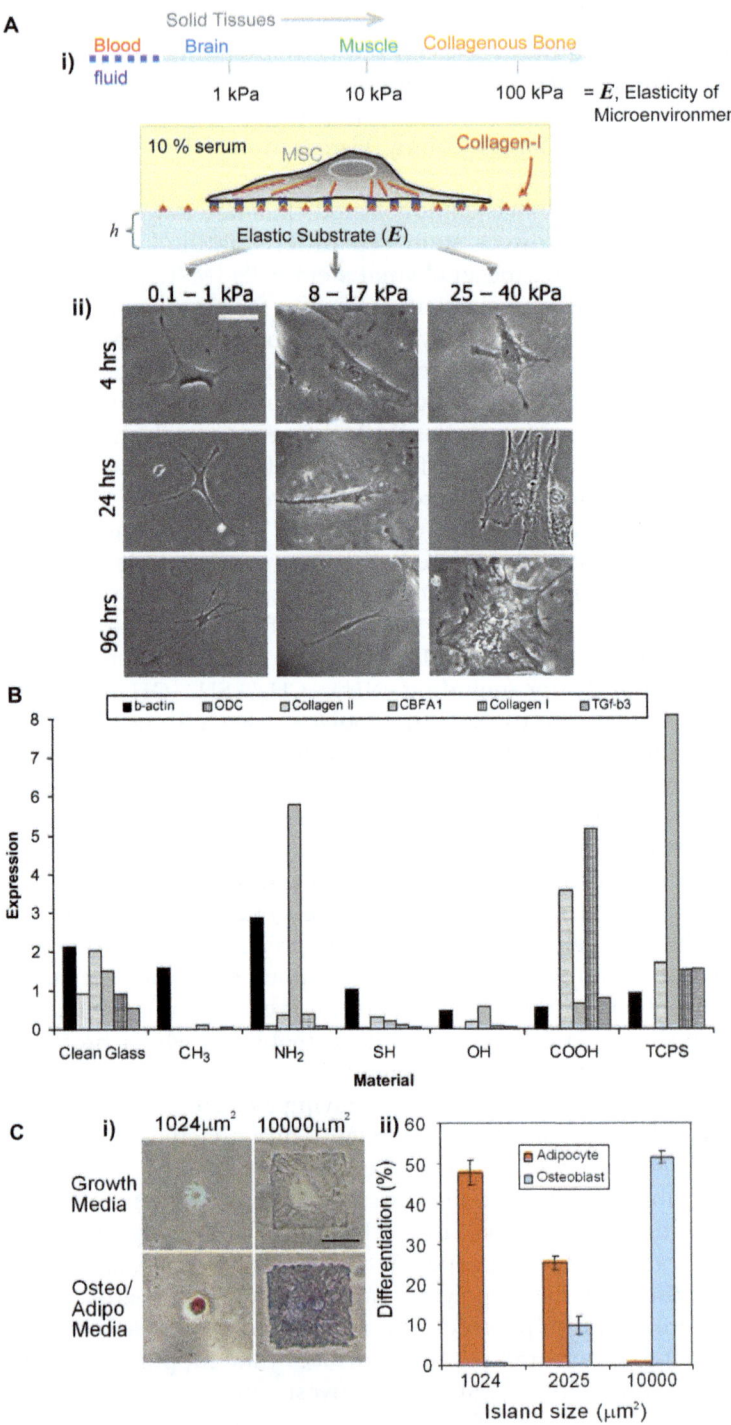

A

i) Solid Tissues →

Blood fluid   Brain   Muscle   Collagenous Bone

1 kPa   10 kPa   100 kPa = *E*, Elasticity of Microenvironment

10 % serum   MSC   Collagen-I

h { Elastic Substrate (*E*)

ii)   0.1 – 1 kPa   8 – 17 kPa   25 – 40 kPa

4 hrs

24 hrs

96 hrs

B

■ b-actin ▨ ODC ▨ Collagen II ▨ CBFA1 ▨ Collagen I ▨ TGf-b3

Expression

Clean Glass   CH₃   NH₂   SH   OH   COOH   TCPS

Material

C

i)   1024μm²   10000μm²

Growth Media

Osteo/ Adipo Media

ii)   Differentiation (%)

■ Adipocyte ▫ Osteoblast

1024   2025   10000

Island size (μm²)

and of those five, all were able to maintain the cells in an undifferentiated state.

10.1.3   Screening of Cell–Surface Interactions: An Overview of High-throughput Techniques

Screening of cell–surface interactions can be a time-consuming process given the vast array of topographical, chemical and biological cues available for investigation, which in turn translates into spiralling costs associated with such experiments. Additionally, systematic errors associated with analysing individual samples can lead to uncertainties in results and hence a requirement for increased numbers of replicates. The advent of high-throughput screening techniques including protein and cell microarrays, encoded microparticles, microfluidic devices as well as surface-bound gradients has precipitated a paradigm shift in the field (Figure 10.3).[22,56,65,74–78]

Pregibon *et al.*[78] have developed a high-throughput screening method using encoded microparticles. Here, multi-functional particles displaying encoding elements and analyte detection regions were developed using microfluidics and continuous flow lithography (Figure 10.3A). In the example shown, researchers loaded the particles with DNA oligomer probes and incubated them with fluorescently labelled targets. Using flow-through particle reading under a fluorescence microscope, different DNA oligomers could be detected without the need for signal amplification. Kothapalli *et al.*[77] investigated neuron-guided growth using microfluidic devices (Figure 10.3). Here, hippocampal or dorsal root ganglion neuronal cells were incubated inside a microfluidic device on a collagen gel. Upon induction of a chemical gradient of various biomolecules known to influence neurite growth including netrin-1, brain pump and slit-2, neurite turning was observed to varying degrees depending on the applied cue and its concentration.

The use of a microarray format, whereby many 'spots' of differing properties and chemistries can be immobilised onto a single substrate at a given

---

Figure 10.2   (A) Substrate elasticity influences stem cell fate,[2] (i) range of stiffnesses observed by solid tissues, (ii) images of MSCs grown on surfaces with a range of stiffnesses for 4, 24 and 96 h, demonstrating the change in morphology with substrate stiffness; (B) surface functional groups influence MSC phenotype, graph showing expression of various expression markers: β-actin (increase in expression), ornithine decarboxylase (ODC, indication of proliferation), collagen II (chondrocytes), CBFA1 (bone transcription factor), collagen I (MSC marker), TGF-β3 (ECM production) on a variety of different functionalised surfaces;[62] (C) cell shape drives hMSC commitment, (i) bright field images of MSCs grown on different sized Fn islands (1024 μm^2 and 10 000 μm^2), (ii) % differentiation of hMSC on different sized islands.[72]
Figures A, B and C adapted from Engler *et al.*,[2] Curran *et al.*[62] and McBeath *et al.*,[72] respectively.

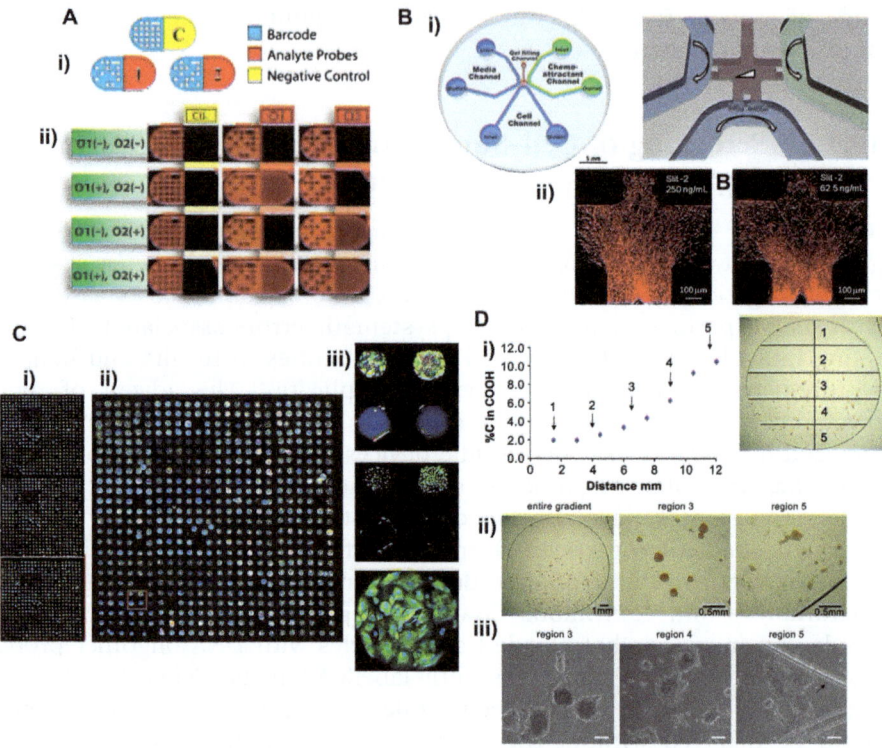

Figure 10.3   Examples of high-throughput methods for the optimisation on cellular response: (A) encoded microparticles, (i) schematic of synthesised particles, (ii) fluorescence image of DNA oligomer loaded particles, after incubation in fluorescently labelled targets; (B) microfluidic devices, (i) schematic of microfluidic device, (ii) confocal microscopy images showing effect of neuron repellent precursors on neuron response; (C) protein/cell microarrays, (i) overview of entire array showing many different polymer concentrations and compositions, (ii) human mesenchymal stem cells grown on polymer arrays, (iii) close up of individual spots; (D) gradient surfaces, for example plasma polymer gradients of acrylic acid to octadiene, (i) XPS % COOH content and photo showing the cell attachment of mouse embryonic stem cells across the AA–OD gradient, (ii) bright field images showing alkaline phosphatase stained colonies, (iii) phase contrast images showing different colony morphologies.
Figures A, B, C and D adapted from Pregibon *et al.*,[78] Kothapalli *et al.*,[77] Anderson *et al.*[56] and Wells *et al.*,[65] respectively.

time, allows for many different polymers, polymer compositions, chemical functionalities or biomolecules to be investigated on a single substrate.[75,79] Hence, thousands of samples to be screened without any knowledge or expectation for the outcome, with spots indicating an interesting response to be investigated further in greater detail. For example, Anderson *et al.*[80] have investigated the effect of polymer composition on the behaviour of human mesenchymal stem cells (hMSCs) (Figure 10.3C). One thousand, seven hundred spots of differing polymer compositions and molecular weights were deposited onto poly(hydroxyethyl methacrylate) (PHEMA) coated slides. PHEMA has been shown to inhibit cell binding and hence was expected to stop cell binding between polymer spots. Polymer compositions were identified that promoted cell attachment and spreading of hMSCs and polymer properties were identified which allowed for high levels of differentiation into cytokeratin-positive cells. Mant *et al.*[81] also investigated cellular response to a variety of polyurethane polymers. One hundred and twenty different polyurethane combinations were prepared in a microarray format to identify polymers that promoted cell attachment of bone marrow dendritic cells. All polymer compositions that promoted cell attachment were found to contain poly(tetramethylene glycol). Flaim *et al.*[82] have also prepared a microarray to investigate different combinations of five extracellular matrix proteins (collagen I, collagen III, collagen IV, Fn and laminin) on the attachment of rat hepatocytes and mouse embryonic stem cells, observing that collagen III showed the highest cell attachment, whilst laminin showed the lowest cell attachment.

Although a broad spectrum of surface properties have been investigated using the microarray format, the majority of experiments were conducted using different samples for every variant of every parameter, *e.g.*, polymer composition. However, an alternative high-throughput technique, gradient surfaces, allows for the optimisation of a limited set of parameters in a single experiment. Surface chemistry and topography gradients are a format that is particularly well suited to study subtleties in cell response to surface topography or chemistry. The format permits one variable, such as functional group density, to be continuously changed with respect to the position on a test surface whilst other parameters are held constant.[11,22,83,84] This platform also requires significantly lower cell numbers, lower quantities of culture medium and sample materials in comparison to the testing of surface chemical properties with discrete samples.

10.2   Gradient Surfaces

A gradient can be defined as a surface with continuously varying chemical composition along one dimension, or a transition from one surface property to another.[11,22,85] Gradient surfaces were first described in 1987 by Elwing *et al.*[86] who created methylsilane wettability gradients to investigate the effect of wettability on protein adsorption. However, since this time many gradients based on topography, polymer compositions, elasticity and biomolecule density have been formed.[5,22,87,88] There have been several detailed review articles published surrounding the preparation of topography and chemical gradients in recent years.[11,22,79,83,84] A summary of some of the gradient preparation techniques is shown in Figure 10.4. This section highlights some of the fundamental and recent developments in the field of gradient surface preparation.

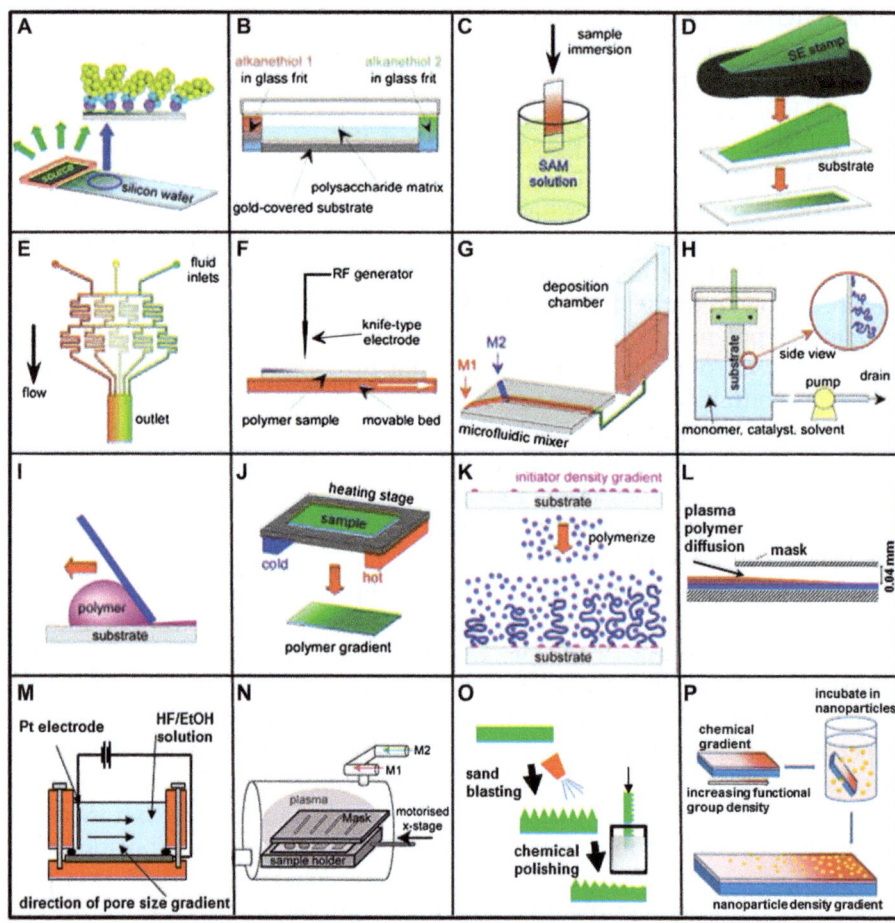

Figure 10.4 Various methods of creating chemical and topographical gradients: (A) vapour diffusion of organosilanes;[89] (B) diffusion of alkanethiols in a polysaccharide matrix;[90] (C) immersion technique applied to self-assembled monolayers;[91] (D) printing alkanethiols from stamps of variable thickness;[92] (E) solution and surface gradients using micro-fluidics;[93,94] (F) corona discharge;[95,96] (G) co-polymer gradients through microfluidic mixing of two monomers followed by chamber filling method;[97] (H) solution draining for creating polymer brushes;[98] (I) knife-edge coating technology;[99–101] (J) temperature gradient heating of substrate;[102,103] (K) molecular gradient of initiator followed by grafting from polymerisation;[104] (L) plasma polymer diffusion under a stationary mask;[7,105] (M) porous silicon gradient by anodic etching (also used for chemical gradients *via* electrografting);[46,106–108] (N) plasma polymerisation using a moving slit mask whilst simultaneously varying monomer ratio;[109,110] (O) sand blasting followed by chemical polishing;[111] (P) immersion of a functional group chemical gradient in a nanoparticle solution.[51,112]
Figures A to K adapted from Genzer and Bhat,[83] Figure L from Zelzer *et al.*,[7] Figure M from Khung *et al.*,[106] Figure N from Murray *et al.*,[110] Figure O from Kunzler *et al.*,[111] and Figure P from Goreham *et al.*[51]

10.2.1 Topography Gradients

Research has shown that surface topography influences cellular response and hence several methods have been developed for the generation of physical (topographical and mechanical) gradients.[25,103,113] One example of topography gradients are pore size gradients. The osteogenic differentiation of MSCs on porous surfaces has been reported to respond differently to the feature size of nanopores[114] and even symmetry and disorder of nanopores.[70] Zhao *et al.*[115] showed that etched titania surfaces, with pore sizes around 100 nm and a pore wall width of 20–50 nm promotes osteogenic differentiation in rMSCs. In another study, it was shown that the ordering of nanopits (120 nm in diameter and 100 nm deep) in polymethylmethacrylate (PMMA) significantly influenced the attachment and differentiation of MSCs.[70] Interestingly, it was observed that the more disordered topography arrays allowed for significantly increased osteospecific differentiation.

Porous silicon (pSi) gradients[46,108,116] have been designed for applications ranging from optical sensors[107] to modulation of cell behaviour[116] and stem cell differentiation.[117] A decrease in pore size has been shown to increase the attachment of rat mesenchymal stem cells (rMSCs).[108,118] In another study, the pore size of >1 μm and <50 nm in diameter was shown to be optimal for the attachment of neuron-like cells.[116]

10.2.1.1 Porous Silicon Gradients

pSi is increasingly being chosen in biomedical applications due to is biocompatibility, degradability and the ease in tailoring its structural properties.[119] pSi is a modified format of the chemical element silicon (Si) whereby micro- and nanoscale pores are formed in the surface structure. Hence, pSi exhibits a large surface area to volume ratio and hence is highly desirable as a substrate in biomaterials applications such as drug delivery.[120,121] pSi is formed through anodisation with hydrofluoric acid (HF) as the electrolyte. The pore structure can be accurately controlled through varying the etching conditions. The current density and the etching time can be modified to control the pore size and depth of the pores, respectively. Various forms of pSi, such as films, micro- particles and membranes, can be formed through adjusting the etching con- ditions and further processing of the etched structures (Figure 10.5).

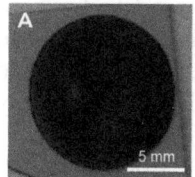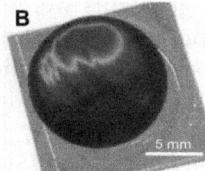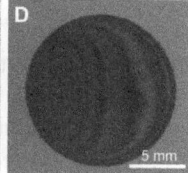

Figure 10.5 Various forms of pSi including (A) films, (B) membranes, (C) micro- particles[122] and (D) gradients.
Image B courtesy of Yit-Lung Khung, Voelcker Research Group, Flinders University.

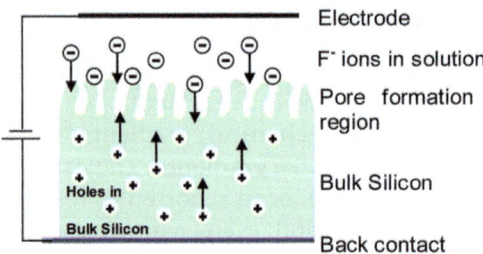

Figure 10.6   Schematic of pore formation.

Pores are readily formed in the Si surface of highly doped substrates. Boron is electron deficient and hence positive holes are present in the bulk silicon. These holes act as charge carriers and facilitate the interaction with F^- ions in the electrolyte solution (Figure 10.6). These positive holes move towards the negative electrode and in doing so, weaken the bonds in the surface silicon layer. This allows for the F^- ions to attack and displace Si on the surface. Once the first Si has been dislodged, a pit is formed in the surface and the electric field is concentrated at the tip of these pits. As a result, the positive holes are attracted to this tip and hence the pores are deepened, rather than widened to make larger pores. The depth of the pores scales with the etching time.

A gradient porous surface can be achieved through placing the platinum (Pt) electrode at one end of the exposed surface as demonstrated in Figure 10.7A. The current density decreases at increasing distances from the electrode and hence pores are the largest in the region closest to the electrode and smallest in the region furthest from the electrode. The pore range that can be achieved with this technique varies with the set etching conditions. However, it has been shown that gradients with pores ranging from 1.5 mm to 2 nm[46] or 3 mm to 70 nm[5] can be achieved over a distance of 1.5 cm. A detailed description of pSi gradient preparation can be found below. Khung *et al.*[5] has produced laterally graded pSi as shown in Figure 10.7. In such experiments, the pore range could be customised through the adjustment of the current density. Collins *et al.*[46] have also prepared lateral pSi gradients. Using profilometry, they demonstrated that film thickness decreased at increasing distances from the electrode. In the examples presented above, the produced gradients were found to be non-linear,[5,46,123,124] whereby the pore size rapidly decreased near the electrode followed by a steady plateau in the pore size. The non-linear nature of the gradient could be attributed to the non-linear decrease in current density across the substrate.

Freshly etched pSi is terminated by silicon–hydride (Si–H) bonds which are unstable and prone to oxidation and corrosion in biological fluids.[41,124] For some applications, such as drug delivery, this may be a desired effect. However, for other biomaterials applications, such as transplantable scaffolds, retaining the stability of the underlying porous layer is critical.

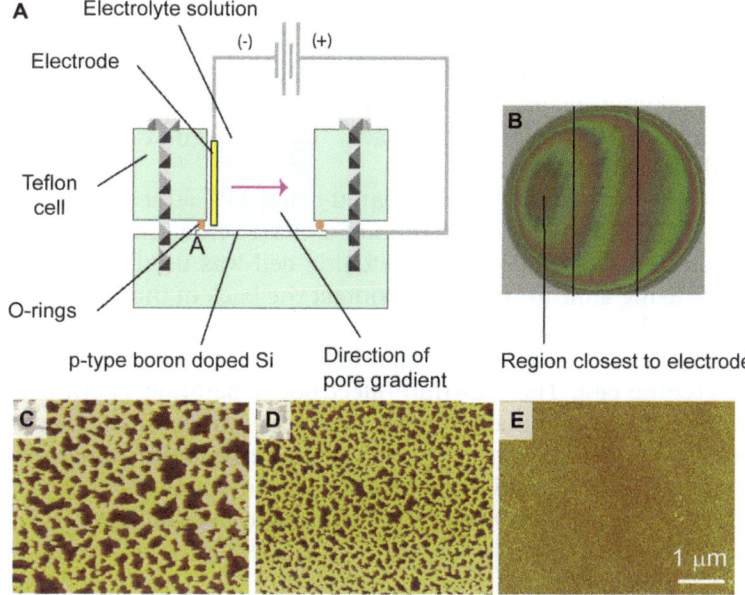

Figure 10.7   (A) Schematic of an etching cell used to create pSi gradients; (B) photograph of laterally etched pSi, the lines indicate the gradient divided into three regions for analysis; (C–E) atomic force microscopy (AFM) images of laterally graded pSi at different regions along a pSi gradient.
Figure adapted from Khung *et al.*[5]

In order to utilise this material in biological applications, surface modification of pSi is required to slow down the degradation process. Considerable research has been conducted in recent years to modify pSi surfaces to increase stability and create chemical moieties on the surface.[41,125,126] As a result, it has been established that the Si–H functionality of pSi can be modified in various ways including thermal or ozone oxidation, silanisation, electrografting, and other chemical modifications.[124,127–129]

The hydride-terminated pSi surface (Si–H) has a short lifetime[124] (a few weeks in air to a few hours in aqueous media) and is gradually oxidised, which converts the Si–H groups to Si–O–Si bonds. pSi can be intentionally oxidised *via* a stream of ozone for a short period of time, whereby the Si–H groups are converted to Si–OH groups.[130] These Si–OH groups cause the surface to become hydrophilic and the surface becomes far more stable in air and aqueous media.[131,132] The conversion from Si–H to Si–O–Si bonds can be monitored easily using infrared spectroscopy. Alternatively, thermal oxidation produces a deeper oxidised layer and produces Si–O–Si bonds rather than the Si–OH bonds seen in ozone oxidation.[132] Thermal oxidation is normally performed by heating the pSi sample above 600 °C for at least 90 min.[133] The oxidation method used however depends on the intention in regard to subsequent modifications of the surface.

A general method for pSi gradient etching is:

1. Highly doped silicon wafers need to be used to generate pSi gradients. In our laboratory, we use p^{++}-type silicon wafers (0.0005–0.001 Ω cm, ⟨100⟩, boron doped) purchased from Virginia Semiconductors (Fredericksburg, VA, USA).

2. HF electrolyte solutions are prepared using 49% aqueous HF and 100% ethanol as a surfactant at a ratio of 1 : 1 HF : ethanol.

3. A custom-built circular Teflon etching cell was used for these experiments using aluminium foil to contact the back of the Si wafer and a Pt mesh as the counter-electrode.

4. The Pt electrode is placed perpendicular to the substrate at one end of the etching cell. The substrate–electrode separation is critical for the formation of pSi gradients. If the electrode is too far from the substrate, a shallow pore-size gradient with a small maximum pore size will be formed. However, positioning the substrate too close to the substrate will result in electropolishing and complete pore collapse.

5. A current density of 115 mA cm^{-2} is applied for 60 s using a Keithley 2425 sourcemeter.

6. Following anodisation, samples should be rinsed with ethanol, methanol, acetone and DCM and subsequently dried with a gentle stream of nitrogen.

7. Following pSi formation, the hydride-terminated surface is highly susceptible to oxidation. Subsequent reactions utilising Si–H chemistry (*e.g.*, hydrosilylations or electrografting) should be performed immediately.

8. Thermal oxidation of the pSi gradients significantly increases their stability in aqueous media. We found that a two-stage thermal oxidation process resulted in the most stable pSi gradients, oxidising them in a tube furnace at 400 °C for 30 min followed by an additional oxidation step at 800 °C for 60 min.

Research by Khung *et al.*,[5] Low *et al.*,[41] Wang *et al.*[117,134] and Karlsson *et al.*[123] has shown that cells are sensitive to the pore size in which they are grown on. Khung *et al.*[5] produced a laterally graded pSi surface on p^{++}-type Si and found that the attachment and morphology of SK-N-SH neuroblastoma cells was critically dependent on pore size. They observed that on 100–300 nm pores, cells adhered poorly to the surface but could still interact through cell–cell contacts, reducing the need for cell–substratum contact. Length and thickness of neurite processes formed by the cells adherent on pSi were also dependent on pore size. Khung *et al.*[5] observed that the cells growing on the 1000–3000 nm pore size region had the largest cell size and area, and were larger than what was observed on flat Si. The smallest length and area of the cells was observed in the 50–100 nm pore range. This study demonstrated that topography can have an effect on cell adhesion and morphology. Clements *et al.*[108] observed that attachment of rat mesenchymal stem cells on pSi gradients spanning pore sizes of 900 nm to 10 nm was optimal at a pore diameter of 19 ± 11 nm. In a complementary experiment, Wang *et al.*[135] investigated the response of three different MSC lines on pSi gradient. Rat bone marrow-derived MSCs were found to be more sensitive to pore size than human bone marrow-derived MSCs. Additionally, human adipose-derived MSCs were shown to attach more strongly to a pore size of ~300 nm than on flat Si substrates. These studies clearly demonstrate that cells derived from different sources display remarkably different cellular response to the same topography cues and highlight the complexity of defining specific topographical cues to drive a specific cellular response. Karlsson *et al.*[123] investigated the effect of pore size of pSi on the attachment of human serum albumin. A non-linear curve was observed for the adsorption of HSA on pSi with preferential attachment found in the largest pore size region of the gradient. Wang *et al.*[134] also found that a decrease in the ridge roughness of n-type pSi gradients lead to an increase in the attachment and spreading of rat MSCs. These results demonstrate the utility of pSi to quickly and simply study the effect of nanotopography on cell behaviour. Whilst the use of pSi gradients has many advantages in terms of its ease of fabrication and further functionalisation, the stability of the films is clearly an issue for long-term cell culture studies. For this reason, a stable substrate with tuneable porous gradients is highly desirable.

10.2.1.2   Porous Alumina Gradients

Porous alumina (pAl) is formed through anodisation of aluminium (Al) foil in aqueous acids. A current is applied between the anode (Al substrate) and cathode (usually lead or platinum) and traditionally the electrodes are positioned parallel to each other to create an evenly dispersed electric field and therefore a uniform pore size across the surface. Phosphoric acid, oxalic acid and sulfuric acid are the most commonly used electrolytes for pAl formation. Based on previous research, smaller pore diameters are achieved when using oxalic acid compared to phosphoric acid.[47] The pore diameter, interpore distance and pore wall width can be optimised through adjusting the anodisation conditions. Ordered hexagonal patterns formed under certain pAl etching conditions are thought to be attributed to the stress-driven interface caused by repulsive forces between neighbouring pores.[136] In recent years, two-step anodisations have become increasingly popular as a means to create intricate patterned architectures including triangular or square shaped pores, checkerboard patterns and complex internal pore structures.[136]

The first pAl gradient was described by Kant *et al.*[47] whereby the electrodes were positioned at an angle of 45° to each other to achieve nonlinear anodisation. The formation of pAl gradients can be explained by the gradual change in current density, similarly observed in pSi gradient formation, arising from the non-parallel electrodes as well as a gradual temperature change across the surface during the anodisation process. Two different

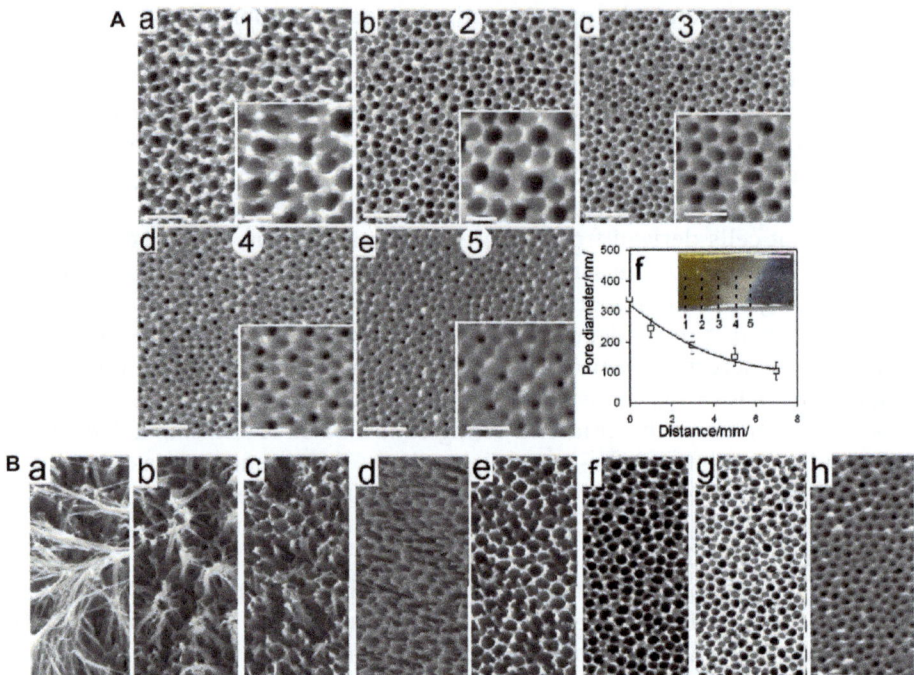

Figure 10.8   SEM images of pAl gradient formed through anodic etching of Al
substrates in the presence of (A) oxalic acid and (B) phosphoric as the
electrolyte at increasing distances from the electrode. Numbers 1 to 5 in
(A) represent locations on the optical image (inset f). Scale bar = 1 μm
and 500 nm (insets).
Figure adapted from Kant *et al.*[47]

electrolytes, oxalic acid and phosphoric acid, were used in this study to
achieve different pore geometries and features. Brush-like structures were
observed in the region with the shortest separation from the cathode, which
can be explained as remnants of pore wall structures that have collapsed
(Figure 10.8). As the distance between the anode and cathode was increased,
the pore size decreased in a non-linear manner. Whilst the technique
demonstrated the successful formation of a pAl gradient, there are obvious
limitations with this method including reproducibility, extending the gradient beyond 10 mm in length and varying the pore size distribution.

Dronov *et al.*[137] explored the possibility of changing the pore size of uniform pAl using a chemical etching technique. Here, a uniform pAl surface
was generated with a small pore size and large interpore distance. Substrates
were then immersed in 5% phosphoric acid solutions for different time
periods to achieve varying degrees of pore enlargement. An extension of this
study was described by Wang *et al.*[138] who prepared gradient pore size pAl
substrates *via* a two-step process. A uniform pAl substrate of a small pore
size was first prepared. pAl gradients were subsequently prepared by dipping

the uniform pAl layer in 5% aqueous phosphoric acid using a vertical motion stage at a constant velocity, resulting in pore sizes ranging from 5 nm to 3 μm. A detailed description of this pAl gradient preparation method is described below. This controlled immersion technique is arguably more versatile than other pAl fabrication techniques described above, given the ease of extending the length and steepness of the pAl gradient.

A general method for pAl gradient formation is:

1. Al foil should be cleaned by sonication in acetone for 30 min followed by submersion in a 5% NaOH solution for 10 s and rinsing with copious amounts of MilliQ water and drying in a nitrogen stream.
2. Porous alumina (pAl) with a uniform pore size is first generated by anodic etching in a custom-built reaction cell. The Al foil and cell are immersed in an oxalic acid solution (0.3 M, 0.0 ± 0.1 °C) and vigorously stirred. A voltage is applied between the gold back contact and the lead counter-electrode (100 V) for 2 min.
3. Following etching, the pAl was rinsed with MilliQ water and then incubated in toluene for 20 min and dried with nitrogen. This step significantly decreased the degree of wicking observed by the phosphoric acid solution.
4. Membranes were then converted into pAl gradients *via* pore enlargement using chemical etching.[138] This chemical etching was performed by fixing the pAl membrane to a vertical motion stage and subsequently slowly dipping them into a solution of phosphoric acid (5%, 25 °C, gentle stirring) over a period of 180 min at a constant velocity (0.0012 mm s^{-1}).
5. Following complete immersion, the surfaces were immediately washed with copious amounts of MilliQ water and dried with a gentle nitrogen stream.
6. pAl substrates were found to be stable in aqueous conditions for at least 21 days following incubation in PBS at 37 °C without any further surface modifications.
7. The steepness and pore size range of the pAl gradient can be controlled through adjustment of the dipping velocity into the phosphoric acid solution, with a velocity of 1.2 μm s^{-1} (corresponding to a maximum incubation time of 180 min) proving to be optimal to cover the largest pore range. Longer incubation times (>200 min) led to complete pore collapse. Conversely, whilst a shorter maximum incubation time (<150 min) led to the successful preparation of a gradient surface, a smaller pore range was achieved.
8. One of the key advantages of this method is the ability to extend the length of the gradient by varying the dipping velocity in accordance with the increased length of the gradient desired.

It has been shown that the pore size not only influences the attachment of cells but also the morphology and differentiation of cells.[116,118,139–143]

This has also been demonstrated on pAl in an extensive review recently published by Brüggemann.[141] As an example, Popat et al.[140] found that pAl surfaces with a pore size of ~72 nm enhance the spreading and proliferation of MSCs over 1 day and 7 days, respectively. Additionally, higher alkaline phosphatase activity as well as increased levels of calcium and phosphorous were observed compared to flat Al surfaces. Human osteoblast cells were also shown to attach, spread and proliferate at a higher rate on pAl surfaces (pore size ~160 nm) than on control samples.[142,143] Furthermore, filipodia were found to readily attach to the pAl surface. Additionally, Swan et al.[139] demonstrated that the attachment and spreading of osteoblast cells could be enhanced by the adsorption of vitronectin or immobilisation of an RGD peptide on pAl substrates.

Whilst there has been increased demand for the use of pAl to investigate the influence of topographic features on cell attachment, there has only been two reports of pAl gradients to investigate cellular response. The attachment of SK-N-SH neuroblastoma cells on pAl gradients was investigated by Kant et al.[47] Interestingly, the highest cell attachment was observed for the collapsed pore brush-like region of the gradient, with many interconnecting networks. At smaller pore sizes, the cells displayed a more neuronal phenotype, with long bipolar extensions. Wang et al.[138] demonstrated that human MSC attachment and cell density decreased with increasing pore size. Additionally, after 2 weeks, osteogenesis of hMSCs was found to be optimal at a pore size of 120–230 nm and a 10 nm pore width. Given the increased stability of pAl over pSi, it is envisaged that the use of pAl gradients will become increasingly popular for longer-term cell culture experiments with the purpose of investigating the effect of porous materials on cell differentiation.

10.2.1.3   Surface Roughness Gradients

Methods that have been developed to generate roughness or other topographical gradients in addition to those described previously include nanoparticle gradients based on an underlying chemical gradient[51,112,144,145] or electrostatic interactions,[146,147] annealing of polymer surfaces via a temperature gradient,[148–150] photolithography to create grooved surfaces[151] and sandblasting coupled with subsequent chemical polishing.[25,111] Several studies have demonstrated the influence of surface roughness on cellular response.[25,52,103] The obvious advantage of using gradient surfaces for such experiments is that a wide range of surface roughness values can be investigated simultaneously under identical experimental conditions.

One approach for generating topography gradients is to use a chemical gradient such as plasma polymer,[51] silane,[112,145] or brush co-polymer[112,144,152] gradients as a template to adhere nanoparticles to a surface in a gradient fashion. This method may be advantageous as the size of the topographical features can be controlled by selecting the size of the nanoparticles. Plasma polymer gradients were used by Goreham et al.[51] to

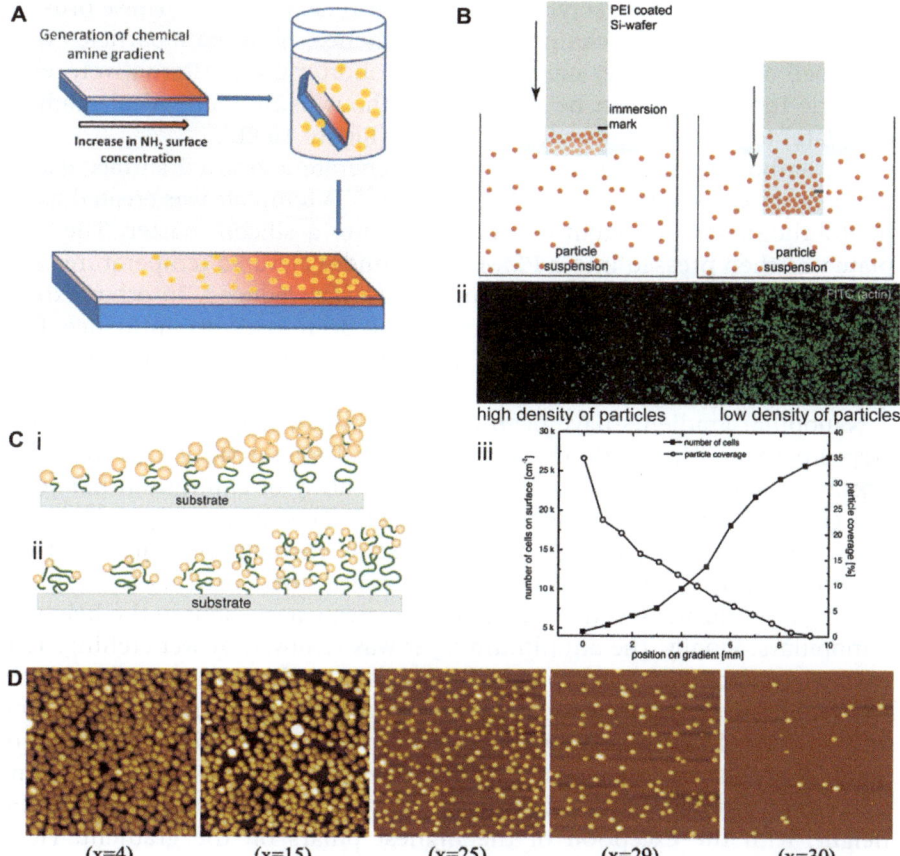

Figure 10.9   Schematic showing the methods used to create nanoparticle density gradients: (A) plasma polymer chemical template with subsequent nanoparticle attachment;[51] (B) electrostatic attachment of nanoparticles,[146] (i) schematic for gradient formation, (ii) osteoblast cell response to nanoparticle gradient, (iii) graph demonstrating that as the density of the nanoparticles decreases the cell density increases; (C) polymer brush chemical gradient template for nanoparticle gradient formation,[144] through adjusting the (i) polymer chain length, (ii) grafting density; (D) AFM images of nanoparticle gradient prepared through using an APTES silane gradient template.[145]

generate a gold nanoparticle gradient (Figure 10.9A). Gold nanoparticle adhesion correlated well with the amine functional groups in the allylamine–octadiene chemical gradient. Bhat *et al.*[112,145] used an organosilane vapour deposition technique to form a concentration gradient of aminopropyltriethoxysilane (APTES) on flat silicon substrates (Figure 10.9D).[89] The negatively charged gold nanoparticles were then electrostatically attracted to the positively charged APTES, resulting in the binding of particles to the regions covered by APTES.[112] Nanoparticles can also adsorbed onto grafted polymer brushes of poly(acryl amide) (Figure 10.9C).[112,144,152] These brushes displayed a gradient of chain length and, hence, more nanoparticles were adsorbed onto the longer chains than the shorter ones. Alternatively, electrostatic interactions can be used (*i.e.*, without using a chemical gradient template) to create nanoparticle gradients (Figure 10.9B).[146,147]

Hot embossing has been employed to generate a grooved surface with a gradient in terms of the pitch of the grooves.[151] A template was created using photolithography and etching the pattern into a silicon master. The template was then pressed into a PMMA layer under elevated temperatures and then slowly released whilst the temperature was reduced. In other experiments, a gradient temperature stage had been used to influence film microstructure, phase transitions, polymer de-wetting and crystallisation rates across the surface.[148–150]

Nanopillar gradients have been prepared by Reynolds *et al.*[153] in a four-step process to investigate the attachment and migration of a co-culture of fibroblast and endothelial cells. Initially, an array of aluminium nanodots was generated on a quartz slide through electron beam lithography.[154] A thin plasma polymer thickness gradient of hexane was then deposited over the nanodot surface.[7] This thickness gradient was then used as a sacrificial etching layer during reactive ion etching to create a gradient in the height of nanopillars. Finally, the aluminium layer was removed by wet etching. This master gradient was then replicated in polystyrene by injection moulding to create identical replicates for cell culture experiments. Fibroblast and endothelial cell co-cultures were incubated onto the nanopillar gradients for 96 h. Fibroblast density was found to steadily decrease with increasing nanopillar height, whilst endothelial cell density increased with pillar height, with the exception of the highest pillars on the gradient. Here, automated detection, measurement and classification of co-cultured cells was employed for high-throughput analysis of cell phenomena.

Al roughness gradients produced by the use of sandblasting and chemical polishing which systematically removed small features from the surface were first described by Kunzler *et al.*[25,111] on an aluminium substrate. However, this technique can be extended to many other substrates through the use of templating,[111] whereby a negative impression is created using polyvinylsiloxane and subsequently a positive replica using an epoxy is formed. These epoxy roughness gradients can then be used as is or coated with metals or a specific chemical functionality can be imparted on the surface. Al roughness gradients were subsequently incubated with osteoblast and fibroblast cells for 1–7 days to determine preferential cell response based on surface roughness.[25] In this study, it was found that osteoblasts showed preferential treatment towards a surface roughness of around 6 mm after 4 and 7 days incubation. The fibroblasts however were found to favour the smoother end of the gradient (roughness ~1 mm) which is also in agreement with published literature.[155,156] From these examples, it is obvious that a generic assumption can not be made as to the optimum surface characteristics of a surface for maximum cell response, and that each cell type will have to be investigated individually. Lampin *et al.*[52] have prepared PMMA roughness substrates using alumina sandblasting whereby the sand grain size dictated the roughness of the surface. Topography characterisation was carried out using profilometry. Although these surfaces were not gradients, it was found that attachment of vascular and corneal cells increased with increasing surface roughness. Whilst this result contradicts the findings of Kunzler *et al.*, it should be noted that the surfaces prepared by Lampin *et al.* showed an increase in hydrophobicity with roughness which may have influenced cell attachment. Additionally, the surface chemistry in the two studies was completely different, which, given the sensitivity of cells to surface chemistry, would undoubtedly influence the cellular response.

10.2.1.4   Surface Elasticity Gradients

The elasticity of a substrate surface has also been shown to influence cellular responses.[2,54,87,93,157–160] There have been several methods employed to generate elasticity gradients including gradient co-polymerisation of two monomers,[87] a gradient in the concentration of curing agent[159,161] and applying a temperature gradient during PDMS curing.[158]

Tse *et al.*[93] created a stiffness gradient of polyacrylamide by UV curing through a photomask. A radial gradient photomask was positioned in between the UV source and monomer–substrate assembly, with the degree of cross-linking decreasing moving from the centre to the outside of the substrate. This decrease in cross-linking leads to an increase in the elasticity of the hydrogel. A uniform layer of collagen was covalently bound to the gradient hydrogels to help facilitate cell attachment. MSCs were found towards the stiffer region of the gradient. Using PDMS stiffness gradients, it was found that that osteogenesis of rat mesenchymal stem cells (rMSCs) was strongly dependent on the PDMS stiffness as indicated by the calcium mineralisation observed in the stiffer regions of the gradient.[158] Several modifications to this method described by Vincent *et al.*[159] involved using a linear gradient photomask to vary the degree of curing, or a custom-built microfluidic gradient maker to vary the cross-linker concentration. Following deposition of a uniform Fn adsorption layer, an in depth analysis of MSC durotaxis (directed migration of cells up a stiffness gradient) was performed. Whilst uniform attachment was observed within hours of cell seeding, after 3 days cells were found to migrate towards stiffer regions of the gradient.

Poly(L-lactic acid) (PLLA) and poly(DL-lactic acid) (PDLLA) gradients were prepared by Simon *et al.*[87] These polymers differ only in their tacticity (*i.e.*, their stereochemistry) leading to PLLA being crystalline and PDLLA being amorphous. Hence, PLLA has a higher modulus (amount of force required to deform a material). PLLA–PDLLA gradients were prepared using a syringe system. PDLLA monomer was slowly introduced into a beaker containing PLLA, whilst a syringe was simultaneously withdrawing a solution of the mixed monomer composition in the beaker. In this way, the concentration of

PDLLA in the syringe gradient gradually increased as the syringe was filled. Whilst the 24 h attachment of osteoblast-like MC3T3-E1 cells was similar across the entire gradient, cell proliferation was found to be faster on the PDLLA-rich end. This could be attributed to the fact that the PDLLA-rich end of the gradient was smoother than that of the PLLA-rich end. Similarly, van Sliedregt *et al.*[157] found that epithelial cells, osteosarcoma cells and fibroblasts proliferated faster on PDLLA surfaces as opposed to PLLA surfaces. Conversely, Ishaug-Riley *et al.* found that human articular cartilage chondrocyte attachment was faster on PLLA compared to PDLLA.[162] As highlighted previously, the elasticity of a substrate has the ability to direct the lineage of stem cells.[2,54,67] Through altering the cross-linking density to obtain a soft or rigid surface, cells could be directed towards neuronal cells or to bone cells.[2]

10.2.1.5 Gradients of Coating Thickness

The thickness of a coating has also been shown to affect the attachment of cells.[7,99,100] Whilst a thickness gradient is not strictly a topography gradient, it is discussed in this section for simplicity.

Smith *et al.*[99,100] and Crosby *et al.*[163] have produced thickness gradients *via* the use of a flow coater. A Si wafer was mounted onto a motorised stage. A knife-edge was then placed just above the surface at a 5° angle to the substrate, and a drop of polymer solution was placed under the knife. As the stage was moved, the knife spread the polymer solution over the substrate. Through increasing the stage velocity and upon solvent evaporation (which occurs in seconds) a thickness gradient was formed. The velocity of the stage determined the thickness of the polymer layer, with film thicknesses ranging from 25–500 nm. Poly(styrene-*b*-methyl methacrylate) co-polymer thickness gradients were formed using this technique on silicon wafers. It has previously been observed that the methacrylate preferentially segregates to the substrate polymer interface whereas the styrene preferentially migrates to the air interface.[164,165] In this example, whilst the researchers were intentionally altering the thickness of the polymer film, a secondary effect, namely phase separation between the two polymers was also observed. As the thickness changed across the gradient surface, the surface changed from smooth, to discreet islands arising from phase separation, and then back to smooth again. Film thickness was measured with a UV-visible interferometer, whereby a contour map was formed through recording measurements at discreet intervals across the surface.

10.2.2 Chemical Gradients

Chemical gradients allow for a diverse range of surface properties to be created through the use of different functional groups. Polymers are increasingly being used for biomaterial and biomedical applications due to their cheap synthesis and ease of structural manipulation, which allows for the creation of a vast selection of compositions and geometries. The difficulty with some gradients is that several parameters are altered in the course of the investigation.[87] For example, the chemical composition and/or phase separation can alter with a variation in film thickness.[99] In such cases it can be difficult to assign a single surface property to the observed cell–surface interactions, with a combination of factors potentially leading to the observed result. Therefore, adequate characterisation of the deposited surface is essential to determine the main parameters influencing the cellular response. The vast array of chemical gradient preparation techniques existing today have been expertly reviewed by Genzer and Bhat,[83] Morgenthaler *et al.*[84] and others.[11,79]

10.2.2.1   Self-assembled Monolayer Gradients

There are two distinct categories of self-assembled monolayer (SAM) gradient preparation techniques; those that use silane chemistry on glass or Si substrates, and those that use alkanethiols on gold or silver substrates. The first reported preparation of a chemical gradient by Elwing *et al.*[86] in 1987 utilised silane chemistry to create gradients of methylsilanes on Si. The methyl silane was bedded under a xylene solution containing an immersed silicon wafer and the methyl silane slowly diffused through a xylol region and was simultaneously bound to the surface. A modification of this procedure was described by Chaudhry and Whitesides,[89] whereby decyltrichlorosilane in paraffin oil was positioned next to a Si wafer with the vapours being allowed to diffuse along the surface, adsorbing and therefore generating a gradual change in silane coverage.

This technique has been readily employed to create silane gradients with semi-fluorinated,[166] terminal alkene,[167] chlorine,[167] amine,[89] hydrocarbon,[86] and specially modified protecting groups for patterned UV irradiation.[168–170] Many of these functional groups can be further modified or irradiated to generate COOH terminated surfaces, peptide gradients and even carbon nanotube density gradients.[167,171,172] In most of these instances, backfilling was employed, either as the second step to fill vacant spaces or as the primary step to create a fill template for subsequent silanisation of the functional groups of interest.

Another method for generating silane chemistry gradients is through UV irradiating a uniform silane coating for varying periods of time as described by Gallant *et al.*[173] (Figure 10.10). In the first step, uniform *n*-octadimethylchlorosilane coatings were deposited on a Si wafer. The sample was then placed under a motorised UV lamp. Varying the UV exposure time across the sample led to variations in the extent of oxidation of the hydrocarbon chains. Such method produced gradients with contact angles ranging from 25° to 95°. Subsequent functionalisation using 'click chemistry'[174] led to the generation of an RGD density gradient. As an extension to this study, Kennedy *et al.*[175] adsorbed Fn on the UV irradiated surfaces. It was found that Fn-mediated cell attachment was faster on the more hydrophobic surface than on the hydrophilic surfaces. This finding was also supported by

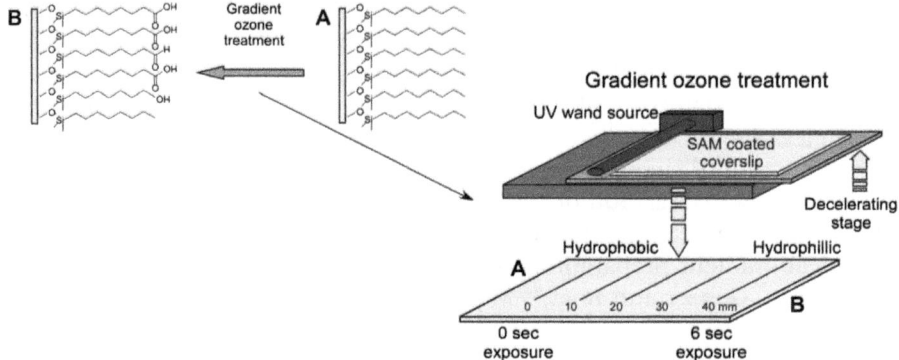

Figure 10.10   Silane gradient preparation through UV irradiation of *n*-octadimethyl-chlorosilane (A). Ozone treatment for increasing time periods across the gradient lead to a gradual increase the concentration of carbonyl functional groups (B).
Modified from Gallant *et al.*[173]

Elwing *et al.*[86] and Bhat *et al.*[176] Allen *et al.*[26] also observed an increase in Fn adsorption on hydrophobic surfaces. However, in a study by Wei *et al.*[9] Fn was found to bind to the hydrophilic end of the gradient, with contact angles ranging from 3° to 40°. This discrepancy highlights that, unlike popular belief in the past, the contact angle alone is not believed to be the sole contributor for the attachment of proteins and that surface chemistry itself also plays an important role. Other techniques for preparing chemical gradients with silanes include microcontact printing and oxidation of a uniform silane coating through UV irradiation through a gradient mask.

> **NOTE:** *The term 'click chemistry' was originally coined by Kolb et al.*[174] *to describe a specific range of chemical processes that quickly and reliably join small units together.*

Semi-fluorinated SAMs have been shown to prevent the attachment of some biomolecules.[177] Using surface plasmon resonance (SPR), the immobilisation of immunoglobulin was monitored, with adsorption only observed on non-SAM coated regions. *N*-Octadecyltrichlorosilane (OTS) is commonly used in chemical modifications of a surface. It can be easily deposited and has been shown to be effective in preventing cell attachment due to its hydrophobic nature.[178] Upon UV exposure, the hydrocarbon chain degrades, leaving a terminal OH region. The hydroxyl-terminated regions can then be further functionalised. Through the use of a gradient mask or shutter system, a density gradient could be prepared. Matsuzawa *et al.*[178] deposited OTS onto glass and silicon surfaces. The surface was exposed to UV light through a mask, patterning the surface. *N*-(2-Aminoethyl-3-aminopropyl)-trimethoxysilane (EDA) was then selectively immobilised onto the surface followed by seeding with SK-N-SH neuroblastoma cells. It was found that the cell attachment was confined to the EDA functionalised regions, which demonstrates the effectiveness of OTS in preventing cell attachment.

Alkanethiols were first used in gradient preparation in 1995 by Liedberg and Tengvall,[90] through a diffusion based method. Two different alkanethiols were loaded at each end of a polysaccharide covered gold substrate and allowed to diffuse for several hours, forming a densely packed monolayer of gradually changing end functional group.

A simpler method of generating alkanethiol gradients is through a two-step immersion process as described by Morgenthaler *et al.*[91] Gold-coated samples were slowly immersed into the thiol solution, generating a density gradient of the alkane thiol (Figure 10.11). The concentration and immersion time were tailored such that a densely packed monolayer is achieved on the end of longest immersion time down to only a few molecules adsorbing at the end of the shortest incubation time. The substrate was backfilled with a second alkane thiol, (*e.g.*, hydroxyl terminated) so that a complete densely packed monolayer was formed over the entire length of the gradient. Whilst there may be an island sub-micron structure arising from the two different thiols, a linear change in surface wettability can be achieved.

10.2.2.2   Plasma Polymerisation Gradients

Plasma polymers can be deposited onto virtually any substrate, affording a smooth and pinhole-free coating.[179] Typically, plasma polymerisation is

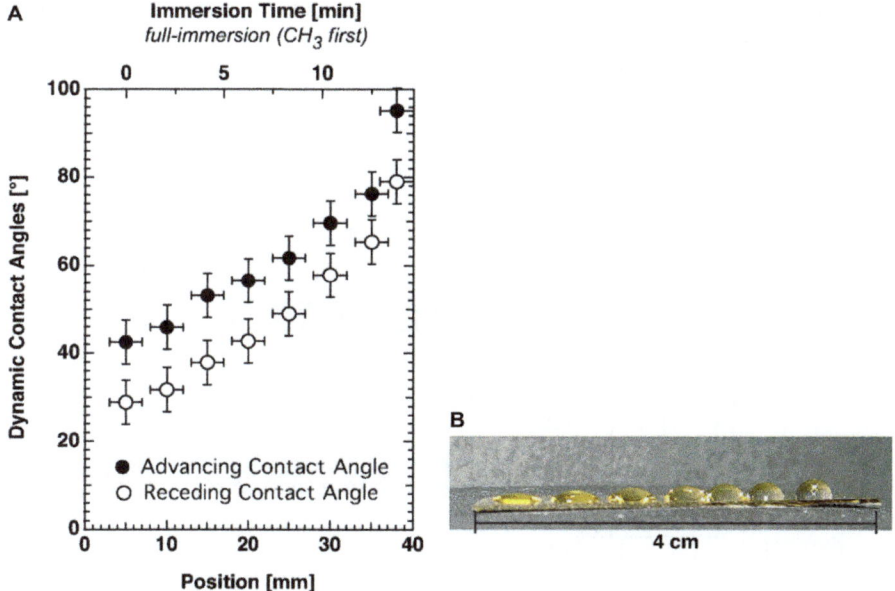

Figure 10.11   Alkanethiol immersion method to generate chemical gradients: (A) advancing and receding contact angles along the gradient; (B) water droplets along the alkanethiol gradient.
Figure adapted from Morgenthaler *et al.*[91]

conducted at room temperature. In this method, a vapour of monomer molecules are electrically excited to form a plasma phase; components of the plasma phase (ions, radicals, and neutrals) then may oligomerise and deposit on the substrate placed in contact with the plasma.[180–182] Through the use of low power and low pressure, functional groups in the monomer are retained in the final deposited film, for example, carboxylic acid groups from acrylic acid.[183] Chemical gradients displaying a single or several functional groups can be easily prepared using plasma polymerisation. There have been several methods reported for the preparation of plasma polymerisation gradients in recent years including using an instrument termed a 'gradientiser', where a mask shields part of the surface whilst the monomer composition continuously changes,[109,110,184] using a knife edge electrode whereby the chemistry changes with increasing distance from the electrode,[185] and diffusion based depositions by shielding the surface with a stationary mask,[7,186] A variety of terminal groups can be incorporated into the gradient surface including, amino, carboxylic acid, methyl, aldehyde and epoxy using these methods.

Two versions of the gradientiser have been produced. In the first method, the sample was placed into a drawer as shown in Figure 10.12A.[109,184] Initially, a thin allylamine plasma polymer coating was deposited over the entire substrate. A second monomer, acrylic acid, was then introduced into the chamber while simultaneously reducing the flow of allylamine. During this process the drawer was slowly retracted. This resulted in a chemical composition gradient from amine to carboxyl over a distance of 11 mm.

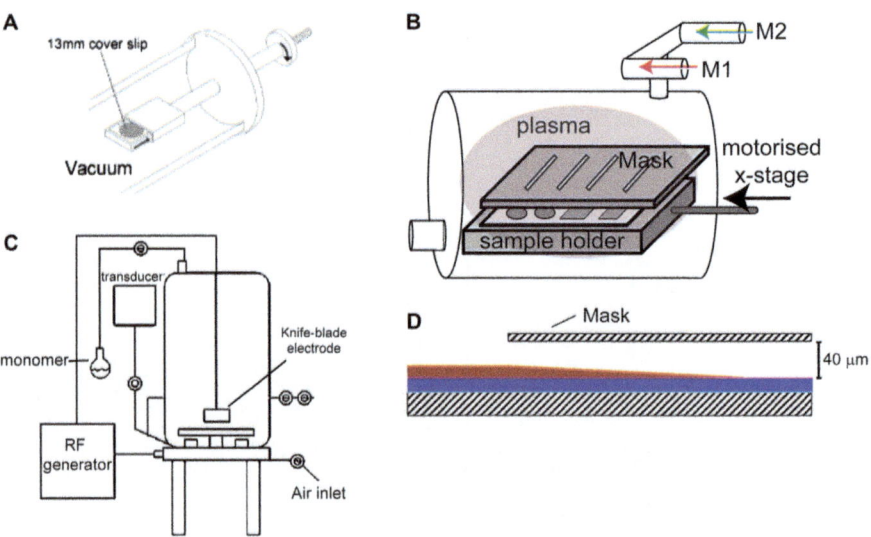

Figure 10.12   Various plasma polymerisation reactors for gradient preparation: (A) gradientiser drawer;[109,184] (B) gradientiser slit mask;[110] (C) knife-blade electrode;[185] (D) diffusion under a mask.[7,105]

This gradient preparation method is versatile in that by changing the monomer solutions, many functional groups can be deposited onto the surface in this manner.

In the second version, the sample is placed into the reaction chamber with a thin slit mask placed over the top at one side of the sample (Figure 10.12B).[51,65,110,187,188] The monomer composition is gradually changed from one monomer to another, whilst the slit is very slowly moved across the sample. This method has advantages over the first generation of its type as the thickness of the plasma polymer is relatively consistent. Parry *et al.*[184] prepared a graded acrylic acid (AA) to octadiene (OD) chemical gradient. Immunoglobin G (IgG) was adsorbed onto the plasma polymer gradient and it was observed that more IgG was adsorbed at the octadiene end (>200 ng cm^{-2}) compared to the acrylic acid end (<20 ng cm^{-2}).

Another approach is the use of a stationary knife blade electrode as described by Menzies *et al.*[185] (Figure 10.12C). A knife blade was positioned ~ 1 mm above the substrate and a gradient of diethylene glycol dimethyl ether (DG) plasma polymer was deposited due to the non-uniform plasma glow discharge. A variation in both the thickness and chemical composition was achieved, with the ether component (O–C–O) gradually decreasing at increasing distances from the knife-blade electrode. The DG plasma polymer produces a PEO-like coating minimising protein and cell attachment.[189] Hence, a gradient from a high cell-binding surface to a low cell-binding surface could be formed in this manner. Adsorption of BSA and lysozyme onto the PEO-like plasma polymer gradients was assessed by XPS and TOF-SIMS analysis.[17]

Zelzer *et al.*[7] prepared thickness gradients *via* diffusion-controlled plasma polymerisation (Figure 10.12D). Gradients ranging from chemistry based on allylamine plasma polymers to hexane plasma polymers were prepared on glass through diffusion under a fixed mask. Sharp and shallow gradients were prepared depending on the distance between the mask and substrate. Once again, a sigmoidal curve of film thickness was observed. The wettability ranged from hydrophobic at the hexane plasma polymer end to hydrophilic at the allylamine plasma polymer end of the gradient. NIH 3T3 fibroblasts were seeded on the surface with the highest cell attachment observed at the hydrophilic acrylic acid end of the gradient. One concerning observation was that a significant difference was observed in the cell density per surface area between the allylamine end of the gradient and a uniform allylamine plasma polymer control. This result demonstrates the need for simultaneous uniform substrates in conjunction with gradient materials. Using a similar technique, Harding *et al.*[105] generated a plasma polymer gradient varying form cell adhesive acrylic acid to cell repellent diethylene glycol dimethyl ether (DG). In this instance, a glass mask was tilted at 12° to the substrate with the sides enclosed. Detailed methods for the preparation of plasma polymer gradients are described below. Mouse embryonic stem cells were incubated on the AA-DG gradients for 3–5 days with both colony size and colony morphology correlating with the continued expression of stem cell markers. Cell response was found to be influenced strongly by the plasma film depth, which mediated the adsorption of proteins from the culture medium.

A method for plasma polymer gradient formation using a diffusion-based technique is:

1. Diffusion-based plasma polymer gradients as described here, can be performed in a wide variety of plasma reactors, and is not limited to expensive and specialised reactor systems.
2. Thermanox plastic coverslips or p^+-type Si wafer (Si, 3–6 Ω cm resistivity) have traditionally been used as substrates in this process. However, this gradient deposition method is adaptable to a wide variety of substrates.
3. All monomers were freeze–pump–thawed degassed three times to remove traces of oxygen.
4. Samples were placed in the centre of the chamber on a glass plate and the chamber was evacuated to a base pressure of 1.2×10^{-3} mbar.
5. For all monomers, a constant flow rate of the monomer was set to 2 sccm (standard cubic centimeters per minute).
6. A base coat of 1,7-octadiene was applied to the substrates for 10 min at generator RF output power measured at 2 W to act as an adhesion layer between the substrate and subsequent plasma layers.
7. Subsequently, an acrylic acid plasma polymer layer was deposited on top of the OD plasma polymer layer.
8. In order to achieve a thickness gradient, a glass slide, tilted 12° to the substrate surface with the sides enclosed, is placed over the substrate as a mask.[105] The outer edge of the substrate protrudes 5 mm from the opening of the glass slide.
9. A gradient of diethylene glycol dimethyl ether (DG) plasma polymer layer is then deposited for 5 or 10 min at an output power of 4 W.
10. This process is not limited to the monomers described in this method, and can be adapted to generate a wide variety of difference chemical gradient functionalities.

In another study, a gradient of oxygen containing functional groups was prepared through corona treatment on a polyethylene surface.[95,96] Polyethylene sheets were first exposed to an air plasma to introduce oxygen-containing functional groups.[22,96] The sample was then moved gradually under the knife-like electrode whilst simultaneously increasing the corona power. The functional group density gradually increased, and hence, a gradient was formed. Once again, the prepared gradient was found to be non-linear. The surface was then treated with polyethyleneimine and then biotin was attached through a NHS assisted esterification. Fluorescein conjugated streptavidin adsorption increased with increasing biotin concentration.

The thickness of polymer films are most commonly determined through atomic force microscopy (AFM)[190] or profilometry.[87,105] In order to measure the step height using these methods the surface can be masked with a flexible polymer prior to polymer deposition. Following deposition the mask can be peeled off and the step height measured. Alternatively a scratch can be made on the deposited polymer surface. Although AFM is potentially slightly more accurate, the experimental effort is higher than using profilometry. Techniques such as ellipsometry and quartz crystal microgravimetry could also be used to determine the thickness of the polymer; however, prior knowledge such as the refractive index or density of the polymer is required to use these techniques.[7,99,100] Interferometric reflectance spectroscopy, UV-visible interferometry and XPS can also be used to determine film thickness within limitations.

10.2.2.3 Electrografting Gradients

Electrografting is very useful for generating surfaces of a wide variety of functional groups through electrochemical attachment onto pSi. Electrochemical reactions of pSi to form Si–C bonds on pSi can be performed using organohalides, alkynes or Grignard reagents.[191]

Organohalides, most commonly bromides or iodides, react with pSi upon electrochemical reduction to form Si–C bonds as summarised in Figure 10.13. The lithium iodide (LiI) present in the reaction reduces the organo halide to form a radical in the first step of the reaction.[192] This alkyl radical then removes a H$^\bullet$ radical from the Si–H surface, thereby creating Si radicals.[193] Subsequently, the Si radicals react with the alkyl radicals to form a Si–C bond. This cathodic electrografting reaction is prone to surface oxidation arising from moisture or oxygen in the environment and hence, the reaction must take place under an inert atmosphere.

A wide variety of functional groups can be attached to pSi using this method (summarised by Gurtner *et al.*[192]) including: trifluoroacetylamino,[126] ethyl ester,[126] pentyl acetate,[107] bipyridines[194] and hydrocarbon-based molecules.[107,195] Some functional group terminated alkyl halides, however, can react with the Si–H surface. These groups include phenol groups, alcohols or terminal acids, and they compete with the Si–C bond formation.[192] These unwanted side reactions could be overcome by using an ester functional group, which can be later reduced to a terminal acid or alcohol.[192] The stability of a pSi surface can be further improved by methylating or 'end-capping' the unreacted Si–H functional groups with

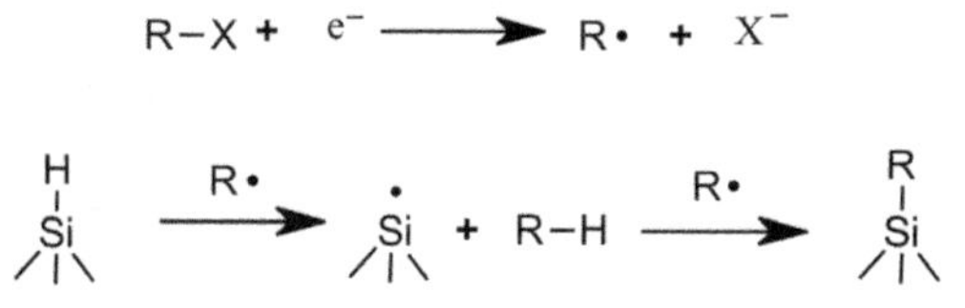

$$R\text{–}X + e^- \longrightarrow R\bullet + X^-$$

Figure 10.13 Mechanism for the attachment of organic functional groups onto pSi by the electrochemical reduction of organohalides.
Adapted from Gurtner *et al.*[192]

methyl iodide or similar.[107,126,195] The functionalisations described above can easily be confirmed by infrared (IR) spectroscopy or X-ray photoelectron spectroscopy (XPS). Terminal alkynes can also be immobilised onto pSi through the use of either an anodic or cathodic electrografting.[194,196,197] Using a cathodic current, alkynes are attached directly onto the surface, still retaining the triple bond functionality. Using an anodic current however, the triple bond functionality is not retained, instead yielding an alkyl surface.

All of the examples listed above describe the modification of a pSi surface to create uniform surface functionality; however, there are few reports to date where this technique has been employed to generate chemical gradients on pSi. The preparation of gradients by electrografting is a relatively straightforward method and uses the same electrochemical cell that is used for the generation of pSi pore size gradients as described previously. Additionally, the electrografting technique is arguably simpler and quicker in terms of generating chemical gradients than many of the techniques described in this chapter and hence, constitutes an efficient approach to prepare substrate materials for a wide range of applications. Clements *et al.*[108,198] generated chemical gradients of ethyl-6-bromohexanoate on porous silicon substrates *via* electrografting. The chemical gradient was prepared using an asymmetrical electrode placement, followed by end-capping of unreacted silicon–hydride groups with methyl groups using a parallel electrode placement. The resulting ester moieties were hydrolysed and activated to produce a gradient of carboxylic acid functional groups. After subsequent immobilisation of cyclic Arg–Gly–Asp–D-Phe–Lys (c(RGDfK)), the surfaces were used to screen the extent of rat mesenchymal stem cell (rMSC) attachment. Detailed methods and tips for the preparation of electrografted and peptide density gradients are described below. Mapping of the surface chemistry was carried out by means of infrared microscopy (IRm) and X-ray photoelectron spectroscopy (XPS). rMSC culture studies showed that short-term cell attachment responded to the c(RGDfK) density gradient present on the surface.

Thompson *et al.*[107] created gradients of methyl, pentyl acetate and decyl groups on pSi substrates. Interestingly, gradients were generated using a magnesium anode rather than the standard Pt electrode which lead to reduced oxidation of the surface. Subsequent cleavage of the pentyl acetate groups to generate a hydroxy terminated surface allowed researchers to investigate the ethanol concentration in water samples.

Electrografting and peptide density gradient formation:

1. Uniform pSi or gradient pSi substrates should be prepared as described above.
2. Freshly etched pSi substrates were immediately transferred to an inert reaction cell, brought under vacuum and then purged with argon. Whilst under an argon atmosphere, 2 mL of alkyl halide solution was transferred to the reaction cell and the cell was again sealed under an argon atmosphere.

3. A flask containing a solution of ethyl-6-bromohexanoate (EBH, 0.4 M) and LiI (0.2 M) in acetonitrile was degassed *via* three FPT cycles.

4. For gradient surface preparation, the electrode was positioned perpendicular to the substrate at a height of 4 mm in a Teflon cell, similar to that used for pSi gradient fabrication. Subsequently, cathodic electrografting was carried out at a linearly ramped current density from 1–8 mA cm^{-2} for 90 s. The samples were illuminated with 80 mW cm^{-2} white light to supply sufficient photocurrent. Unreacted silicon hydride (Si–H) sites were subsequently capped with methyl iodide (MI) (0.4 M MI and 0.2 M LiI in 2 mL acetonitrile, applying a current density of 5 mA cm^{-2} for 120 s) using the same procedure described above. However, for capping the unreacted Si–H sites, the electrode was positioned parallel to the substrate surface rather than perpendicular. Following derivatisation, samples were washed with glacial acetic acid, acetonitrile and ethanol and dried with a light stream of nitrogen gas.

5. The EBH functionalised surfaces were converted into free acids by the addition of boiling sulfuric acid (3 M, 90 °C, 4 h). The resulting hexanoic acid (HA) functionalised surface was activated with a solution containing 1-ethyl-3-(3-dimethylaminopropyl) carbodiimide (EDC, 125 mM) and *N*-hydroxysuccinimide (NHS), both at 125 mM for 20 min followed by a brief washing step with MilliQ water. The amino-terminated cyclic Arg–Gly–Asp–D-Phe–Lys (c(RGDfk) was then added (200 µg mL^{-1} in PBS) and reacted at room temperature for 3 h. Samples were then rinsed with PBS and MilliQ water and dried with a gentle stream of nitrogen gas.

TIPS

- *The electrode height is very important for successful gradient formation. Too close and the surface will oxidise, and too far away will result in poor coverage.*
- *All solutions should be anhydrous and/or distilled prior to use.*
- *All solutions must be freeze–pump–thaw degassed to remove traces of oxygen.*
- *The reaction must take place in an inert atmosphere (we found argon to produce most favourable results).*

10.2.2.4 Polymer Brush Gradients

There have been many methods developed in recent years to prepare polymer brush gradients. Some techniques utilise SAM methods to generate a gradient of functional groups or initiator molecules, whilst others rely on immersion times or microfluidic based techniques. Initiator gradients have been prepared using namely silane or alkanethiol diffusion methods, or UV irradiation.[104,173,199–201] These gradients can then act as initiator sites, *e.g.*, for atom transfer radical polymerisation (ATRP), or click chemistry reactions.

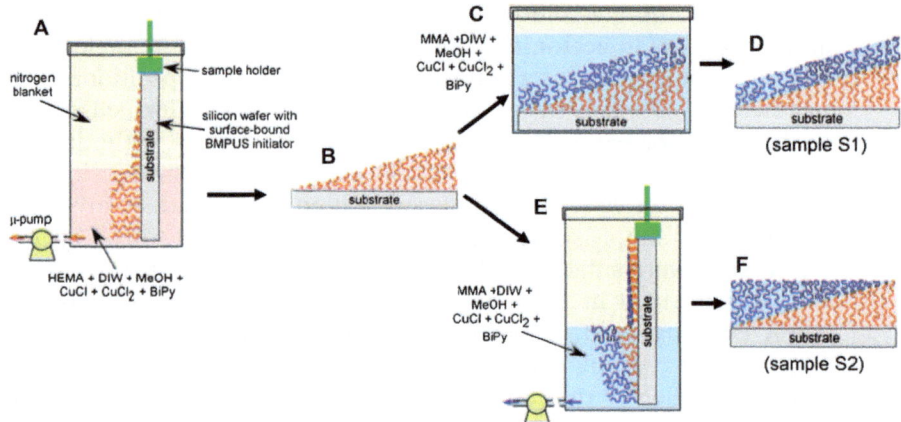

Figure 10.14 Schematic for the preparation of co-polymer brush gradients: (A) solution draining method; (B) resultant polymer molecular weight gradient; (C and D) complete submersion to generate co-polymer brush gradient of varying lengths; (E and F) second solution draining procedure to create a co-polymer brush gradient of varying polymer ratios but the same thickness.
Figure adapted from Tomlinson *et al.*[98]

This work was pioneered by Tomlinson and Genzer who developed several methods of preparing polymer brush-like gradients, as summarised in Figure 10.14.[202,203] A solution draining method was prepared to induce a gradual change in molecular weight, involving submersing a uniformly deposited initiator substrate into a monomer solution and slowly draining the monomer solution (Figure 10.14A and B). Subsequently, the gradient was completely submerged in a second monomer solution to create a co-polymer brush gradient of varying lengths (Figure 10.14C and D). Alternatively, a second solution draining procedure could be implemented to generate a co-polymer brush gradient of varying polymer ratios (Figure 10.14E and F). These methods have been readily utilised to prepare 2-dimensional (2D) gradient surfaces.

Several modifications to these processes were developed. Xu *et al.*[204] described a method whereby a monomer solution was slowly injected into one end of the initiator functionalised substrate using a syringe pump and microchannel system (Figure 10.15A), with polymerisation times varying according to the monomer incubation time. This method resulted in a homopolymer molecular weight or a block co-polymer gradient. In another experiment, a microfluidic static mixer was employed to mix two monomer solutions with constantly ratios of monomer A and monomer B (Figure 10.15B).[97] Following UV polymerisation, the result was a gradient in the composition of the polymer brush, with a gradual variation from monomer A to monomer B.

An alternative method, originally designed for the preparation of alkanethiol gradients involves a simple dip-coating process into a

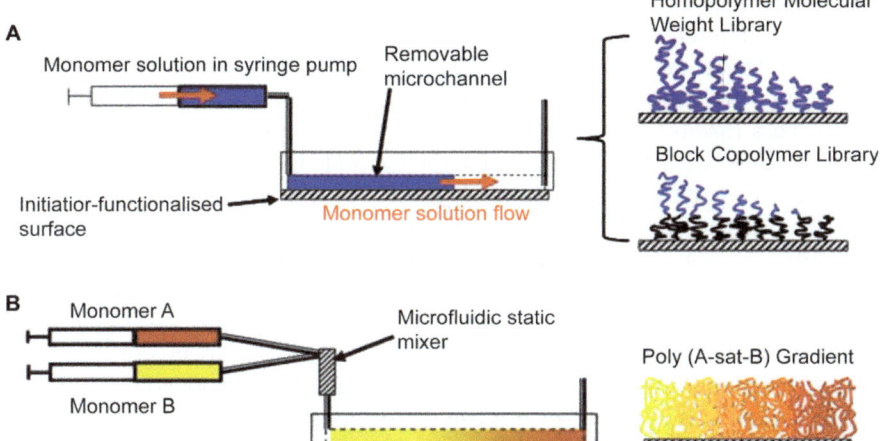

Figure 10.15 Schematic for the preparation of brush-*co*-polymer gradients: (A) homopolymer molecular weight gradient using a syringe pump and microchannel; (B) microfluidic static mixer with two different monomers to create a compositional gradient from monomer A to monomer B.
Figure modified from Fasolka *et al.*[148]

poly(L-lysine)-*g*-poly(ethylene glycol) solution.[205] By controlling the adsorption time across the substrate, a gradient in polymer density could be created. Subsequent backfilling with a second polymer or different chemistry, or additional functionality, provided a saturated polymer gradient.

Larsson and Liedberg[206] have produced PEG thickness gradients ranging from 0 to 500 Å through graft polymerisation. Firstly, a cyclo-olefin polymer is exposed to an air plasma to create peroxides and other reactive species on the surface.[207] Subsequent UV irradiation leads to the generation of surface radicals, which interact with the PEG monomers in the solution. During UV irradiation a shutter is gradually slid across the surface. This results in the formation of a polymer brush gradient, with the thickness depending upon the length of UV exposure. Human Fn was found to penetrate into the PEG layer for ~200 Å whereas human serum albumin was unable to penetrate into the PEG layer at all.

10.2.2.5 Gradients of Surface Wettability

As discussed earlier, it is too simplistic to view the contact angle as a major parameter that determines cellular response as there are often multiple factors (functional group chemistry, surface topography, *etc.*) influencing the contact angle observed. Nonetheless, many researchers have investigated wettability gradients using a wide variety of functional groups and polymer compositions. Hydrophobic groups are most commonly comprised of carbon chains, ranging from a single carbon to long chains. Hydrophilic groups however can be charged groups such as anions (carboxylates, sulfates, sulfonates, phosphates) or cations (amines) or polar uncharged groups such as alcohols. There are several methods that have been employed to create wettability gradients including plasma polymerisation,[9] self-assembled monolayers (SAMs),[175] corona discharge,[8,208] and photodegradation[209] just to name a few, with a comprehensive review on this subject by Ruardy *et al.*[22] Choee *et al.*[8] produced a wettability gradient on polyethylene (PE) through oxidation of the polymer with increasing power *via* corona discharge. The resulting carbon-based radicals including metastable peroxides and hydroperoxides reacted further to produce various oxygen-containing functionalities such as ketones, aldehydes, hydroxyls and carboxylic acids that are more hydrophilic compared to PE.[210] A resulting gradient with contact angles ranging from 95° to 18° was then exposed to NIH/3T3 fibroblast cells. It was found that the fibroblasts grew best on a moderately hydrophilic surface (contact angle $\sim 55°$). Wei *et al.*[9] and Yoshinari *et al.*[10] have prepared hexamethyldisiloxane (HMDSO) coatings, displaying a hydrophobic surface, *via* plasma polymerisation. Subsequent O_2-plasma treatment with varying duration across the surface gave a gradient of wettability. The adsorption of Fn increased with increasing wettability, whilst albumin adsorption occurred preferentially on more hydrophobic regions of the gradient.

10.2.3 Biological Density Gradients

The ECM changes comprises many proteins including collagen, vitronectin, Fn, elastin and laminin. These proteins have been used extensively in the investigation of cellular responses to surfaces through either deliberately adsorbing a particular protein to the surface or in many cases the inadvertent deposition of proteins (*e.g.*, from serum in culture media).[1,3,15,82,123,211,212]

Gradient surfaces displaying a gradually decreasing density of Fn were prepared by Bhat *et al.*[176] PHEMA gradients were prepared *via* atom transfer radical polymerisation (ATRP).[202] Fn adsorption was found to decrease across a gradient displaying a linear increase in PHEMA density. For very low thicknesses of PHEMA, a thick layer of Fn was adsorbed. However, the Fn thickness decreased rapidly as the thickness of the PHEMA increased until no Fn was adsorbed approximately half way along the gradient. Fn serves as an anchor for the attachment of osteoblastic cells used in this study.[31-33] Therefore as expected, the concentration of the cells decreased as the thickness (density) of the PHEMA increased (*i.e.*, as the density of Fn decreased). Variances in cell morphology was also observed along the gradient. Well spread cells were observed on the thin PHEMA region whereas rounded cells were observed on the thick PHEMA. In an PHEMA/Fn gradient prepared by Mei *et al.*[213] a sigmoidal response was observed for both Fn adsorption and cell attachment as the PHEMA density increased with maximum cell attachment observed at a Fn density of 50 ng cm^{-2}.

Fn gradients have also been generated by Smith *et al.*[14] using an alkanethiol SAM diffusion approach as described previously. Fn was then covalently bound through EDC–NHS coupling to the carboxy-terminated regions of the surface. Similarly to other examples described above, bovine endothelial cells were found to attach or migrate towards the high Fn density region of the gradient. A gradient of heparin, a glycosaminoglycan known to interact with a wide variety of proteins crucial for a range of cellular processes,[214] was developed by Robinson *et al.*[215] using a plasma polymer gradient template. Chemical gradients of allylamine to 1,7-octadiene (OD) functionality were deposited using the gradientiser, with heparin subsequently adsorbed onto the plasma polymer surface. Using the same allylamine–OD plasma polymer gradient described above, Vasilev *et al.*[187] created grafted PEG gradients and subsequently fibrinogen and lysosome protein gradients exploiting the varying allylamine concentration across the surface.

As discussed previously, cell attachment can be mediated by integrin receptors present on the surface of anchorage dependent cells on the one hand and ECM proteins, *e.g.*, Fn, or fragments of ECM proteins on the other. The RGD motif discovered by Erkki Ruoslahti[216] has been a particularly popular target for immobilisation on biomaterial surfaces[27,176] as it is recognised by integrins that are expressed by a variety of cell types.[76,173,217,218] In addition to enhancing cell attachment, the display of ECM-based ligands affects the migration, apoptosis and differentiation of adherent cells.[75,176,184,219,220] The ability to optimise the presentation and density of such signals is vital for the advancement of implantable devices and regenerative medicine. Using short peptide sequences has advantages over whole ECM proteins because the former are fully chemically defined, are not of animal origin and do not pose any risk of disease transmission. In addition, small peptides are less susceptible to degradation during sterilisation processes.

Many techniques have been developed to prepare gradients in RGD peptide (or similar) concentration including functionalised SAM gradients,[173,221] hydrogels,[222,223] microfluidic mixing,[94] polymer brushes,[217] and biotin–streptavidin coupling.[224,225] Burdick *et al.*[222] developed a gradient in RGD peptide concentration using a microfluidic device (Figure 10.16A). Two monomer/initiator solutions (one containing just the PEG monomer, the other containing a modified monomer containing an RGD peptide) were injected into the microfluidic device, with the gradient formed *via* mixing within the microfluidic device through the channels repeatedly splitting and mixing the solutions. The density of endothelial[222] and fibroblast[223] cells was found to increase with increasing RGD concentration. Likewise, Petty *et al.*[94] have utilised microfluidics to create gradients of RGD peptides along the substrate. The gradients varied from the active RGD peptides (GRGDSC) to an inactive peptide (GGRDGSC), which meant that the topography and surface chemistry of the surface was uniform over the whole substrate. Mouse melanoma cells attached preferentially to the high-density active RGD sites. Lagunas *et al.*[225] developed a RGD gradient using biotin streptavidin coupling (Figure 10.16B). A uniform PMMA surface was converted by

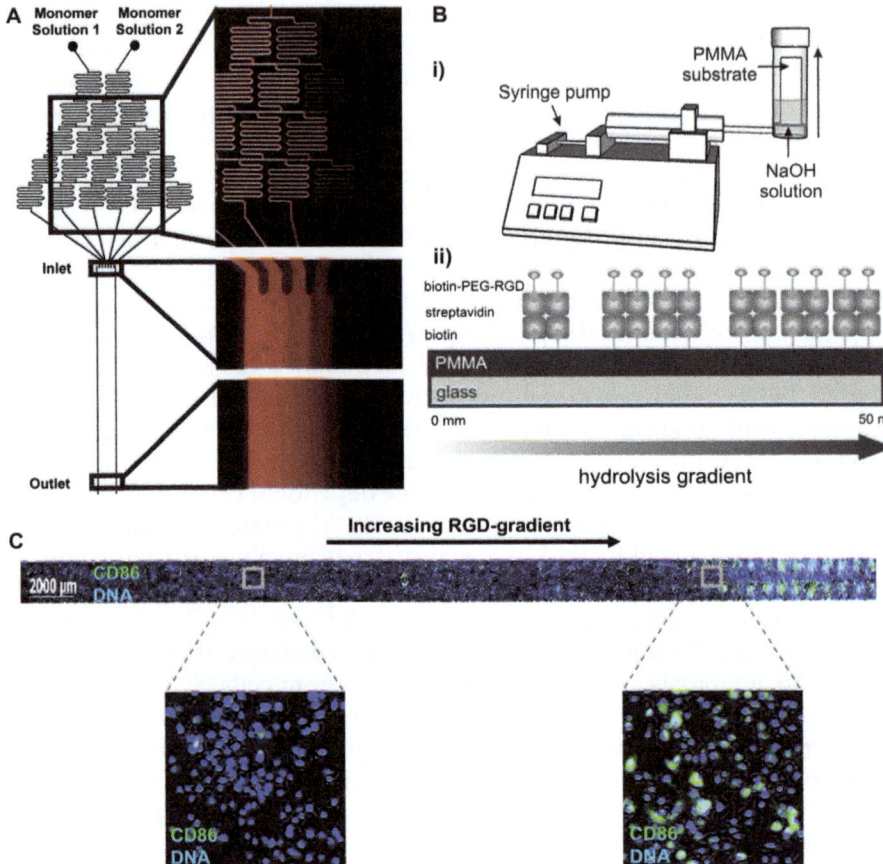

Figure 10.16 Examples of several RGD gradients: (A) schematic of the microfluidics device used to create hydrogel gradients; (B) schematic for the preparation of (i) hydrolysed PMMA gradients, (ii) subsequent biotin–streptavidin–biotin–PEG–RGD binding; (C) investigation of dendritic cell response to RGD density gradients (gradient fabrication method described in Figure 10.10), stained for CD86 (activation marker, green) and cell nuclei (blue), demonstrating an increase in CD86 activation with increasing RGD concentration.
Figures A, B and C adapted from Burdick *et al.*,[222] Lagunas *et al.*[224,225] and Acharya *et al.*,[76] respectively.

controlled hydrolysis *via* gradual immersion into a sodium hydroxide solution. Biotin was then immobilised on the reactive COOH groups followed by subsequent coupling with RGD functionalised streptavidin.[224]

Gallant *et al.*[173] demonstrated that in the presence of a surface-bound RGD concentration gradient, the attachment of smooth muscle cells increased with increasing RGD concentration (Figure 10.16C). As an extension to this study, the up-regulated expression of activation markers and production of cytokines in dendritic cells was shown to correlate with increasing surface-bound RGD concentrations (Figure 10.16C).[76] These studies highlight an important conceptual advantage of the gradient format in that it allows for rapid screening of a large number of surface properties that are important in the bio-interface context, whilst determining effective limits of immobilised biomolecules required for the observation of a given cellular phenomena, thereby enabling optimisation. Earlier studies of peptide gradient surfaces have been limited to the linear form of RGD. However, the cyclic form of RGD has demonstrated significantly enhanced bioactivity compared to the linear version.[226–228]

10.2.4   Two-dimensional Chemical Gradients

The preparation of 2D chemical gradients has been pursued by several Research Groups in recent years, with applications ranging from surface dewetting and nanoparticle attachment to protein adsorption and optimisation of cellular response.[23,101,144,152,167,176,229–234] Orthogonal polymer brush gradients displaying variations in grafting density and molecular weight have been reported by Bhat *et al.* (Figure 10.17A).[23,144,152] Firstly, a concentration gradient of initiator molecules was deposited onto the substrate through vapour deposition. Secondly, the substrate was placed into a polymerisation chamber with the initiator gradient running in the 10 direction. The monomer, poly(dimethyl aminoethyl methacrylate) (PDMAEMA) solution was slowly drained from the polymerisation chamber which in turn lead to the formation of a molecular weight gradient in the *y*-direction. PHEMA orthogonal gradients were also prepared using this procedure

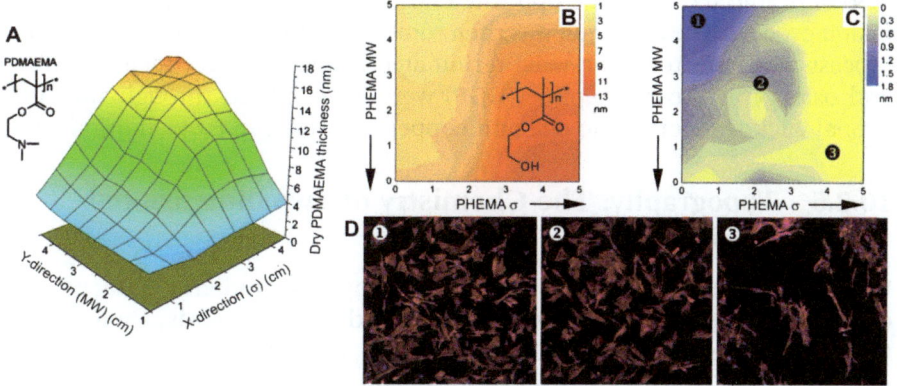

Figure 10.17   (A) Thickness of PHEMA in MW/grafting density (σ) orthogonal PHEMA gradient; (B) thickness of PHEMA MW/σ orthogonal gradient (*z*-scale in nm); (C) Fn density in MW/σ orthogonal gradient (*z*-scale in nm). Scales for position on the substrate in parts (B) and (C) are in cm, (D) images of MC3T3-E1 cells cultured on the PHEMA/Fn gradient substrates. Images 1 to 3 refer to different regions of the gradient as shown in (C).
Figure adapted from Bhat *et al.*[23]

(Figure 10.17B), and the adsorption of Fn as well as the growth of osteoblast cells was investigated. Fn was found to bind to the region on the orthogonal gradient displaying the lowest density and lowest molecular weight of PHEMA (Figure 10.17C). Osteoblast cell density was also the highest in this region (Figure 10.17D).

Dual molecular weight polymers have also been prepared using PHEMA and poly(methyl methacrylate) (PMMA).[23] Firstly, a monolayer of the polymerisation initiator was deposited onto the surface. A thickness gradient of polymer 'A' was prepared through slowly draining the solution from the beaker so that longer polymer chains would be present in the region immersed in the polymer solution for the longest period of time. Since this polymer has been grown *via* ATRP the chains can act as initiators for the polymerisation of polymer B. The substrate was rotated 90° and placed in the reaction vessel containing polymer B. following the same solution draining method, a chain length gradient of polymer B was created orthogonal to that of polymer A. Polymer thickness was found to be linear along both axes, a phenomenon that has rarely been seen in gradient preparation before.[7,25,123] This technique has allowed the authors to determine the optimal surface conditions for both Fn adsorption and cell attachment on PHEMA and PMMA substrates.

'Click chemistry' has been employed to generate orthogonal peptide concentration gradients.[167] In the first of several steps, a silane chemistry gradient with a terminal chlorine functional group was prepared through a confined channel vacuum deposition. The sample and a silane reservoir was placed in a confined chamber and the vacuum was pulled away from the side nearest the silane reservoir and away from the substrate (Figure 10.18), with a stream of methanol injected form the opposite end of the chamber to quench the diffusion process. The substrate was then rotated 90° and the vacuum deposition process was repeated for a vinyl terminated silane. Peptides including RGD and osteogenic growth peptide (OGP) were immobilised through two orthogonal click reactions (thiolene and copper-free Huisgen cycloaddition).

10.2.5 Topography: The Chemistry of Two-dimensional Gradients

Whilst there are several examples depicting 2D gradient surfaces with changes in chemical properties along both the *x*- and *y*-axis, there are far fewer examples detailing 2D gradients of topography and chemistry[101,102,108,151,235–237] or topography and topography.[238–240] Meredith *et al.*[101,102] reported on an orthogonal gradient with variations in polymer composition (*x*-axis) and annealing temperature (*y*-axis). Poly(DL-lactide) (PDLA) and poly(ε-caprolactone) (PCL) co-polymer gradients were prepared through a syringe pump, as detailed in Figure 10.19A.[101] Although molecular diffusion will lead to a uniform composition throughout the gradient over time, the time scale for which the gradient is in the syringe is too short to reach equilibrium. The polymer gradient was deposited in a strip along the

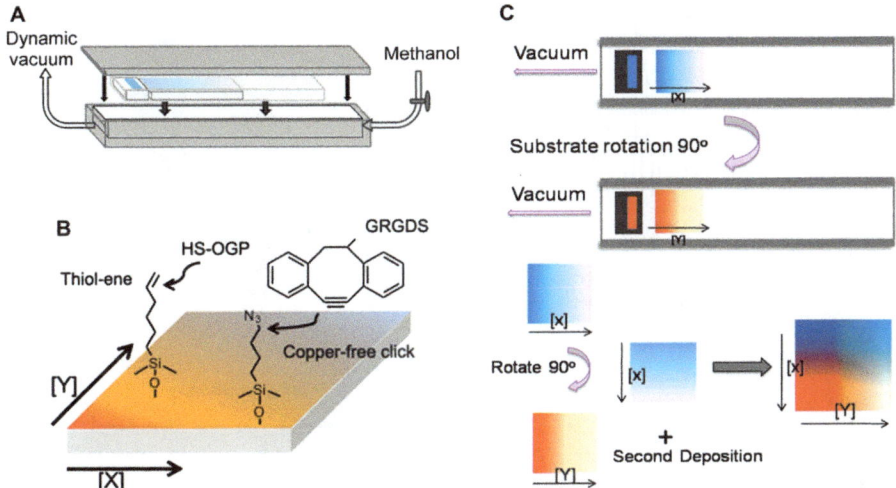

Figure 10.18 (A) Schematic for the fabrication of silane concentration gradients using a confined channel diffusion method; (B) surface chemistry to generate gradients of RGD and OGD; (C) two-step sequential deposition: following the first silane deposition, the sample was rotated 90°, and then a second silane was deposited.
Figure adapted from Ma *et al.*[167]

substrate and a knife was used to spread the polymer solution over the substrate by flow coating (Figure 10.19C and D). An annealing temperature gradient was prepared orthogonal to the polymer composition gradient through the use of a gradient temperature heat block (Figure 10.19E). Osteoblast attachment turned out to be greatest near the boundaries of the polymer islands and at annealing temperatures of ~ 110 °C.

Yang *et al.*[151] prepared a 2D gradient in groove width and surface chemistry (from amine to hydrocarbon) (Figure 10.20). The groove width topography gradient was prepared by hot embossing, pressing a silicon master (prepared by reactive-ion etching) into a thin PMMA substrate. A chemical gradient was prepared by plasma polymerisation orthogonal to the topography.[7] Contact angles measured from $\sim 55°$ at the amine-rich end to $\sim 95°$ at the hydrocarbon end. Whilst there was little difference in cell attachment after 1 and 2 days, after 3 days in culture a hot spot was observed for a contact angle of $\sim 73°$ and groove width of ~ 50 μm. There was little cell attachment on the extreme hydrophilic and hydrophobic ends of the gradient regardless of the groove width, suggesting the cells were more sensitive to changes in the surface chemistry/charge than topography. Another point to note is that the groove width appeared to influence the alignment of cells regardless of the surface chemistry for all 1–3 days of culture. The findings here further highlight the importance of 2D gradient surfaces when investigating influences on cellular response.

2D gradient samples combining topography and peptide density gradients were prepared by combining a pSi topography gradient orthogonal to a

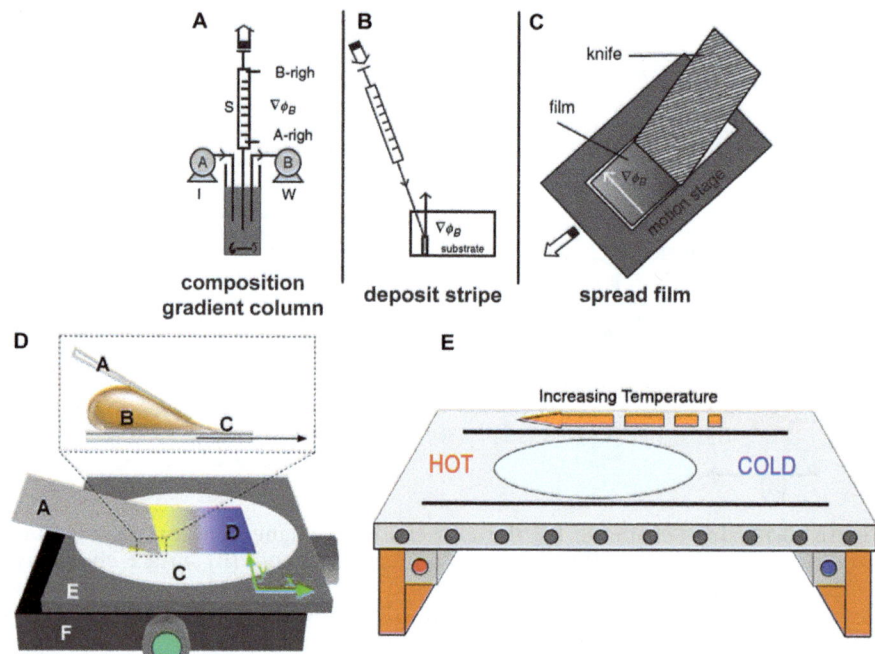

Figure 10.19 Preparation of 2D polymer composition and film thickness gradients: (A) a composition gradient column through altering the bulk monomer concentration whilst simultaneously loading a syringe; (B) deposition of a 'stripe' of the varying polymer composition; (C and D) flow coating where the substrate is moved beneath a knife blade and liquid is spread across a substrate. Accelerating the substrate movement during this process leads to a thickness gradient; (E) temperature stage with increasing temperature.[101,148]

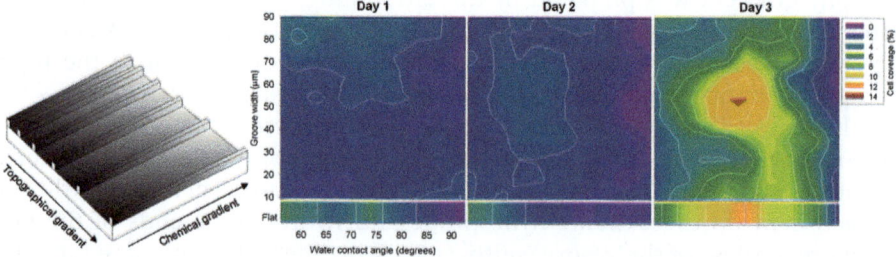

Figure 10.20 Fibroblast cell response to groove pitch and amine density (topography-chemistry) 2D gradient after 1, 2 and 3 days culture. Heat maps show % cell coverage with amine density (represented as contact angle) on the x-axis and groove width on the y-axis. A 'hot spot' was observed after 3 days culture, and the surface chemistry was found to more strongly influence cell attachment and proliferation than the topography.
Adapted from Yang et al.[151]

c(RGDfK) density gradient on the same substrate were created by Clements *et al.*[108] The orthogonal gradient platform revealed that rMSCs respond more favourably to changes in c(RGDfK) density than to changes in pore size. In these experiments, since the same apparatus is used for the preparation of both gradient surfaces, generation of 2D gradients is arguably more straight-forward than other reported 2D gradient methods. The 2D format illustrates that surface-bound gradients are useful for the combinatorial screening of cell–material surface interactions where one can optimise the effects of vary-ing two surface properties in regard to the cellular response on a single sample and in a single experiment. Additionally, this study highlights the importance of screening multiple factors that influence cellular response simultaneously.

10.2.6 Two-dimensional Topography Gradients

Recently, several groups have developed 2D topography gradients.[238–240] Zink *et al.*[238] developed micro-roughness and particle nano-roughness density gradients (Figure 10.21A). Roughness gradients were prepared *via*

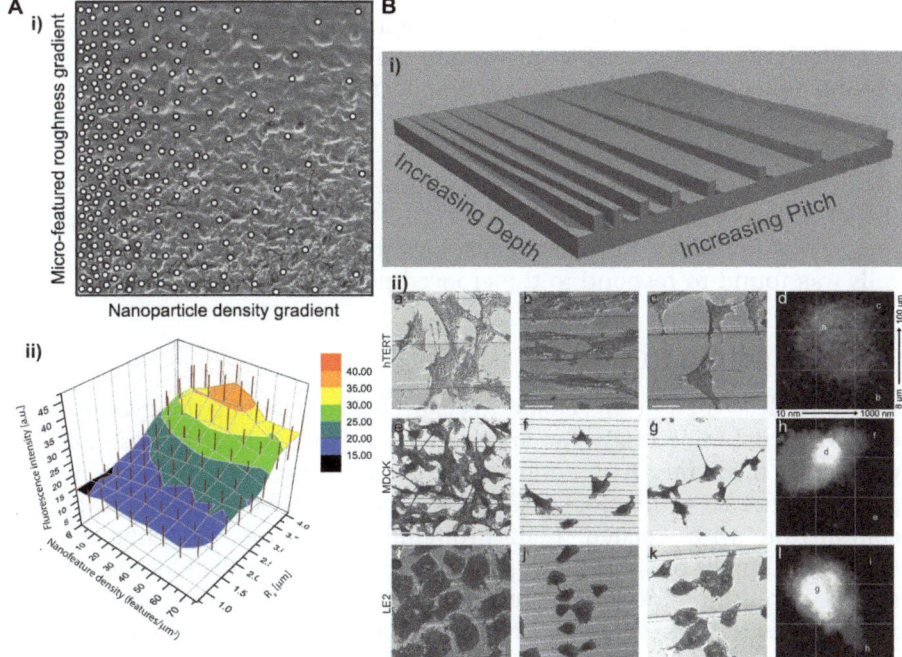

Figure 10.21   Examples of two topography *vs.* topography gradients: (A) micro-roughness *versus* nanoparticle nano-roughness (i) schematic of 2D gradient, (ii) graph showing osteopontin expression (indicator for osteogenesis); (B) groove pitch *vs.* depth, (i) schematic of 2D gradient, (ii) cell response of fibroblasts, epithelial cells and endothelial cells with heat maps (d, h, l) showing optimum surface properties on 2D gradients for epithelial and endothelial cells.
Figure adapted from Reynolds *et al.*[239] and Zink *et al.*[238]

sandblasting and subsequent chemical etching. Subsequently, a nanoparticle gradient was prepared orthogonally by gradual immersion into a nanoparticle solution followed by sintering to firmly attach nanoparticles. From here, the 2D gradient was replicated using PDMS to create a negative replica, followed by a positive replica fabrication using an epoxy resin. The final step was to sputter coat the epoxy replica with titanium. In general, an increase in micro-roughness led to an increase in osteopontin expression (an osteogenesis marker). This investigation of the combined effects of micro- and nano-roughness in a single experiment revealed that whilst there was a relatively linear response of osteopontin expression with increasing roughness, there was an optimum where high micro surface roughness coupled with moderate nano-roughness gave highest osteopontin expression. Had these topographies been investigated separate from one another, this optimal may not have been discovered.

A 2D gradient with variations in both groove pitch and depth was developed by Reynolds *et al.*[239] (Figure 10.21B). Dual layer etching masks were created using a combination of micropatterning and plasma polymer deposition. A gradient pitch was generated through the use of a gradient photomask and photolithography. The depth axis was generated using the same technique as described in Section 1.4.1.3. Briefly, a gradient hexane plasma polymer was deposited orthogonally to the groove pitch gradient as a sacrificial layer, followed by etching of the surface.[153] The attachment of fibroblast, endothelial and epithelial cells was investigated on these 2D gradients whereby an optimum was found for both epithelial and endothelial cells as shown in Figure 10.21B(ii). No such increased cell response region was found for fibroblast cells, however the orientation of fibroblast cells was found to respond to the grooved/pitch 2D gradient.

10.3   Conclusions

The work described in this chapter is expected to facilitate the development of advanced biomaterials for applications in biomedical devices and regenerative medicine. It is anticipated that the gradient approach, and in particular the 2D orthogonal gradient approach, will be used increasingly in conjunction with related high-throughput platforms to enhance our understanding of bio-interfacial phenomena and to rapidly identify optimal surface conditions for cellular responses such as attachment, proliferation and differentiation. Once identified, the optimal characteristics can be scaled up and used in the manufacturing of new and improved biomaterials and biomedical devices.

With the rapid advancement of regenerative medicine and cell therapy in recent years, for example advancements in skin regeneration and human tissue grown in a Petri dish, growth in this field will rely on the effective isolation and scale up of specific stem cell populations. Additionally, reliable procedures for the controlled and directed differentiation of stem cells will be required. The use of 1D, 2D and 3D gradients in conjunction with other high-throughput techniques including microarray and microfluidics technologies will be vital for the identification and optimisation of reliable surface properties for the large-scale maintenance and differentiation of stem cells. These advancements bring feasible regenerative medicine applications closer to reality.

References

1. D. G. Castner and B. D. Ratner, *Surf. Sci.*, 2002, **500**, 28–60.
2. A. J. Engler, S. Sen, H. L. Sweeney and D. E. Discher, *Cell*, 2006, **126**, 677–689.
3. B. Kasemo, *Surf. Sci.*, 2002, **500**, 656–677.
4. X. L. Zhu, J. Chen, L. Scheideler, R. Reichl and J. Geis-Gerstorfer, *Biomaterials*, 2004, **25**, 4087–4103.
5. Y. L. Khung, G. Barritt and N. H. Voelcker, *Exp. Cell Res.*, 2008, **314**(4), 789–800.
6. D. E. Discher, P. Janmey and Y. l. Wang, *Mater. Biol.*, 2005, **310**, 1139–1143.
7. M. Zelzer, R. Majani, J. W. Bradley, F. Rose, M. C. Davies and M. R. Alexander, *Biomaterials*, 2008, **29**, 172–184.
8. J. H. Choee, S. J. Lee, Y. M. Lee, J. M. Rhee, H. B. Lee and G. Khang, *J. Appl. Polym. Sci.*, 2004, **92**, 599–606.
9. J. H. Wei, M. Yoshinari, S. Takemoto, M. Hattori, E. Kawada, B. L. Liu and Y. Oda, *J. Biomed. Mater. Res., Part B*, 2007, **81**, 66–75.
10. M. Yoshinari, T. Hayakawa, K. Matsuzaka, T. Inoue, Y. Oda and M. Shimono, *J. Oral Tissue Eng.*, 2004, **1**, 69–79.
11. M. S. Kim, G. Khang and H. B. Lee, *Prog. Polym. Sci.*, 2008, **33**, 138–164.
12. H. Thissen, J. P. Hayes, P. Kingshott, G. Johnson, E. C. Harvey and H. J. Griesser, *Smart Mater. Struct.*, 2002, **11**, 792–799.
13. C. S. Chen, M. Mrksich, S. Huang, G. M. Whitesides and D. E. Ingber, *Biotechnol. Prog.*, 1998, **14**, 356–363.
14. J. T. Smith, J. K. Tomfohr, M. C. Wells, T. P. Beebe, T. B. Kepler and W. M. Reichert, *Langmuir*, 2004, **20**, 8279–8286.
15. J. D. Whittle, N. A. Bullett, R. D. Short, C. W. Ian Douglas, A. P. Hollander and J. Davies, *J. Mater. Chem.*, 2002, **12**, 2726–2732.
16. M. A. Cole, H. Thissen, D. Losic and N. H. Voelcker, *Surf. Sci.*, 2007, **601**, 1716–1725.
17. D. J. Menzies, M. Jasieniak, H. J. Griesser, J. S. Forsythe, G. Johnson, G. A. McFarland and B. W. Muir, *Surf. Sci.*, 2012, **606**, 1798–1807.
18. D. S. W. Benoit, A. R. Durney and K. S. Anseth, *Biomaterials*, 2007, **28**, 66–77.
19. A. L. Hook, H. Thissen, J. P. Hayes and N. H. Voelcker, *Proc. SPIE*, 2006, **6413**, Article 64130C, 1–11.
20. C. W. Xu, *Genome Res.*, 2002, **12**, 482–486.
21. B. Angres, *Expert Rev. Mol. Diagn.*, 2005, **5**, 769–779.

22. T. G. Ruardy, J. M. Schakenraad, H. C. van der Mei and H. J. Busscher, *Surf. Sci. Rep.*, 1997, **29**, 3–30.

23. R. R. Bhat, M. R. Tomlinson and J. Genzer, *J. Polym. Sci., Part B: Polym. Phys.*, 2005, **43**, 3384–3394.

24. S. I. Stupp, J. J. J. M. Donners, L. S. Li and A. Mata, *MRS Bull.*, 2005, **30**, 864–873.

25. T. P. Kunzler, T. Drobek, M. Schuler and N. D. Spencer, *Biomaterials*, 2007, **28**, 2175–2182.

26. L. T. Allen, M. Tosetto, I. S. Miller, D. P. O'Connor, S. C. Penney, I. Lynch, A. K. Keenan, S. R. Pennington, K. A. Dawson and W. M. Gallagher, *Biomaterials*, 2006, **27**, 3096–3108.

27. S. L. Bellis, *Biomaterials*, 2011, **32**, 4205–4210.

28. R. O. Hynes, *Cell*, 1992, **69**, 11–25.

29. N. J. Boudreau and P. L. Jones, *Biochem. J.*, 1999, **339**, 481–488.

30. S. Drotleff, U. Lungwitz, M. Breunig, A. Dennis, T. Blunk, J. Tessmar and A. Gopferich, *Eur. J. Pharm. Biopharm.*, 2004, **58**, 385–407.

31. E. Rusoslahti and M. D. Pierschbacher, *Cell*, 1986, **44**, 517–518.

32. M. D. Pierschbacher and E. Ruoslahti, *Nature*, 1984, **309**, 30–33.

33. G. Maheshwari, G. Brown, D. A. Lauffenburger, A. Wells and L. G. Griffith, *J. Cell Sci.*, 2000, **113**, 1677–1686.

34. C. R. Witmer, J. A. Phelps, W. M. Saltzman and P. R. van Tassel, *Biomaterials*, 2007, **28**, 851–860.

35. H. Thissen, G. Johnson, P. G. Hartley, P. Kingshott and H. J. Griesser, *Biomaterials*, 2006, **27**, 35–43.

36. P. Kingshott, H. Thissen and H. J. Griesser, *Biomaterials*, 2002, **23**, 2043–2056.

37. M. A. Cole, N. H. Voelcker and H. Thissen, *Smart Mater. Struct.*, 2007, **16**, 2222–2228.

38. P. Kingshott and H. J. Griesser, *Curr. Opin. Solid State Mater. Sci.*, 1999, **4**, 403–412.

39. J. M. Harris, *Poly(ethylene glycol) chemistry: biotechnical and biomedical applications*, Plenum Press, New York, 1992.

40. S. R. Sheth and D. Leckband, *Proc. Natl. Acad. Sci. U. S. A.*, 1997, **94**, 8399–8404.

41. S. P. Low, K. A. Williams, L. T. Canham and N. H. Voelcker, *Biomaterials*, 2006, **27**, 4538–4546.

42. J. Tan and W. M. Saltzman, *Biomaterials*, 2002, **23**, 3215–3225.

43. A. I. Teixeira, P. F. Nealey and C. J. Murphy, *J. Biomed. Mater. Res., Part A*, 2004, **71**, 369–376.

44. M. Schindler, I. Ahmed, J. Kamal, A. Nur-E-Kamal, T. H. Grafe, H. Y. Chung and S. Meiners, *Biomaterials*, 2005, **26**, 5624–5631.

45. C. L. Klein, M. Scholl and A. Maelicke, *J. Mater. Sci.: Mater. Med.*, 1999, **10**, 721–727.

46. B. E. Collins, K. P. S. Dancil, G. Abbi and M. J. Sailor, *Adv. Funct. Mater.*, 2002, **12**, 187–191.

47. K. Kant, S. P. Low, A. Marshal, J. G. Shapter and D. Losic, *ACS Appl. Mater. Interfaces*, 2010, **2**, 3447–3454.
48. J. Y. Lim, A. D. Dreiss, Z. Zhou, J. C. Hansen, C. A. Siedlecki, R. W. Hengstebeck, J. Cheng, N. Winograd and H. J. Donahue, *Biomaterials*, 2007, **28**, 1787–1797.
49. B. S. Zhu, Q. H. Lu, J. Yin, J. Hu and Z. G. Wang, *J. Biomed. Mater. Res., Part B*, 2004, **70**, 43–48.
50. P.-Y. Wang, W.-T. Li, J. Yu and W.-B. Tsai, *J. Mater. Sci.: Mater. Med.*, 2012, **23**, 3015–3028.
51. R. V. Goreham, R. D. Short and K. Vasilev, *J. Phys. Chem. C*, 2011, **115**, 3429–3433.
52. M. Lampin, R. WarocquierClerout, C. Legris, M. Degrange and M. F. SigotLuizard, *J. Biomed. Mater. Res.*, 1997, **36**, 99–108.
53. A. Chaubey, K. J. Ross, R. M. Leadbetter and K. J. L. Burg, *J. Biomed. Mater. Res., Part B*, 2008, **84**, 70–78.
54. C. Chai and K. W. Leong, *Mol. Ther.*, 2007, **15**, 467–480.
55. S. R. Ghaemi, F. J. Harding, B. Delalat, S. Gronthos and N. H. Voelcker, *Biomaterials*, 2013, **34**, 7601–7615.
56. D. G. Anderson, S. Levenberg and R. Langer, *Nat. Biotechnol.*, 2004, **22**, 863–866.
57. B. G. Keselowsky, D. M. Collard and A. J. Garcia, *Proc. Natl. Acad. Sci. U. S. A.*, 2005, **102**, 5953–5957.
58. Y. Mei, K. Saha, S. R. Bogatyrev, J. Yang, A. L. Hook, Z. I. Kalcioglu, S.-W. Cho, M. Mitalipova, N. Pyzocha, F. Rojas, K. J. Van Vliet, M. C. Davies, M. R. Alexander, R. Langer, R. Jaenisch and D. G. Anderson, *Nat. Mater.*, 2010, **9**, 768–778.
59. J. Yang, Y. Mei, A. L. Hook, M. Taylor, A. J. Urquhart, S. R. Bogatyrev, R. Langer, D. G. Anderson, M. C. Davies and M. R. Alexander, *Biomaterials*, 2010, **31**, 8827–8838.
60. D. S. W. Benoit, M. P. Schwartz, A. R. Durney and K. S. Anseth, *Nat. Mater.*, 2008, 7, 816–823.
61. J. M. Curran, R. Stokes, E. Irvine, D. Graham, N. A. Amro, R. G. Sanedrin, H. Jamil and J. A. Hunt, *Lab Chip*, 2010, **10**, 1662–1670.
62. J. M. Curran, R. Chen and J. A. Hunt, *Biomaterials*, 2005, **26**, 7057–7067.
63. J. E. Phillips, T. A. Petrie, F. P. Creighton and A. J. Garcia, *Acta Biomater.*, 2010, **6**, 12–20.
64. S. Neuss, C. Apel, P. Buttler, B. Denecke, A. Dhanasingh, X. Ding, D. Grafahrend, A. Groger, K. Hemmrich, A. Herr, W. Jahnen-Dechent, S. Mastitskaya, A. Perez-Bouza, S. Rosewick, J. Salber, M. Woeltje and M. Zenke, *Biomaterials*, 2008, **29**, 302–313.
65. N. Wells, M. A. Baxter, J. E. Turnbull, P. Murray, D. Edgar, K. L. Parry, D. A. Steele and R. D. Short, *Biomaterials*, 2009, **30**, 1066–1070.
66. R. Peerani, K. Onishi, A. Mahdavi, E. Kumacheva and P. W. Zandstra, *PLoS One*, 2009, **4**, Article 460870, 1–18.
67. S. Even-Ram, V. Artym and K. M. Yamada, *Cell*, 2006, **126**, 645–647.

68. S. Ding and P. G. Schultz, *Nat. Biotechnol.*, 2004, **22**, 833–840.
69. G. H. Underhill and S. N. Bhatia, *Curr. Opin. Chem. Biol.*, 2007, **11**, 357–366.
70. M. J. Dalby, N. Gadegaard, R. Tare, A. Andar, M. O. Riehle, P. Herzyk, C. D. W. Wilkinson and R. O. C. Oreffo, *Nat. Mater.*, 2007, **6**, 997–1003.
71. J. M. Curran, R. Chen and J. A. Hunt, *Biomaterials*, 2006, **27**, 4783–4793.
72. R. McBeath, D. M. Pirone, C. M. Nelson, K. Bhadriraju and C. S. Chen, *Dev. Cell*, 2004, **6**, 483–495.
73. R. Derda, L. Y. Li, B. P. Orner, R. L. Lewis, J. A. Thomson and L. L. Kiessling, *ACS Chem. Biol.*, 2007, **2**, 347–355.
74. A. L. Hook, H. Thissen and N. H. Voelcker, *Trends Biotechnol.*, 2006, **24**, 471–477.
75. A. L. Hook, D. G. Anderson, R. Langer, P. Williams, M. C. Davies and M. R. Alexander, *Biomaterials*, 2010, **31**, 187–198.
76. A. P. Acharya, N. V. Dolgova, N. M. Moore, C.-Q. Xia, M. J. Clare-Salzler, M. L. Becker, N. D. Gallant and B. G. Keselowsky, *Biomaterials*, 2010, **31**, 7444–7454.
77. C. R. Kothapalli, E. van Veen, S. de Valence, S. Chung, I. K. Zervantonakis, F. B. Gertler and R. D. Kamm, *Lab Chip*, 2011, **11**, 497–507.
78. D. C. Pregibon, M. Toner and P. S. Doyle, *Science*, 2007, **315**, 1393–1396.
79. J. C. Meredith, *J. Mater. Chem.*, 2009, **19**, 34–45.
80. D. G. Anderson, D. Putnam, E. B. Lavik, T. A. Mahmood and R. Langer, *Biomaterials*, 2005, **26**, 4892–4897.
81. A. Mant, G. Tourniaire, J. J. Diaz-Mochon, T. J. Elliott, A. P. Williams and M. Bradley, *Biomaterials*, 2006, **27**, 5299–5306.
82. C. J. Flaim, S. Chien and S. N. Bhatia, *Nat. Methods*, 2005, **2**, 119–125.
83. J. Genzer and R. R. Bhat, *Langmuir*, 2008, **24**, 2294–2317.
84. S. Morgenthaler, C. Zink and N. D. Spencer, *Soft Matter*, 2008, **4**, 419–434.
85. H. B. Lee, M. S. Kim, Y. H. Cho and G. Khang, *Polymer*, 2005, **29**, 423–432.
86. H. Elwing, S. Welin, A. Askendal, U. Nilsson and I. Lundstrom, *J. Colloid Interface Sci.*, 1987, **119**, 203–210.
87. C. G. Simon, N. Eidelman, S. B. Kennedy, A. Sehgal, C. A. Khatri and N. R. Washburn, *Biomaterials*, 2005, **26**, 6906–6915.
88. C. D. McFarland, C. H. Thomas, C. DeFilippis, J. G. Steele and K. E. Healy, *J. Biomed. Mater. Res.*, 2000, **49**, 200–210.
89. M. K. Chaudhry and G. M. Whitesides, *Science*, 1992, **256**, 1539–1541.
90. B. Liedberg and P. Tengvall, *Langmuir*, 1995, **11**, 3821–3827.
91. S. Morgenthaler, S. Lee, S. Zurcher and N. D. Spencer, *Langmuir*, 2003, **19**, 10459–10462.
92. T. Kraus, R. Stutz, T. E. Balmer, H. Schmid, L. Malaquin, N. D. Spencer and H. Wolf, *Langmuir*, 2005, **21**, 7796–7804.
93. J. R. Tse and A. J. Engler, *PLoS One*, 2011, **6**, 1, Article e15978.
94. R. T. Petty, H. W. Li, J. H. Maduram, R. Ismagilov and M. Mrksich, *J. Am. Chem. Soc.*, 2007, **129**, 8966–8967.

95. M. S. Kim, K. S. Seo, G. Khang and H. B. Lee, *Bioconjugate Chem.*, 2005, **16**, 245–249.
96. M. S. Kim, K. S. Seo, G. Khang and H. B. Lee, *Langmuir*, 2005, **21**, 4066–4070.
97. C. Xu, S. E. Barnes, T. Wu, D. A. Fischer, D. M. DeLongchamp, J. D. Batteas and K. L. Beers, *Adv. Mater.*, 2006, **18**, 1427–1430.
98. M. R. Tomlinson and J. Genzer, *Chem. Commun.*, 2003, 1350–1351.
99. A. P. Smith, J. F. Douglas, J. C. Meredith, E. J. Amis and A. Karim, *J. Polym. Sci., Part B: Polym. Phys.*, 2001, **39**, 2141–2158.
100. A. P. Smith, J. F. Douglas, J. C. Meredith, E. J. Amis and A. Karim, *Phys. Rev. Lett.*, 2001, **87**, 15503–15506.
101. J. C. Meredith, A. Karim and E. J. Amis, *Macromolecules*, 2000, **33**, 5760–5762.
102. J. C. Meredith, J. L. Sormana, B. G. Keselowsky, A. J. Garcia, A. Tona, A. Karim and E. J. Amis, *J. Biomed. Mater. Res., Part A*, 2003, **66**, 483–490.
103. N. R. Washburn, K. M. Yamada, C. G. Simon, S. B. Kennedy and E. J. Amis, *Biomaterials*, 2004, **25**, 1215–1224.
104. B. Zhao, *Langmuir*, 2004, **20**, 11748–11755.
105. F. J. Harding, L. R. Clements, R. D. Short, H. Thissen and N. H. Voelcker, *Acta Biomater.*, 2012, **8**, 1739–1748.
106. Y. L. Khung, G. Barritt and N. H. Voelcker, *Exp. Cell Res.*, 2008, **314**, 789–800.
107. C. M. Thompson, M. Nieuwoudt, A. M. Ruminski, M. J. Sailor and G. M. Miskelly, *Langmuir*, 2010, **26**, 7598–7603.
108. L. R. Clements, P.-Y. Wang, W.-B. Tsai, H. Thissen and N. H. Voelcker, *Lab Chip*, 2012, **12**, 1480–1486.
109. J. D. Whittle, D. Barton, M. R. Alexander and R. D. Short, *Chem. Commun.*, 2003, 1766–1767.
110. P. Murray, K. Vasilev, C. F. Mora, E. Ranghini, H. Tensaout, A. Rak-Raszewska, B. Wilm, D. Edgar, R. D. Short and S. E. Kenny, *Biochem. Soc. Trans.*, 2010, **38**, 1062–1066.
111. T. P. Kunzler, T. Drobek, C. M. Sprecher, M. Schuler and N. D. Spencer, *Appl. Surf. Sci.*, 2006, **253**, 2148–2153.
112. R. R. Bhat, J. Genzer, B. N. Chaney, H. W. Sugg and A. Liebmann-Vinson, *Nanotechnology*, 2003, **14**, 1145–1152.
113. M. G. Kutty, S. Bhaduri and S. B. Bhaduri, *J. Mater. Sci.: Mater. Med.*, 2004, **15**, 145–150.
114. M. J. Dalby, N. Gadegaard, A. S. G. Curtis and R. O. C. Oreffo, *Curr. Stem Cell Res. Ther.*, 2007, **2**, 129–138.
115. L. Z. Zhao, L. Liu, Z. F. Wu, Y. M. Zhang and P. K. Chu, *Biomaterials*, 2012, **33**, 2629–2641.
116. Y. L. Khung, G. Barritt and N. H. Voelcker, *Exp. Cell Res.*, 2008, **314**, 789.
117. P.-Y. Wang, L. R. Clements, H. Thissen, A. Jane, W.-B. Tsai and N. H. Voelcker, *Adv. Funct. Mater.*, 2012, **22**, 3414–3423.
118. P.-Y. Wang, L. R. Clements, H. Thissen, A. Jane, W.-B. Tsai and N. H. Voelcker, *Adv. Funct. Mater.*, 2012, **22**, 3414–3423.

119. L. Canham, *Properties of Porous Silicon*, INSPEC, The Institution of Electrical Engineers, London, UK, 1997.

120. X. Li, J. John, J. L. Coffer, Y. Chen, R. F. Pinizzotto, J. Newey, C. Reeves and L. T. Canham, *Biomed. Microdevices*, 2000, **2**, 265–272.

121. J. Salonen, L. Laitinen, A. M. Kaukonen, J. Tuura, M. Bjorkqvist, T. Heikkila, K. Vaha-Heikkila, J. Hirvonen and V. P. Lehto, *J. Controlled Release*, 2005, **108**, 362–374.

122. E. C. Wu, J.-H. Park, J. Park, E. Segal, F. Cunin and M. J. Sailor, *ACS Nano*, 2008, **2**, 2401–2409.

123. L. M. Karlsson, P. Tengvall, I. Lundstrom and H. Arwin, *Phys. Status Solidi A*, 2003, **197**, 326–330.

124. M. J. Sailor and E. J. Lee, *Adv. Mater.*, 1997, **9**, 783–793.

125. K. A. Kilian, T. Boecking and J. J. Gooding, *Chem. Commun.*, 2009, 630–640.

126. I. N. Lees, H. Lin, C. A. Canaria, C. Gurtner, M. J. Sailor and G. M. Miskelly, *Langmuir*, 2003, **19**, 9812–9817.

127. M. P. Stewart and J. M. Buriak, *Adv. Mater.*, 2000, **12**, 859–869.

128. J. M. Buriak, *Adv. Mater.*, 1999, **11**, 265–267.

129. D. J. Guo, S. J. Xiao, B. Xia, W. Shuai, J. Pei, Y. Pan, X. Z. You, Z. Z. Gu and Z. H. Lu, *J. Phys. Chem. B*, 2005, **109**, 20620–20628.

130. S. E. Letant and M. J. Sailor, *Adv. Mater.*, 2000, **12**, 355–359.

131. J. Gao, T. Gao, Y. Y. Li and M. J. Sailor, *Langmuir*, 2002, **18**, 2229–2233.

132. T. Gao, J. Gao and M. J. Sailor, *Langmuir*, 2002, **18**, 9953–9957.

133. V. Petrovakoch, T. Muschik, A. Kux, B. K. Meyer, F. Koch and V. Lehmann, *Appl. Phys. Lett.*, 1992, **61**, 943–945.

134. P. Y. Wang, L. R. Clements, H. Thissen, A. Jane, W. B. Tsai and N. H. Voelcker, *Adv. Funct. Mater.*, 2012, **22**, 3414–3423.

135. P. Y. Wang, L. R. Clements, H. Thissen, S. C. Hung, N. C. Cheng, W. B. Tsai and N. H. Voelcker, *RSC Adv.*, 2012, **2**, 12857–12865.

136. A. M. M. Jani, D. Losic and N. H. Voelcker, *Prog. Mater. Sci.*, 2013, **58**, 636–704.

137. R. Dronov, A. Jane, J. G. Shapter, A. Hodges and N. H. Voelcker, *Nanoscale*, 2011, **3**, 3109–3114.

138. P. Y. Wang, L. R. Clements, H. Thissen, W. B. Tsai and N. H. Voelcker, *Biomater. Sci.*, 2013, **1**, 924–932.

139. E. E. L. Swan, K. C. Popat and T. A. Desai, *Biomaterials*, 2005, **26**, 1969–1976.

140. K. C. Popat, K. I. Chatvanichkul, G. L. Barnes, T. J. Latempa, C. A. Grimes and T. A. Desai, *J. Biomed. Mater. Res., Part A*, 2007, **80**, 955–964.

141. D. Brüggemann, *J. Nanomater.*, 2013, Article 460870, 1–18.

142. E. P. Briggs, A. R. Walpole, P. R. Wilshaw, M. Karlsson and E. Palsgard, *J. Mater. Sci.: Mater. Med.*, 2004, **15**, 1021–1029.

143. A. R. Walpole, E. P. Briggs, M. Karlsson, E. Palsgard and P. R. Wilshaw, *Materialwiss. Werkstofftech.*, 2003, **34**, 1064–1068.

144. R. R. Bhat and J. Genzer, *Appl. Surf. Sci.*, 2006, **252**, 2549–2554.

145. R. R. Bhat, D. A. Fischer and J. Genzer, *Langmuir*, 2002, **18**, 5640–5643.
146. C. Huwiler, T. P. Kunzler, M. Textor, J. Voros and N. D. Spencer, *Langmuir*, 2007, **23**, 5929–5935.
147. A. Lundgren, M. Hulander, J. Brorsson, M. Hermansson, H. Elwing, O. Andersson, B. Liedberg and M. Berglin, *Part. Part. Syst. Charact.*, 2014, **31**(2), 173–177.
148. M. J. Fasolka, C. M. Stafford and K. L. Beers, in *Polymer Libraries, Advances in Polymer Science*, Springer-Verlag, Berlin Heidelberg, 2009, vol. 225, pp. 63–105.
149. A. Sehgal, A. Karim, C. M. Stafford and M. J. Fasolka, *Microsc. Today*, 2003, **11**, 26–29.
150. K. L. Beers, J. F. Douglas, E. J. Amis and A. Karim, *Langmuir*, 2003, **19**, 3935–3940.
151. J. Yang, F. Rose, N. Gadegaard and M. R. Alexander, *Adv. Mater.*, 2009, **21**, 300–304.
152. R. R. Bhat, M. R. Tomlinson and J. Genzer, *Macromol. Rapid Commun.*, 2004, **25**, 270–274.
153. P. M. Reynolds, R. H. Pedersen, J. Stormonth-Darling, M. J. Dalby, M. O. Riehle and N. Gadegaard, *Nano Lett.*, 2013, **13**, 570–576.
154. R. H. Pedersen, M. Hamzah, S. Thoms, P. Roach, M. R. Alexander and N. Gadegaard, *Microelectron. Eng.*, 2010, **87**, 1112–1114.
155. M. Kononen and M. Hormian, *J. Biomed. Mater. Res.*, 1992, **26**, 1325–1341.
156. U. Joos, 2001, US6554867 B1.
157. A. van Sliedregt, A. M. Radder, K. de Groot and C. A. van Blitterswijk, *J. Mater. Sci.: Mater. Med.*, 1992, **3**, 365–370.
158. P. Y. Wang, W. B. Tsai and N. H. Voelcker, *Acta Biomater.*, 2012, **8**, 519–530.
159. L. G. Vincent, Y. S. Choi, B. Alonso-Latorre, J. C. del Alamo and A. J. Engler, *Biotechnol. J.*, 2013, **8**(4), 472–484.
160. K. Chatterjee, S. Lin-Gibson, W. E. Wallace, S. H. Parekh, Y. J. Lee, M. T. Cicerone, M. F. Young and C. G. Simon, Jr., *Biomaterials*, 2010, **31**, 5051–5062.
161. E. A. Wilder, S. Guo, S. Lin-Gibson, M. J. Fasolka and C. M. Stafford, *Macromolecules*, 2006, **39**, 4138–4143.
162. S. L. Ishaug-Riley, L. E. Okun, G. Prado, M. A. Applegate and A. Ratcliffe, *Biomaterials*, 1999, **20**, 2245–2256.
163. A. J. Crosby, M. J. Fasolka and K. L. Beers, *Macromolecules*, 2004, **37**, 9968–9974.
164. P. F. Green, T. M. Christensen, T. P. Russell and R. Jerome, *Macromolecules*, 1989, **22**, 2189–2194.
165. T. P. Russell, G. Coulon, V. R. Deline and D. C. Miller, *Macromolecules*, 1989, **22**, 4600–4606.
166. R. Bos, J. H. de Jonge, B. van de Belt-Gritter, J. de Vries and H. J. Busscher, *Langmuir*, 2000, **16**, 2845–2850.
167. Y. R. Ma, J. K. Zheng, E. F. Amond, C. M. Stafford and M. L. Becker, *Biomacromolecules*, 2013, **14**, 665–671.

168. A. Kros, J. A. Jansen, S. J. Holder, R. J. M. Nolte and N. Sommerdijk, *J. Adhes. Sci. Technol.*, 2002, **16**, 143–155.

169. A. del Campo, D. Boos, H. W. Spiess and U. Jonas, *Angew. Chem., Int. Ed.*, 2005, **44**, 4707–4712.

170. N. Herzer, S. Hoeppener and U. S. Schubert, *Chem. Commun.*, 2010, **46**, 5634–5652.

171. B. C. Berry, C. M. Stafford, M. Pandya, L. A. Lucas, A. Karim and M. J. Fasolka, *Rev. Sci. Instrum.*, 2007, **78**, 072202.

172. C. J. Shearer, A. V. Ellis, J. G. Shapter and N. H. Voelcker, *Langmuir*, 2010, **26**, 18468–18475.

173. N. D. Gallant, K. A. Lavery, E. J. Amis and M. L. Becker, *Adv. Mater.*, 2007, **19**, 965–969.

174. H. C. Kolb, M. G. Finn and K. B. Sharpless, *Angew. Chem., Int. Ed.*, 2001, **40**, 2004–2021.

175. S. B. Kennedy, N. R. Washburn, C. G. Simon and E. J. Amis, *Biomaterials*, 2006, **27**, 3817–3824.

176. R. R. Bhat, B. N. Chaney, J. Rowley, A. Liebmann-Vinson and J. Genzer, *Adv. Mater.*, 2005, **17**, 2802–2807.

177. E. Klein, P. Kerth and L. Lebeau, *Biomaterials*, 2008, **29**, 204–214.

178. M. Matsuzawa, R. S. Potember, D. A. Stenger and V. Krauthamer, *J. Neurosci. Methods*, 1993, **50**, 253–260.

179. T. Gengebach, Z. R. Vasic, R. C. Chatelier and H. J. Griesser, *J. Polym. Sci., Part A: Polym. Chem.*, 1994, **32**, 1399–1414.

180. H. Yasuda, *Plasma Polymerization*, Academic Press, New York, NY, USA, 1985.

181. M. A. Lieberman and A. J. Lichtenberg, *Principles of Plasma Discharges and Materials Processing*, John Wiley and Sons, Chicester, UK, 1994.

182. H. Biederman, *Plasma Polymer Films*, Imperial College Press, London, UK, 2004.

183. A. J. Beck, J. D. Whittle, N. A. Bullett, P. Eves, S. Mac Neil, S. L. McArthur and A. G. Shard, *Plasma Processes Polym.*, 2005, **2**, 641–649.

184. K. L. Parry, A. G. Shard, R. D. Short, R. G. White, J. D. Whittle and A. Wright, *Surf. Interface Anal.*, 2006, **38**, 1497–1504.

185. D. J. Menzies, B. Cowie, C. Fong, J. S. Forsythe, T. R. Gengenbach, K. M. McLean, L. Puskar, M. Textor, L. Thomsen, M. Tobin and B. W. Muir, *Langmuir*, 2010, **26**, 13987–13994.

186. M. Zelzer, D. Scurr, B. Abdullah, A. J. Urquhart, N. Gadegaard, J. W. Bradley and M. R. Alexander, *J. Phys. Chem. B*, 2009, **113**, 8487–8494.

187. K. Vasilev, A. Mierczynska, A. L. Hook, J. Chan, N. H. Voelcker and R. D. Short, *Biomaterials*, 2010, **31**, 392–397.

188. A. Michelmore, L. Clements, D. A. Steele, N. H. Voelcker and E. J. Szili, *J. Nanomater.*, 2012, **2012**, 1–7.

189. D. S. Kumar, M. Fujioka, K. Asano, A. Shoji and A. Jayakrishnan, *J. Mater. Sci.: Mater. Med.*, 2007, **18**, 1831–1835.

190. P. G. Hartley, H. Thissen, T. Vaithianathan and H. J. Griesser, *Plasmas Polym.*, 2000, **5**, 47–60.

191. S. S. Vegunta, J. N. Ngunjiri and J. C. Flake, *Langmuir*, 2009, **25**, 12750–12756.
192. C. Gurtner, A. W. Wun and M. J. Sailor, *Angew. Chem., Int. Ed.*, 1999, **38**, 1966–1968.
193. P. Yang, *The Chemistry of Nanostructured Materials*, World Scientific Publishing Co. Pty. Ltd, Singapore, 2003.
194. A. Aurora, F. Cattaruzza, C. Coluzza, C. Della Volpe, G. Di Santo, A. Flamini, C. Mangano, S. Morpurgo, P. Pallavicini and R. Zanoni, *Chem. - Eur. J.*, 2007, **13**, 1240–1250.
195. C. A. Canaria, I. N. Lees, A. W. Wun, G. M. Miskelly and M. J. Sailor, *Inorg. Chem. Commun.*, 2002, **5**, 560–564.
196. E. G. Robins, M. P. Stewart and J. M. Buriak, *Chem. Commun.*, 1999, 2479–2480.
197. P. T. Hurley, A. E. Ribbe and J. M. Buriak, *J. Am. Chem. Soc.*, 2003, **125**, 11334–11339.
198. L. Clements, P. Y. Wang, F. Harding, W. B. Tsai, H. Thissen and N. H. Voelcker, *Phys. Status Solidi A*, 2011, **208**, 1440–1445.
199. T. Wu, K. Efimenko and J. Genzer, *J. Am. Chem. Soc.*, 2002, **124**, 9394–9395.
200. T. Wu, K. Efimenko, P. Vlcek, V. Subr and J. Genzer, *Macromolecules*, 2003, **36**, 2448–2453.
201. Y. S. Lin, V. Hlady and C. G. Golander, *Colloids Surf., B*, 1994, **3**, 49–62.
202. M. R. Tomlinson and J. Genzer, *Macromolecules*, 2003, **36**, 3449–3451.
203. M. R. Tomlinson and J. Genzer, *Langmuir*, 2005, **21**, 11552–11555.
204. C. Xu, T. Wu, C. M. Drain, J. D. Batteas and K. L. Beers, *Macromolecules*, 2005, **38**, 6–8.
205. S. Morgenthaler, C. Zink, B. Städler, J. Vrs, S. Lee, N. D. Spencer and S. G. P. Tosatti, *Biointerphases*, 2006, **1**, 156–165.
206. A. Larsson and B. Liedberg, *Langmuir*, 2007, **23**, 11319–11325.
207. A. Larsson, T. Ekbald, O. Andersson and B. Liedberg, *Biomacromolecules*, 2007, **8**, 287–295.
208. S. J. Lee, G. Khang, Y. M. Lee and H. B. Lee, *J. Colloid Interface Sci.*, 2003, **259**, 228–235.
209. Y. Ito, M. Heydari, A. Hashimoto, T. Konno, A. Hirasawa, S. Hori, K. Kurita and A. Nakajima, *Langmuir*, 2007, **23**, 1845–1850.
210. G. Khang, S. J. Lee, Y. M. Lee, J. H. Lee and H. B. Lee, *Korea Polym. J.*, 2000, **8**, 179–185.
211. F. J. O'Brien, B. A. Harley, I. V. Yannas and L. J. Gibson, *Biomaterials*, 2005, **26**, 433–441.
212. C. Streuli, *Curr. Opin. Cell Biol.*, 1999, **11**, 634–640.
213. Y. Mei, J. T. Elliott, J. R. Smith, K. J. Langenbach, T. Wu, C. Xu, K. L. Beers, E. J. Amis and L. Henderson, *J. Biomed. Mater. Res., Part A*, 2006, **79**, 974–988.
214. D. J. Mahoney, J. D. Whittle, C. M. Milner, S. J. Clark, B. Mulloy, D. J. Buttle, G. C. Jones, A. J. Day and R. D. Short, *Anal. Chem.*, 2004, **330**, 123–129.

215. D. E. Robinson, A. Marson, R. D. Short, D. J. Buttle, A. J. Day, K. L. Parry, M. Wiles, P. Highfield, A. Mistry and J. D. Whittle, *Adv. Mater.*, 2008, **20**, 1166–1169.
216. E. Ruoslahti and M. D. Pierschacher, *Science*, 1987, **238**(4826), 491–497.
217. B. P. Harris, J. K. Kutty, E. W. Fritz, C. K. Webb, K. J. L. Burg and A. T. Metters, *Langmuir*, 2006, **22**, 4467–4471.
218. A. A. Sawyer, K. M. Hennessy and S. L. Bellis, *Biomaterials*, 2005, **26**, 1467–1475.
219. Y. Mei, M. Goldberg and D. Anderson, *Curr. Opin. Chem. Biol.*, 2007, **11**, 388–393.
220. L. Y. Santiago, R. W. Nowak, J. P. Rubin and K. G. Marra, *Biomaterials*, 2006, **27**, 2962–2969.
221. N. M. Moore, N. J. Lin, N. D. Gallant and M. L. Becker, *Acta Biomater.*, 2011, **7**, 2091–2100.
222. J. A. Burdick, A. Khademhosseini and R. Langer, *Langmuir*, 2004, **20**, 5153–5156.
223. D. Guarnieri, A. De Capua, M. Ventre, A. Borzacchiello, C. Pedone, D. Marasco, M. Ruvo and P. A. Netti, *Acta Biomater.*, 2010, **6**, 2532–2539.
224. A. Lagunas, J. Comelles, E. Martinez, E. Prats-Alfonso, G. A. Acosta, F. Albericio and J. Samitier, *Nanomedicine*, 2012, **8**, 432–439.
225. A. Lagunas, J. Comelles, E. Martinez and J. Samitier, *Langmuir*, 2010, **26**, 14154–14161.
226. S. Verrier, S. Pallu, R. Bareille, A. Jonczyk, J. Meyer, M. Dard and J. Amedee, *Biomaterials*, 2002, **23**, 585–596.
227. S. J. Bogdanowich-Knipp, D. S. S. Jois and T. J. Siahaan, *J. Pept. Res.*, 1999, **53**, 523–529.
228. R. Haubner, R. Gratias, B. Diefenbach, S. L. Goodman, A. Jonczyk and H. Kessler, *J. Am. Chem. Soc.*, 1996, **118**, 7461–7472.
229. E. Beurer, N. V. Venkataraman, A. Rossi, F. Bachmann, R. Engeli and N. D. Spencer, *Langmuir*, 2010, **26**, 8392–8399.
230. E. Beurer, N. V. Venkataraman, M. Sommer and N. D. Spencer, *Langmuir*, 2012, **28**, 3159–3166.
231. A. del Campo, T. Sen, J. P. Lellouche and I. J. Bruce, *J. Magn. Magn. Mater.*, 2005, **293**, 33–40.
232. V. S. Khire, D. S. W. Benoit, K. S. Anseth and C. N. Bowman, *J. Polym. Sci., Part A: Polym. Chem.*, 2006, **44**, 7027–7039.
233. D. M. Lentz, A. M. Rhoades, R. A. Pyles, K. W. Haider, M. S. Angelone and R. C. Hedden, *J. Nanopart. Res.*, 2011, **13**, 4795–4808.
234. K. M. Ashley, D. Raghavan, J. F. Douglas and A. Karim, *Langmuir*, 2005, **21**, 9518–9523.
235. L. Wei, E. A. Vogler, T. M. Ritty and A. Lakhtakia, *Mater. Sci. Eng., C*, 2011, **31**, 1861–1866.
236. J. L. Zhang and Y. C. Han, *Langmuir*, 2008, **24**, 796–801.
237. J. C. Meredith, A. P. Smith, A. Karim and E. J. Amis, *Macromolecules*, 2000, **33**, 9747–9756.

238. C. Zink, H. Hall, D. M. Brunette and N. D. Spencer, *Biomaterials*, 2012, **33**, 8055–8061.
239. P. M. Reynolds, R. H. Pedersen, M. O. Riehle and N. Gadegaard, *Small*, 2012, **8**, 2541–2547.
240. T. W. H. Oates, Y. Shiratori and S. Noda, *J. Phys. Chem. C*, 2009, **113**, 9588–9594.

Grafting of Functional Monomers on Biomaterials

LISBETH GRØNDAHL* AND JING ZHONG LUK

The University of Queensland, School of Chemistry and Molecular Biosciences, 4072 St Lucia, QLD, Australia
*Email: l.grondahl@uq.edu.au

11.1 Introduction

Biointerfaces are seen as a key component when designing new generation biomaterials. The events that take place at the biointerface upon implantation or body fluid exposure of a biomaterial have been investigated extensively over the past decades. Furthermore, the means of changing the surface properties to elicit a specific biological response has received ample attention.[1–4] Yet, most commercial medical implants have been designed based mainly on their bulk properties (strength, stability, macroscopic structure) and only a few medical implants exist on the market for which the material's surface properties have subsequently been optimised. One such example is CARMEDA® BioActive Surface in which heparin is end-point attached to an expanded polytetrafluoroethylene (ePTFE) vascular graft fabricated by GORE-TEX®.[5] It is apparent that at present an ideal solution cannot be 'dialled up' and much fundamental research is still required before medical implants will have ideal surface properties for their intended application.

This chapter aims to illustrate the current status with regards to grafting of functional monomers on biomaterials and will be restricted to grafting on polymeric substrates. Since optimisation of the biointerface is important for both biostable and biodegradable polymers examples from both classes will

RSC Smart Materials No. 10
Biointerfaces: Where Material Meets Biology
Edited by Dietmar Hutmacher and Wojciech Chrzanowski
© The Royal Society of Chemistry 2015
Published by the Royal Society of Chemistry, www.rsc.org

be included. Four commonly used methods are described highlighting their advantages and disadvantages as well as specific factors that affect the grafting outcome. This chapter will not detail the 'grafting to' approach in which a pre-formed polymer is attached to a functional group on a surface nor will it include syn-grafting of monomers adsorbed to a substrate prior to plasma treatment; for these topics the reader is referred to excellent reviews.[2,6,7] Furthermore, the focus will be on monomers carrying functional groups that can impart a charge to the surface and/or be used for subsequent attachment of biomolecules. When grafting of functional monomers to biomaterials it is important to properly characterise the modified substrate before evaluation of the *in vitro* or *in vivo* responses. Without such characterisation the interpretation of biological data is not warranted. Thus, the second half of this chapter is dedicated to various aspects of characterisation of grafted substrates as well as the grafted chains themselves and their position in/on the polymer substrate.

11.2   Methods for Grafting of Functional Monomers

Surface grafting of functional monomers can be employed to tailor the surface properties leading to new molecular functionalities incorporated onto an activated polymer surface. The four methods considered are: (1) simultaneous grafting method (Scheme 11.1A); (2) peroxy/hydroxyl grafting (Scheme 11.1B); (3) grafting from a surface using CRP (Scheme 11.1C); and (4) plasma polymerisation (Scheme 11.1D). Each method has its advantages and disadvantages, and in addition, different polymer topology (architecture) and polymer chemistry of the grafted chains will result from the different methods. Detailed descriptions of these methods are given in Sections 11.2.1 to 11.2.4. Furthermore, grafting of functional monomers to porous substrates such as membranes and tissue engineering scaffolds require an additional set of considerations and will be described separately in Section 11.2.5.

Depending on the purpose of the surface modification, different monomers can be used to introduce various functionalities to a substrate surface. The most commonly explored charged functional groups of carboxyl (–COOH), amine (–NH$_2$), phosphate (–O–P(O)(OH)$_2$), and sulfate (–O–S(O)(OH)$_2$) will be discussed. In addition, included will be functional groups suitable for attachment of biological molecules, *i.e.*, aldehyde [–CH(O)] and epoxy (–CHOCH$_2$).[4] Examples of monomers employed for grafting of these functional groups on polymer substrates will be given in the following individual sections.

11.2.1   Simultaneous Grafting

In the simultaneous grafting technique (Scheme 11.1A), the polymer substrate and monomer (a vinyl compound) are subjected simultaneously to radiation. The monomer is typically dissolved in a solvent which does not dissolve the polymer substrate. Oxygen, which is an inhibitor, is eliminated

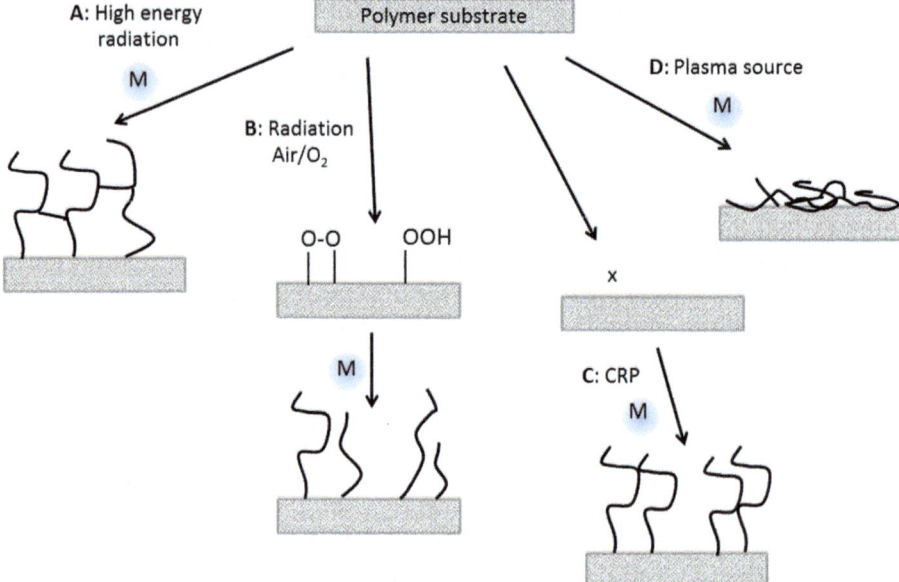

Scheme 11.1 Four methods for grafting of functional monomers on biomaterials considered in this chapter: (A) simultaneous grafting method; (B) peroxy/hydroxyl grafting (a two-step process); (C) grafting from a surface using CRP (may involve an initial step of introducing a functional group); (D) plasma polymerisation.

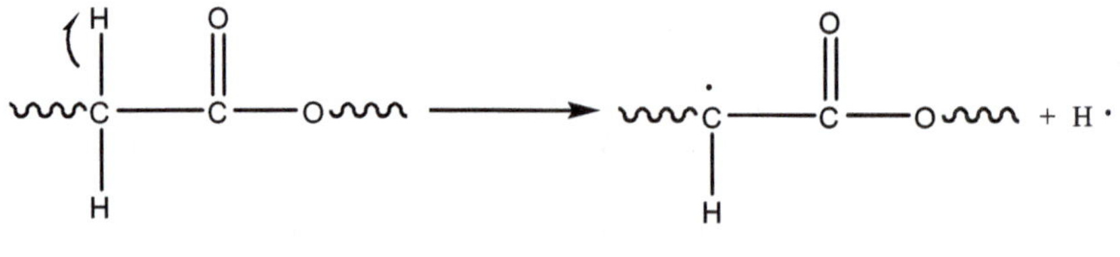

Scheme 11.2 Schematic illustration of hydrogen abstraction from a polyester polymer as a result of high-energy radiation.

from the grafting solution either by nitrogen degassing or by applying a vacuum.[8,9] This radiation process causes hydrogen (or in the absence of hydrogen, other element) abstraction from the polymer substrate's backbone and formation of polymer centred radicals as illustrated for a polyester in Scheme 11.2. These radicals can subsequently undergo graft co-polymerisation.

The grafting process is a radical polymerisation reaction from the surface; however, due to the continuous radiation process side reactions such as hydrogen abstraction from the growing graft co-polymer may take place concomitantly with the grafting process and as such branched or crosslinked grafted chains of a broad molecular mass distribution but with well-defined polymer chemistry will result as illustrated in Scheme 11.1A. In addition, homopolymer will form in the grafting solution as a result of the monomer radicals that form. This, however, can be minimised by using a homopolymer inhibitor such as Mohr's salt (*e.g.*, ammonium iron(II) sulfate hexahydrate, typically 1%). The extent of grafting can be controlled by varying the radiation conditions such as dose rate and radiation dose (*e.g.*, time exposed to radiation), as well as monomer concentration and solvent as illustrated in Box 1. In addition to the effects listed in Box 1, the polymer substrate's crystallinity will affect the grafting outcome as the monomer solution will diffuse into an irradiated substrate of low crystallinity more readily and thus the monomer will potentially come in contact with radicals throughout the polymer substrate. This effect has been observed for AA grafting onto the highly crystalline PLLA and semi-crystalline PCL with the former resulting in surface grafting and the latter in grafting throughout the bulk.[19]

Box 1: Effect of grafting parameters on grafting outcome for simultaneous grafting

Dose rate	*The degree of grafting may either increase or decrease with increased dose rate.[10]*
Dose	*Higher degree of grafting with higher dose is frequently observed[11,12a,13] but radiation at higher dose also affects bulk properties (see Table 1 for examples of doses used for various polymers).*
Monomer conc.	*Higher degree of grafting with increase in concentration generally observed[8,11,12,14,15] and can be rationalised based on radical polymerisation kinetics, i.e. the monomer concentration affects the chain length.*
	At a certain concentration a plateau or decrease in degree of grafting with monomer concentration may be observed (exact concentration depends on monomer and solvent).[16] This phenomenon is attributed to an increase in solution viscosity at high monomer concentration.[17]
Solvent	*Affects chain length depending on chain transfer constant which reflects its ability to transfer a radical from a growing chain thereby causing chain termination.[14]*
	Affects grafting extent depending on ability to swell the polymer substrate: the higher the degree of swelling of the polymer substrate by the solvent the further the graft copolymer will penetrate into the substrate.[10,14,18]
	Affects graft morphology depending on its ability to solvate the graft copolymer.[10]

The source of radiation can be either ultraviolet (UV) or high-energy ions such as an electron beam or gamma rays. While UV radiation is confined to the surface, electron beam and gamma rays will penetrate the entire material. Upon exposure of polymeric materials to radiation the radicals that form in the material will, if they are not initiating a graft co-polymer, either undergo radical combination to form cross-links or disproportionation to give scission and thus the bulk properties will be altered. The overall effect of ionising radiation depends on the polymer substrate and the radiation dose and these can be characterised with a *G*-value that quantifies the yield of an event. In general, *G*-values are described for cross-linking *G*(X), scission *G*(S) and gas evolution *G*(g).[20] Many polymers can undergo both chain scissoring and cross-linking but depending on the polymer type as well as the absorbed dose, one mechanism dominates.[19,21]

The advantage of the simultaneous radiation method, when using gamma ray or electron beam, is that it is a simple one-pot, one-step process. In contrast to gamma rays, electron beam needs much shorter time of exposure of the substrate to the irradiation source and as such has less of an effect on the bulk properties.[17] When UV is used as the radiation source, the grafting of monomers is a two-step process. In the first step a photo-initiator, such as benzophenone (BP; a Norrish type II photo-initiator) is irradiated with the substrate and in the second step, this pre-treated and washed substrate is irradiated in the presence of a monomer. Since this process uses a photo-initiator as well as other chemicals to obtain high efficiency of the grafting reaction concerns of complete removal of such additives post-grafting arise.

The radical concentration in the polymer substrate is affected by the irradiation process and depends on the type of radiation employed. The reader is referred to comprehensive reviews on UV radiation[22,23] as well as on high energy irradiation[19,24,25] for in-depth details on this topic. The radical concentration generated on (and in) a polymer substrate will furthermore be dependent on the polymer chemistry and crystallinity as radicals form preferentially in the amorphous or amorphous/crystalline interface regions.[24,25]

While electron beam radiation has been used for simultaneous grafting, most studies involve UV or gamma radiation induced grafting and examples of these are listed in Table 11.1. It is evident that the acrylate acrylic acid (AA) has been studied extensively for grafting onto various substrates; however, the phosphate-containing monomers monoacryloxyethyl phosphate (MAEP) and 2-(methacryloyloxy) ethyl phosphate (MOEP) have also been the focus of a number of studies. The monomer reactivity affects the grafting outcome with some being highly reactive (*e.g.*, AA) while other are more difficult to graft [*e.g.*, 2-aminoethyl methacrylate (AeMA)[18]]. There are a couple of important characteristics to note about some of the monomers listed in Table 11.1:

- AeMA can undergo rearrangement (intra- and inter-molecular amidation) and degradation (hydrolysis) reactions in alkaline media,[26] and it is thus important that the grafting process is carried out at neutral or weakly acidic pH.

Table 11.1 Examples of monomers that have been employed in the simultaneous radiation grafting method as well as the polymer substrates that were functionalised and the radiation source used.

Functional group	Monomer (concentration)	Polymer substrate	Radiation source (dose)	Ref.
Carboxyl	AA (2–40 w/v% in water, 10 g L^{-1} Mohr's salt)	PTFE	Gamma ray (2 or 10 kGy)	31
	AA (2–15% in methanol)	PHBV	Gamma ray (2–9 kGy)	11
	AA (2–15% in water, 1% Mohr's salt)	PCL	Gamma ray (5 or 20 kGy)	14
	AA (1, 3, 5, 10 wt% in water, 0.01 M Mohr's salt)	PLCL	Gamma ray (5–15 kGy)	12
	AA (5–25% in isopropanol–water)	PP	UV (BP as photo-initiator)	15
	AA (10% in ethanol)	PLA	UV (BP as photo-initiator)	32
Epoxy	GMA (1, 3, 5% in water–methanol mixtures)	PE	Gamma ray (0.8–3 kGy)	13
Amine	AeMA (2.5 or 4.5 w/v% in either water or methanol)	PHBV	Gamma ray (4.5 or 9 kGy)	18
	Acrylamide (4 wt%) converted amide to amine *via* Hoffmann degradation	PHBV	UV (BP as photo-initiator)	33
	AeMA (20–160 g L^{-1})	PP	UV (BP as photo-initiator)	16
Phosphate	MAEP (20–40% in water, CuCl$_2$ homopolymer inhibitor)	PTFE and FEP	Gamma ray (25–150 kGy)	34
	MOEP (1–40% in MEK or methanol)	PTFE	Gamma ray (8 or 10 kGy)	8, 10
	MOEP (10% in one and two-phase solvent systems from water, methanol and DCM)	PTFE	Gamma ray (10 kGy)	28

AA, acrylic acid; PTFE, polytetrafluoroethylene; PHBV, poly(3-hydroxybutyrate-*co*-3-hydroxyvalerate); PCL, polycaprolactone; PLCL, poly(L-lactide-*co*-caprolactone); PP, polypropylene; BP, benzophenone; PLA, poly(lactic acid); GMA, glycidyl methacrylate; PE, polyethylene; AeMA, 2-aminoethyl methacrylate; MAEP, monoacryloxyethyl phosphate; FEP, poly(tetra fluoroethylene-*co*-hexafluoropropylene); MOEP, 2-(methacryloyloxy) ethyl phosphate; MEK, methyl ethyl ketone; DCM, dichloromethane.

- The phosphate-containing monomers MAEP and MOEP contain a large percentage of the corresponding diene in the commercial monomers (25%) and as a result highly branched or cross-linked graft co-polymers form in most solvent systems.[27] However, a linear polymer topology of the grafted chains can be achieved in a two-phase mixed solvent system.[28] An additional complexity of using these phosphate-containing monomers in graft polymerisation reactions is the instability of the ester bonds during polymerisation (the extent of ester hydrolysis is dependent on the solvent used) and detailed analysis of the graft co-polymers (as well as soluble polymers produced by RAFT mediated polymerisation[29]) revealed that the MOEP graft co-polymer is best described as MOEP-*co*-HEMA while the MAEP graft co-polymers are MAEP-*co*-AA.[8–10,28,30]

11.2.2 Peroxy/hydroxyl Grafting

The peroxy/hydroxyl grafting method is a two-step method. In the first step reactive species are formed upon exposure of the polymer substrate to radiation and in the second step the monomer is grafted onto the activated surface. In the first step, hydroperoxides or diperoxides species are introduced onto the polymer substrate as shown in Scheme 11.1B. As above for the simultaneous radiation method, abstraction of a hydrogen (or other element) from the polymer backbone must take place before such reactive oxygen species can be introduced. This can be achieved in a number of ways as listed below and the first four methods will be considered in detail:

- Exposure of the polymer substrate to high energy radiation (gamma ray or electron beam) in vacuum or in the presence of an inert gas followed by exposure to air
- Exposure of the polymer substrate to high energy radiation (gamma ray or electron beam) in the presence of oxygen or air
- Plasma treatment in an inter atmosphere followed exposure to air
- Plasma treatment in an oxygen, water or air atmosphere
- Hydrogen peroxide or ozone treatment (sometimes combined with UV radiation)[23]

The peroxy/hydroxyl method is by some authors referred to as the 'pre-irradiation method'. However, this is confusing as that term (together with 'post-irradiation grafting') refers to a process where there is no exposure to an oxidising atmosphere at any point in the surface modification process. Indeed, the first radiation step involves exposure of the polymer substrate to radiation in vacuum or in the presence of an inert gas and in the second step the polymer substrate is exposed to a monomer either in the gas phase or in solution. The degassed monomer solution must be transferred to the polymer substrate through an air-tight seal in order to avoid exposure to oxygen. The pre-irradiation method is not commonly used in biomaterials science and instead, the peroxy/hydroxyl grafting method is preferred since it does not require exclusion of oxygen between the irradiation step and the grafting step; in fact, exposure to oxygen (*e.g.*, air) is required. When the radiation source used to introduce the peroxy radicals is plasma (typically Ar or O_2 plasma) the process is often referred to as 'plasma-induced graft polymerisation'. The surface concentration of peroxy radicals can be measured by the diphenylpi-crylhydrazyl (DPPH) method[35] and have been reported for various substrates to be up to 2.9×10^{-11} mol cm^{-2} for PTFE films,[36] up to 3.5×10^{-10} mol cm^{-2} for polyethylene (PE) films[35] and up to 2.5×10^{-8} mol cm^{-2} for poly(vinylidene fluoride) membranes.[37] The concentration of these peroxy radicals will be dependent on the polymer substrate (as discussed above in Section 11.2.1) and the source of radiation, the radiation dose and dose rate. With respect to high energy radiation details were provided in Section 11.2.1 while for plasma-induced grafting additional aspects of the type of plasma reactor and position of the sample relative to the plasma zone will also affect the incorporation of radicals and the reader is referred to articles on plasma modification by Desmet *et al.*[2] and Morent *et al.*[38] for further details.

It is important to realise that the different irradiation sources have different penetration depths into the polymer substrate with gamma rays and electron beam penetrating the entire material as described above while a plasma source will affect a depth of up to 60 nm when using Ar or O_2. This has impact not only on the position of the graft co-polymer (it will form where activated species, *e.g.*, radicals, are present in contact with the monomer) but also on the bulk properties of the polymer. While the use of a plasma source will leave the bulk properties practically intact, ionising radiation will lead to molecular changes as discussed above in Section 11.2.1.

The produced hydroperoxides or diperoxides species decompose to oxygen-centred radicals in the second step after immersion of the activated polymer substrate in a degassed monomer solution either at elevated temperature or upon exposure to UV radiation resulting in grafting. As for the simultaneous radiation method, the grafting process is essentially a radical polymerisation reaction from the surface. However, since it is a two-step process and the monomer (solution) is not exposed to radiation, linear (rather than branched or crosslinked) grafted chains of a broad molecular mass distribution and with well-defined polymer chemistry will result from this approach as illustrated in Scheme 11.1B. In the second grafting step the solvent used to prepare the monomer solution and the monomer concentration will impact on the grafting outcome as outlined in Box 1. In addition, the reaction temperature and reaction time potentially affect the grafting outcome when grafting at elevated temperature as indicated in Box 2. As an alternative to performing the second grafting step in solution, a monomer vapour can be exposed to the activated surface. This can be done simply by leaving the sample in the plasma reactor and at reduced pressure, bleeding in the monomer in the gaseous state. In this type of grafting process the reaction time or monomer pressure will affect the grafting yield.[44] While the monomer AA has most extensively been grafted on a large variety of

---

Box 2: **Effect of grafting parameters of the second step on the grafting outcome for the peroxy/hydroxyl method when the grafting step is done at elevated temperature**

Grafting time *Longer grafting time increases the graft yield which can be rationalised based on radical polymerisation kinetics, i.e. the grafting time affects the number of grafting reactions initiated and thus the density. A maximum graft yield is observed after 2–6 hours after which no further increase is achieved.[39-41]*

Temperature *An increase in grafting temperature (in water above 70 °C[42,43]) generally increases the degree of grafting and this parameter can be optimised for hydrolytically unstable polymer substrates to reduce concomitant degradation as demonstrated for samples irradiated by electron beam.[19]*

---

substrates including PTFE, polyethylene (PE), polycaprolactone (PCL) and poly(3-hydroxybutyrate-*co*-3-hydroxyvalerate) (PHBV) using this method, monomers with other functional groups have also been successfully employed as illustrated in Table 11.2.

11.2.3 Controlled Radical Polymerisation Grafting from a Surface

The main advantage of using controlled radical polymerisation (CRP) grafting from a surface (often referred to as surface-initiated polymerisation) is that it allows control over not only the chemistry of the grafted chains but also greater control of polymer topology and chain length with a narrow size distribution resulting from this approach as shown in Scheme 11.1C. Comparing grafting from and grafting to approaches where CRP is used it is found that the grafting to approach will give the best control over chain length (*i.e.*, low polymer dispersity) while the grafting from approach will allow more densely packed polymer brushes to be formed[53,54]; however, only the grafting from approach will be considered here. An additional advantage of CRP is that well-defined co-polymer brushes can be produced allowing for a wealth of chemistry to be explored. There are a number of different CRP techniques including atom transfer radical polymerisation (ATRP) and reversible addition-fragmentation chain transfer (RAFT) polymerisation using an R-group (grafting from) approach with the former being the most extensively used for surface-initiated reactions.[54,55] When employing the method of grafting from a surface using CRP, functional groups on the polymer substrate are required

Table 11.2 Examples of monomers that have been employed in the peroxy/hydroxyl grafting method as well as the polymer substrates that were functionalised and the pre-treatment conditions used to introduce the peroxy radicals.

Functional group	Polymer substrate	Step 1: pre-irradiation treatment	Step 2: monomer grafting	Ref.
Carboxyl	PTFE	(iii) Argon plasma (1200/2500 W, 30–360 s, 0.27 mbar)	AA (40 wt%, 80 °C)	45
	PTFE	(iv) Water plasma (400 W, 60 s, water pressure 2×10^{-3} mbar)	AA (vapour, 0.5–4 mbar, 30 min)	44
	PCL	(iii) Argon plasma (37 W, 10 s, Ar pressure 0.67 mbar)	AA (0.03–0.11 g L^{-1}, UV radiation)	46
	PHBV	(iv) Oxygen plasma (120 W, 30 s, O$_2$ pressure 0.27 mbar)	AA (10 wt%, 75 °C)	47
	HDPE	(iii) Argon plasma (40 W, 3 min, Ar pressure 0.33 mbar)	AA (10–30 wt% in alcohol, 70 °C)	39
	PTFE	(ii) Gamma ray (5–20 kGy, irradiation in air)	AA (50%)	48
	PP	(ii) Gamma ray (5–20 kGy, irradiation in air)	AA (40% in water, 50 °C)	49
	PCL	(i) or (ii) Electron beam (50–700 kGy, irradiated in air or nitrogen)	AA (30% in water, 25 °C)	41
Epoxy	PE	(i) Electron beam (200 kGy, irradiation in nitrogen)	GMA (10% in methanol, 313 K)	40
Amine	PCL	(iii) Argon plasma (100 W, 30 s, Ar pressure 0.8 mbar)	AeMA (0.1–1.0 M, UV radiation)	50
Phosphate	PE	(iii) Argon plasma (24 W, 10 s)	MOEP (water–acetic acid solution, UV irradiation)	51
	SPCL	(iv) Oxygen plasma (30 W, 15 min, O$_2$ pressure 0.2 mbar)	VPA (0.1 M in 2-propanol, RT)	52
Sulfate	SPCL	(iv) Oxygen plasma (30 W, 15 min, O$_2$ pressure 0.2 mbar)	VSA (10 vol% in water, RT)	52

HDPE, high density polyethylene; VPA, vinyl phosphoric acid; VSA, vinyl sulfonic acid; SPCL, Starch/PCL blend 30/70 wt%.

for the attachment of an ATRP initiator (commonly 2-bromoisobutyrate bromide, BIBB) or a chain transfer agent (CTA) for ATRP and RAFT polymerisation, respectively. The polymerisation reaction can subsequently take place from these surface-bound molecules. In some polymer substrates, functional groups exist as part of the structure enabling direct attachment of an initiating molecule to the polymer substrate. Cellulose is an example of such a polymer for which a comprehensive review by Roy *et al.* on modification by polymer grafting serves as an excellent reference.[56] In other polymer substrates, however, an initial step of introducing functional groups onto the surface is required before CRP can take place. This is the case for most of the polymer substrates covered in this chapter (both biostable and biodegradable) and many methods exist which allow introduction of functional groups as illustrated in Table 11.3. The factors which affect the grafting outcome when using CRP are similar to those affecting chain length in solution-based CRP and include not only monomer concentration but also concentration of other reagents. The density of the grafted chains will of course be dependent on the density of the functional group and subsequently the density of the attached initiating molecule. This can be controlled to some extent by using a mixture of initiating molecules and blocking agents attached to the surface.[57] In addition, for RAFT polymerisation using the R-group approach it has been found that the presence of a CTA in the grafting solution (and not only attached to the substrate surface) has a significant effect on the grafting outcome as evident in recent reviews.[54,56]

In addition to the use of monomers already carrying functional groups, it is common to carry out post polymerisation reactions[54,58] to introduce some of the functional moieties which are the focus of this chapter. Box 3 gives examples of such post polymerisation reactions. This approach is often used for CRP due to the low tolerance of CRP to certain functional groups and examples of grafting monomers suitable for post polymerisation reactions are thus included in Table 11.3. It should be noted, however, that while such approaches are useful for chemically stable polymer substrates they may not be able to be transferred to hydrolytically labile substrates such as polyesters. Furthermore, the approach is not applicable if the surface-bound initiating molecule contain, for example, ester bonds as some of the grafted chains may be lost during the post polymerisation reaction. As an alternative to initiating the polymerisation reaction from an initiating molecule-functionalised surface, radiation or plasma mediated CRP from a surface can be used without the requirement for functional groups to be present on the substrate's surface. Most work in this area has used monomers which require post polymerisation reactions to introduce the functional groups of interest as illustrated in Table 11.3.

11.2.4 Plasma Polymerisation

Plasma polymerisation, also known as plasma deposition, is a means to create a thin film (termed a plasma polymer) on a substrate. A monomer in

Table 11.3 Examples of monomers that have been employed in the CRP grafting method as well as the polymer substrates that were functionalised and the method use to introduce functional groups if required as well as the type of CRP technique employed.

Functional group	Monomer	Polymer substrate (incl. introduction of functional group)	Technique (conditions)	Ref.
Carboxyl	AA	Cellulose (contains OH groups)	ATRP (water pH 10.2)	59
	MAA (sodium salt)	PCL (aminolysis; NH_2 groups introduced)	ATRP (water, 30 °C)	60
	tBA	PE-co-PP	Bimolecular RAFT (CTA was PEPDA; gamma ray 0.2–0.8 kGy)	61
Epoxy	GMA	Cellulose (contains OH groups)	ATRP (EtOAc, RT)	62
	GMA	PCL (aminolysis; NH_2 groups introduced)	ATRP (methanol–water, RT)	63
	GMA	PCL (hydrolysis; OH groups introduced)	ATRP (methanol–water, RT)	64
	GMA	PTFE (Ar plasma and exposure to air)	Bimolecular RAFT (CTA was CDB; DMF, 85 °C)	65
Amine	DMAEMA	Cellulose (contains OH groups)	RAFT (CTA was MCPDB; initiator AIBN, toluene, 60 °C)	66
	DMAEMA	Cellulose (contains OH groups)	ATRP (dichloro benzene, 80 °C)	57
	EGMA; DMAEMA	PP (O_3; reduction to OH groups)	ATRP (DMF, 80 °C)	67
Sulfate	StS (sodium salt)	PTFE (H_2 plasma/O_3 treatment; introduction of OH groups)	ATRP (DMSO, 40 °C)	68
	St	PP	Bimolecular RAFT (CTA was CPDA; gamma ray 0.1–14 kGy)	69

MAA, methacrylic acid; tBA, tert-butylacrylate; PEPDA, 1-phenylethyl phenyldithioacetate; EtOAc, ethyl acetate; CDB, cumyl dithiobenzoate; DMF, dimethylformamide; DMSO, dimethylsulfoxide; DMAEMA, N,N-dimethyl aminoethyl methacrylate; MCPDB, S-methoxycarbonyl phenylmethyl dithiobenzoate; AIBN, azobisisobutyronitrile; EGMA, ethylene glycidyl methacrylate; CPDA, cumylphenyl dithioacetate; St, styrene; StS, styrene sulfonate.

Box 3:   Examples of post polymerisation reactions for introduction of ionic functional groups[54,58]

Carboxylate   Hydrolysis of tert-butyl acrylate (tBA) or tert-butyl methacrylate (tBMA) using strong acid (e.g., 10 % HCl or CF₃COOH).

Pyrolysis of tBA or tBMA at 190–200 °C.

Esterfication of 2-hydroxyethyl methacrylate (HEMA) with succinic anhydride in the presence of pyridine.

Amine   CDI activation of HEMA followed by reaction with ethylene diamine.

Amination of glycidyl methacrylate (GMA) by reaction with excess ethylene diamine.

Sulfate   Sulfation of GMA using sodium sulfite.

Sulfation of styrene (St).

---

the vapour phase, upon exposure to plasma, is converted into reactive species as a result of fragmentation reactions. These species in turn recombine to produce the plasma polymer on the substrate surface exposed to the plasma. The polymers formed in the plasma do not necessarily have similar chemical composition or topology to the polymers obtained by the conventional polymerisation techniques discussed in the sections above.[1,4] Generally, plasma polymers are pin-hole free and highly cross-linked as illustrated in Scheme 11.1D. They are therefore insoluble, chemically inert, thermally stable and mechanically tough. They are not attached to the polymer substrate *via* covalent bonds as the graft co-polymers described in the sections above but they are rather coated onto the substrate. One of the main advantages of this approach to surface grafting of functional monomers is that the same outcome is achieved regardless of the substrate which is in contrast to the previous methods discussed above where optimisation of the grafting conditions is required for each polymer substrate. In plasma polymerisation, once the conditions for producing the plasma polymer have been optimised for a particular monomer, it can be applied to any substrate.[1,4] However, the reaction conditions do need to be optimised for a particular plasma reactor. Another advantage of this technique is that it is solvent-free eliminating concerns of residual toxic solvents in the modified substrate. However, due to the processes of fragmentation and recombination taking place in the plasma reactor loss of functional groups can be a drawback. There are many parameters that affect the outcome of a plasma polymerisation reaction including the plasma conditions (power, frequency, pressure, monomer flow rate, exposure time, position of sample in plasma chamber) and the monomer. It is generally found that low radio-frequency power and/or pulsing result in better retainment of the functional groups; however, too low a power will lead to low cross-linking density of the plasma polymer causing dissolution upon exposure to a solution.[4] It is therefore essential to optimise the conditions for each monomer. Another concern regarding plasma polymers is their instability in air and ageing phenomena have been studied extensively; the reader is referred to an excellent review covering the topic.[1]

In contrast to the previously described methods the monomers used during plasma polymerisation do not need to contain a vinyl group. For example, for the introduction of amine groups heptylamine (an aliphatic amine) and allylamine (a vinyl monomer) can both be applied. While heptylamine undergoes fragmentation and recombination reactions only, allylamine display higher deposition rates and this is attributed to additional conventional radical polymerisation taking place concomitantly for this monomer.[1] Likewise, the vinyl monomer AA has a significantly higher deposition rate than propanoic acid, again attributed to the additional conventional radical polymerisation. Moreover, this leads to plasma polymers produced from AA displaying a more linear chain structure than that obtained from saturated monomers like propanoic acid.[1] Optimisation of the density of functional groups can be achieved by careful selection of the monomer. For introduction of amine groups it has thus been found that the use of diaminocyclohexane yields a higher density than heptylamine.[1] In addition to introduction of carboxyl[70,71] and amine[72-74] groups, plasma polymerisation also provides a convenient avenue for the introduction of epoxide[75] and aldehyde[72,76] functionalities. The reader is referred to specific papers given above for a guide to optimal plasma parameters for each monomer considering the required optimisation for each plasma reactor type.

11.2.5 Grafting of Functional Monomers to Porous Substrates

There is a great interest in membranes and 3D scaffolds for biomedical applications.[77-79] 3D scaffolds can be fabricated by a range of methods including supercritical fluid technology, rapid prototyping, porogen leaching and phase separation techniques (*e.g.*, thermally induced phase separation (TIPS)). In addition, membranes can be fabricated by different techniques with electrospinning being one of the popular approaches used for tissue engineering constructs. Each of these methods allow for fabrication of scaffolds/membranes with different overall porosity and different pore sizes and shapes. The pore tortuosity (a description of how many turns the pores have) will also depend on the fabrication method used. While studies investigating grafting of functional monomers on membranes are common, similar studies on 3D scaffolds have mainly appeared in recent years.

One aspect that must be taken into account when grafting of functional monomers to porous substrates is whether the method and conditions chosen will allow modification throughout the interior of the porous material or if it will only allow for modification of the external surface. In some applications it may be desirable to graft throughout the matrix while for other applications the goal may be to have surface-confined modification. Gamma and electron beam irradiation will produce reactive species used for grafting throughout the entire material. Modification throughout the entire porous material will then occur if the monomer solution contacts the internal pore walls. This in turn will be affected both by the method use for degassing the polymer substrate solution[9] and by the solvent system used to dissolve the monomer[9,14] when using any of the 'solvent' methods described above. For UV radiation or plasma methods described in this chapter the ability to modify the interior will depend on the dimensions of the porous substrate and on the pore tortuosity. Since the scaffold morphology is affected by the processing techniques this information has been included in Table 11.4 together with the method used to modify the porous material. Regardless of the intended application and regardless of the method used it is important to analyse the position of the grafted chains. An indication of the characterisation technique used for this analysis is included in Table 11.4 and it is clear that most of the studies on 3D scaffolds do investigate chemical changes in the interior using some type of spectroscopic technique. Studies which have evaluated the chemical change at various positions for membranes which have been grafted with functional monomers are included in Table 11.4, however the list is by no means exhaustive. A final important aspect of grafting of functional monomers to any substrate but particularly for porous ones which do not have as high strength as their solid counterparts, is the evaluation of changes in mechanical properties due to the grafting process. This aspect is often overlooked and has only been included in a few of the studies included in Table 11.4.

11.3   Characterisation of Grafted Materials

Grafted materials are characterised with respect to the surface chemistry to evaluate the overall change in chemistry upon grafting as well as to evaluate to which extent the graft co-polymer resembles the theoretical structure. The most commonly employed techniques are described in Section 11.3.1. For 2D substrates physical properties including the surface wettability, roughness and viscoelastic properties of the grafted materials can be characterised by methods described in Section 11.3.2. For 3D substrates, however, these properties are often inferred from model 2D substrates. In addition to these commonly used characterisation methods, evaluation of the grafted polymer chains with respect to topology and chain length will be discussed in Section 11.3.3; methods for the evaluation the lateral distribution of the graft co-polymer across a surface will be detailed in Section 11.3.4 while the penetration depth of the graft co-polymer both into solid substrate and into the pores of porous membranes and scaffolds will be discussed in Section 11.3.5. If a true evaluation of a biomaterial performance

Table 11.4 Examples of 3D scaffolds and membranes that have been surface-modified by methods discussed in this chapter.

Fabrication technique and dimensions of matrix	Surface modification method	Evaluation of position of graft co-polymer[a]	Compressive modulus	Ref.
Supercritical CO_2 produced PLA scaffold: 3 mm thick, 10 mm diameter	Plasma polymerisation	XPS (N 1s signal throughout the cross-section)	No	80
Gas foaming, porogen leached PLLA scaffold: 5 mm thick, 10 mm diameter	Plasma polymerisation	No	Unmodified: 0.4–0.7 MPa	81
Osteofoam™ PLGA scaffold: 10×10×10 mm	Plasma polymerisation	XPS (N 1s signal both on surface and cross-section)	No	82
Porogen leached PHBV scaffold. Dimension not stated.	Simultaneous grafting (UV irradiation)	ATR-FTIR (N–H stretching vibration in spectrum of cross-section)	Unmodified: 0.3–0.6 MPa Modified: 1.3–2.3 MPa	83
Porogen leached PHBV scaffold. Dimension not stated.	Simultaneous grafting (UV irradiation)	EDX (presence of nitrogen across the cross-section)	No	84
3D printing PCL scaffold: 5×5 mm, 8 mm diameter	Plasma polymerisation	XPS (N 1s across cross-section)	Unmodified: 34 MPa Modified: 33 MPa	85
TIPS PCL scaffold. 5 mm thick, 13 mm diameter	Simultaneous grafting (gamma irradiation)	XPS (presence of N 1s signal throughout the cross-section)	Unmodified: 1.2 MPa Modified: 0.8 MPa	14
ePTFE membrane (thickness of 1 mm)	Simultaneous grafting (gamma irradiation)	MRI (presence of water signal associated with grafted chains)	No	9
ePTFE membrane (thickness of 3.5 mm)	Peroxy/hydroxy (Ar plasma)	XPS (each side of membrane)	Unmodified: 1.9 MPa Modified: 1.8 MPa	86

[a]Full names for all techniques are given in Section 11.3.1.

with respect to introduction of grafted chains is to be achieved it is crucial to understand both the exact chemistry introduced, the physical properties that result from introduction of the grafted chains and the position of the graft co-polymer in the substrate.

11.3.1   Chemical Characterisation of Surfaces

The most commonly used techniques for evaluation of chemical changes to the surface of a polymeric substrate are Fourier transform infrared spectroscopy (FTIR) and X-ray photoelectron spectroscopy (XPS). While FTIR spectra are recorded at ambient conditions XPS requires vacuum conditions. These two techniques are somewhat complementary with XPS giving information regarding the elemental composition (as well as the element's 'oxidation state') while FTIR will give information regarding the functional groups present. It should be noted, that when the chemistry of the substrate and the grafted chains are chemically distinct, *e.g.*, PAA chains on PTFE, then it is fairly straight forward to characterize the grafted material using either technique.[31] However, when the chemistry is similar, *e.g.*, PAA chains on PCL, then it is more challenging and this is illustrated in Box 4. A limitation of XPS is that it cannot be used directly to assess for example, carboxylate groups on a polyester substrate due to the similar binding energies of COOH and COOR (289.2-289.2 eV and 288.6-289.2, respectively[90]) and FTIR cannot easily be used to *e.g.*, quantify different nitrogen-containing groups due to overlap in band positions.[33] Some of these shortcomings can be overcome using derivatisation methods, most commonly in conjunction with XPS although it has also successfully been used with Raman spectroscopy.[88] Examples of commonly used probes and the functional groups with which they react are given in Table 11.5. Some of these probes are used in a gas phase reaction in which a sealed container with the substrate as well as the probe molecule (sometimes at elevated temperature) is reacted. Other probes react with the functional group in a solution phase reaction, in some cases assisted by other reagents (*e.g.*, some probes require carbodiimide chemistry for coupling). Some studies have found that probes in the gas phase will react only with the functional groups exposed to the surface[88] while solution based probes may penetrate further in to the substrate if the solution is capable of swelling the grafted chains and/or the polymer substrate. Some of these probes display high selectivity towards a specific functional group; however, others are reactive towards many different groups, *e.g.*, trifluoroacetic anhydride (TFAA) is reactive towards amines, carboxyl, epoxides and hydroxyl groups. An example of a highly selective probe is trifluoromethyl benzaldehyde (TFBA) used to detect amine groups. This probe was found to yield the same information with regards to percent amine groups relative to all nitrogen species as that obtained from narrow scan fitting of the N 1s spectrum.[88]

A technique which is gaining increasingly more use is time-of-flight secondary ion mass spectrometry (ToF-SIMS).[99] This technique provides

---

Box 4: Information gained from FTIR and XPS for examples of grafted substrates

Carboxylate *Carboxylate groups on a polyester can be detected by FTIR (example of an AA grafted PLGA film is shown in Figure 11.1A and 1B). In the 4000–2000 cm^{-1} region, the very broad band from the OH vibrational mode appears clearly after grafting while in the 2000 to 600 cm^{-1} region the most distinct change is from a single carbonyl stretch from the ester group at 1750 cm^{-1} in the unmodified polymer substrate to an additional carbonyl stretch at lower wavenumber (1700 cm^{-1}) for the carboxylic acid group in the grafted substrate.[11,14,32]*

Relative amounts of acid and ester moieties can be obtained by FTIR or Raman spectroscopy.[14]

Amine *Amine groups can be detected by FTIR by the presence of a amine vibrational modes at 1610 and 750 cm^{-1}.[18,33]*

Amine groups on a polyester substrate are detected and quantified by XPS as a new element is introduced during the grafting process.[18,33,50]

XPS can be used to evaluate the relative amounts of various nitrogen-containing functional groups such as amine, amide and protonated amines[18] and thus is a useful tool to assess the level of e.g., surface oxidation of amine groups.[87,88]

Sulfate *Sufate groups on a polyester substrate can be detected by XPS as a new element is introduced during the grafting process.[52]*

Sulfate groups display a strong band due to the S=O symmetric stretching in the 1230–1250 cm^{-1} region.[89]

---

chemical information regarding the surface layer and structural information based on the molecular fragments observed. Like XPS, this technique requires the sample to be in a vacuum and no volatiles can be present. The data sets obtained from ToF-SIMS are very complex and are routinely coupled with detailed principal component analysis to evaluate chemical differences between samples.[72] Energy dispersive X-ray (EDX) spectroscopy coupled to scanning electron microscopy (SEM) analysis has also been applied in a few studies. However, it is restricted to information regarding elemental composition and it is difficult to obtain quantitative data for samples that are not smooth.

It should, of course, be noted that while these different techniques used for the characterisation of the surface chemistry are indeed complementary, they do probe the surface to different depths. ToF-SIMS is a true surface analysis technique evaluating only the outmost 1 nm of a surface while XPS (limited by the escape of the core electrons from the surface) will analyse the top 5–10 nm. It is possible, through angle resolved XPS, to probe a more

Table 11.5 Examples of XPS probes used for derivatisation of functional groups.

Functional group	Reactant	Comments	Ref.
Carboxyl –COOH	CF$_3$CH$_2$OH TFE	Gas phase reaction. Reaction mixture contains di-*t*BuC and py.	91–93
	 BTFBAM	Solution phase reaction. Sample first activated by NHS.	94
	CF$_3$CH$_2$NH$_2$ TFEA	Solution phase reaction. Reaction mixture contains EDS and NHS.	95
	 PFP	Solution phase reaction. Reaction mixture contains EDC.	11, 14
	NaOH	Solution phase reaction. Produces the carboxylate from the carboxylic acid.	91
Aldehyde –CHO	H$_2$NNH$_2$ Hydrazine	Gas phase reaction. Also reactive towards ketones ($>$C$=$O).	92
	 TFPH	Solution reaction. Also reactive towards ketones ($>$C$=$O).	91

Table 11.5 (*Continued*)

Functional group	Reactant	Comments	Ref.
Amine –NH$_2$	PFBA	Solution phase reaction. Reaction product is an imine.	91, 93
	TFBA	Gas phase reaction. Reaction product is an imine.	88, 96–98
	(CF$_3$CO)$_2$O TFAA	Solution phase reaction. Reported to also react with carboxyl, epoxide, and hydroxyl groups.	91, 93

TFE, trifluoroethanol; di-*t*BuC, di-*tert*-butylcarbodiimide; py, pyridine; BTFBAm, 3,5-bis-(trifluoromethyl)benzyl amine; TFPH, (pentafluorophenyl)hydrazine; PFBA, pentafluorobenzaldehyde; TFBA, trifluoromethyl benzaldehyde; TFEA, 2,2,2-trifluoroethylamine; TFAA, trifluoroacetic anhydride.

shallow depth. EDX and ATR-FTIR are often categorised as surface analysis techniques although their penetration depth into a substrate can be as much as microns. Other FTIR techniques will allow a more shallow depth to be analysed (*e.g.*, grazing angle specular reflectance technique); however, such techniques are not used extensively in the literature for the analysis of bio-materials as they require a very smooth substrate. It is important to recognise the different penetration depths of the characterisation techniques and often by combining date from more than one technique detailed evaluation of the penetration depth of the graft co-polymer can be obtained, as discussed further in Section 11.3.5.

Furthermore, colorimetric methods can also be used to determine the surface density of functional groups.[1,100] Such techniques make use of various dyes, both those that covalently attach to a functional group (similar to XPS probes) and those that interact *via* electrostatic interaction between oppositely charged groups on the surface of the substrate and the dye molecule. For the latter approach an assumption that a 1 : 1 interaction occurs must be made. Furthermore, for acidic and basic functional groups potentiometric titrations may be used to evaluate surface density of the functional group.[101]

11.3.2   Physical Characterisation of Surfaces

A challenge when evaluating the biological effect of the grafting of functional monomers on biomaterials is that not only is the surface chemistry altered, but many physical properties are also changed concomitantly making it difficult to discern the effect of any one property. A wide range of physical properties of a material's surface are relevant to the ultimate performance of the material in a biological setting.[3] It is thus important to evaluate as many of these as practically possible and also to recognise that some techniques are only suitable for the analysis of 2D and not 3D substrates. A list of the most commonly studied properties and suitable techniques for their analysis are given below:

- *Wettability*: obtained by contact angle measurement; suitable for non-porous substrates although it is also used for porous membranes for evaluation of relative changes, for example; affected by surface roughness and heterogeneity of grafted chains[3,23]
- *Surface energy*: obtained by contact angle measurements using a range of solvents with different surface tension; suitable only for non-porous and topographically smooth surfaces[102]
- *Surface roughness and topography*: micrometer level using SEM, nanometre level using AFM
- *Zeta potential*: electrokinetic measurements by streaming potential
- *Viscoelastic properties*: obtained from AFM nano-indentation, suitable to analysis of 2D substrates[14,103] but may be applied to 3D membranes with detailed analysis[86]
- *Surface crystallinity*: evaluated using *e.g.*, FTIR, suitable for 2D substrates only[104]

In addition to these surface properties, it is important to evaluate the bulk properties especially if the method used for grafting of functional monomers uses high energy radiation. In such instances, bulk crystallinity is an important property as changes caused by the grafting process indicates that, for example, if grafting takes place not only in the amorphous region as expected but also in the crystalline region.[41] In addition, changes in bulk crystallinity are likely to affect the bulk mechanical properties as are chain scissoring or cross-linking of the polymer substrate. It is therefore also important to evaluate such mechanical bulk properties in particular for porous substrates as highlighted in Section 11.2.5.

11.3.3   Characterisation of Grafted Chains

It is difficult to characterise the polymer topology and chain length of the grafted chains. None of the techniques described in the previous sections can provide this detail. FTIR is able to evaluate relative degree of cross-linking between different samples based on the fact that the presence of hydrogen bonding will affect the band width. Broader bands will arise for linear chains while the functional groups of cross-linked graft co-polymers are less available for hydrogen bonding resulting in sharper bands.[27] A common approach used when the grafted chains have been fabricated using the simultaneous grafting method is to isolate the homopolymer that forms concomitantly near the grafted substrate as it is inferred that the polymerisation reaction is similar for both the grafted chains and the homopolymer. The homopolymer can be characterised by traditional methods used for soluble polymers. These include GPC to evaluate the chain length[66,105] and NMR to evaluate chemistry and topology.[27] One such study compared the molecular mass of the homopolymer formed in solution and that which formed occluded at the surface of a grafted substrate.[105] It was found that the occluded homopolymer had a significantly higher molecular mass (approximately one order of magnitude) than that which formed in the surrounding solution and this was regardless of the solvent used during the grafting reaction.[105] This was rationalised based on the different viscosity near the substrate and in solution with the higher viscosity region decreasing chain mobility and therefore termination.

The only way, however, to obtain exact information regarding polymer topology and chain length is by cleaving off the grafted chains and analysing these by traditional methods used for soluble polymers. This approach is viable when the grafted chains are attached using CRP and the grafted chains can be cleaved from the substrate post-grafting. An example of a study which has used this approach is the work by Perrier and co-workers.[66] They evaluated the molecular mass and dispersity of polymers produced in solution as well as grafted on a cellulose substrate using RAFT incorporating CTA in solution (in addition to that which was surface-bound). In this system it was found that the reaction kinetics differed from solution to surface-grafted polymerisation. The polymer in solution displayed faster growth resulting in higher molecular mass,[66] while the higher CTA concentration at the surface and its rate retardation effect resulting in lower molecular mass for the surface initiated polymer chains.[66]

11.3.4 Lateral Distribution of Grafted Chains

The polymer crystallinity of semi-crystalline polymers can affect the lateral distribution of the graft co-polymer and prevent full surface coverage.[8,9,25,106] In addition, the solvent used in the grafting process can also affect the lateral distribution.[9] Since a heterogeneous surface is expected to result in a 'mixed' biological response which can be difficult to interpret, it is important to analyse the lateral distribution of the graft co-polymer across the surface. The lateral distribution is commonly assessed by analysing several positions across the surface and using the data to generate a surface map or image of the distribution of a specific chemical functionality. Evaluation of surface crystallinity has been done using μ-ATR-FTIR[100] and both Raman and μ-ATR-FTIR mapping can be used to evaluate the lateral distribution of the graft co-polymer.[9,88,107] An illustration of the principle of μ-ATR-FTIR mapping is shown in Figure 11.1. The μ-ATR objective makes contact with the sample and records a spectrum (Figure 11.1C). Each spectrum is then analysed (*e.g.*, area ratios of specific bands calculated) and each of these obtained values are then displayed as an area of the map which is produced (Figure 11.1D). The resolution of μ-ATR-FTIR mapping is controlled by the aperture size and typically results in 30×30 μm being analysed at each position.[9] Furthermore, mapping of functional groups across a surface can be acheived using XPS and ToF-SIMS which provide high resolution images of the spatial distribution of elements or fragments.[9,99] An example of ToF-SIMS images (analysing a total area of 200×200 μm) of the CF^+ ion from the surface of an ePTFE membrane before and after grafting is displayed in Figure 11.2. It can be seen that while the lateral distribution of the CF^+ ion on the untreated membrane is homogeneous, the grafted membrane displays a patchy distribution. It should be noted that the techniques described here have different penetration depths into the grafted

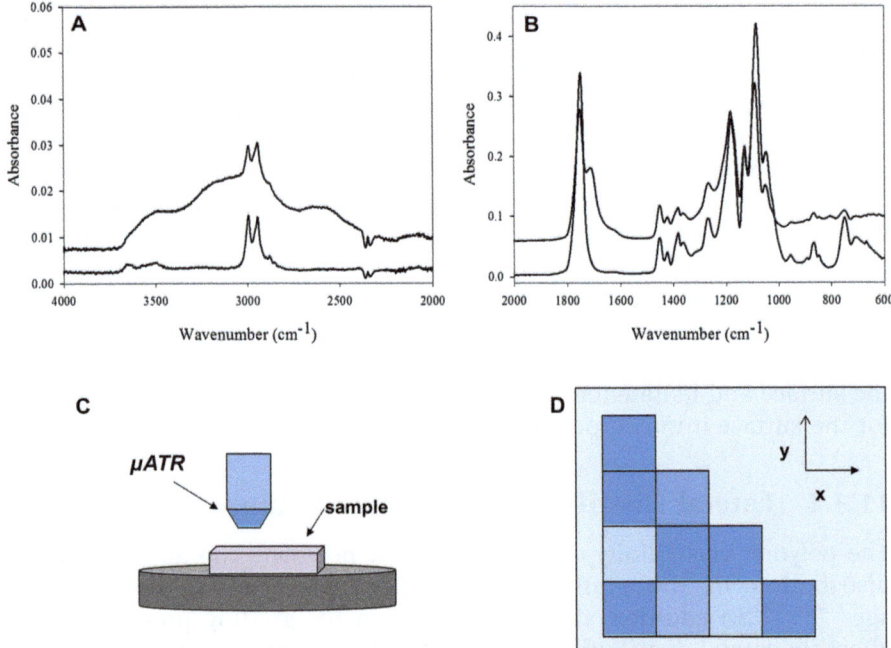

Figure 11.1   (A) FTIR spectra of PLGA film (bottom) and AA grafted PLGA film (top) in the 4000–2000 cm^{-1} region; (B) FTIR spectra of the same samples in the 2000 to 600 cm^{-1} region; (C) schematic illustration of the μ-ATR objective making contact with a sample; (D) each spectrum recorded by μ-ATR is analysed to obtain a parameter (area or intensity ratio) and each of these obtained values are displayed as an area in the map where the relative colour intensity indicates the magnitude of the value.

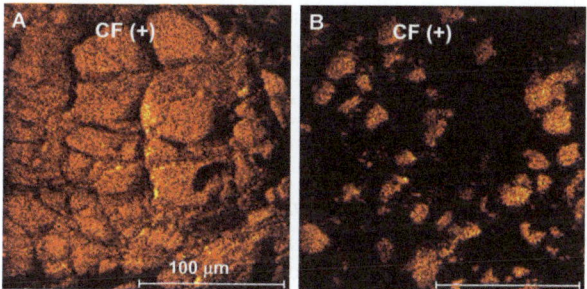

Figure 11.2 Positive ion ToF-SIMS images. Bright red areas show the presence of the CF^+ ion on the surface on (A) unmodified ePTFE, and (B) MOEP grafted ePTFE.

substrate as detailed in Section 11.3.1 and this must be kept in mind when analysing the lateral distribution of a graft co-polymer.

11.3.5 Penetration Depth of Grafted Chains

Considering the two scenarios in Figure 11.3, both polymer substrates (A and B) have the same degree of grafting. In one grafted sample (A) the depth of the graft co-polymer is shallow, while in the other grafted sample (B) the graft co-polymer has penetrated deep into the substrate; however, it is presented at the surface in a patchy manner. Depending on the penetration depth of the technique used to analyse the surface different information will be obtained. If, on the one hand, a technique that evaluates the entire depth of the grafted region is used (*e.g.*, ATR-FTIR) then the outcome of the analysis would be that both substrates were similar. If, on the other hand, a technique that examines only the outmost surface is used (*e.g.*, ToF-SIMS or contact angle) then the outcome would be that the substrates are very different. In essence, none

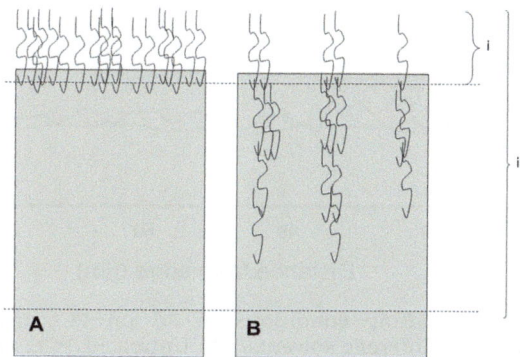

Figure 11.3 Schematic illustration of two polymer substrates (A and B) with the same degree of grafting. (A) The depth of the graft co-polymer is shallow; (B) the graft co-polymer has penetrated deep into the substrate, but at the surface it is presented in a patchy manner.

of the conclusions are completely wrong, nor are they completely right in the sense that none of the techniques provides the full picture. It is thus desirable to use multiple techniques for the analysis in order to get a full picture of the position of the grafted chains.[8] Alternatively, the penetration of the grafted chains can be probed directly by cross-section analysis.

Cross-sections of grafted polymer substrates may be analysed using a variety of techniques as evident from the data in Table 11.4 for porous substrates. Such an approach is also used for solid substrates, and for this purpose the use of Raman spectroscopy is common.[14,105] An example of Raman mapping across the cross-sections for a series of PCL films is shown in Figure 11.4. The carbonyl band of the ester from the PCL substrate and the carbonyl band from the introduced grafted PAA chains were used in the analysis and it is clearly evident that the penetration depth of the graft co-polymer is highly dependent on the solvent used during the grafting process.[14] The disadvantage of using any technique that requires destructive sample preparation, such as resin embedding or cross-sectioning, is that it can cause the introduction of artefacts. Cross-sectioning of highly porous substrates, particularly when soft in nature, can also be a challenge because this can cause deformation of the sample.

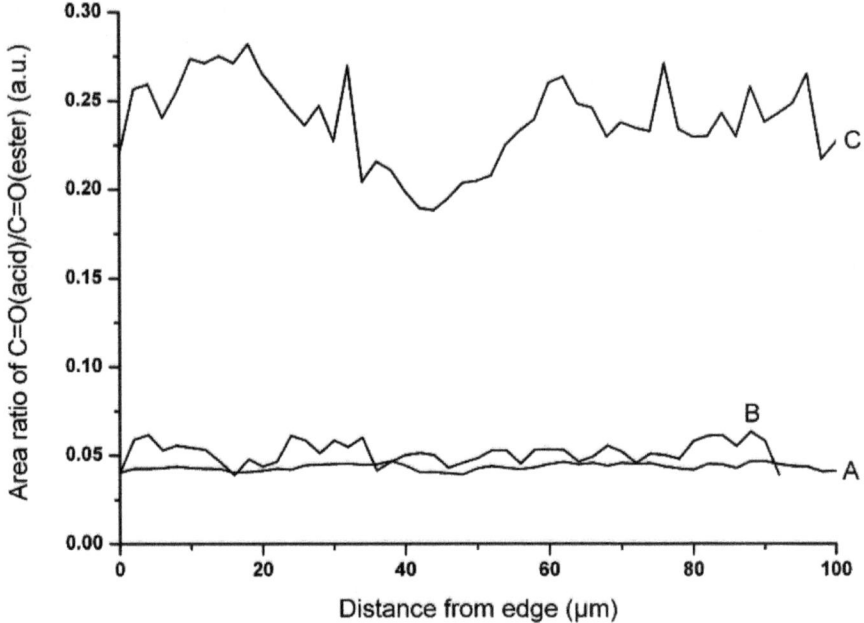

Figure 11.4   Effect of grafting conditions of AA on PCL using simultaneous grafting in different solvents. (A) Untreated PCL; (B) sample grafted in water; (C) sample grafted in a mixed solvent (water–methanol) solvent system. Data represent graft penetration depth as evaluated from the area ratios of the C=O stretch of the acid and the C=O stretch of the ester from Raman spectroscopy.
Data obtained from ref. 14.

A range of techniques which do not require destructive sample preparation have been employed to determine the depth of penetration of grafted chains including dye assays,[11] angle-dependent XPS[9,99] and MRI.[9] In angle-dependent XPS the stage position is altered allowing a more shallow depth of the sample to be analysed. While, in general, angle dependent XPS can be used to investigate the outermost sample surfaces, dye assays and MRI are very useful for examining the position of the grafted chains for the entire depth of the substrate up to hundreds of microns in thickness, albeit at a lower resolution. Both techniques require that the dye solution or water will be able to penetrate into areas where the graft co-polymer is present. MRI relies on the polymer substrate being hydrophobic, thus water will only penetrate the substrate when interacting with the hydrophillic graft co-polymer. This has been shown to provide a useful method of visualising the graft co-polymer distribution within a porous membrane.[9]

11.4 Concluding Remarks

Grafting of functional monomers on biomaterials offers a wealth of opportunities for tailoring the surface properties for specific medical applications. As illustrated in this chapter it is important to carefully select an appropriate grafting method and grafting conditions based on the polymer substrate and the required position of the grafted chains. It is equally important to thoroughly characterise the modified material if a sound analysis of a resultant biological response is to be made. It was noted in this chapter that for grafting in 3D scaffolds the methods that dominated were plasma polymerisation and simultaneous grafting. It is envisaged that the future will see CRP methods being used in this regard. Finally, it is hoped that through careful selection of grafting methods coupled with thorough analysis of grafted substrates that more optimised materials will make their way to the medical implant market.

Acknowledgements

Professor Sebastien Perrier (The University of Warwick, UK) and Doctor Adrienne Chandler-Temple are thanked for helpful suggestions. Doctor Imelda Keen (The University of Queensland, Australia) is acknowledged for providing the spectra for the FTIR figure and Doctors Adrienne Chandler-Temple and Marek Jasieniak for obtaining the ToF-SIMS images.

References

1. K. S. Siow, L. Britcher, S. Kumar and H. J. Griesser, *Plasma Process Polym.*, 2006, **3**, 392.
2. T. Desmet, R. Morent, N. De Geyter, C. Leys, E. Schacht and P. Dubruel, *Biomacromolecules*, 2009, **10**, 2351.
3. E. A. Dubiel, Y. Martin and P. Vermette, *Chem. Rev.*, 2011, **111**, 2900.

4. B. R. Coad, M. Jasieniak, S. S. Griesser and H. J. Griesser, *Surf. Coat. Tech.*, 2013, **233**, 169.
5. (a) P. C. Begovac, R. C. Thomson, J. L. Fisher, A. Hughson and A. Gällhagen, *Eur. J. Vasc. Endovasc.*, 2003, **25**, 432; (b) J. S. Lindholt, B. Gottschalksen and N. Johannesen, *et al.*, *Eur. J. Vasc. Endovasc.*, 2011, **41**, 668.
6. J. Rühe and W. Knoll, *J. Macromol. Sci-Pol. R*, 2002, **C42**, 91.
7. T. Jacobs, R. Morent, N. De Geyter, P. Dubruel and C. Leys, *Plasma Chem. Plasma P*, 2012, **32**, 1039.
8. A. Chandler-Temple, E. Wentrup-Byrne, A. K. Whittaker and L. Grøndahl, *J. Appl. Polym. Sci.*, 2010, **117**, 3331.
9. A. Chandler-Temple, E. Wentrup-Byrne, A. K. Whittaker, H. J. Griesser, M. Jasieniak and L. Grøndahl, *Langmuir*, 2010, **26**, 15409.
10. E. Wentrup-Byrne, L. Grøndahl and S. Suzuki, *Polym. Int.*, 2005, **54**, 1581.
11. L. Grøndahl, A. Chandler-Temple and M. Trau, *Biomacromolecules*, 2005, **6**, 2197.
12. (a) Y. M. Shin, K. S. Kim, Y. M. Lim, Y. C. Nho and H. Shin, *Biomacromolecules*, 2008, **9**, 1772; (b) Y. M. Shin, H. Shin and Y. M. Lim, *Macromol. Res.*, 2010, **18**, 472.
13. M. Grasselli, A. A. Navarro del Gañizo, S. A. Comperi, F. J. Wolman, E. E. Smolko and O. Cascone, *Radiat. Phys. Chem.*, 1999, **5**, 203.
14. J. Z. Luk, J. J. Cooper-White, L. Rintoul, E. Taran and L. Grøndahl, *J. Mater. Chem. B*, 2013, **1**, 4171.
15. S. Li, J. Wei, H. Yang, A. Wang, Y. Zhang and Y. Nie, *Polym-Plast Technol.*, 2013, **52**, 427.
16. Q. Yang, Z.-K. Xu, M.-X. Hu, J.-J. Li and J. Wu, *Langmuir*, 2005, **21**, 10717.
17. B. Gupta and N. Anjum, *Adv. Polym. Sci.*, 2003, **162**, 35.
18. J. Z. Luk, E. Rondeau, M. Trau, J. J. Cooper-White and L. Grøndahl, *Polymer*, 2011, **52**, 3251.
19. A. Södergård, *J. Bioact. Compat. Pol.*, 2004, **19**, 511.
20. L. Pruitt, H. Kausch, N. Anjum, Y. Chevolot, B. Gupta, D. Léonard, H. Mathieu, L. Pruitt, L. Ruiz-Taylor and M. Scholz, *The Effects of Radiation on the Structural and Mechanical Properties of Medical Polymers: Radiation Effects on Polymers for Biological Use.* Springer Berlin/Heidelberg, 2003, 162, 63.
21. J. H. O'Donnell, *Radiation chemistry of polymers. ACS Sym. Ser.*, 1989, **381**, 1.
22. D. He, H. Susanto and M. Ulbricht, *Prog. Polym. Sci.*, 2009, **34**, 62.
23. K. Kato, E. Uchida, E.-T. Kang, Y. Uyama and Y. Ikada, *Prog. Polym. Sci.*, 2003, **28**, 209.
24. J. S. Forsythe and D. J. T. Hill, *Prog. Polym. Sci.*, 2000, **25**, 101.
25. T. R. Dargaville, G. A. George, D. J. T. Hill and A. K. Whittaker, *Prog. Polym. Sci.*, 2003, **28**, 1355.
26. J. H. He, E. S. Read, S. P. Armes and D. J. Adams, *Macromolecules*, 2007, **40**, 4429.

27. L. Grøndahl, S. Suzuki and E. Wentrup-Byrne, *Chem. Commun.*, 2008, 3314.
28. E. Wentrup-Byrne, S. Suzuki, J. J. Suwanasilp and L. Grøndahl, *Biomed Mater.*, 2010, **5**, Article Number: 045010.
29. S. Suzuki, M. R. Whittaker, L. Grøndahl, E. Wentrup-Byrne and M. J. Monteiro, *Biomacromolecules*, 2006, **7**, 3178.
30. A. Chandler-Temple, P. Kingshott, E. Wentrup-Byrne, A. I. Cassady and L. Grøndahl, *J. Biomed. Mater. Res. A*, 2013, **101A**, 1047.
31. N. M. Hidzir, D. J. T. Hill, D. Martin and L. Grøndahl, *Polymer*, 2012, **53**, 6063.
32. A. V. Janorkar, S. E. Proulx, A. T. Metters and D. E. Hirt, *J. Polym. Sci. Pol. Chem.*, 2006, **44**, 6534.
33. Y. Wang, Y. Ke, L. Ren, G. Wu, X. Chen and Q. Zhao, *J. Biomed. Mater. Res.*, 2008, **88A**, 616.
34. L. Grøndahl, F. Cardona, K. Chiem and E. Wentrup-Byrne, *J. Appl. Polym. Sci.*, 2002, **86**, 2550.
35. M. Suzuki, A. Kishida, H. Iwata and Y. Ikada, *Macromolecules*, 1986, **19**, 1804.
36. C. Wang and J.-R. Chen, *Appl. Surf. Sci.*, 2007, **253**, 4599.
37. Y. M. Lee and J. K. Shim, *J. Appl. Polym. Sci.*, 1996, **61**, 1245.
38. R. Morent, N. De Geyter, T. Desmet, P. Dubruel and C. Leys, *Plasma Process Polym.*, 2011, **8**, 171.
39. C.-Y. Huang, W.-L. Lu and Y-C. Feng, *Surf. Coat. Tech.*, 2003, **167**, 1.
40. M. Kim and K. Saito, *Radiat. Phys. Chem.*, 2000, **57**, 167.
41. A. Södergård, *J. Polym. Sci. Pol. Chem.*, 1998, **36**, 1805.
42. J. Wang, X. Lui and H.-S. Choi, *J. Polym. Sci. Part B: Polym. Phys.*, 2008, **46**, 1594.
43. H.-S. Choi, Y.-S. Kim, Y. Zhang, S. Tang, S.-W. Myung and B.-C. Shin, *Surf. Coat. Tech.*, 2004, **182**, 55.
44. U. König, M. Nitschke, A. Menning, G. Eberth, M. Pliz, C. Arnold, F. Simon, G. Adam and C. Werner, *Colloid Surface B*, 2002, **24**, 63.
45. S. Turmanova, M. Minchev, K. Vassilev and G. Danev, *J. Polym. Res.*, 2008, **15**, 309.
46. Z. Y. Cheng and S. H. Teoh, *Biomaterials*, 2004, **25**, 1991.
47. I.-K. Kang, S.-H. Choi, D.-S. Shin and S. C. Yoon, *Int. J. Biol. Macromol.*, 2001, **28**, 205.
48. E. Bucio and G. Burillo, *Radiat. Phys. Chem.*, 2007, **76**, 1724.
49. Y. S. Ramirez-Fuentes, E. Bucio and G. Burillo, *Nucl. Instrum. Meth. B*, 2007, **265**, 183.
50. T. Desmet, T. Billiet, E. Berneel, R. Cornelissen, D. Schaubroeck, E. Schacht and P. Dubruel, *Macromol. Biosci.*, 2010, **10**, 1484.
51. O. N. Tretinnikov, K. Kato and Y. Ikada, *J. Biomed. Mater. Res.*, 1994, **28**, 1365.
52. P. M. Lopez-Perez, R. M. P. Da Silva, R. A. Sousa, I. Pashkuleva and R. L. Reis, *Acta Biomater.*, 2010, **6**, 3704.

53. S. Edmondson, V. L. Osborne and W. T. S. Huck, *Chem. Soc. Rev.*, 2004, **33**, 14.
54. R. Barbey, L. Lavanant, D. Paripovic, N. Schüwer, C. Sugnaux, S. Tugulu and H-A. Klok, *Chem. Rev.*, 2009, **109**, 5437.
55. H. Jiang and F-J. Xu, *Chem. Soc. Rev.*, 2013, **42**, 3394.
56. D. Roy, M. Semsarilar, J. T. Guthrie and S. Perrier, *Chem. Rev.*, 2009, **38**, 2046.
57. S. B. Lee, R. R. Koepsei, S. W. Morley, K. Matyjaszewski, Y. Sun and A. J. Russell, *Biomacromolecules*, 2004, **5**, 877.
58. C. J. Galvin and J. Genzer, *Prog. Polym. Sci.*, 2012, **37**, 871.
59. N. Singh, J. Wang, M. Ulbricht, S. R. Wickramasinghe and S. M. Husson, *J. Membrane Sci.*, 2008, **309**, 64.
60. S. Yuan, G. Xiong, X. Wang, S. Zhang and C. Choong, *J. Mater. Chem.*, 2012, **22**, 13039.
61. K. Kiani, D. T. J. Hill, F. Rasoul, M. Whittaker and L. Rintoul, *J. Polym. Sci. Pol. Chem.*, 2007, **45**, 1074.
62. J. Lindqvist and E. Malmström, *J. Appl. Polym. Sci.*, 2006, **100**, 4155.
63. S. Yuan, G. Xiong, A. Roguin and C. Choong, *Biointerphases*, 2012, **7**, 30.
64. F. J. Xu, Z. H. Wang and W. T. Yang, *Biomaterials*, 2010, **31**, 3139.
65. W. H. Yu, E. T. Kang and K. G. Neoh, *Langmuir*, 2005, **21**, 450.
66. (a) D. Roy, J. T. Guthrie and S. Perrier, *Aust. J. Chem.*, 2006, **59**, 737; (b) D. Roy, J. T. Guthrie and S. Perrier, *Soft. Matter*, 2008, **4**, 145; (c) D. Roy, J. S. Knapp, J. T. Guthrie and S. Perrier, *Biomacromolecules*, 2008, **9**, 91.
67. F. Yao, G.-D. Fu, J. Zhao and E.-T. Kang, *J. Membrane Sci.*, 2008, **319**, 149.
68. Y.-L. Liu, M.-T. Luo and J.-Y. Lai, *Macromol. Rapid Comm.*, 2007, **28**, 329.
69. L. Barner, N. Zwaneveld, S. Perera, Y. Pham and T. P. Davis, *J. Polym. Sci. Pol. Chem.*, 2002, **40**, 4180.
70. G. Chen, M. Zhou, Z. Zhang, G. Lv, S. Massey, W. Smith and M. Tatoulian, *Plasma. Process Polym.*, 2011, **8**, 701.
71. D. B. Haddow, R. F. France, R. D. Sort, J. W. Bradley and D. Barton, *Langmuir*, 2000, **16**, 5654.
72. A. M. Sandstrom, M. Jasieniak, H. J. Griesser, L. Grøndahl and J. J. Cooper-White, *Plasma. Process Polym.*, 2013, **10**, 19.
73. X. J. Dai, J. D. Plessis, I. L. Kyratzis, G. Maurdev, M. G. Huson and C. Coombs, *Plasma. Process Polym.*, 2009, **6**, 490.
74. N. Guerrouani, A. Baldo, A. Bouffin, C. Drakides, M.-F. Guimon and A. Mas, *J. Appl. Polym. Sci.*, 2007, **105**, 1978.
75. C. Vreuls, G. Zocchi, B. Thierry, G. Garitte, S. S. Griesser, C. Archambeau, C. Van de Weerdt, J. Martial and H. J. Griesser, *J. Mater. Chem.*, 2010, **20**, 8092.
76. X. Gong and H. J. Griesser, *Plasmas. Polym.*, 1997, **2**, 261.

77. Y. Cao, T. I. Croll, A. J. O'Connor, G. W. Stevens and J. J. Cooper-White, *J. Biomat. Sci-Polym. E*, 2006, **17**, 369–402.
78. Q. P. Pham, U. Sharma and A. G. Mikos, *Tissue Eng.*, 2006, **12**, 1197.
79. M. Martina and D. W. Hutmacher, *Polym. Int.*, 2007, **56**, 145.
80. J. J. A. Barry, M. M. C. G. Silva, K. M. Shakesheff, S. M. Howdle and M. R. Alexander, *Adv. Funct. Mater.*, 2005, **15**, 1134.
81. Y. M. Ju, K. Park, J. S. Son, J. J. Kim, J. W. Rhie and D. K. Han, *J. Biomed. Mater. Res. B*, 2008, **85B**, 252.
82. I. Djordjevic, L. G. Britcher and S. Kumar, *Appl. Surf. Sci.*, 2008, **254**, 1929.
83. Y. Ke, Y. J. Wang, L. Ren, Q. C. Zhao and W. Huang, *Acta. Biomater.*, 2010, **6**, 1329.
84. Y. Ke, Y. Wang and L. Ren, *J. Biomat. Sci-Polym. E*, 2010, **21**, 1589.
85. M. Domingos, F. Intranuovo, A. Gloria, R. Gristine, L. Ambrosio, P. J. Bartolo and P. Favia, *Acta. Biomater.*, 2013, **9**, 5997.
86. N. M. Hidzir, D. J. T. Hill, E. Taran, D. Martin and L. Grøndahl, *Polymer*, 2013, **54**, 6536.
87. T. R. Gengenbach, R. C. Chatelier and H. J. Griesser, *Surf Interface Anal.*, 1996, **24**, 271.
88. I. Keen, P. Broota, L. Rintoul, P. Fredericks, M. Trau and L. Grøndahl, *Biomacromolecules*, 2006, 7, 427.
89. I. Freeman, A. Kedem and S. Cohen, *Biomaterials*, 2008, **29**, 3260.
90. G. Beamson, D. Briggs, *The scienta ESCA300 database*, New York: Wiley, 1992.
91. D. S. Everhart and C. N. Reilley, *Anal. Chem.*, 1981, **53**, 665.
92. A. Chilkoti, B. D. Ratner and D. Briggs, *Chem. Mater.*, 1991, **3**, 51.
93. Y. Nakayama, T. Takahagi and F. Soeda, *J. Polym. Sci. Pol. Chem.*, 1988, **26**, 559.
94. C. Salvagnini, A. Roback, M. Momtaz, V. Pourcelle and J. Marchand-Brynaert, *J. Biomat. Sci-Polym. E*, 2007, **18**, 1491.
95. T. Ameringer, F. Ercole, K. M. Tsang, B. R. Coad, X. Hou, A. Rodda, N. R. Nisbet, H. Thissen, R. A. Evans, L. Meagher and J. Forsythe, *Biointerphases*, 2013, **8**, 16.
96. J. G. Terlingen, L. M. Breneisen, H. T. Super, A. P. Pijpers, A. S. Hoffman and J. Feijen, *J. Biomat. Sci-Polym. E*, 1993, **4**, 165.
97. P. Favia, M. Stendardo and R. d'Agostino, *Plasmas Polym*, 1996, **1**, 91.
98. M. Nitschke, G. Schmack, A. Janke, F. Simon, D. Pleul and C. Werner, *J. Biomed. Mater. Res.*, 2002, **59**, 632.
99. P. Kingshott, G. Andersson, S. L. McArthur and H. J. Griesser, *Curr. Opin. Chem. Biol.*, 2011, **15**, 667.
100. Z. Ma, Z. Mao and C. Gao, *Colloid Surface B*, 2007, **60**, 137.
101. C. Werner and H.-J. Jcobasch, *Int. J. Artif. Organs*, 1999, **22**, 160.
102. D. Y. Kwok and A. W. Neumann, *Adv. Colloid Interfac.*, 1999, **81**, 167.

103. G. Francius, J. Hemmerle, J. Ohayon, P. Schaaf, J. C. Voegel, C. Picart and B. Senger, *Microsc. Res. Techniq.*, 2006, **69**, 84.
104. I. Keen, L. Raggatt, S. Cool, V. Nurcombe, P. Fredericks, M. Trau and L. Grøndahl, *J. Biomat. Sci-Polym. E*, 2007, **18**, 1101.
105. T. Dargaville, D. Hill and S. Perera, *Aust. J. Chem.*, 2002, **55**, 439.
106. B. Gupta, F. N. Büchi, G. G. Scherer and A. Chapiro, *Polym. Advan. Technol.*, 1994, **5**, 493.
107. (a) I. Keen, L. Rintoul and P. M. Fredericks, *Macromol Symp*, 2002, **184**, 287; (b) I. Keen, L. Rintoul and P. M. Fredericks, *Appl. Spectrosc.*, 2001, **55**, 984.

Design of Mobile Supramolecular Biointerfaces for Regulation of Biological Responses

JI-HUN SEO[a,b] AND NOBUHIKO YUI[*a,b]

[a] JST-CREST, Chiyoda, Tokyo 102-0076, Japan; [b] Institute of Biomaterials and Bioengineering, Tokyo Medical and Dental University, Chiyoda, Tokyo 102-0062, Japan
*Email: yui.org@tmd.ac.jp

12.1 Introduction

As a result of the development of materials engineering, various types of artificial materials are now being used in biological environments for the purpose of repairing organs, regenerating tissues, or slowing down of diseases.[1] The artificial materials that directly contact with the living body to replace or augment the faulted places are called 'biomaterials', and understanding interfacial phenomenon on the biomaterials surfaces in the biological environments is crucial to the design of effective and safe biomaterials.

When artificial materials are placed in biological environments large amounts of protein molecules interact immediately with biomaterials surfaces.[2] Because around 80% of dried body fluid is composed of protein molecules, protein material surface interactions are responsible for the most

RSC Smart Materials No. 10
Biointerfaces: Where Material Meets Biology
Edited by Dietmar Hutmacher and Wojciech Chrzanowski
Published by the Royal Society of Chemistry, www.rsc.org

of the biological reactions on the biomaterials. After initial contact of the surface with proteins they adsorbed, and continuously migrating on the surfaces until they determine thermodynamic standpoint for the final conformation on the materials surfaces.[3] Because the conformational states on the materials surfaces is different from those in the body fluidics, hindered bioactive domain of protein molecules is easily exposed to the biological environments *via* strong interaction with the materials surfaces.[4] At this stage, the artificial materials are eventually recognized by a biological system, and various biological responses are triggered on their surfaces by the exposed bioactive domain. Therefore, the total amount of protein adsorption and the following degree of conformational change, *i.e.*, adsorption state of the surface proteins, has been considered as a criterion to estimate the degree of biological responses on the materials surfaces.

The critical factors affecting the amount of protein adsorption on the biomaterials are relatively well known. For example, it is well established that electrically neutral and chemical groups of non-hydrogen binding donor but acceptor are more favorable to reduce the amount of protein adsorption on the biomaterials surfaces.[5] Surface free energy, one of the traditional physicochemical factors explaining the polarity of the surfaces, is also known as a useful factors impacting on the amount of protein interactions on the biomaterials.[6] Biomaterials surface with high value of surface free energy, *i.e.*, polar surface, are known to reduce protein adsorption induced by hydrophobic interaction between protein and materials surfaces. Therefore, much effort has been made to provide a hydrophilic nature on the materials surfaces to reduce total amount of protein adsorption for the development of 'bio-inert' surfaces.[7–10]

The physicochemical factors affecting the adsorption state of the surface proteins is, however, relatively not yet clearly understood. Three dimensional conformations of protein molecules are normally stabilized by the hydrophobic collapse between hydrophobic cores in protein molecules.[11] This intermolecular hydrophobic interaction is greatly diminished when the protein molecules get a chance to develop hydrophobic interaction with extra molecules such as hydrophobic material surfaces. Therefore, it is believed that the protein molecules adsorbed onto more hydrophobic surface show higher degree of conformational changes.[12] Ishihara *et al.* reported the importance of hydrated states near polymer surfaces in discussion of the conformational change of the surface proteins.[13] A hydrated layer containing higher fraction of free water layer could induce reverse contact of the protein molecules with the hydrated polymer surfaces, thus it could preserve the three dimensional conformation of proteins on the surfaces. This result suggests that the hydrated states of the polymer surfaces is one of the considerable factors dominating the adsorption state of proteins. Recently, it was suggested that the surface elasticity could also make an effect on the adsorption state of protein molecules. The surface elasticity of elastomer surfaces has been known to affect the adhesion behavior of cells, such as adhering morphologies, proliferation rate, or stem cell lineages.[14,15]

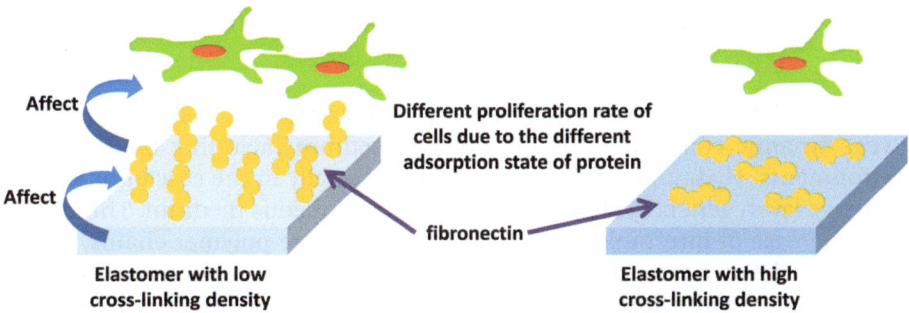

Figure 12.1 Schematic explanation of different adsorption states of protein molecules on elastomer surfaces with different surface stiffness.

However, the importance of the states of mediating ECM has always been out of discussion. We proposed the hypothesis that the adsorption state of proteins on elastomer surfaces is responsible for different proliferation rate of adhering fibroblasts on the elastomer surfaces having different stiffness (Figure 12.1). Indeed, the strong liner relationship between the hydrated viscoelasticity of the adsorbed protein layer estimated by quartz crystal microbalance-dissipation (QCM-D) measurement, and the number of adhering fibroblasts was observed.[16] This result suggests the possibility that the surface stiffness makes an effect on the adsorption states of the protein prior to induce the different adhesion responses of the cells. Recently, a similar result was also reported by Amsden's group.[17] They also proposed the similar hypothesis that the elastomer surfaces having different cross-linking density could induce different amount of protein adsorptions. These results indicate that the adsorption state of proteins is greatly affected by physicochemical factors of materials surfaces, and this could be a possible reason why cells showed different adhesion behavior on different materials surfaces.

Although these several possibilities have been suggested, still only a limited number of reports have been dealing with the critical factors relating to the adsorption states of the surface proteins. The dynamic nature of hydrated polymer surfaces in aqueous media is one of the important factors that are anticipated to impact on the states of the surface proteins. The degree of molecular interaction between the protein molecules and the materials surfaces, *e.g.*, hydrophobic interaction, is one of the criteria determining the adsorption states of the surface proteins as mentioned above. In this point of view, the dynamic nature of the hydrated polymer surfaces is anticipated to make an effect on the degree of protein–material interactions. Based on this hypothesis, several case studies have recently been conducted to estimate the effect of the dynamic properties of polymer surfaces on the adsorption states of protein molecules. In this chapter, the basic concept in designing the dynamic polymer surfaces and the adsorption behavior of protein molecules will be introduced.

12.2 Hydrated Viscoelastic Factor of Polymeric Biomaterials

Being different from metallic or ceramic materials, polymeric materials reveal dynamic conformations in aqueous medium including biological environments. Hydrated polymer materials, especially in the case of outermost surface layer, generally show a swelled state in aqueous medium. This causes the increase of intermolecular distance between the polymer chains, thus it induces the dynamic dimensional change resulting in highly viscoelastic properties.[18] The effect of these dynamic changes of polymer surface influences the protein/cell responses and it has been suggested as one of the important factors in designing biomaterials.[19,20] However, a systematic study to understand the effect of surface dynamics on the protein/cell responses has not yet conducted because of its difficulties in measuring dynamic properties of the outermost surfaces of polymer materials, and in modeling the polymer materials.

Recently, a quartz crystal microbalance-dissipation (QCM-D) technique has been shown to allow measuring changes in hydrated viscoelastic property at the molecular level.[21] The quartz crystal between a pair of electrodes in the sensor chip is oscillated as showing acoustic resonance frequency (f_A) when the AC voltage is given to the electrodes. When the substance is adsorbed on the sensor surface, a change in the frequency will be observed in proportion to the amount of adsorption. When the AC voltage is removed, the oscillation of the quartz crystal is exponentially decayed and provides the dissipation factor of oscillating energy (D). When a highly viscous substance is adsorbed on the surface, the energy dissipation of the sensor surface will be rapidly increased (high ΔD) due to the energy transfer to the substance (Figure 12.2a). Namely, a change in the f (Δf) is generally related with the mass changes on the outermost surfaces denoting the release or adsorption of substance, whereas a change in the D (ΔD) represents the change in viscoelastic properties of the surfaces. Therefore, the hydrated viscoelastic factor (Mf) on the outermost cast polymer surfaces on the sensor surface could be obtained in the normalized value using eqn (12.1).

> **TIP:** *It is recommended to cast around 10–20 mg of polymers on sensor surface, because the ΔD value will be saturated and show no significant change along different polymer surfaces. A polymer solution prepared by a vaporish solvent should be carefully cast to prevent segregation of solutes on the center of the sensor surface.*

$$\text{Mf} = \frac{D_{\text{polymer,wet}} - D_{\text{gold,wet}}}{f_{\text{gold,dry}} - f_{\text{polymer,dry}}} \qquad (12.1)$$

where Δf indicates the amount of coated polymer on the electrodes, whereas ΔD indicates the total amount of changes in dissipation energy on the

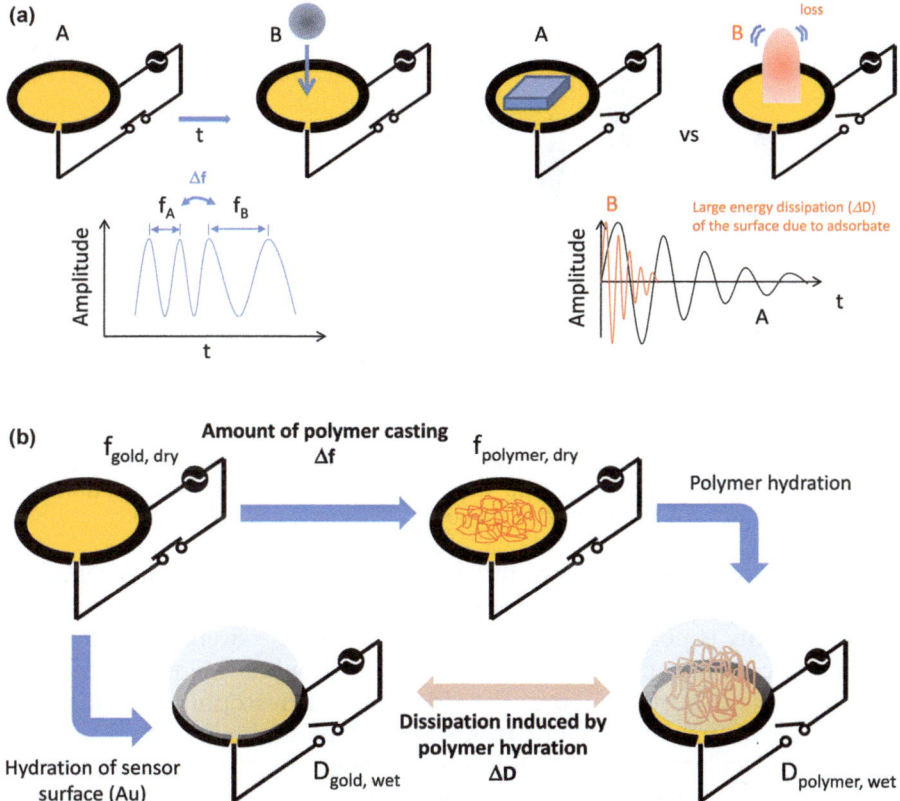

Figure 12.2 (a) Principle of measuring using QCM-D equipment; (b) schematic explanation of measuring the hydrated viscoelastic factor (Mf).

polymer surfaces during the hydration process (Figure 12.2b). The comparison results of Mf values and the traditional factor, *i.e.*, surface free energy, of five different conventional homopolymer surfaces, poly(methyl methacrylate) (pMMA), poly(*n*-butyl methacrylate) (pBMA), poly(isobutyl methacrylate) (piBMA), poly(methyl acrylate) (pMEA), and poly(2-hydroxy-ethyl methacrylate) (pHEMA), are shown in Figure 12.3. The surface free energy shown in this figure was calculated by the following Owens–Wendt equation based on the result of contact angle measurement by means of sessile drops of water and methylene iodide:[22]

$$1 + \cos\theta = 2\sqrt{\gamma_s^d}\left(\frac{\sqrt{\gamma_l^d}}{\sqrt{\gamma_{lv}}}\right) + 2\sqrt{\gamma_s^h}\left(\frac{\sqrt{\gamma_l^h}}{\sqrt{\gamma_{lv}}}\right) \tag{12.2}$$

where γ_s^d, γ_s^h and γ_{lv}, indicate the non-polar and polar components of the surface free energy on the polymer surface, and the surface free energy of liquid, respectively.

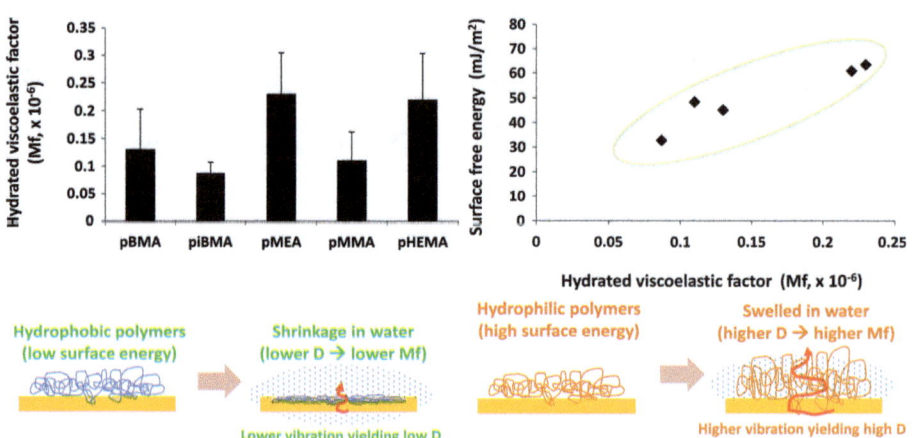

Figure 12.3   Plot of surface free energy and the Mf values of the conventional polymer surfaces. The reason for the straightforward relationships between two factors is explained schematically.

As shown in Figure 12.2, two independent factors show similar tendency (linear co-relationship) on the five different conventional polymer surfaces. This similarity is possibly due to the difference in the swelled states on the hydrophilic (high surface free energy) and hydrophobic (low surface free energy) polymer surfaces. Since the side chain of hydrophilic polymer surfaces (*e.g.*, hydroxyl group in pHEMA) might be highly hydrated, a more swollen molecular layer will be formed at the outermost surface. This might show a higher level of energy dissipation (larger ΔD) when the oscillation is turned off, and this will give a large value of Mf as depicted in Figure 12.3. In the case of hydrophobic polymer surface (*e.g.*, *n*-butyl group in pBMA), the side chain group might be rapidly shrunk when it is immersed in aqueous medium. The shrunk surface molecules might show a low level of energy dissipation (lower ΔD), and this will provide a low value of Mf. This linear correlation between the Mf value and the traditional surface free energy provides useful information. That is, the Mf value obtained by means of QCM-D measurement could be an alternative method to estimate the polarity of polymer surfaces. However, this also indicates that the effect of Mf value on the protein/cell responses on polymer biomaterials could not be independently discussed from those of surface free energy. Namely, designing a novel concept of polymer surfaces is essential to clearly understand the relationship between the Mf value and biological responses.

12.3   Designing Dynamic Surface for Regulation of Hydrated Viscoelastic Factor

In order to specifically understand the effect of the Mf value on protein/cell responses, polymer surfaces with a much broader range of Mf values but a narrow range of surface free energy must be considered as a model surface.

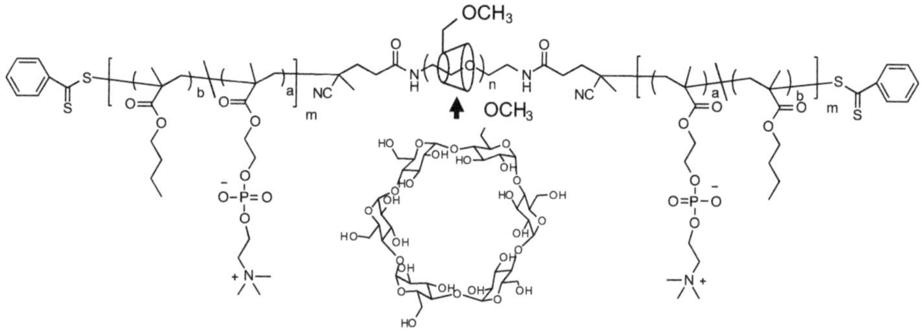

Figure 12.4 Molecular structure of a methoxy group-containing PRX block co-polymer.

From this point of view, a polyrotaxane (PRX) is considered as a proper materials to provide dynamic nature on the polymer surfaces. PRX is a molecular assembly consisting of a host molecule, *e.g.*, α-cyclodextrin (α-CD), threaded by a linear guest molecule, *e.g.*, polyethylene glycol (PEG).[23] α-CD and PEG form a non-covalent inclusion complex due to intermolecular hydrogen bonding of neighboring CD molecules as well as van der Waals interaction between the cavities of CD with PEG chains. For this reason, threaded α-CD molecules in PRX are expected to freely slide and rotate along a linear PEG backbone after diminishing the intermolecular hydrogen bonding of CD by chemical modification.[24] By taking advantage of this highly movable inclusion complex, it is expected that a polymer surface with a broad range of Mf values could be achieved. To this end, an ABA-type block co-polymer was recently reported to develop dynamic polymer surfaces. Figure 12.4 shows the representative molecular structure of the PRX block co-polymer. A random co-polymer composed with 2-methacryloyloxyethyl phosphorylcholine (MPC) and *n*-butyl methacrylate (*n*-BMA) (PMB) is introduced at both ends of the PRX segment for the stable forming PRX segment on the substrate surfaces. The PMB is a well-known surface anchoring group which can be immobilized on the various surfaces by hydrophobic interaction induced by *n*-butyl segment.[25] Furthermore, effective suppression of non-specific protein adsorption could be achieved due to the cell membrane mimic anti-fouling phosphorylcholine group.[26] Because α-CD molecules in PRX segment is covered with 18 hydroxyl groups, various functional groups could be introduced on the dynamic PRX segment by chemical modification such as carbonyldiimidazole-mediated chemistry.[27] The hydrophobic methoxy group is one of the most commonly used chemical groups to provide hydrophobic property onto PRX segment. The change in Mf value is observed when the hydrophobic methoxy group was introduced on PRX segment (mPRX) (Figure 12.5). Of particular interest is the independent manner of the Mf values on PRX block co-polymer surfaces with those of the surface free energy. While most of the prepared polymer surfaces show similar level of surface free energy, *i.e.*, within a narrow range of surface free

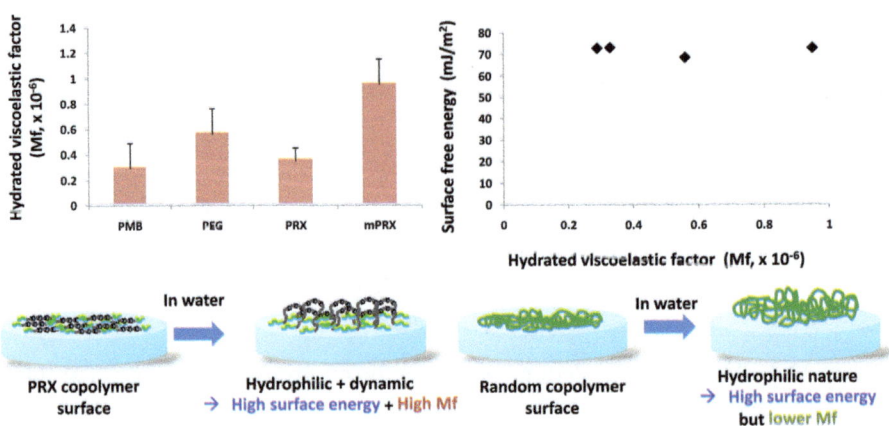

Figure 12.5 Plot of surface free energy and the Mf values of the mPRX with the control co-polymer surfaces. The reason for the independent manner of two factors is explained schematically.

energy near 70 mJ m^{-2}, a broad range of Mf values are observed on the PRX surfaces. Because hydrated viscoelastic properties of the conventional methacrylic polymer surface induce similar tendencies to that of the surface free energy, this broad range of Mf values is possibly due to the dynamic property of the PRX segment. In any case, this suggests that the PRX block co-polymer surfaces are useful as a platform to independently discuss the effect of Mf values on the various surface phenomena including protein/cell responses.

12.4 Biological Responses on the Dynamic Surfaces

12.4.1 The Effect of Hydrated Viscoelastic Properties on Fibrinogen Adsorption

The total amount of protein adsorption on biomaterials surfaces is generally determined by the surface free energy. However, the critical factor affecting the conformational change of adsorbed proteins is not yet broadly studied. One considerable factors that affects the adsorption state of a protein, *i.e.*, dynamic property of the polymer surface, was recently studied on a series of PRX surfaces using a fibrinogen as a model protein.[28] Fibrinogen is a glycoprotein abundantly present in the blood plasma (~ 3 mg mL^{-1}) and is responsible for blood coagulation and thrombus formation.[29] Fibrinogen molecules have two bioactive motifs responsible for cellular adhesion and aggregation, one is a dodecapeptide sequence (H12, HHLGGADQAGDV) existing close to the C-terminus of the γ-chain (400–411 peptide sequence), which is a binding site to GPIIb/IIIa on platelet to finally trigger the activation of platelet.[30] The other is two RGD sequences existing on the α-chain (RGDF and RGDS on 95–98 and 572–575 peptide sequences) which are binding sites to GPIIb/IIIa of platelets as well as the $\alpha_v\beta_3$ integrin on various cells.[31] These two active motifs are generally hidden in hydrophobic cores in fibrinogen molecules, and become exposed when the adsorbed fibrinogen undergoes significant conformational changes along the physicochemical properties of the polymer surfaces.[32] Therefore, different adsorption states on various materials surfaces might induce different degree of exposure level of each active motif. Figure 12.6a shows the total amount of fibrinogen adsorption on various polymer surfaces including dynamic PRX surfaces. Hydrophilic control samples (PMB, PEG and PRX surfaces) showed relatively low fibrinogen adsorption. In contrast, a significantly large amount of fibrinogen adsorption was observed on the mPRX surface. Because mPRX containing large amount of hydrophobic methoxy group (over 90% of hydroxyl groups was substituted) on PRX segment, hydrophobic methoxy

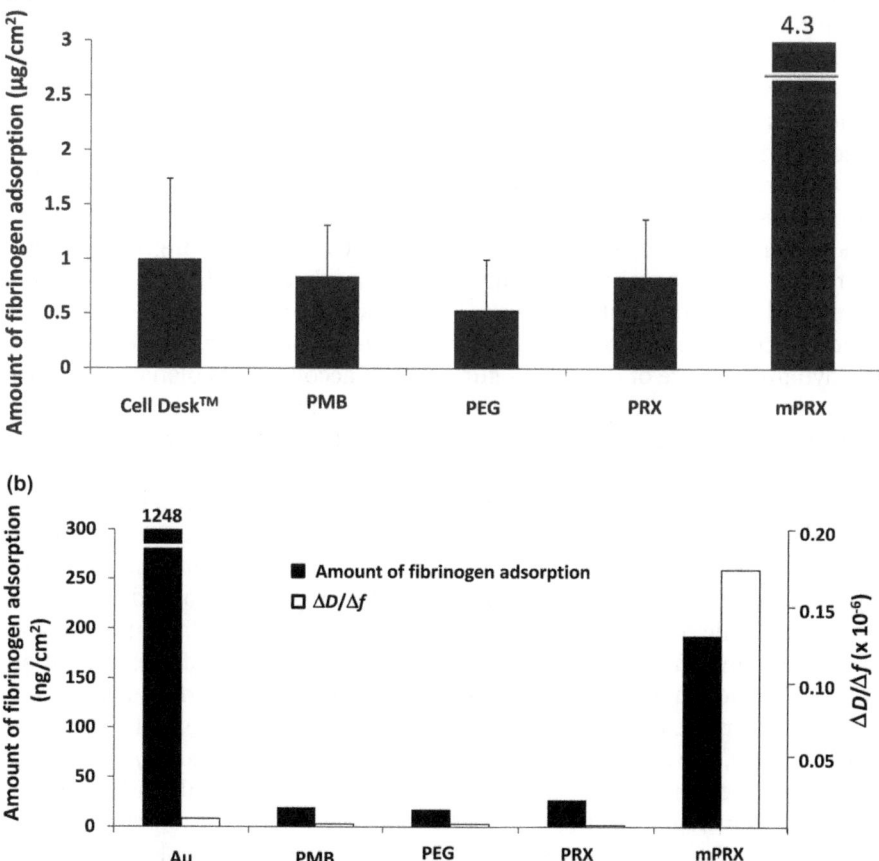

Figure 12.6 (a) The amount of fibrinogen adsorption on the mPRX and the control surfaces. (b) The amount of fibrinogen adsorption calculated by Δf of QCM-D measurement, and the values of $\Delta D/\Delta f$ of adsorbed fibrinogen. Reproduced/adapted from ref. 28.

groups can be sufficiently exposed to the outermost surface during the contact with plasma proteins, and leads to the protein accumulation.

The adsorption state of fibrinogen was analyzed using QCM-D measurement and the result is shown in Figure 12.6b. As mentioned above, a change in the *f* value during the adsorption test indicates the mass change occurred on the polymer surfaces. The amount of adsorbed proteins could be calculated by using the simplified Sauerbrey equation with overtone value 7 and $C = 17.7$ ng cm^{-2},[33] as follows:

$$\Delta m = -C \Delta f / n \qquad (12.3)$$

where m is adsorption mass, C is a proportional constant for the 5 MHz AT-cut crystal used, and n is overtone value. As clearly shown, a significantly large value of fibrinogen adsorption was observed during the contact with fibrinogen only on the hydrophobic Au and mPRX surfaces. This result is consistent with the result of micro-BCA™, which is a well-known protein quantification kit. Although both Au and mPRX surfaces induce a large amount of fibrinogen adsorption, the change in D values was quite different between two surfaces. In the case of a bare Au surface, adsorbed fibrinogen molecules showed very low value of $\Delta D / \Delta f$, which indicates that the state of adsorbed fibrinogen molecules is very rigid due to strong interaction with the Au surface. In contrast to this, adsorbed fibrinogen on the mPRX surface showed a significantly higher value of $\Delta D / \Delta f$, which indicates that the energy dissipation of adsorbed fibrinogen molecules are quite high. This phenomenon could be observed only when adsorbed substances are weakly interacted with materials surfaces.[34] The reason why this type of soft interaction occurred only on the mPRX surface is still unclear. However, taking the dynamic nature of the mPRX surface into account, it is plausible that the dynamic nature of the mPRX surface is responsible for the soft adsorption behavior of a fibrinogen molecule. Generally, the strong hydrophobic interaction of a protein molecule with the materials surfaces is known to induce significant conformational changes of adsorbed proteins.[35] Therefore, this weakly interacting protein layer on the dynamic PRX surface is anticipated to reveal moderate conformational change comparing with a rigid-uncoated surface. The conformational change of adsorbed fibrinogen on the dynamic polymer surfaces was quantified using ELISA test. Two types of primary antibodies which can specifically bind to α- and γ-chains of adsorbed fibrinogen were used to determine the degree of surface exposure of each chain. In the case of non-coated (Cell Desk™, a commercial cell culture dish) surface, the amount of binding antibody on both α- and γ-chains was higher than other hydrophilic polymer surfaces. This result indicates that adsorbed fibrinogen on the Cell Desk™ surface underwent significant conformational change. In contrast to this, other hydrophilic polymer surfaces showed significantly low level of antibody binding in both α- and γ-chains. This is presumably due to the low level of adsorbed proteins as confirmed in the micro-BCA™ and QCM-D measurement. However, fibrinogen molecules on the mPRX surface showed quite interesting tendency. In spite of the

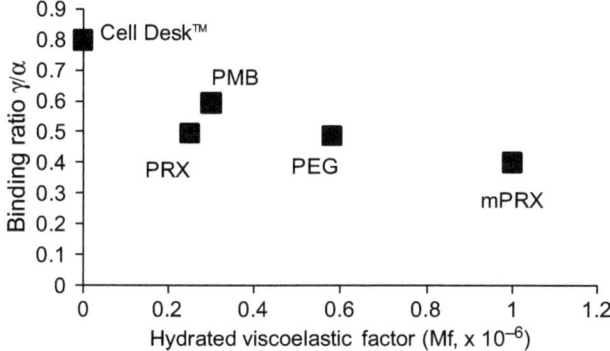

Figure 12.7 Relative binding ratio of second antibody to γ- and α-chain-binding primary antibodies plotted with the Mf values.
Reproduced/adapted from ref. 28.

significant amount of adsorbed fibrinogen, the amount of secondary antibody for the C-terminus γ-chain binding antibody on the mPRX surface was quite a low level comparing with non-coated Cell Desk™ surface. This result suggests the possibility that weak interaction of fibrinogen molecules with the dynamic surface induces moderate conformational change of adsorbed fibrinogen. To confirm this theory, the antibody binding ratio (C-terminus γ-chain binding to N-terminus α-chain binding) was plotted with the Mf values and the result is shown in Figure 12.7. The C-terminus γ-chain binding ratio comparing the N-terminus α-chain was decreased as the Mf value increased. Namely, it is considered that molecularly dynamic surfaces with high Mf (high viscoelastic surface) value can prevent conformational changes of adsorbed fibrinogen molecules.

The similar effect of the dynamic property of the PRX on the preservation of the protein nature was also found in solution states. We recently reported the preservation of the enzymatic activity of β-galactosidase before and after forming the complex with the PRX derivatives (Figure 12.8).[36] End group-cleavable PRX was aminated to induce electrostatic interaction with negatively charged β-galactosidase (pI = 4.6) at pH 7.4. PRX complex on β-galactosidase surface induce the steric hindrance to obstruct the enzymatic activity of β-galactosidase. When the complex was disassembled, the enzymatic activity of β-galactosidase was almost 100% recovered, and the circular dichroism/fluorescence study verified the preservation of the protein conformation after the significant contact (forming a complex) with the PRX molecules. This result is other evidence to support the fact that the dynamic PRX segment is useful to preserve the conformation of protein molecules during the interaction.

In any event, these series of studies suggested that the dynamic PRX interface could induce strong protein interactions, but prevent significant conformational changes of protein molecules, as being different with the traditional materials interfaces.

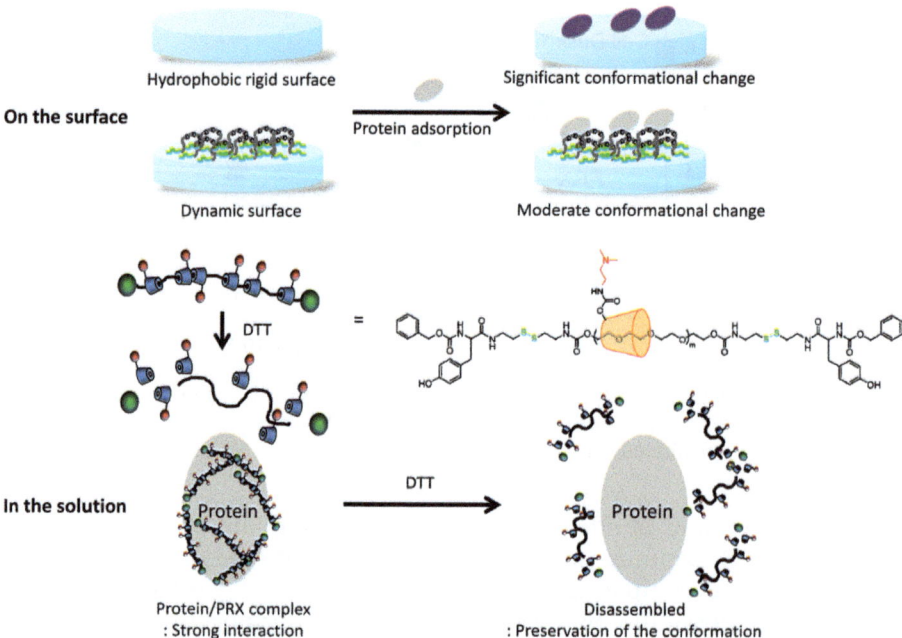

Figure 12.8   Schematic explanation of properties of the dynamic PRX interfaces contacting with protein molecules as preserving their conformation.

12.4.2   Platelet Responses on the Dynamic Surfaces

Previous studies on adsorbed fibrinogen on polymer surfaces demonstrated that C-terminus of γ-chain presenting dodecapeptide sequence in fibrinogen is essential to induce platelet–fibrinogen interaction.[37] Therefore, how well suppress the presentation of platelet GPIIb/IIIa binding site in the C-terminus γ-chain could be a key to developing 'bio-inert' blood-contacting materials. To estimate the effect of adsorbed plasma proteins on the cellular responses, a platelet adhesion test was carried out by using human platelet-rich-plasma (PRP). Figure 12.9 shows the result of platelet adhesion on the prepared polymer surfaces. Obviously, the Cell Desk™ surface induced a large amount of platelet adhesion on its surface, suggesting that the platelet adhesion is derived from significant conformational change of adsorbed proteins including fibrinogen as confirmed by ELISA. In contrast, the mPRX surface showed eliminated platelet adhesion even though it induced significant amount of fibrinogen adsorption. Figure 12.9 also shows the quantitative result of adhering platelets along with the relative amount of C-terminus γ-chain binding antibody. Obviously, an almost straightforward relationship between the number of adhering platelets and the relative amount of exposed γ-chain was observed. Namely, the platelet adhesion was increased with the extent of exposing C-terminus γ-chain of adsorbed fibrinogen on the surface, and this tendency is consistent with previous report.[38–40] It has been already reported that amount of platelet adhesion is

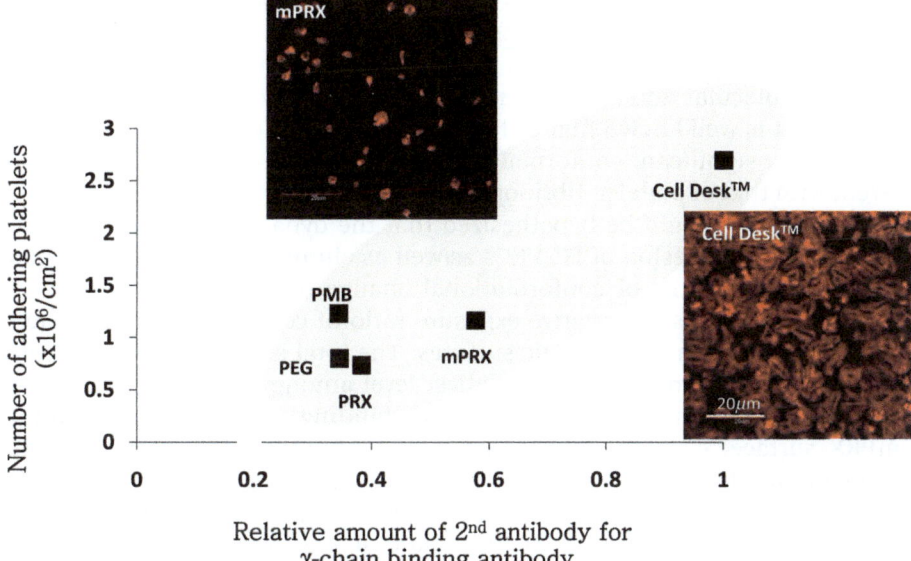

Figure 12.9 Morphology of adhering platelets and the number of adhering platelets plotted with γ-chain exposure of adsorbed fibrinogen. Reproduced/adapted from ref. 28.

normally increased in proportion to the amount of adsorbed fibrinogen. However, the mPRX surface showed different tendency with those of the other materials surfaces. Namely, it is considered that eliminating conformational change of adsorbed proteins including fibrinogen on the dynamic mPRX surface can prevent platelet adhesion, in spite of significant amount of adsorbed proteins.

12.4.3 Adsorption and Adhesion Behavior of Fibronectin and Human Umbilical Vein Endothelial Cell on the Dynamic Surfaces

Adhesion behavior of the different type of cell on the dynamic PRX surface was explored using human umbilical vein endothelial cells (HUVECs). Being different to the platelet responses promoted adhesion of endothelial cells on the biomaterial surfaces is preferable in designing blood-contacting biomaterials for the regeneration of the damaged tissues.[41] However, nonspecifically adsorbed surface protein, which is required for the adhesion of HUVECs, also triggers undesirable biological responses such platelet adhesion and activation. This paradoxical problem of the surface protein has been a fundamental barrier to developing ideal biomaterials.

Fibronectin, a major cell adhesive protein in serum, exposes the RGD and synergetic pentapeptide sequence (PHSRN) when it adsorbs on biomaterials.[42] These exposed peptide sequences specifically bind to the integrin β and α molecules, respectively, which exists on the most of the plasma membrane. Cell adhesion on biomaterial surfaces is induced by this type of molecular interaction between adsorbed protein and membrane proteins. It is well known that cell adhesion *via* fibronectin molecules is not initiated by significant conformational change of fibronectin, which is different with that of platelet–fibrinogen interaction.[43] By taking this advantage into account, it could be hypothesized that the dynamic surfaces could induce selective adhesion of HUVECs as well as eliminating platelet adhesion by regulating degree of conformational change of the surface proteins.[44] Figure 12.10 shows the relative exposure ratio of cell-binding motif of surface fibronectin on the dynamic surfaces. The total amount of exposed cell-binding motifs remained at the highest level among the polymer surfaces, which indicates that there are many cell-binding motifs on the dynamic mPRX surface, even though the protein layer underwent moderate conformational changes. The relative amount of exposed cell-binding motifs was then plotted with the amount of adhering HUVECs, as shown in Figure 12.11. Being similar to the result of platelet adhesion, a strong straightforward relationship is observed, which indicates that amount of HUVEC adhesion is dominated by the density of the cell-binding motifs on the polymer surfaces. An interesting point is that highly dynamic PRX surface revealed a large amount of HUVEC adhesion while the adhesion level of platelet was quite low. Taking the result of low platelet adhesion on the mPRX surface into account, this cellular response seems out of the ordinary. Usually, cell adhesive surfaces also induce strong platelet adhesion. Because

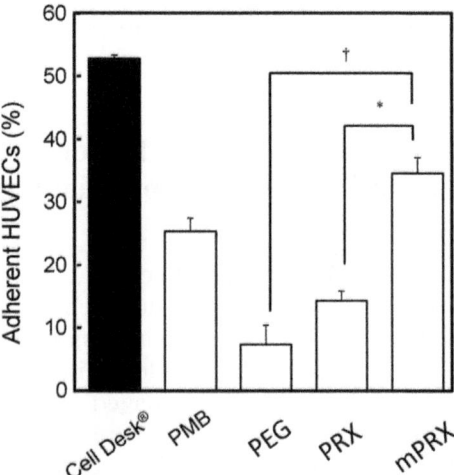

Figure 12.10   Relative surface density of the cell-binding motifs on fibronectin adsorbed surfaces. All the values are normalized to that of Cell Desk™. Reproduced/adapted from ref. 44 with permission from Taylor & Francis.

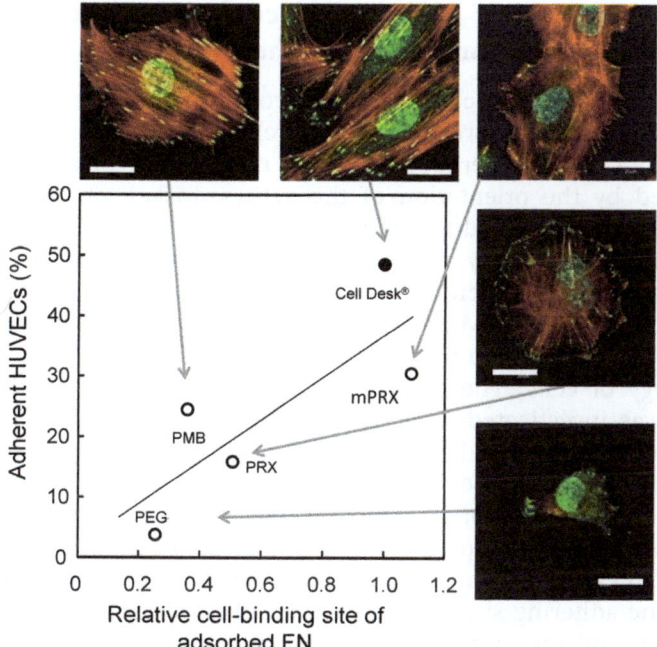

Figure 12.11 Plot of percent adhering number of HUVECs and relative surface density of the cell-binding motifs on fibronectin.
Reproduced/adapted from ref. 44 with permission from Taylor & Francis.

materials surfaces that easily induce fibrinogen adsorption also induce adsorption of other cell adhesive proteins such as fibronectin and vitronectin at the same time on its surfaces.[45] The reason why the selective cellular responses could be achieved on the dynamic PRX surface could be explained in the point of view of the different adsorption states of proteins. In the case of adsorbed fibrinogen on the mPRX surface, the specific binding site for platelet GPIIb/IIIa was eliminated in much lower level than Cell Desk™ even though the amount of adsorbed proteins was much higher level than Cell Desk™. However, the cell-binding motifs of the surface fibronectin on the mPRX revealed almost similar level to Cell Desk™. This suggests that the modulated conformational change of the cell adhesive proteins in serum is one of the reasons why the mPRX surface showed moderate fibroblast adhesion as eliminating platelet adhesion.

Of particular interest in Figure 12.11 is the different morphology of adhering HUVECs on various polymer surfaces having different value of Mf. Because the morphology of adhering cells is determined by the continuous communication between cells and the microenvironment of the adhering surfaces,[46] this result suggests the possibility that dynamic nature of the polymer surfaces make an effect on not only the adsorption states of proteins but also the adhering morphology of cells.

12.4.4 The Effect of Hydrated Viscoelastic Properties on the Adhering Morphology of the Fibroblasts

Adhesion morphology of cells plays a key role in the maintenance process of a living body such as directing tissue architecture, tissue repair, and the regulation of cell proliferation.[47–49] The morphology of adhering cells is determined by the orientation of the actin cytoskeleton linked with dynamically distributed endoplasmic proteins such as paxillin, vinculin, and tensin, *etc.*[50] The density and distribution of these phosphorylated endoplasmic proteins are determined by dynamic contact of integrin with extracellular matrix (ECM).[51] Therefore, the adsorption states of ECM on materials surfaces is also considered as a critical factor affecting adhesion morphology of cells. The morphology of adhering cells on the dynamic surfaces was investigated by using NIH3T3 fibroblasts. In order to clearly estimate the effect of the dynamic properties on the adhesion morphology of fibroblasts, several types of PRXs containing different number of α-CDs and different compositions of hydrophobic methoxy groups were synthesized. A fibroblast is a key cell to maintain the structure of connective tissues, and is known to easily change its morphology depending on the microenvironment of the adhering sites.[52] The number of adhering fibroblasts, the projected area, and the morphology of adhering fibroblasts were sequentially considered in terms of the density of the cell-binding motifs on ECM and the degree of hydrated viscoelastic factor (Mf), respectively.[53] Figure 12.12 summarizes the relationships between each adhesion behaviors of fibroblasts and the density of cell-binding motifs, or Mf. The projected cell area is linearly increased as the amount of the surface fibronectin. Because interaction between materials and cells are induced by the surface proteins, this result is well-consistent with the fact that protein–materials interaction is the primary concern to induce cellular responses on the materials surfaces. The number of adhering fibroblasts after longer-term adhesion test (1 day) also shows the similar relationship to the fibronectin density on the surface. As a result, it could be confirmed that the number of adhering fibroblasts is strongly dependent on the projected cell area at the early stage of cell adhesion as shown in Figure 12.12. However, no significant relationship between the adhesion morphology (aspect ratio) and the projected cell area. It is known that the morphology of adhering cells is not directly related to the quantity of surface proteins. Instead, physical or geometrical characteristics such as stiffness or surface morphology are known to contribute to the morphology of adhering cells.[54] The projected cell area and the aspect ratio of adhering fibroblasts were plotted with the Mf value as shown in Figure 12.12. Although, the projected cell area revealed a strong relationship with the density of fibronectin and the number of adhering cells, no significant relationship is found with the Mf value. This result indicates that the dynamic nature of the polymer surfaces is not directly related to the amount of protein adsorption or cell adhesion. Instead, a strong linear relationship between the aspect ratio and the Mf value was found. The exact

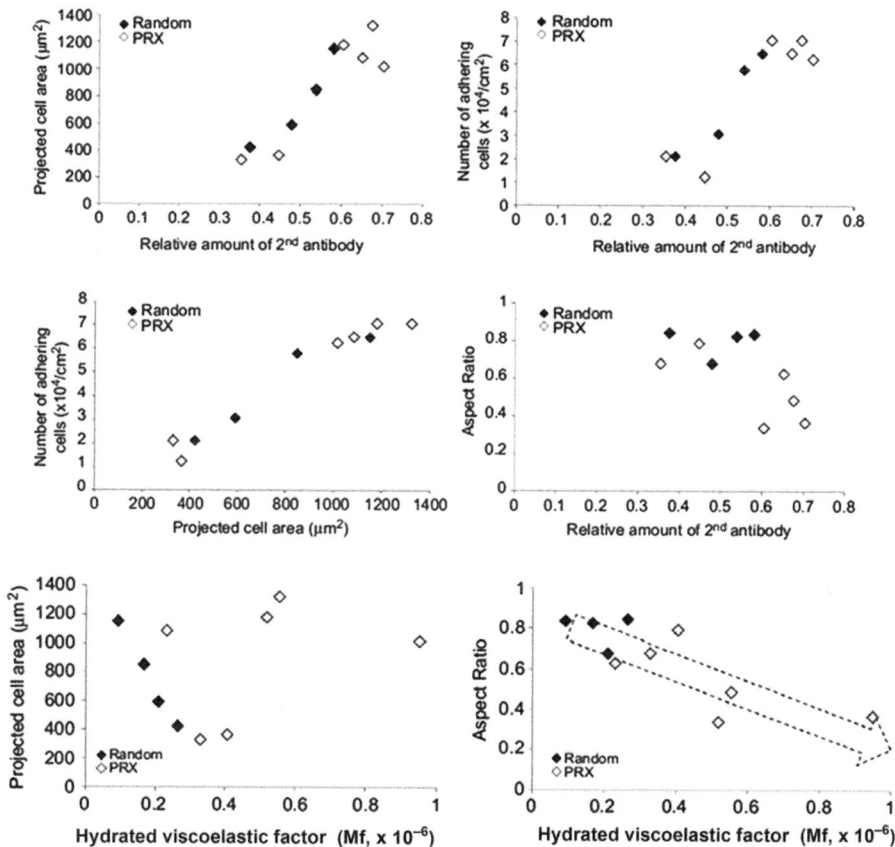

Figure 12.12 Plots of the number, projected area, and the aspect ratio of adhering fibroblasts with the relative amount of the cell-binding motifs on the polymer surfaces as well as with the Mf values. The aspect ratio was calculated by dividing the short axis of the best-fit ellipse of adhering fibroblast with the long axis. The best-fit ellipse was modeled by using Image-J.
Reproduced/adapted from ref. 53 with permission from Elsevier.

reason why this relationship was found is still unclear. However, the following hypothesis was postulated: 'Because cell surfaces are dynamically changing and reordering their surfaces, the materials surfaces can conduct dynamic ECM adsorption are considered to induce the specific morphology of adhering cells'.

Acknowledgements

The authors are grateful for the many collaborations, discussions and experimental support for the various research studies cited here. Special thanks are due to Prof. Tetsuji Yamaoka, Prof. Kazuhiko Ishihara, Dr Sachiro Kakinoki, Dr Yuuki Inoue and Dr Atsushi Tamura.

References

1. B. D. Ratner, A. S. Hoffman, F. J. Schoen and J. E. Lemons, *Biomaterials Science*, Elsevier Academic Press, London, UK, 2nd edn, 2004, pp. 2–5.
2. M. Amiji and K. Park, *Biomaterials*, 1992, **13**, 682.
3. P. Roach, D. Farrar and C. C. Perry, *J. Am. Chem. Soc.*, 2005, **127**, 8168.
4. M. Tirrell, E. Kokkoli and M. Biesalski, *Surf. Sci.*, 2002, **500**, 61.
5. E. Ostuni, R. G. Chapman, R. E. Holmlin, S. Takayama and G. M. Whitesides, *Langmuir*, 2001, **17**, 5605.
6. A. Baszkin and D. J. Lyman, *J. Biomed. Mater. Res., Part A*, 1980, **14**, 393.
7. F. Rupp, L. Scheideler, N. Olshanska, M. de Wild, M. Wieland and J. Geis-Gerstorfer, *J. Biomed. Mater. Res., Part A*, 2006, **76**, 323.
8. A. Roosjen, W. Norde, H. C. van der Mei and H. J. Busscher, *Prog. Colloid Polym. Sci.*, 2006, **132**, 138.
9. B. Jansen and G. Ellinghorst, *J. Biomed. Mater. Res.*, 1984, **18**, 655.
10. K. S. Kim, K. H. Lee, K. Cho and C. E. Park, *J. Membr. Sci.*, 2002, **199**, 135.
11. K. A. Walther, F. Grater, L. Dougan, C. L. Badilla, B. J. Berne and J. M. Fernandez, *Proc. Natl. Acad. Sci. U. S. A.*, 2007, **104**, 7916.
12. P. Roach, D. Farrar and C. C. Perry, *J. Am. Chem. Soc.*, 2005, **127**, 8168.
13. K. Ishihara, H. Nomura, T. Mihara, K. Kurita, Y. Iwasaki and N. Nakabayashi, *J. Biomed. Mater. Res.*, 1998, **39**, 323.
14. A. J. Engler, S. Sen, J. L. Sweeney and D. E. Discher, *Cell*, 2006, **126**, 677.
15. M. Prager-Khoutorsky, A. Lichtenstein, R. Krishnan, K. Rajendran, A. Mayo, Z. Kam, E. Geiger and A. D. Bershadsky, *Nat. Cell Biol.*, 2011, **13**, 1457.
16. J.-H. Seo, K. Sakai and N. Yui, *Acta Biomater.*, 2013, **9**, 5493.
17. M. C. Vyner, L. Liu, H. D. Sheardown and B. G. Amsden, *Biomaterials*, 2013, **34**, 9287.
18. D. M. Ebenstein and L. A. Pruitt, *J. Biomed. Mater. Res., Part A*, 2004, **69**, 222.
19. S. Tugulu, P. Silacci, N. Stergiopulos and H.-A. Klok, *Biomaterials*, 2007, **28**, 2536.
20. N. Singh, X. Cui, T. Boland and S. M. Husson, *Biomaterials*, 2007, **28**, 763.
21. F. Hook, J. Voros, M. Rodahl, R. Kurrat, P. Boni, J. J. Ramsden, M. Textor, N. D. Spencer, P. Tengvall, J. Gold and B. Kasemo, *Colloids Surf., B*, 2002, **24**, 155.
22. D. K. Owens and R. C. Wendt, *J. Appl. Polym. Sci.*, 1969, **13**, 1741.
23. A. Harada, J. Li and M. Kamachi, *Nature*, 1992, **356**, 325.
24. T. Ooya, M. Eguchi and N. Yui, *J. Am. Chem. Soc.*, 2003, **125**, 13016.
25. K. Ishihara, H. Oshida, Y. Endo, T. Ueda, A. Watanabe and N. Nakabayashi, *J. Biomed. Mater. Res.*, 1992, **26**, 1543.
26. K. Ishihara, E. Ishikawa, Y. Iwasaki and N. Nakabayashi, *J. Biomater. Sci., Polym. Ed.*, 1999, **10**, 1047.
27. Y. Yamada, T. Nomura, H. Harashima, A. Yamashita and N. Yui, *Biomaterials*, 2012, **33**, 3952.

28. J.-H. Seo, S. Kakinoki, Y. Inoue, T. Yamaoka, K. Ishihara and N. Yui, *Soft Matter*, 2012, **8**, 5477.
29. P. J. Sims, T. Wiedmer, C. T. Esmon, H. J. Weiss and S. J. Shattil, *J. Biol. Chem.*, 1989, **264**, 17049.
30. M. B. Gorbet and M. V. Sefton, *Biomaterials*, 2004, **25**, 5681.
31. J. M. Gfunkemeier, W. B. Tsai, C. D. McFarland and T. A. Horbet, *Biomaterials*, 2000, **21**, 2243.
32. D. R. Lu and K. Park, *J. Colloid Interface Sci.*, 1991, **144**, 271.
33. A. G. Hemmersam, K. Rechendorff, F. Besenbacher, B. Kasemo and D. S. Sutherland, *J. Phys. Chem. C*, 2008, **112**, 4180.
34. J. Malmstrom, H. Agheli, P. Kinshott and D. S. Sutherland, *Langmuir*, 2007, **23**, 9760.
35. H. Elwing, B. Nilsson, K.-E. Svensson, A. Askendahl, U. R. Nilsson and I. Lundstrom, *J. Colloid Interface Sci.*, 1988, **125**, 139.
36. A. Tamura, G. Ikeda, J.-H. Seo, K. Tsuchiya, H. Yajima, Y. Sasaki, K. Akiyoshi and N. Yui, *Sci. Rep.*, 2013, **3**, 2252.
37. M. M. Rooney, L. V. Parise and S. T. Lord, *J. Biol. Chem.*, 1996, **271**, 8553.
38. D. A. Cheresh, S. A. Berliner, V. Vicente and Z. M. Ruggeri, *Cell*, 1989, **58**, 945.
39. Y. Wu, F. I. Simonovsky, B. D. Ratner and T. A. Horbett, *J. Biomed. Mater. Res., Part A*, 2005, **74**, 722.
40. W.-B. Tsai, J. M. Grunkemeier and T. A. Horbett, *J. Biomed. Mater. Res., Part A*, 2003, **67**, 1255.
41. Y. Zhu, C. Gao, X. Liu, T. He and J. Shen, *Tissue Eng.*, 2004, **10**, 53.
42. W. J. Kao, D. Lee, J. C. Schense and J. A. Hubbell, *J. Biomed. Mater. Res.*, 2001, **55**, 79.
43. A. J. Garcia, M. D. Vega and D. Boettiger, *Mol. Biol. Cell*, 1999, **10**, 785.
44. S. Kakinoki, J.-H. Seo, Y. Inoue, K. Ishihara and N. Yui, *J. Biomater. Sci., Polym. Ed.*, 2013, **24**, 1320.
45. P. A. Underwood and F. A. Bennett, *J. Cell Sci.*, 1989, **93**, 641.
46. M. D. Mager, V. LaPointe and M. M. Stevens, *Nat. Chem.*, 2011, **3**, 582.
47. A. M. Pizzo, K. Kokini, L. C. Vaughn, B. Z. Waisner and S. L. Voytik-Harbin, *J. Appl. Physiol.*, 2005, **98**, 1909.
48. R. J. Petrie, A. D. Doyle and K. M. Yamada, *Nat. Rev. Mol. Cell Biol.*, 2009, **10**, 538.
49. H. Yamaguchi, J. Wyckoff and J. Condeelis, *Curr. Opin. Cell Biol.*, 2005, **17**, 559.
50. E. Zamir, B. Z. Katz, S. Aota, K. M. Yamada, B. Geiger and Z. Kam, *J. Cell Sci.*, 1999, **112**, 1655.
51. E. Zamir and B. Geiger, *J. Cell Sci.*, 2001, **114**, 3583.
52. J. A. Green and K. M. Yamada, *Adv. Drug Delivery Rev.*, 2007, **59**, 1293.
53. J.-H. Seo and N. Yui, *Biomaterials*, 2013, **34**, 55.
54. Y. S. Choi, L. G. Vincent, A. R. Lee, K. C. Kretchmer, S. Chirasatitsin, M. K. Dobke and A. J. Engler, *Biomaterials*, 2012, **33**, 6943.

Section E
Interfaces that Control Bacterial Responses

CHAPTER 13

Bacterial Adhesion and Interaction with Biomaterial Surfaces

LI-CHONG XU*[a] AND CHRISTOPHER A. SIEDLECKI[a,b]

[a] Department of Surgery, Biomedical Engineering Institute, The Pennsylvania State University, College of Medicine, Hershey, PA 17033, USA; [b] Department of Bioengineering, Biomedical Engineering Institute, The Pennsylvania State University, College of Medicine, Hershey, PA 17033, USA
*Email: lxx5@psu.edu

13.1 Introduction

Microbial infection is one of the main problems for the use of medical implants, and is induced by bacterial adhesion and biofilm development on surfaces of medical devices. Microbial infections have been observed on most medical devices including: intravascular catheters, prosthetic heart valves, left ventricular assist devices, urinary catheters, orthopedic implants, and intrauterine contraceptive devices.[1-4] In many cases, surgical removal and replacement of the implanted devices is often the only treatment for microbial infection, causing a significant increase in morbidity and mortality.[5,6] Nosocomial (hospital-acquired) infections have been the fourth leading cause of death in U.S. with 2 million cases annually leading to $5 billion medical cost per annum.[7]

RSC Smart Materials No. 10
Biointerfaces: Where Material Meets Biology
Edited by Dietmar Hutmacher and Wojciech Chrzanowski
© The Royal Society of Chemistry 2015
Published by the Royal Society of Chemistry, www.rsc.org

The difficulty in treatment of device-centered infections is primarily related to biofilm formation. Bacteria adhere, colonize on the surface, and produce a polysaccharide slime forming a biofilm, which can protect bacteria from host microbicidal systems and antimicrobial agents.[8–10] In addition, antibiotic resistant strains including *Staphylococcus aureus* and *Staphylococcus epidermidis* have proven difficult to be treated by standard antibiotic therapy after colonization on the prosthetic surface.[11,12] In the case of blood-contacting devices, a biofilm may continuously release bacteria into the blood stream and interact with platelets resulting in platelet activation[13] and thrombus formation.[14] Despite intense research efforts in the last decades, the problems associated with biofilms as well as the increasing levels of antibiotic resistance pose severe challenges in health care-associated infections.

A broad range of biomaterials, both natural and synthetic, encompassing metals, ceramics, synthetic polymers, biopolymers, self-assembled systems, nanoparticles, carbon nanotubes and quantum dots, are widely applied in medical devices.[15] The application of newly developed biomaterials must satisfy certain performance criteria such as biocompatibility and long-term stability. When a biomaterial is implanted into the body, biomaterials induce different biological responses.[16,17] Generally, proteins are adsorbed onto the biomaterial surfaces immediately upon implantation in a range of conformations from native to denatured, and mediate the subsequent biological reactions. A number of different cells such as monocytes, leukocytes, platelets, and bacteria arriving at the biomaterial surfaces will interact with the adsorbed protein layer as well as materials. Thus the initial protein adsorption onto a biomaterial surface plays a key role in how the body responds to an implanted biomaterial.[18] Understanding the host responses of implants to bacterial adhesion is important for improving the biocompatibility of biomaterials.

The most common bacteria leading to nosocomial infection include *Staphylococcus, Enterococci, Pseudomonas* and *Candida* species.[7,19] Particularly, staphylococci are the most commonly diagnosed microorganisms in microbial infection on blood-contacting devices.[7] *S. aureus* and *S. epidermidis* are two major opportunistic pathogens of this genus. *S. aureus*, a coagulase-positive staphylococci, is much more virulent and aggressive, and synthesizes an array of toxins and other virulence factors, causing a range of acute and pyogenic infections. *S. aureus* has received significant attention and been intensively studied.[20–23] *S. epidermidis*, a coagulase-negative staphylococci, ranks first among the causative agents of nosocomial infections and represents the most common source of infections on indwelling medical devices such as prosthetic heart valves and joint prostheses.[24,25] The main defined virulence factor associated with *S. epidermidis* is its ability to colonize on biomaterials and form biofilms which are recalcitrant to the deleterious action of antibiotics and impedes the host immune response.[26]

13.2 Bacterial Adhesion and Biofilm Formation on Biomaterial Surfaces

Microbial biofilms have been defined as complex consortia of adherent microorganisms encased by a polymeric matrix. The exact mechanism to form a functional, mature bacterial biofilm is still unknown; however, it is generally proposed as a four-stage model for the development of staphylo-coccal biofilms: adherence, accumulation, maturation, and dispersal,[27] as shown in Figure 13.1. The molecular mechanisms of staphylococcal biofilm formation have been reviewed.[26,28,29]

13.2.1 Adherence of Planktonic Bacterial Cells to Surfaces

The first stage of biofilm formation includes the initial instantaneous and reversible physical phase which is followed by a time-dependent and irreversible molecular and cellular phase.[30] The planktonic bacteria make contact with a surface due to the initial attraction of the cells to the biomaterial surface followed by adsorption and attachment. Bacterial movement to the material is driven by physical forces and chemical factors. The former is non-specific and acts between cell and surface, including hydrophobic, electrostatic, van der Waals forces, *etc.* The latter, referred to as chemo-attractants, is due to interaction with amino acids, sugars and oligopeptides.[18] In the molecular and cellular phase, molecular reactions between surface structure and material surface become predominant. A firmer adhesion of bacteria to a surface is formed by the bridging function of bacterial surface polymeric structures including capsules, fimbriae or pili, and slime.[31] The polysaccharide adhesins, clumping factors and proteins are involved in this process.[32,33]

When a biomaterial is implanted in contact with blood, plasma proteins can rapidly adsorb onto the material surface to form a 'conditioning film'.

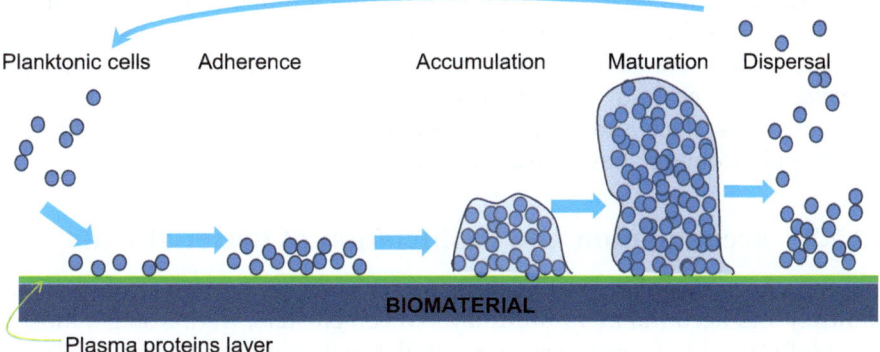

Figure 13.1 The schematic representation of bacterial adhesion and biofilm formation on biomaterial surface.
Adapted from Otto[75] with permission from Nature Publishing Group.

This adsorbed protein layer may minimize the effect of biomaterial surface properties on bacterial adhesion[34,35] so that interactions between the bacteria and the proteins mediate bacterial adhesion. The serum proteins generally suppress initial bacterial adhesion due to the lack of a specific interaction between albumin and bacteria[36,37] while adhesive plasma proteins such as fibrinogen (Fg) and fibronectin (Fn) were reported to contribute to increase of bacterial adhesion.[35,38,39] This increase in bacterial adhesion is believed to be due to specific ligand–receptor events between plasma proteins and bacterial cell surface proteins known as the microbial surface components recognizing adhesive matrix molecules (MSCRAMM).[32,40–42] Multiple MSCRAMM found on the *S. epidermidis* surface promote adhesion of bacteria including SdrG,[43,44] SdrF,[42,45] and extracellular matrix binding protein (Embp),[41,46] which were identified to bind Fg, collagen, and Fn, respectively. It has also been demonstrated that SdrG promotes platelet adhesion/activation and aggregation.[47]

Additional protein and polysaccharide components are also involved in primary attachment to biomaterial surface. One of the most prominent adhesion molecules is autolysin. Heilmann *et al.* first described autolysin, AtlE, a 148 kDa protein in *S. epidermidis* which mediated attachment to polystyrene surface.[48] It is highly homologous to the autolysin/adhesin AtlA, a 137 kDa protein found in *S. aureus*.[49] Later, the other novel autolysin/adhesin proteins were found in *S. epidermidis* (Aae)[50] and in *S. aureus* (Aaa),[51] respectively. Both were identified as a 35 kDa surface-associated protein that has bacteriolytic activity and binds vitronectin, Fg, and Fn. Not only do these proteins have specific adherence functions, they also function to release extracellular DNA (eDNA) which is found to be an important adherence/aggregation factor in both *S. epidermidis* and *S. aureus* biofilm formation.[52,53] Studies have shown that extracellular DNA is released by different components on cell depending on species, and functions as a structural support to maintain biofilm architecture.[54] Extracellular DNA has also been demonstrated to chelate cations and induce antibiotic resistance of biofilms.[55] In addition, non-proteinaceous adhesins, cell wall teichoic acids, were also found to function as a bridging molecule between *S. epidermidis* and Fn adsorbed polymer surfaces.[56] The features and functions of surface and surface-associated protein adhesins as well as of non-proteinaceous adhesins were recently reviewed.[57]

13.2.2 Accumulation and Proliferation of Bacterial Cells

Once the bacteria adhere to the surface of the foreign body, microorganisms multiply and accumulate in multi-layered cell clusters. In this stage, most of the biofilm bacteria produce extracellular polymeric substances (EPS) to form the biofilm matrix. A specific polysaccharide antigen, termed polysaccharide intercellular adhesion (PIA), has been identified to be a major functional component for intercellular adhesion and biofilm accumulation in staphylococci.[58–60] Its importance in the biofilm production of

S. epidermidis has been demonstrated in two animal models, a rat central venous catheter-associated infection model and a mouse foreign-body infection model.[61,62] Homologues of PIA are also found and identified in many other staphylococcal species including *S. aureus* suggesting that it might represent a more general pathogenicity principle in biofilm-related infections.[63]

PIA is a homoglycan composed of β-1,6-linked *N*-acetylglucosamine residues, synthesized by enzymes encoded in the icaADBC operon,[59] and confers hydrophobic character to the *Staphylococcus* bacterial capsule.[64] Secretion of PIA mediates initial adherence to material surfaces, interbacterial adhesion, and facilitates biofilm formation.[65] Furthermore, PIA also mediates the biocide resistance and protects against major components of human innate host defense.[66] Therefore, the synthesis of PIA has been a general mechanism employed in biofilm formation by a diverse group of Gram-positive and Gram-negative eubacteria. Its structure, function, and contribution to *S. epidermidis* biofilm formation and pathogenesis of biomaterial-associated infections have been reviewed.[67]

Although PIA is a highly important factor in staphylococcal biofilm formation, the biofilm-positive *S. epidermidis* strains have been isolated from clinical relevant infections that produced PIA-independent biofilms, in part mediated by the accumulation associated protein (Aap).[68] Such biofilm accumulation proteins include Embp, Aap and BhP in *S. epidermidis* and Bap in *S. aureus*.[69] Aap, a 140 kDa antigen, was shown to be a factor essential in *S. epidermidis* accumulation on polymer surfaces. An antiserum specific for Aap inhibited accumulation by up to 98% of the strains, suggesting the novel immunotherapeutic strategies for prevention of foreign body infection.[70] In fact, an immunotherapeutic approach is becoming the new tool to combat staphylococcal infection, although no anti-staphylococcal vaccines have yet successfully passed clinical trials.[71] So far, a limited number of targets have been examined for their immunotherapeutic potentials. One group of targeted compounds is MSCRAMM that play an important role in bacterial adhesion to the biomaterial surface. Other potential targets include factors that influence cell–cell interactions, such as polysaccharide intercellular adhesin and proteinaceous intercellular adhesin. The development and challenges of these vaccine targets for immunotherapeutic approaches have been excellently reviewed recently.[72,73]

13.2.3 Maturation and Dispersal of Biofilm

After attachment to a surface, bacteria undergo further adaptation and form a mature biofilm. The synthesis of EPS and antibiotic resistance are further developed in this process. Biofilm maturation comprises adhesive processes that link bacteria together during proliferation and disruptive processes that form channels in the biofilm structure.[74] The latter is necessary for nutrients to reach cells in deeper biofilm layer, and also causes the detachment of cell clusters from a biofilm. Biofilm maturation and detachment processes are of key importance for the formation of vital biofilms *in vivo* with the capacity of bacterial dissemination to secondary sites of infection.[75]

It is well accepted that the bacteria growing within a biofilm are unique from those growing in the planktonic phase. Microarray studies have demonstrated the differential gene expression profiling of staphylococci cultivated under biofilm and planktonic conditions.[76] As the cells grow in biofilm, they also have the spatial and temporal responses to the immediate environment, *e.g.*, availability of nutrients and oxygen, resulted in the shift of physiology toward to anaerobic or microaerobic metabolism and cell wall, DNA synthesis.[26] Rani *et al.* demonstrated that staphylococcal biofilms contain cells in at least four distinct states: growing aerobically, growing fermentatively, dormant, and dead. The variety of activity states represented in a biofilm may contribute to the special ecology and tolerance to antimicrobial agents of biofilms.[77]

Although the molecular mechanisms for the adhesion/accumulation of staphylococcal cells to material surfaces have been investigated in great detail, little is known for the forces and molecular determinants behind the structuring of biofilms and the detachment of cellular clusters from biofilms. Molecular proteins and polysaccharides are involved in biofilm maturation. PIA is one of the most important adhesive biofilm molecules.[61,78] The growing list of biofilm adhesive proteins comprises accumulation associated protein (Aap),[70] extracellular matrix binding protein (Embp),[46] protein A,[79] Fg-binding protein[80] and others. Other polymers are also implicated in staphylococcal biofilm formation. For example, teichoic acids contribute to *S. aureus* biofilm formation[81] and eDNA is an essential part of the network.[82] The *agr* (accessory gene regulator) quorum-sensing system has been recognized to govern biofilm maturation.[83,84]

As a biofilm matures, individual cells or intact section of biofilm can detach and metastasize to other sites. It is important for the dissemination of infection from surface of an indwelling medical device to other sites *via* the lymph and blood stream.[75] Because staphylococcal biofilms are embedded in an extracellular matrix composed of proteins, polysaccharides, eDNA and other host factors, degradation of these matrix components can disrupt established biofilms or prevent the biofilm formation. The production of compounds such as proteases,[85] DNases,[86,87] and surfactants[88] that can degrade and solubilize adhesive components in biofilm matrix is the primary mechanism of biofilm dispersal and detachment.[89] Recent work demonstrated that phenol-soluble modulins (PSMs), α-helical peptides with surfactant properties, secreted by *S. epidermidis*, promote biofilm structuring and detachment *in vitro* and dissemination from colonized catheters in a mouse model of device-related infection.[90] PSMs are regulated by the *agr* quorum-sensing system, and all PSMs have been identified in *S. aureus* and *S. epidermidis*, and known to participate in biofilm structuring and detachment processes.[91–93]

13.3 Factors Influencing Bacterial Adhesion and Interaction with Biomaterial Surfaces

Bacterial adhesion is the critical step in the pathogenesis of biomaterial associated infection. Factors that influence bacterial adhesion on a biomaterial surface include the nature of environment, type of microorganism, and properties of material, and each one of these factors is in turn affected by several other parameters.[18]

13.3.1 Biomaterial Surface Properties Affect Bacterial Adhesion

Biomaterial surface characteristics including surface chemistry, roughness, surface energy, and charge are known to be the major factors to determine initial bacterial adhesion and biofilm formation on implant surfaces.[94–96] Surface material composition determines physico- and chemical properties, and moderates the bacterial adhesion. On a series of gold film surfaces with self-assembled monolayers (SAMs) formed, adhesion of *S. aureus* was found to be lowest on the ethylene oxide-bearing surfaces, followed by the hydroxyl surfaces, while adhesion on the carboxylic- and methyl-terminated SAMs was much higher.[35] The Anderson group studied the adhesion of *S. epidermidis* on polymeric biomaterials with different charged surfaces and hydrophobicity, and found surface chemistry significantly impact the initial adhesion and aggregation of bacteria on biomaterials.[97] Modification of polyurethane with polyethylene oxide significantly inhibited *S. epidermidis* biofilm formation over 48 h *in vitro*.[37]

Binding of bacteria to surfaces is influenced by the properties of all three phases involved (substrate, bacteria, and medium). From the thermodynamic theory for the adhesion of small particles from a suspension onto a solid substratum, it is predicated that bacterial cells preferentially attach to hydrophilic substrata of relatively high surface energy, when the surface energy of the bacteria is larger than that of the medium in which they are suspended. When the surface energy of the suspending medium is greater than that of the bacteria the opposite pattern of behavior prevails.[98] However, the application of thermodynamic theory has not been entirely successful in explaining or predicting all various attachment behaviors in bacterial system.[30] An adhesion study of *S. epidermidis* to plasma modified polyethylene terephthalate films demonstrated that the increase in the free energy of polyethylene surfaces by He and He/O$_2$ plasma treatments significantly reduced the adhesion of *S. epidermidis* under static condition.[99] Similar experiments were carried out on the model glass surfaces coated with SAMs. Bacterial adhesion was found to depend on the monolayer's terminal functionality. Adhesion was higher on the CH$_3$ followed by the positively charged NH$_2$, the non-charged NH$_2$ groups, the COOH, and minimal on the OH-terminated glass, in accordance with the predictions of the thermodynamic theory.[100,101] However, both studies showed that the increase in the shear rate restricted the predictability of the theory, suggesting the importance of nature of environment on bacterial adhesion.[99,100]

Surface roughness is a two-dimensional parameter of a material surface and is generally characterized as the value of arithmetical mean deviation of the height profile (R_a). It has been reported that increases in surface roughness promotes the bacterial adhesion.[102,103] The reason for this phenomenon may include that a rough surface has a greater surface area and provides more favorable site for colonization. However, the effect of surface roughness on adhesion is apparently related to the degree of roughness and topographic features of surface. Bollen *et al.* suggested a 'threshold R_a' that a reduction in surface roughness on titanium surface below $R_a = 0.2$ μm had no effect on microbial adhesion or colonization.[104,105] Tang *et al.* reported that surface roughness significantly affects the adhesion of *S. epidermidis* on silicon surfaces only when the root-mean-square roughness was higher than 200 nm.[106] The studies of influence of physical structuring of surface nanofeatures on bacterial adhesion further revealed the role of surface roughness in bacterial adhesion. Whitehead *et al.*[107] measured the bacterial adhesion on Si wafers coated with titanium having featured surface pit sizes of 0.2, 0.5, 1, 2 μm in diameter and different depths, resulting in R_a values ranging from 45 to 220 nm. Results confirmed that adhesion generally increased with surface roughness, however, surfaces with 500 nm diameter pits showed the lowest number of bacterial adhesion. This is probably because the 500 nm pits had a low contact area and reduced the opportunity for bacteria interaction. Reduction in bacterial adhesion by surface topography will be further discussed in the Section 13.4.

13.3.2   Effect of Plasma Proteins on Bacterial Adhesion on Biomaterial Surface

When a foreign material is implanted in the body, plasma proteins rapidly interact with the surface and form a protein layer. The nature of adsorbed proteins is affected by the physicochemical properties of surface such as surface chemistry and charge, roughness, and hydrophobicity, and those in turn moderate the initial bacterial adhesion.

Through specific interaction between the cell surface receptors and adsorbed proteins ligands, staphylococci have the ability to bind many proteins found in the blood including Fg,[35] Fn,[108] collagen, vitronectin,[109,110] and von Willebrand factor.[111] The specific adhesion to these ligands is mediated by the bacterial cell wall receptors known as MSCRAMM (see Section 13.2.1).

Fg is the most third abundant protein plasma in blood and plays a prominent role in development of surface-induced thrombosis. It serves as a ligand binding to the platelet integrin receptor $\alpha_{IIb}\beta_3$, leading to platelet immobilization, activation, and aggregation.[112,113] It is also found to promote the adherence of bacteria. For example, Tegoulia and Cooper found pre-adsorption of Fg on SAMs coated surface minimized the effect of the surface properties of the substrate and adherence of *S. aureus* was increased on all SAMs surfaces.[35] The increase in adhesion of *S. aureus* by Fg is due to the receptor–ligand interactions between protein and Fg-binding MSCRAMM clumping factors on cell surface.[32,33] A Fg-binding adhesin gene was found on *S. epidermidis* cell surfaces from central venous catheters-associated and orthopedic implant-associated infections, demonstrating that Fg also increased the adhesion of *S. epidermidis* to biomaterial surfaces.[114]

Fn is another main plasma protein responsible for forming a condition film on implanted biomaterials. It can bind a variety of extracellular molecules including fibrin, heparin, and collagen, and plays a key role in cell adhesion and proliferation.[115,116] However, the effect of Fn on adherence of bacteria is highly variable. On the one hand, numerous studies have found that Fn facilitates bacterial adhesion to biomaterials including *S. aureus* and *S. epidermidis*.[38,117–119] Similar to the mechanism of enhancement of bacterial adhesion by Fg, Fn-binding proteins were found on *S. aureus*[120] and *S. epidermidis*[46] cell surfaces, respectively, and contributed to initial bacterial adhesion. On the other hand, several studies showed that Fn had no effect or even inhibited the adhesion of *S. epidermidis* to protein-coated surfaces.[121–123] Inhibition of adhesion is likely due to the production of an exopolysaccharide by *S. epidermidis* that influences interactions with protein-coated surfaces[121] or the adhesion of *S. epidermidis* to Fn-coated surface is not a specific adhesion.[123] The conflicting results indicate the most basic mechanisms of binding between *S. epidermidis* and Fn-coated surface are unknown yet, and more work should be done to examine the role of Fn in adhesion of bacteria to surfaces, especially in light of interactions of protein–cell at molecular level on substrata with different physicochemical properties.

Conformational structure and orientation of plasma proteins may influence bacterial adhesion. Fn is a dimer of two similar polypeptides linked by disulfide bonds at the carboxyl terminus, and possesses several functional domains that binds to a variety of extracellular molecules such as heparin and collagen.[124] There are two physiologically relevant binding sites to adhere to bacterial cell surface, which are located at the N-terminus and the C-terminus of Fn. *S. epidermidis* was shown to exhibit a different adhesion response when Fn was oriented with the C-terminus *versus* the N-terminus bound to the surface.[125] Holmes *et al.* studied the bacterial interactions with intact or fragments of Fn by surface plasmon resonance technique and found *S. aureus* preferentially binds to the Fn N-terminal region whereas *S. epidermidis* binds to the C-terminal domain.[126] Vadillo-Rodriguez *et al.* found that Fn molecules adopt a more extended conformation on hydrophobic than hydrophilic Ti6A14V surfaces and the extended conformation of the proteins facilitates the exposure of specific sites for adhesion, resulting in higher bacterial attachment.[127] We used immuno-AFM (atomic force

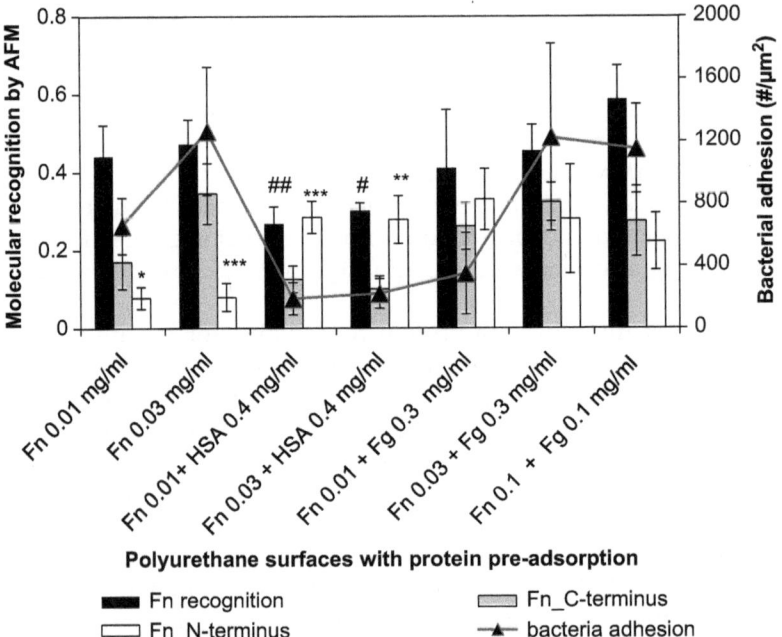

Figure 13.2   Correlation of molecular Fn recognition, and amino (N)- or carboxy (C) termini of Fn, and bacterial adhesion on polyurethane surfaces. Statistical analysis symbol # denotes the comparison of Fn recognition to surfaces adsorbed from pure Fn (0.01 mg mL^{-1}) solution, and * denotes comparing N- and C-termini on same sample.
Reproduced from Xu and Siedlecki[128] with permission from Scientific Research Publishing.

microscopy) to assess the Fn adsorption/orientation on polyurethane surfaces at the molecular scale. The orientation of Fn adsorbed on polyurethane surface was detected by the monoclonal antibodies (MAb) coupled AFM probes. Results show that Fn adsorbed on polyurethane surfaces from pure Fn solutions showed more C-terminus available than N-terminus in same sample, and corresponded to higher bacterial adhesion (Figure 13.2), indicating that *S. epidermidis* bacterial cell surface prefers to bind at the domain of C-terminus of Fn. Furthermore, the presence of albumin influences the orientation of Fn and more N-terminus is available although the amount of Fn adsorption is lower in this case.[128]

13.3.3   Interaction of Bacteria and Platelets on Biomaterial Surfaces

Pathogenic bacteria can occasionally enter the human circulatory system and lead to the interaction of bacterial pathogens with platelets, resulting in life-threatening diseases such as stroke. Interaction of bacteria and platelets can induce platelet activation, which is a necessary step in thrombus formation. On the other hand, the activated platelets can also promote bacterial adhesion on biomaterial surface. Wang *et al.* studied the adhesion of *S. epidermidis* on hydrophobic polystyrene and found that the presence of platelet increased adhesion more than on a protein adsorbed surface by at least one order of magnitude.[129]

Numerous studies showed that bacterial surface proteins and plasma proteins are involved in platelet–bacteria interactions.[13,130–132] The former includes the clumping factors, Fn-binding proteins of *S. aureus*,[133,134] and Fg-binding protein SdrG of *S. epidermidis*.[47] Cox *et al.* reviewed the bacteria–platelet interactions and summarized that interactions can be characterized by the binding of bacteria to platelets either directly through a bacterial surface protein or indirectly by a plasma bridging molecule that links bacteria and platelet surface receptors.[135,136] Platelets are activated by interaction with bacteria *via* either direct interaction or indirectly through bridging molecules or the secreted bacterial products. Bacteria may be enhanced to adhere to biomaterial surfaces.

Plasma proteins play an important role in interaction of bacteria–platelet on biomaterial surface. Figure 13.3 shows the images of bacteria–platelet interactions on the polyurethane surfaces which were incubated in phosphate buffered solution containing *S. epidermidis* (1×10^8 CFU mL^{-1}) and platelets (2.5×10^8 mL^{-1}) for 1 h with shaking at 250 rpm. Results show that bacteria adhered on surface and aggregated to form clusters. Although fewer platelets were observed compared to the numbers of bacteria, most of the platelets were found to be either entrapped in bacterial aggregates (green arrows) or co-adherent with bacteria (red arrows), suggesting the formation of platelet–bacteria aggregates. To study the effect of plasma proteins on bacteria–platelet interactions, human serum albumin (HSA), Fg and Fn were

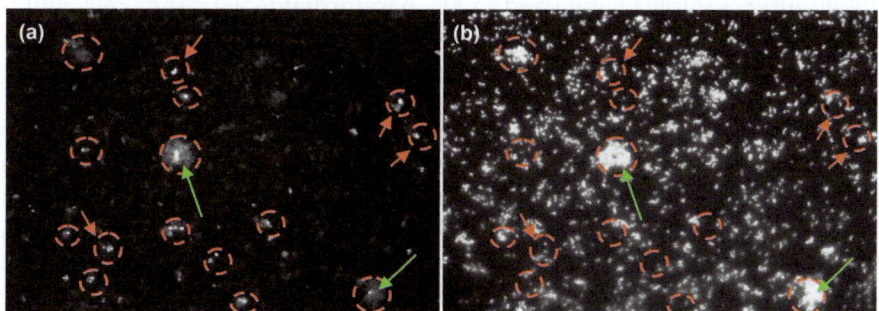

Figure 13.3   Fluorescent images of *S. epidermidis* RP62A bacteria and platelets interactions on polyurethane surface: (a) platelets and (b) bacteria. Red circles are drawn for comparison of areas between images. Image size: 226 μm × 169 μm.
Reproduced from Xu and Siedlecki[128] with permission from Scientific Research Publishing.

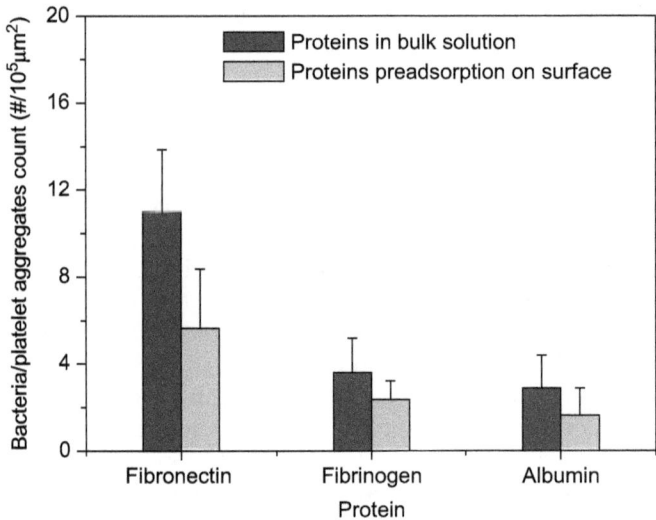

Figure 13.4   Counts of bacteria–platelet aggregates on polyurethane surfaces when proteins were present.
Reproduced from Xu and Siedlecki[128] with permission from Scientific Research Publishing.

added in bulk solution or pre-adsorbed on polyurethane surfaces. The aggregates were counted as illustrated in Figure 13.4. Results show that Fn leads to more aggregates than either fibrinogen or albumin in both cases, and more aggregates formed when the protein was present in bulk solution as opposed to just pre-adsorbed on the polymer surface.

Interactions of platelets and bacteria were further imaged by AFM. Similar as above, the polyurethane films were incubated in solution containing platelets and bacteria for 1 h. After rinsing with phosphate-buffered saline (PBS) buffer, samples were fixed in 1% paraformaldehyde and 2.5% glutaraldehyde for 1 h. The fixed samples were rinsed with pure water and dried in air, and imaged by AFM with tapping mode. The images were collected as the height, amplitude, and phase data at scan size of 50 μm×50 μm or 25 μm×25 μm and scan rate of ∼0.6 Hz. Figure 13.5 shows the morphology of platelets and bacteria adhered on polyurethane surfaces. The activation of platelets can be distinguished from the shape change of platelets adhered on surface, where the round shape is a non-activated platelet and the spread, pseudopodia shape indicates the platelet is activated. Both non-activated (round) and activated (spread) platelets were found on surface when only platelets were present in solution (Figure 13.5a); however, platelets were found to be activated and spread on surface when bacteria were present with platelets. Bacteria were seen to be adhered with activated platelets and formed aggregates (Figure 13.5b), suggesting that bacteria increased the platelets activation and formed bacteria–platelet aggregates. When plasma proteins were present in solution, HSA significantly decreased the amount of

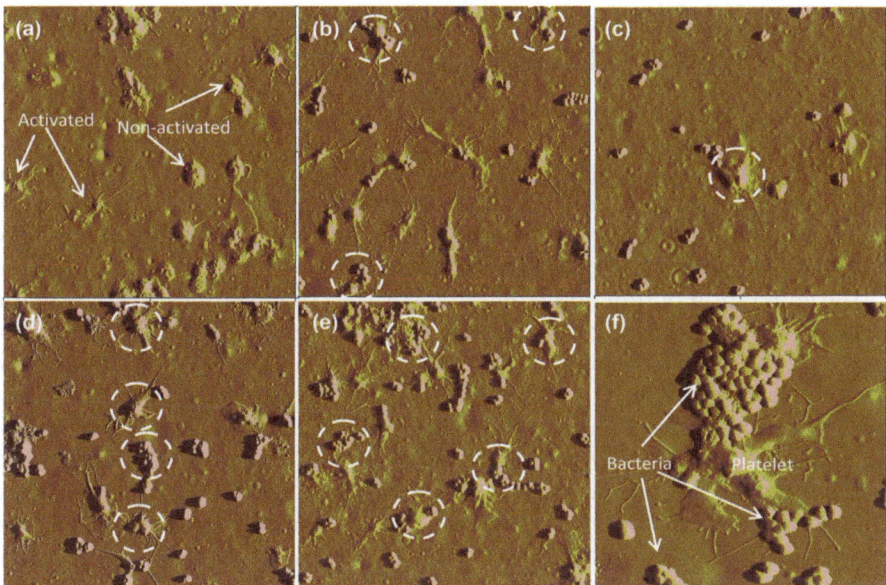

Figure 13.5 AFM images of platelet–bacteria interactions on polyurethane surfaces: (a) platelet only, (b) platelet and bacteria without plasma proteins, and platelets and bacteria in the presence of (c) human serum albumin, (d) fibrinogen and (e) fibronectin (a–e, image size: 50 μm×50 μm), (f) enlarged image showing platelets adherent or entrapped with bacteria (Image size: 20 μm×20 μm). The bacteria–platelet aggregates are indicated by circles.
Reproduced from Xu and Siedlecki[128] with permission from Scientific Research Publishing.

bacteria–platelet aggregates on polyurethane surface (Figure 13.5c) while Fg and Fn increased the formation of aggregates (Figure 13.5d and 13.5e). Furthermore, non-activated platelets were observed on surface in the case of HSA while all platelets were activated in the cases of Fg and Fn. The images with higher magnification show that the activated platelets are either adhered with bacteria or entrapped in bacterial cluster, forming platelet–bacteria aggregates (Figure 13.5f).

As described previously, bacteria–platelet interaction can be characterized by direct or indirect interactions between bacteria and platelet. *S. epidermidis* induced the platelet activation and aggregation of bacteria and platelets without addition of plasma proteins, suggesting a direct mechanism is involved. When plasma proteins (Fg and Fn) were added in solution, more bacteria–platelet aggregates were observed on biomaterial surface, suggesting both direct and indirect mechanisms involved. More aggregations of bacteria–platelet were measured on polyurethane surface when Fn or Fg are added in bulk solution compared to the case of proteins only pre-adsorbed on surface. This suggests that Fg or Fn may serve as linker in interaction of bacteria and platelets.

13.4 Inhibition of Bacterial Adhesion by Textured Biomaterial Surfaces

13.4.1 Background

Although addition of antibiotics is still the main strategy for prevention of biomaterial-associated infections, reduction of bacterial adhesion to an implant without the use of drugs has been increasingly attractive to biomaterial scientists. One promising approach to prevent microbial infection is the generation of functional material surfaces that significantly reduce the initial attachment of microorganisms and the number of persistent pathogens.[137] Numerous surface modification techniques, such as anti-adhesive and antibiotic coatings, surface grafting, chemical modification, and biological methods have been applied to biomaterials for inhibition of microbial adhesion.[138–140] Along these strategies, altering surface topography with micro- or nano-textured patterns is highly attractive since the textured surface applied to biomaterials does not require modification of surface chemistry and mechanical properties of bulk materials, but does significantly reduce bacterial adhesion and biofilm formation.

Clues for the prevention of biofilm formation by surface topography come from natural antifouling surfaces such as marine organisms (*e.g.*, shark, mussel) or the lotus leaf.[141–144] The endothelium of a healthy artery is another example of a natural antifouling system.[145] These natural surfaces are generally microtopographically structured and also may secret bioactive products that inhibit the adhesion of microorganism or cells. The combination of chemical and physical structures creates an ideal antifouling surface.[146] With interest in this area and the related research increased, biomimetic or bio-inspired materials have been developed over the last two decades.[147] For example, materials with topographical features mimicking the skin of sharks at certain length scales have shown increased resistance to marine biofouling.[146,148] Similar topography applied into a poly(dimethylsiloxane) (PDMS) elastomer was found to disrupt biofilm formation of *S. aureus*, providing the possibility for application in biomedical devices.[149] These bioinspired antifouling materials generally possess micron-size structured surfaces having dimensions ranging from 1 to 300 μm.[150] The creation of engineered nanoforce gradients on this specific patterned surface was believed to contribute to the low attachment of microorganisms in the marine environment.[151]

Topographical structure affects the surface contact area and surface physicochemical properties, and therefore influences the interactions of bacterial cells and material surface, and the subsequent biofilm development. We developed submicron-textured surfaces on polyurethane biomaterials originally as a means to reduce blood platelet adhesion,[152] but later results demonstrated that this structure also decreased the adhesion of *S. epidermidis* and *S. aureus* to polyurethane and inhibited biofilm formation under shear or static conditions.[153] Thus, the *in vitro* success of textured materials is of great interest for the potential clinical use in combating health-care infections and thrombosis.

13.4.2 Principles of the Inhibition of Bacterial Adhesion by Textured Surfaces

Attachment of bacteria onto a material's surface is a result of interactions between the cell surface and materials or molecules adsorbed on a surface. The availability of binding sites from cell surface or the material surface that bacteria can access to contact is important for initial attachment of bacteria. The hypothesis for the inhibition of bacterial adhesion by a textured surface is that surface textures with sub-bacterial dimensions can reduce the material surface area accessible to bacteria, resulting in decreased opportunity for interaction of bacteria with the material surface or adhesive plasma proteins (*e.g.*, Fg and Fn) adsorbed on the material. Thus, the flow of fluid over the material surface removes bacteria from a textured surface more efficiently than it would be from a smooth surface, and resists bacterial adhesion and biofilm formation (Figure 13.6).

In our study, a polyurethane block co-polymer is used as the model biomaterial for modification with a textured surface since it is a primary material used in a variety of blood-contacting medical devices due to its excellent properties. Using a soft lithography two-stage replication molding technique, polyurethane films were prepared with ordered arrays of pillars with submicron dimensions, smaller than the diameter of bacterial cells.[153] One representative AFM three-dimensional (3D) topography image of polyurethane film with pattern of 500/500 nm (diameter/spacing of pillars) is

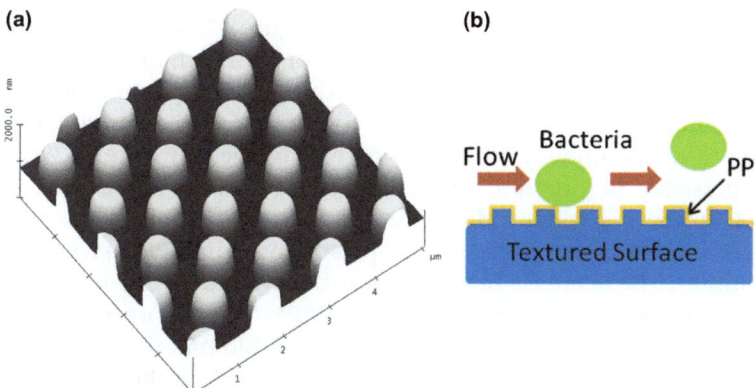

Figure 13.6 (a) AFM 3D topography image of textured polyurethane surfaces with pattern of 500/500 nm (round pillar diameter/separation) (scan size 5×5 μm^2), and (b) schematic diagram for bacterial adhesion inhibition by surface texturing, PP = plasma proteins.
Reproduced from Xu and Siedlecki[153] with permission from Elsevier.

shown in Figure 13.6. The total accessible top pillar surface area for bacteria interaction is only about 24.5% of a smooth surface. The diameter and separation of pillars are smaller than the dimension of bacterial cells, ensuring a reduction in available contact area and minimizing the opportunities for bacteria to become lodged in the spaces between pillars.

The surface texturing greatly increases the overall surface roughness (RMS, R_q) and surface hydrophobicity. For example, pattern 500/500 nm has the roughness of 248.5 ± 3.4 nm and water contact angle of $138.0 \pm 3.1°$, while the smooth polyurethane surface is with roughness of 15.7 ± 2.7 nm and water contact angle of $92.8 \pm 3.0°$. The increase in hydrophobicity of the textured surfaces is believed to be due to air captured in the spaces between pillars, which may prevent the intimate contact of the surface with bacteria resulting in low adhesion.

13.4.3 Bacterial Adhesion and Biofilm Formation on Textured Surfaces

Bacterial adhesion to smooth and textured polyurethane films was assessed across a low shear stress range (0–13.2 dyn cm^{-2}) in phosphate buffered saline (PBS) solutions with or without plasma proteins. A rotating disk system (RDS)[152] was used to produce well-defined dynamic flow conditions across the polyurethane surface. Figure 13.7 illustrates the adhesion of bacterial strains *S. epidermidis* RP62A and *S. aureus* Newman on smooth control and textured polyurethane surface with 500/500 pattern. A significant reduction in bacterial adhesion was observed at nearly all shear stresses for a textured surface for both strains in PBS solution for 1 h. The 500/500 pattern produced a reduction in adhesion of *S. epidermidis* RP62A ranging from 68.1% to 89.2% with an average reduction of $81.4 \pm 7.7\%$ over the shear stress range of 0–13.2 dyn cm^{-2} (Figure 13.7a). *S. aureus* Newman showed lower adhesion to smooth polyurethane films than did *S. epidermidis* RP62A, with reductions in adhesion on the 500/500 pattern polyurethane surface being 82.7% at the near static condition (~ 0.2 dyn cm^{-2}), and decreasing to approximately 48% with increasing shear (Figure 13.7c).

The presence of proteins influences the adhesion of bacteria on biomaterial surfaces. A greater decrease in adhesion was observed on all polyurethane surfaces when platelet poor plasma proteins (PPPs) were added in PBS solution. The decrease in adhesion is believed to be due to the presence of a large amount of serum albumin in PPPs. The adsorption of albumin suppressed the adhesion and interaction with biomaterial surface.[37] However, bacterial adhesion still significantly decreased on the textured surface for both strains, the reduction rates in adhesion range from 44.8% to 81.5% for *S. epidermidis* and from 57.1% to 84.1% for *S. aureus* over the whole shear stress range. It should be noted that adhesion of *S. aureus* was run in 25% human serum (clotting factors removed from plasma) instead of plasma due to the clumping by coagulase positive *S. aureus* in 25% normal plasma.

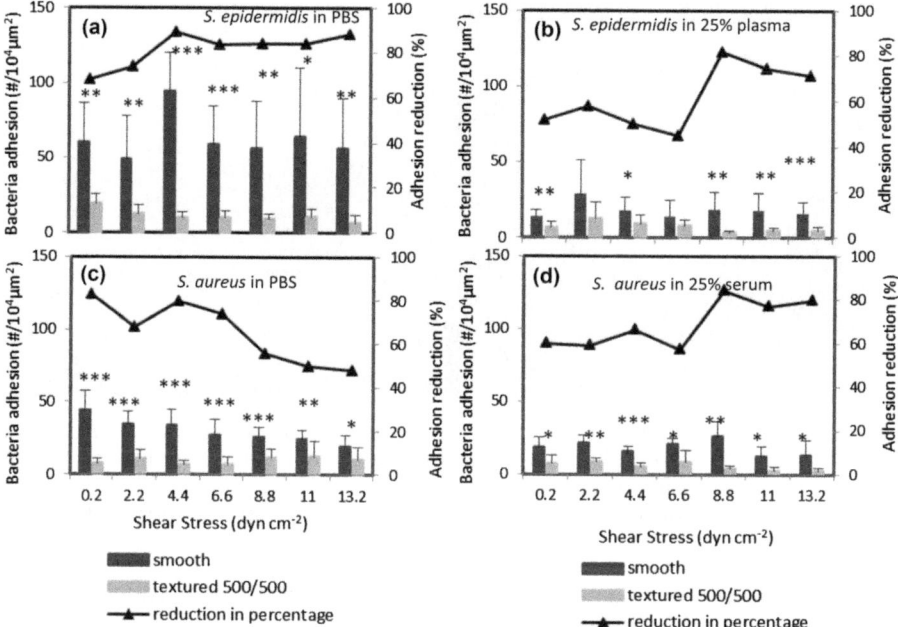

Figure 13.7   Bacterial adhesion on smooth and textured 500/500 polyurethane surfaces (a) *S. epidermidis* RP62A in PBS for 1 h, (b) *S. epidermidis* RP62A in 25% human plasma for 2 h, (c) *S. aureus* Newman in PBS for 1 h, and (d) *S. aureus* Newman in 25% human serum for 2 h. *$P < 0.05$, **$P < 0.01$, and ***$P < 0.001$.
Reproduced from Xu and Siedlecki[153] with permission from Elsevier.

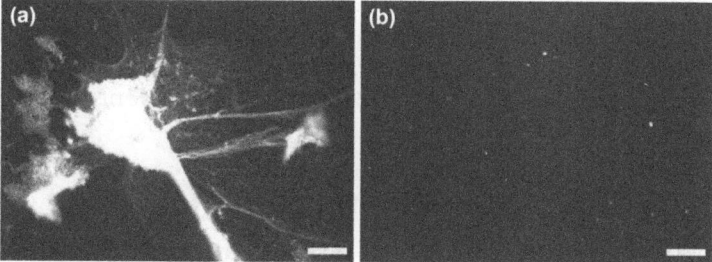

Figure 13.8   Fluorescence microscopy images of *S. aureus* Newman biofilms formed on (a) smooth and (b) textured 500/500 polyurethane surfaces after 5 days incubation in 25% serum and 25% brain heart infusion solution at a shear stress of 2.2 dyn cm^{-2} (scale bar: 100 μm).
Reproduced from Xu and Siedlecki[153] with permission from Elsevier.

Reduction in bacterial adhesion on a textured surface leads to inhibition of biofilm formation. Both *S. epidermidis* and *S. aureus* formed a biofilm and produced slime on smooth control sample surfaces, while only individual bacteria were seen on a textured surface. Figure 13.8 shows an example of a

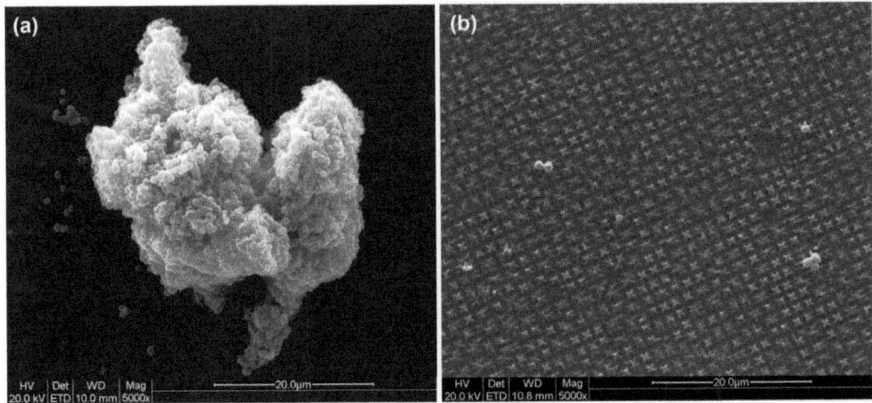

Figure 13.9   Representative scanning electron microscopy images of *S. epidermidis* RP62A on (a) smooth and (b) textured 400/400 nm polyurethane surfaces after 2 days incubation.
Reproduced from Xu and Siedlecki[153] with permission from Elsevier.

biofilm of *S. aureus* on a smooth surface, and individual bacteria/cluster on a 500/500 nm patterned surface.

A similar experiment was also carried out on another textured pattern surface with dimensions of 400/400 nm and the fraction of accessible surface area of 27.5%. Significant reduction in adhesion was obtained on the textured surface.[153] With an incubation of 2 days, biofilm and large clusters of bacteria aggregates were formed on a smooth surface and only individual bacteria were attached on a textured 400/400 nm surface (Figure 13.9). Therefore, it can be concluded that surface texture and the concomitant reduction in accessible surface area leads to inhibition of bacterial adhesion and subsequent biofilm development. Results suggest that surface texture may provide a novel approach to resist staphylococcal bacterial adhesion and biofilm formation on a material's surface and prevent the biomaterial-associated infections.

13.5   Bacterial Adhesion Assay Methods and Techniques Used in Studying Bacteria–Material Interactions

Various different methods and techniques have been used to evaluate bacteria–material interactions in either quantitative or qualitative manners, and can be described as two major categories. One is the macro-scale analysis of bacterial adhesion that utilizes fluid flowing against the adhered cells and counts the number or percentage of cells attached, aided with microscopy techniques and various labeling methods including radio-activity, fluorescence, and crystal violet staining.[154] The other one is the micro- or molecular scale analysis of interaction forces of single cell with material surface providing the basis for theoretical analysis of the receptor-ligand mechanics including AFM[155,156] and optical tweezers.[118] Other methods include total internal reflection microscopy,[157] quartz crystal microbalance,[158] *etc.* The reviews of most techniques and related theories can be found elsewhere;[30,159–161] however, further investigations in techniques for studying bacteria–material interactions are still needed to advance understanding of the mechanisms of bacterial adhesion and interaction on biomaterial surfaces. In this section, the general bacterial adhesion assay methods under static and dynamic conditions as well as the development and application of atomic force microscopy technique used in study of the bacterial–material interactions are briefly reviewed.

13.5.1 Static Fluid Bacterial Adhesion Assay

A static assay is a simple, inexpensive, and straightforward method to evaluate bacterial adhesion on different surfaces. Generally, the prepared surface is soaked or overlaid with a suspension of bacteria for a determined period of time. Afterwards, the non-adherent bacterial cells are removed by fluid exchange and the adhered cells on surface are fixed by appropriate chemicals (*e.g.*, glutaraldehyde) and counted. The adhered bacteria and biofilm can be examined by numerous methods,[162] such as microscopy methods (*e.g.*, epifluorescence microscopy, scanning electron microscopy, AFM, scanning confocal laser microscopy), viable bacterial counting methods (*e.g.*, CFU plate counting, radiolabeling), and other direct or indirect methods (*e.g.*, spectrophotometry, Coulter counter, biomarkers ATP). The apparatus for static adhesion assay may vary with the experimental conditions and size of samples. Linnes *et al.* reported a custom designed adhesion apparatus—a microwell plate—which allowed for carefully controlled experimental adhesion parameters and comparison of many different biomaterials.[123] With this apparatus, they found the adhesion of *S. epidermidis* to biomaterials was inhibited by Fn and albumin under static fluid.

13.5.2 Dynamic Fluid Bacterial Adhesion Assay

A hydrodynamic assay is generally required in examining bacteria–material interactions because bacterial adhesion to indwelling medical devices is associated with the flow of body fluids in most cases. Fluid shear rate influences the interaction of bacteria–material, biofilm formation and detachment of bacteria.[163,164] The most commonly used apparatus and techniques for dynamic fluid bacteria adhesion assays include the parallel-plate flow chamber, radial flow chamber, and RDS.

Flow chambers are commonly used to study microbial adhesion to surfaces under environmentally relevant hydrodynamic conditions. The parallel-plate flow chamber is the most common design since it is simple to construct and the flow can be easily analyzed. However, it should be noted in design that the channel length should be long enough to allow a greater length for the establishment of the desired hydrodynamic conditions.[165] The fluid dynamics in flow chambers have been characterized elsewhere.[160] Another configuration of chambers is that of the radial flow chamber. It consists of two flat disks separated by a thin gap, and the fluid flows in through the center of the fixed disc and flows out radially through the gap between the discs. Such a flow pattern provides a gradient of shear stress becoming progressively lower at the outer edge of the disc. The fluid dynamics can be found elsewhere.[166]

RDS can be used to produce well-defined dynamic flow conditions across the material surface. The details of the system have been described elsewhere.[34,152] When operated under laminar flow conditions, the wall shear stress at the surface of the RDS varies linearly with the radial distance from zero at the center point, while the flux of particles (bacteria) is uniform over the entire surface. By assessing adhesion of bacteria at different area, it is possible to determine the variation in adhesion with wall shear stress.

Figure 13.10 illustrates the schematic diagram of RDS, a technique we and others used extensively.[129,152,153,164,167] Biomaterial samples are mounted on to a stainless steel disk. For convenience, the stainless steel disk to which the biomaterial sample is mounted can be scribed with concentric circles at 1 mm radial increments and with radii at 60° increments. The disk is attached to the shaft of an RDS *via* a stainless steel threaded piece. The shear stress rates across the sample vary from 0 dyn cm^{-2} at the center to the maximum value at largest radial distance (Figure 13.10c). The shear stress range can be adjusted by the control of rotating speed. The fluid dynamics on surface of RDS was described elsewhere.[129]

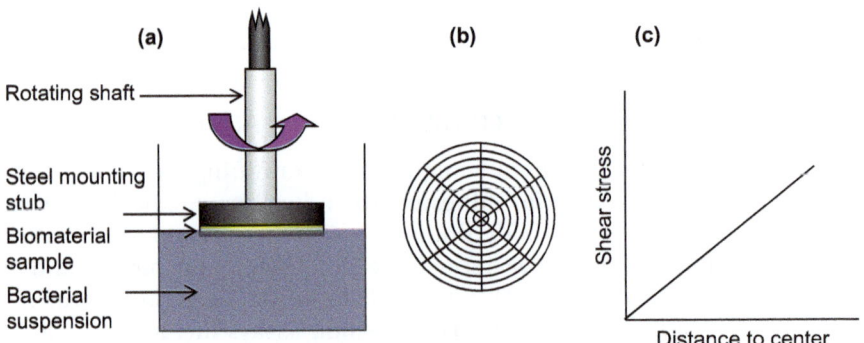

Figure 13.10   Schematic diagram of (a) rotating disk system, (b) the scribed markings metal mounting stub indicating areas for epifluorescence microscopy analysis of bacterial adhesion and (c) shear stresses linearly increased with the distance to center.
Adapted from Milner *et al.*[152] with permission from John Wiley and Sons.

13.5.3 Atomic Force Microscopy

Atomic force microscopy (AFM) has been a powerful tool in research of microbiology.[155,168] It is not only the fact that AFM is useful in imaging the morphology of individual microbial cells and biofilms on biomaterial surfaces both in dried or hydrated states, but also it is used increasingly for mapping interaction forces on microbial surfaces[169–171] and probing the local mechanical properties of bacterial cell surfaces or biofilms.[172] Thus, a major advantage of the AFM over other microscopy techniques is that AFM provides information about the nanoscale surface properties of (living) cells and about the localization and interactions of their individual constituents.[173]

The AFM uses a very sharp probe to measure and map the morphology of nearly any surface in ambient air or liquid environments. As the central component of AFM, the AFM probe is important in obtaining high resolution of images. The standard topographic imaging AFM probe consists of a sharp tip attached at the end of a flexible cantilever. The appropriate choice of cantilever and probe is dependent on the imaging mode and the characteristics of surface features of interest. The parameters for selection of probes include tip material, tip geometry, sharpness, length, stiffness, *etc.* The stiffness of the cantilever should be as low as possible when the AFM is operated in contact mode to avoid damage to the sample, *e.g.*, Si_3N_4 probes, whereas in dynamic imaging modes, the higher spring constant values of the cantilevers may help reduce noise and instabilities, *e.g.*, Si probes. The second important criterion for choosing a probe is minimization or avoidance of the most common artifacts in imaging generated from the tip geometry. For example, the ultra-sharp probe with high aspect ratio, such as a carbon nanotube probe, may be considered for imaging the surface with high aspect ratio features such as holes or pillars. The third important parameter of the AFM probe to be considered is the spring constant of the cantilever, particularly for making measurements of interaction forces. Typical spring constants for commercial AFM cantilevers range from 0.001 to 100 N m^{-1}, enabling force sensitivity down to the pico-newton level. Although manufacturers provide the nominal spring constants, actual spring constants should be calibrated by the appropriate method.[174]

AFM can directly visualize the 3D topography of biomaterial and bacterial cell surfaces at the nanometer scale, and even the structures underneath of the surfaces. This has been extensively studied with polymers, for example, polyurethane co-polymer is the primary material used in a variety of blood-contacting medical devices. It consists of hard and soft segments forming the microphase separation structure. AFM phase imaging technique can be used to visualize the microphase separation structure of polymers and *in situ* dynamic phase restructuring process during hydration.[175] A typical example for the probe and operation parameters for imaging microphase separation structure in polyurethane surface in solution is described in Appendix A.1. Since AFM can be operated in an aqueous environment, it is of great interest for visualizing the high resolution of living bacterial cell surface and characterizing their surface properties.[176] Méndez-Vilas *et al.* characterized two strains of *S. epidermidis* with different slime production by AFM and revealed the nanostructure of cell surfaces and the adhesion correlated to the slime-production zone on surfaces.[177] Dufrêne and colleagues reviewed the application of AFM techniques in exploring the microbial surfaces such as analysis of cell surface proteins and detection of individual cell surface receptors.[178,179] The development and application of AFM greatly contribute to understanding of the structure–function relationship of bacterial cell surfaces.

Apart from visualization of high resolution images of bacterial cell surface and biomaterial surface, an important feature of AFM is the ability to directly measure the interaction forces between bacterial cells and material surface at the nano-newton scale. Force measurements are made by recording the deflection of the cantilever measured as a function of the relative probe motion as it approaches and then retracts, or separates, away from the surface, called as force–distance curve.[174] A force–distance curve records the variation of interaction forces as the bacteria approaches and then separates from the sample surface. Physicochemical forces such as van der Walls and electrostatic forces can be measured from the approaching region of force curve, while the adhesion force or binding strength is measured from the deflection of cantilever during separation. Furthermore, the slope of repulsive section of the approaching curve also provides the information about nano-mechanical properties of bacterial cell surface.[180] In force measurement, bacteria are immobilized either on a substrate[169] or on an AFM probe[181] so that it is possible to measure the cell–material interactions and map the adhesion properties of cell surface by force volume image mode. The immobilization of bacterial cells to AFM probe also makes it possible to measure the interaction forces of cell–cell which is important for biofilm development.[182] The typical values of adhesion forces of cell–material or cell–cell vary in a wide range from pico-newtons to nano-newtons, even in tens of nano-newtons, depending on cell, material surface, environment, and operation conditions. The larger adhesion force, the greater is the tendency towards bacterial adhesion. Therefore, measurement of adhesion forces on single bacterium cell surface, cell–cell interactions, and cell–substratum can lead to better understanding of the mechanism of bacterial adhesion and biofilm formation on biomaterial surfaces.

AFM probing the interaction forces of microbial cells requires them to be physically or chemically attached to a substrate surface or onto AFM probe. Among them, the immobilization of cells on an AFM probe is a challenge for experiments. A protocol for the immobilization of bacterial cells on an AFM probe, which was reported in the literature,[156,181] is described in Appendix A.2. Kang and Elimelech[183] also reported a similar method to attach the live single cell onto polydopamine-coated tipless AFM probe.

Molecular recognition is an important event in immunological reactions and microbial infection. Using biologically modified AFM tips, it is possible to measure the minute forces between biological molecules and recognize the specific receptor–ligand interactions of molecules. Such probing of individual interactions can be used to map cell surface receptors, and to assay the functional states of receptors, binding kinetics and landscapes.[184] For studying bacterial adhesion, it can localize individual proteins and polysaccharides on the surface of living cell, and measure their specific binding strength. Clearly, knowledge of these molecular recognition forces contributes to a refinement in our understanding of the molecular basis of cell–cell and cell–material interactions, and provides insight into the molecular bases of cell adhesion and aggregation.[185,186]

Molecular recognition force measurement records the force curves between modified AFM tips and samples, and then assessing the unbinding force from the adhesion 'pull-off' force curve. A protocol for functionalization of AFM probe with antibodies[187,188] is described in Appendix A.3. An array of such force curves across over the surface form a 3D force map illustrating the distribution of any particular interaction of interest. The measured individual unbinding forces are typically in the range 50–400 pN, depending on experimental conditions (number of interacting molecules and the loading rate).[186] Touhami *et al.* reported the binding strength of 64 pN for interaction of immunoglobulin G and a surface protein adhesin (protein A) on living *S. aureus*.[189] Dupres *et al.* used a heparin-modified AFM tip to measure the adhesion forces between the heparin-binding hemagglutinin adhesin (HBHA) produced by mycobacterium tuberculosis and heparin, then mapped the distribution of single HBHA molecules on the surface of living mycobacteria and found that the adhesin is not homogeneously distributed over the mycobacterial surface, but concentrated into nanodomains that might promote adhesion to target cells.[190] A similar AFM technique was also used to map the interaction forces between Fn and Fn attachment proteins in mucobacteria by a Fn tip and found the adhesion forces were in the range of ~ 50 pN. The interaction force detection revealed that the proteins were widely exposed on the mycobacterial surface.[191] We used Fg and Fn modified tips to map the distributions of Fg- and Fn-receptors on *S. epidermidis* surface, respectively, and found more Fn-receptor recognition events than Fg-receptor recognition events on the cell surface, suggesting Fn is an important plasma protein in bacterial adhesion.[128] AFM can also be used to measure the distribution of cell wall polysaccharides. Using a lectin-modified tip, Francius *et al.* mapped and analyzed individual polysaccharide molecules on the surfaces of live probiotic bacterium *Lactobacillus rhamnosus* GG. This approach offers exciting prospects for analyzing the heterogeneity and diversity of macromolecules constituting cell membranes and cell walls and for understanding their roles in biofilm formation.[192]

13.6 Summary

A larger number of research studies has been carried out and significant achievements have been made in understanding the mechanisms of bacterial adhesion and controlling the bacterial adhesion to biomaterials. However, microbial infection still remains a serious problem for long term use of implanted medical devices. This is partially because bacterial adhesion is a very complicated process that is affected by many factors such as material surfaces and environments. The other reason is that the bacterial surface compositions/structures and their functional roles in bacterial adhesion and biofilm formation are still not clear. For example, future studies may need to address the functions of PIA, AaP, Bhp or Embp in *S. epidermidis* adhesion and biofilm formation, as well as their responses to antibiotics.[26]

Understanding the molecular interaction at the material–bacteria interface is fundamental to controlling bacterial adhesion to biomaterial surfaces. The major point in bacterial adhesion appears to be the role of proteins adsorbed on surfaces. The conformational structure and biological function of proteins adsorbed are affected by material surface properties (*e.g.*, surface chemistry, nanotopography) and they in turn influence bacterial adhesion. Future research in this field will need to better address the correlation of nanoscale surface properties and protein conformation, and correlate to the biological response bacterial adhesion. Certainly, this will rely on a better analysis and understanding of the influence of nanoscale surface properties on quantity and conformational structure of proteins adsorbed, and will require a development of new techniques or an improvement of the existing techniques to obtain this information. One representative example in this field is the increased use of novel, high-resolution AFM techniques for imaging surfaces and mapping the interactions of material and bacteria. Furthermore, combinations of AFM with other characterization methods such as confocal microscopy and Raman spectroscopy[168,193,194] greatly expanded the amount and types of data one can obtain simultaneously, and greatly help to know the molecular mechanisms of bacterial responses to adhering surfaces.

A   Appendix: Methods

A.1   Typical Parameters for AFM Imaging of Microphase Separation Structure in Polyurethane Biomaterial Surfaces in Solution

- *Probe*: Si_3N_4, spring constant $= 0.06$ to 0.58 N m^{-1}, probe curvature radius <10 nm
- *Mode*: tapping
- *Scan size*: 500 nm$\times$500 nm or 1 μm$\times$1 μm
- *Resolution*: 512$\times$512 or 256$\times$256
- *Scan rate*: 0.5 to 1 Hz
- *Moderate tapping force*: $r_{sp} \sim 0.75$ (set point amplitude/free amplitude of oscillation)
- *Data type*: height, phase, amplitude

A.2   Immobilization of Bacterial Cells on an AFM Probe

- *Probe*: commercially available AFM tips can be utilized, including silicon, silicon nitride and gold-coated tips. Curvature of the AFM tips is preferred to be around 50 nm or tipless probe may be used. Before use, AFM probes are cleaned in acetone for 15 min and then treated in glow discharge air plasma at 100 W power for 30 min to remove any organic contaminants.
- *Harvest of bacterial cells*: bacteria are cultured in media at 37 °C with shaking, and harvested in mid-exponential phase by centrifuging at 1360 g for 10 min. Wash the cell pellet with buffer for three times and transfer the pellet onto a clean glass slide, forming a thin film of bacteria.
- *Modification of an AFM probe*: place one drop of 1% polyethyleneimine (PEI, MW = 1200) onto the AFM probe and allow to adsorb for 2–3 h. The probe can also be treated with 0.1% (w/v) poly-L-lysine for 5 min. Remove the excess solution, and rinse the probe with water.
- *Immobilization of cells on the probe*: place the probe in the tip holder; position the probe over the bacterial film using an optical microscope. Set the scan size to 0 nm and scan rate to 0.1 Hz. Engage to allow the tip to approach the bacterial layer. Allow the tip to come into contact with the bacteria for 1–3 min. Withdraw the probe and rinse in buffer.
- *Check the probe*: the presence of bacteria on the AFM tip is checked by scanning electron microscopy or inspection of characteristic force curves for coated and uncoated tips.
- *Note*: bacterial cells may be treated in 2.5% glutaraldehyde for fixation before and after immobilization on the probe, but lose the activity of live cells.

A.3   Functionalization of an AFM Probe with Antibodies

- *AFM probe*: silicon nitride probes with nominal spring constant of 0.06 N m^{-1}.
- Probes are treated by glow discharge plasma at 100 W power for 30 min.
- Incubate the probe in a 1% (v/v) solution of aminopropyltriethoxysilane in ethanol for 1 h to provide reactive amine groups on the tip.
- Thoroughly rinse with Millipore water three times. Then the probes are soaked in 10% glutaraldehyde in aqueous solution for 1 h. Rinse the probes with Millipore water three times.
- Incubate the probes in antibodies solution ($\sim$25 µg mL^{-1}) for 1 h.
- The probes are rinsed with buffer after removal from protein solution and are stored in buffer at 4 °C until use within 2 days.
- Multiple probes should be prepared together to improve consistency between experiments.

Acknowledgements

The authors are grateful for financial support from the National Science Foundation (DMR-0804873) and Surgery Feasibility Grant of Penn State College of Medicine.

References

1. R. M. Donlan, *Emerging Infect. Dis.*, 2001, 7, 277–281.
2. I. Francolini and G. Donelli, *Fems Immunol. Med. Microbiol.*, 2010, **59**, 227–238.
3. L. Zhang, J. Gowardman and C. M. Rickard, *Int. J. Antimicrob. Agents*, 2011, **38**, 9–15.
4. C. von Eiff, B. Jansen, W. Kohnen and K. Becker, *Drugs*, 2005, **65**, 179–214.
5. R. Gutierrez-Gonzalez, G. R. Boto and A. Perez-Zamarron, *Eur. J. Clin. Microbiol. Infect. Dis.*, 2012, **31**, 889–897.
6. B. Johnson, I. Starks, G. Bancroft and P. J. Roberts, *J. Trauma Acute Care Surg.*, 2012, **72**, 1375–1379.
7. R. P. Wenzel, *Clin. Infect. Dis.*, 2007, **45**, 85–88.
8. R. M. Donlan, *ASAIO J.*, 2000, **46**, S47–S52.
9. P. S. Stewart, *Int. J. Med. Microbiol.*, 2002, **292**, 107–113.
10. N. Høiby, T. Bjarnsholt, M. Givskov, S. Molin and O. Ciofu, *Int. J. Antimicrob. Agents*, 2010, **35**, 322–332.
11. S. Delgado, R. Arroyo, E. Jimenez, M. L. Marin, R. del Campo, L. Fernandez and J. M. Rodriguez, *BMC Microbiol.*, 2009, **9**, 82.
12. O. Dumitrescu, O. Dauwalder, S. Boisset, M. E. Reverdy, A. Tristan and F. Vandenesch, *M S-Med. Sci.*, 2010, **26**, 943–949.
13. S. W. Kerrigan, N. Clarke, A. Loughman, G. Meade, T. J. Foster and D. Cox, *Arterioscler., Thromb., Vasc. Biol.*, 2008, **28**, 335–340.
14. M. A. Johnson and J. M. Ross, *Ann. Biomed. Eng.*, 2008, **36**, 349–355.
15. D. F. Williams, *Biomaterials*, 2009, **30**, 5897–5909.
16. J. M. Anderson, *Annu. Rev. Mater. Res.*, 2001, **31**, 81–110.
17. B. D. Ratner and S. J. Bryant, *Annu. Rev. Biomed. Eng.*, 2004, **6**, 41–75.
18. D. Pavithra and M. Doble, *Biomed. Mater.*, 2008, **3**, 034003.
19. D. J. Diekema, M. A. Pfaller, F. J. Schmitz, J. Smayevsky, J. Bell, R. N. Jones, M. Beach and S. P. Grp, *Clin. Infect. Dis.*, 2001, **32**, S114–S132.
20. S. R. Gill, D. E. Fouts, G. L. Archer, E. F. Mongodin, R. T. DeBoy, J. Ravel, I. T. Paulsen, J. F. Kolonay, L. Brinkac, M. Beanan, R. J. Dodson, S. C. Daugherty, R. Madupu, S. V. Angiuoli, A. S. Durkin, D. H. Haft, J. Vamathevan, H. Khouri, T. Utterback, C. Lee, G. Dimitrov, L. X. Jiang, H. Y. Qin, J. Weidman, K. Tran, K. Kang, I. R. Hance, K. E. Nelson and C. M. Fraser, *J. Bacteriol.*, 2005, **187**, 2426–2438.
21. R. J. Gordon and F. D. Lowy, *Clin. Infect. Dis.*, 2008, **46**, S350–S359.

22. L. G. Harris, S. J. Foster and R. G. Richards, *Eur. Cells Mater.*, 2002, **4**, 39–60.
23. C. Arrecubieta, T. Asai, M. Bayern, A. Loughman, J. R. Fitzgerald, C. E. Shelton, H. M. Baron, N. C. Dang, M. C. Deng, Y. Naka, T. J. Foster and F. D. Lowy, *J. Infect. Dis.*, 2006, **193**, 1109–1119.
24. C. von Eiff, G. Peters and C. Heilmann, *Lancet Infect. Dis.*, 2002, **2**, 677–685.
25. M. Otto, *Nat. Rev. Microbiol.*, 2009, 7, 555–567.
26. P. D. Fey and M. E. Olson, *Future Microbiol.*, 2010, **5**, 917–933.
27. P. Stoodley, K. Sauer, D. G. Davies and J. W. Costerton, *Annu. Rev. Microbiol.*, 2002, **56**, 187–209.
28. C. R. Arciola, D. Campoccia, P. Speziale, L. Montanaro and J. W. Costerton, *Biomaterials*, 2012, **33**, 5967–5982.
29. G. Laverty, S. P. Gorman and B. F. Gilmore, *Future Microbiol.*, 2013, **8**, 509–524.
30. M. Katsikogianni and Y. F. Missirlis, *Eur. Cells Mater. J.*, 2004, **8**, 37–57.
31. Y. H. An and R. J. Friedman, *J. Biomed. Mater. Res.*, 1998, **43**, 338–348.
32. O. M. Hartford, E. R. Wann, M. Hook and T. J. Foster, *J. Biol. Chem.*, 2001, **276**, 2466–2473.
33. S. Perkins, E. J. Walsh, C. C. S. Deivanayagam, S. V. L. Narayana, T. J. Foster and M. Hook, *J. Biol. Chem.*, 2001, **276**, 44721–44728.
34. K. Vacheethasanee, J. S. Temenoff, J. M. Higashi, A. Gary, J. M. Anderson, R. Bayston and R. E. Marchant, *J. Biomed. Mater. Res.*, 1998, **42**, 425–432.
35. V. A. Tegoulia and S. L. Cooper, *Colloids Surf., B*, 2002, **24**, 217–228.
36. R. Ardehali, L. Shi, J. Janatova, S. F. Mohammad and G. L. Burns, *J. Biomed. Mater. Res., Part A*, 2003, **66**, 21–28.
37. J. D. Patel, M. Ebert, R. Ward and J. M. Anderson, *J. Biomed. Mater. Res., Part A*, 2007, **80**, 742–751.
38. C. P. Xu, N. P. Boks, J. de Vries, H. J. Kaper, W. Norde, H. J. Busscher and H. C. van der Mei, *Appl. Environ. Microbiol.*, 2008, **74**, 7522–7528.
39. E. Brouillette, G. Grondin, L. Shkreta, P. Lacasse and B. G. Talbot, *Microb. Pathog.*, 2003, **35**, 159–168.
40. E. R. Wann, S. Gurusiddappa and M. Hook, *J. Biol. Chem.*, 2000, **275**, 13863–13871.
41. R. J. Williams, B. Henderson, L. J. Sharp and S. P. Nair, *Infect. Immun.*, 2002, **70**, 6805–6810.
42. C. Arrecubieta, F. A. Toba, M. von Bayern, H. Akashi, M. C. Deng, Y. Naka and F. D. Lowy, *PLoS Pathog.*, 2009, **5**, e1000411.
43. S. L. Davis, S. Gurusiddappa, K. W. McCrea, S. Perkins and M. Hook, *J. Biol. Chem.*, 2001, **276**, 27799–27805.
44. O. Hartford, L. O'Brien, K. Schofield, J. Wells and T. J. Foster, *Microbiology*, 2001, **147**, 2545–2552.
45. C. Arrecubieta, M. H. Lee, A. Macey, T. J. Foster and F. D. Lowy, *J. Biol. Chem.*, 2007, **282**, 18767–18776.

46. M. Christner, G. C. Franke, N. N. Schommer, U. Wendt, K. Wegert, P. Pehle, G. Kroll, C. Schulze, F. Buck, D. Mack, M. Aepfelbacher and H. Rohde, *Mol. Microbiol.*, 2010, **75**, 187–207.

47. M. P. Brennan, A. Loughman, M. Devocelle, S. Arasu, A. J. Chubb, T. J. Foster and D. Cox, *J. Thromb. Haemostasis*, 2009, **7**, 1364–1372.

48. C. Heilmann, M. Hussain, G. Peters and F. Götz, *Mol. Microbiol.*, 1997, **24**, 1013–1024.

49. S. J. Foster, *J. Bacteriol.*, 1995, **177**, 5723–5725.

50. C. Heilmann, G. Thumm, G. S. Chhatwal, J. Hartleib, A. Uekötter and G. Peters, *Microbiology*, 2003, **149**, 2769–2778.

51. C. Heilmann, J. Hartleib, M. S. Hussain and G. Peters, *Infect. Immun.*, 2005, **73**, 4793–4802.

52. Z. Qin, Y. Ou, L. Yang, Y. Zhu, T. Tolker-Nielsen, S. Molin and D. Qu, *Microbiology*, 2007, **153**, 2083–2092.

53. K. C. Rice, E. E. Mann, J. L. Endres, E. C. Weiss, J. E. Cassat, M. S. Smeltzer and K. W. Bayles, *Proc. Natl. Acad. Sci. U. S. A.*, 2007, **104**, 8113–8118.

54. L. Montanaro, A. Poggi, L. Visai, S. Ravaioli, D. Campoccia, P. Speziale and C. R. Arciola, *Int. J. Artif. Organs*, 2011, **34**, 824–831.

55. H. Mulcahy, L. Charron-Mazenod and S. Lewenza, *PLoS Pathog.*, 2008, **4**, e1000213.

56. M. Hussain, C. Heilmann, G. Peters and M. Herrmann, *Microb. Pathog.*, 2001, **31**, 261–270.

57. C. Heilmann and F. Götz, in *Biomaterials Associated Infection*, ed. T. F. Moriarty, S. A. J. Zaat and H. J. Busscher, Springer, New York, 2013, pp. 57–85.

58. D. Mack, M. Haeder, N. Siemssen and R. Laufs, *J. Infect. Dis.*, 1996, **174**, 881–884.

59. D. Mack, W. Fischer, A. Krokotsch, K. Leopold, R. Hartmann, H. Egge and R. Laufs, *J. Bacteriol.*, 1996, **178**, 175–183.

60. P. D. Fey, J. S. Ulphani, F. Götz, C. Heilmann, D. Mack and M. E. Rupp, *J. Infect. Dis.*, 1999, **179**, 1561–1564.

61. M. E. Rupp, J. S. Ulphani, P. D. Fey, K. Bartscht and D. Mack, *Infect. Immun.*, 1999, **67**, 2627–2632.

62. M. E. Rupp, J. S. Ulphani, P. D. Fey and D. Mack, *Infect. Immun.*, 1999, **67**, 2656–2659.

63. H. Rohde, S. Frankenberger, U. Zaehringer and D. Mack, *Eur. J. Cell Biol.*, 2010, **89**, 103–111.

64. T. Maira-Litran, A. Kropec, C. Abeygunawardana, J. Joyce, G. Mark, D. A. Goldmann and G. B. Pier, *Infect. Immun.*, 2002, **70**, 4433–4440.

65. M. E. Olson, K. L. Garvin, P. D. Fey and M. E. Rupp, *Clin. Orthop. Relat. Res.*, 2006, **451**, 21–24, DOI: 10.1097/1001.blo.0000229320.0000245416.0000229320c.

66. C. Vuong, J. M. Voyich, E. R. Fischer, K. R. Braughton, A. R. Whitney, F. R. DeLeo and M. Otto, *Cell. Microbiol.*, 2004, **6**, 269–275.

67. H. Rohde, S. Frankenberger, U. Zahringer and D. Mack, *Eur. J. Cell Biol.*, 2010, **89**, 103–111.
68. H. Rohde, E. C. Burandt, N. Siemssen, L. Frommelt, C. Burdelski, S. Wurster, S. Scherpe, A. P. Davies, L. G. Harris, M. A. Horstkotte, J. K. M. Knobloch, C. Ragunath, J. B. Kaplan and D. Mack, *Biomaterials*, 2007, **28**, 1711–1720.
69. C. Cucarella, C. Solano, J. Valle, B. Amorena, Í. Lasa and J. R. Penadés, *J. Bacteriol.*, 2001, **183**, 2888–2896.
70. M. Hussain, M. Herrmann, C. von Eiff, F. Perdreau-Remington and G. Peters, *Infect. Immun.*, 1997, **65**, 519–524.
71. N. J. Verkaik, W. J. B. van Wamel and A. van Belkum, *Immunotherapy*, 2011, **3**, 1063–1073.
72. M. Otto, *Expert Opin. Biol. Ther.*, 2010, **10**, 1049–1059.
73. L. Van Mellaert, M. Shahrooei, D. Hofmans and J. Van Eldere, *Expert Rev. Vaccines*, 2012, **11**, 319–334.
74. G. O'Toole, H. B. Kaplan and R. Kolter, *Annu. Rev. Microbiol.*, 2000, **54**, 49–79.
75. M. Otto, *Annu. Rev. Med.*, 2013, **64**, 175–188.
76. A. Resch, R. Rosenstein, C. Nerz and F. Gotz, *Appl. Environ. Microbiol.*, 2005, **71**, 2663–2676.
77. S. A. Rani, B. Pitts, H. Beyenal, R. A. Veluchamy, Z. Lewandowski, W. M. Davison, K. Buckingham-Meyer and P. S. Stewart, *J. Bacteriol.*, 2007, **189**, 4223–4233.
78. C. Heilmann, O. Schweitzer, C. Gerke, N. Vanittanakom, D. Mack and F. Gotz, *Mol. Microbiol.*, 1996, **20**, 1083–1091.
79. N. Merino, A. Toledo-Arana, M. Vergara-Irigaray, J. Valle, C. Solano, E. Calvo, J. Antonio Lopez, T. J. Foster, J. R. Penades and I. Lasa, *J. Bacteriol.*, 2009, **191**, 832–843.
80. E. O'Neill, C. Pozzi, P. Houston, H. Humphreys, D. A. Robinson, A. Loughman, T. J. Foster and J. P. O'Gara, *J. Bacteriol.*, 2008, **190**, 3835–3850.
81. M. Gross, S. E. Cramton, F. Gotz and A. Peschel, *Infect. Immun.*, 2001, **69**, 3423–3426.
82. C. B. Whitchurch, T. Tolker-Nielsen, P. C. Ragas and J. S. Mattick, *Science*, 2002, **295**, 1487.
83. C. Vuong, C. Gerke, G. A. Somerville, E. R. Fischer and M. Otto, *J. Infect. Dis.*, 2003, **188**, 706–718.
84. C. Vuong, H. L. Saenz, F. Gotz and M. Otto, *J. Infect. Dis.*, 2000, **182**, 1688–1693.
85. B. R. Boles and A. R. Horswill, *PLoS Pathog.*, 2008, **4**.
86. J. B. Kaplan, S. Jabbouri and I. Sadovskaya, *Res. Microbiol.*, 2011, **162**, 535–541.
87. J. B. Kaplan, K. LoVetri, S. T. Cardona, S. Madhyastha, I. Sadovskaya, S. Jabbouri and E. A. Izano, *J. Antibiot.*, 2012, **65**, 73–77.
88. B. R. Boles, M. Thoendel and P. K. Singh, *Mol. Microbiol.*, 2005, **57**, 1210–1223.

89. B. R. Boles and A. R. Horswill, *Trends Microbiol.*, 2011, **19**, 449–455.

90. R. Wang, B. A. Khan, G. Y. C. Cheung, T.-H. L. Bach, M. Jameson-Lee, K.-F. Kong, S. Y. Queck and M. Otto, *J. Clin. Invest.*, 2011, **121**, 238–248.

91. M. Rautenberg, H.-S. Joo, M. Otto and A. Peschel, *FASEB J.*, 2011, **25**, 1254–1263.

92. Y. F. Yao, D. E. Sturdevant and M. Otto, *J. Infect. Dis.*, 2005, **191**, 289–298.

93. S. Periasamy, H.-S. Joo, A. C. Duong, T.-H. L. Bach, V. Y. Tan, S. S. Chatterjee, G. Y. C. Cheung and M. Otto, *Proc. Natl. Acad. Sci. U. S. A.*, 2012, **109**, 1281–1286.

94. W. Teughels, N. Van Assche, I. Sliepen and M. Quirynen, *Clin. Oral Implan. Res.*, 2006, **17**, 68–81.

95. K. Subramani, R. E. Jung, A. Molenberg and C. H. F. Hammerle, *Int. J. Oral Maxillofac. Pathol.*, 2009, **24**, 616–626.

96. M. Gharechahi, H. Moosavi and M. Forghani, *J. Biomater. Nanobiotechnol.*, 2012, **3**, 541–546.

97. E. E. MacKintosh, J. D. Patel, R. E. Marchant and J. M. Anderson, *J. Biomed. Mater. Res., Part A*, 2006, **78**, 836–842.

98. D. R. Absolom, F. V. Lamberti, Z. Policova, W. Zingg, C. J. van Oss and A. W. Neumann, *Appl. Environ. Microbiol.*, 1983, **46**, 90–97.

99. M. Katsikogianni, E. Amanatides, D. Mataras and Y. F. Missirlis, *Colloids Surf., B*, 2008, **65**, 257–268.

100. M. G. Katsikogianni and Y. F. Missirlis, *J. Mater. Sci.*, 2010, **21**, 963–968.

101. Y. F. Missirlis and M. Katsikogianni, *Materialwiss. Werkstofftech.*, 2007, **38**, 983–994.

102. T. J. Kinnari, J. Esteban, N. Zamora, R. Fernandez, C. Lopez-Santos, F. Yubero, D. Mariscal, J. A. Puertolas and E. Gomez-Barrena, *Clin. Microbiol. Infect.*, 2010, **16**, 1036–1041.

103. M. Quirynen, H. C. Vandermei, C. M. L. Bollen, A. Schotte, M. Marechal, G. I. Doornbusch, I. Naert, H. J. Busscher and D. Vansteenberghe, *J. Dental Res.*, 1993, **72**, 1304–1309.

104. C. M. L. Bollen, P. Lambrechts and M. Quirynen, *Dental Mater.*, 1997, **13**, 258–269.

105. C. M. L. Bollen, W. Papaioanno, J. VanEldere, E. Schepers, M. Quirynen and D. vanSteenberghe, *Clin. Oral Implan. Res.*, 1996, 7, 201–211.

106. H. Y. Tang, T. Cao, X. M. Liang, A. F. Wang, S. O. Salley, J. McAllister and K. Y. S. Ng, *J. Biomed. Mater. Res., Part A*, 2009, **88**, 454–463.

107. K. A. Whitehead, J. Colligon and J. Verran, *Colloids Surf., B*, 2005, **41**, 129–138.

108. H. J. Busscher, B. van de Belt-Gritter, R. J. B. Dijkstra, W. Norde and H. C. van der Mei, *Langmuir*, 2008, **24**, 10968–10973.

109. D. Q. Li, F. Lundberg and A. Ljungh, *Curr. Microbiol.*, 2001, **42**, 361–367.

110. F. Lundberg, S. Schliamser and A. Ljungh, *J. Med. Microbiol.*, 1997, **46**, 285–296.

111. J. Bjerketorp, M. Nilsson, A. Ljungh, J. I. Flock, K. Jacobsson and L. Frykberg, *Microbiology*, 2002, **148**, 2037–2044.

112. A. Chiumiento, S. Lamponi and R. Barbucci, *Biomacromolecules*, 2007, **8**, 523–531.
113. I. Lee and R. E. Marchant, *Ultramicroscopy*, 2003, **97**, 341–352.
114. C. R. Arciola, D. Campoccia, S. Gamberini, M. E. Donati and L. Montanaro, *Biomaterials*, 2004, **25**, 4825–4829.
115. A. J. Garcia, M. D. Vega and D. Boettiger, *Mol. Biol. Cell*, 1999, **10**, 785–798.
116. U. Schwarz-Linek, M. Hook and J. R. Potts, *Mol. Microbiol.*, 2004, **52**, 631–641.
117. S. Kerdudou, M. W. Laschke, B. Sinha, K. T. Preissner, M. D. Menger and M. Herrmann, *Thromb. Haemostasis*, 2006, **96**, 183–189.
118. K. H. Simpson, M. G. Bowden, M. Hook and B. Anvari, *Lasers Surg. Med.*, 2002, **31**, 45–52.
119. A. C. Schroeder, J. M. Schmidbauer, A. Sobke, B. Seitz, K. W. Ruprecht and M. Herrmann, *J. Cataract Refractive Surg.*, 2008, **34**, 497–504.
120. A. L. J. Olsson, P. K. Sharma, H. C. van der Mei and H. J. Busscher, *Appl. Environ. Microbiol.*, 2012, **78**, 99–102.
121. W. M. Dunne and E. M. Burd, *J. Appl. Bacteriol.*, 1993, **74**, 411–416.
122. S. Galliani, M. Viot, A. Cremieux and P. Vanderauwera, *J. Lab. Clin. Med.*, 1994, **123**, 685–692.
123. J. C. Linnes, K. Mikhova and J. D. Bryers, *J. Biomed. Mater. Res., Part A*, 2012, **100**, 1990–1997.
124. B. Henderson, S. Nair, J. Pallas and M. A. Williams, *FEMS Microbiol. Rev.*, 2011, **35**, 147–200.
125. R. A. Jarvis and J. D. Bryers, *J. Biomed. Mater. Res., Part A*, 2005, **75**, 41–55.
126. S. D. Holmes, K. May, V. Johansson, F. Markey and I. A. Critchley, *J. Microbiol. Methods*, 1997, **28**, 77–84.
127. V. Vadillo-Rodriguez, M. A. Pacha-Olivenza, M. Luisa Gonzalez-Martin, J. M. Bruque and A. M. Gallardo-Moreno, *J. Biomed. Mater. Res., Part A*, 2013, **101**, 1397–1404.
128. L. C. Xu and C. A. Siedlecki, *J. Biomater. Nanobiotechnol.*, 2012, **3**, 487–498.
129. I. W. Wang, J. M. Anderson and R. E. Marchant, *J. Infect. Dis.*, 1993, **167**, 329–336.
130. S. W. Kerrigan, I. Douglas, A. Wray, J. Heath, M. F. Byrne, D. Fitzgerald and D. Cox, *Blood*, 2002, **100**, 509–516.
131. A. Loughman, J. R. Fitzgerald, M. P. Brennan, J. Higgins, R. Downer, D. Cox and T. J. Foster, *Mol. Microbiol.*, 2005, **57**, 804–818.
132. L. Obrien, S. W. Kerrigan, G. Kaw, M. Hogan, J. Penades, D. Litt, D. J. Fitzgerald, T. J. Foster and D. Cox, *Mol. Microbiol.*, 2002, **44**, 1033–1044.
133. H. Miajlovic, A. Loughman, M. Brennan, D. Cox and T. J. Foster, *Infect. Immun.*, 2007, **75**, 3335–3343.
134. J. R. Fitzgerald, A. Loughman, F. Keane, M. Brennan, M. Knobel, J. Higgins, L. Visai, P. Speziale, D. Cox and T. J. Foster, *Mol. Microbiol.*, 2006, **59**, 212–230.

135. J. R. Fitzgerald, T. J. Foster and D. Cox, *Nat. Rev. Microbiol.*, 2006, **4**, 445–457.

136. S. W. Kerrigan and D. Cox, *Cell. Mol. Life Sci.*, 2010, **67**, 513–523.

137. A. B. Estrela, M. G. Heck and W. R. Abraham, *Curr. Med. Chem.*, 2009, **16**, 1512–1530.

138. S. F. Chen, L. Y. Li, C. Zhao and J. Zheng, *Polymer*, 2010, **51**, 5283–5293.

139. G. Guerrero, J. Amalric, P. H. Mutin, A. Sotto and J. P. Lavigne, *Pathol. Biol.*, 2009, **57**, 36–43.

140. I. Francolini, L. D'Ilario, E. Guaglianone, G. Donelli, A. Martinelli and A. Piozzi, *Acta Biomater.*, 2010, **6**, 3482–3490.

141. A. J. Scardino, D. Hudleston, Z. Peng, N. A. Paul and R. de Nys, *Biofouling*, 2009, **25**, 83–93.

142. A. J. Scardino, H. Zhang, D. J. Cookson, R. N. Lamb and R. de Nys, *Biofouling*, 2009, **25**, 757–767.

143. W. Barthlott and C. Neinhuis, *Planta*, 1997, **202**, 1–8.

144. B. Bhushan and Y. C. Jung, *Prog. Mater. Sci.*, 2011, **56**, 1–108.

145. A. W. Feinberg, J. F. Schumacher and A. B. Brennan, *Acta Biomater.*, 2009, **5**, 2013–2024.

146. C. M. Magin, S. P. Cooper and A. B. Brennan, *Mater. Today*, 2010, **13**, 36–44.

147. T. Sun, G. Qing, B. Su and L. Jiang, *Chem. Soc. Rev.*, 2011, **40**, 2909–2921.

148. J. F. Schumacher, M. L. Carman, T. G. Estes, A. W. Feinberg, L. H. Wilson, M. E. Callow, J. A. Callow, J. A. Finlay and A. B. Brennan, *Biofouling*, 2007, **23**, 55–62.

149. K. K. Chung, J. F. Schumacher, E. M. Sampson, R. A. Burne, P. J. Antonelli and A. B. Brennan, *Biointerphases*, 2007, **2**, 89–94.

150. A. J. Scardino and R. de Nys, *Biofouling*, 2011, **27**, 73–86.

151. J. F. Schumacher, C. J. Long, M. E. Callow, J. A. Finlay, J. A. Callow and A. B. Brennan, *Langmuir*, 2008, **24**, 4931–4937.

152. K. R. Milner, A. J. Snyder and C. A. Siedlecki, *J. Biomed. Mater. Res., Part A*, 2006, **76**, 561–570.

153. L.-C. Xu and C. A. Siedlecki, *Acta Biomater.*, 2012, **8**, 72–81.

154. S. Vesterlund, J. Paltta, M. Karp and A. C. Ouwehand, *J. Microbiol. Methods*, 2005, **60**, 225–233.

155. L. S. Dorobantu and M. R. Gray, *Scanning*, 2010, **32**, 74–96.

156. A. Razatos, Y. L. Ong, M. M. Sharma and G. Georgiou, *Proc. Natl. Acad. Sci. U. S. A.*, 1998, **95**, 11059–11064.

157. M. A. S. Vigeant, R. M. Ford, M. Wagner and L. K. Tamm, *Appl. Environ. Microbiol.*, 2002, **68**, 2794–2801.

158. J. Strauss, Y. T. Liu and T. A. Camesano, *JOM*, 2009, **61**, 71–74.

159. Y. F. Missirlis and A. D. Spiliotis, *Biomol. Eng.*, 2002, **19**, 287–294.

160. Y. F. Missirlis and M. Katsikogianni, *Materialwiss. Werkstofftech.*, 2007, **38**, 983–994.

161. R. Bos, H. C. van der Mei and H. J. Busscher, *FEMS Microbiol. Rev.*, 1999, **23**, 179–230.

162. Y. H. An and R. J. Friedman, *J. Microbiol. Methods*, 1997, **30**, 141–152.
163. R. B. Dickinson, J. A. Nagel, R. A. Proctor and S. L. Cooper, *J. Biomed. Mater. Res.*, 1997, **36**, 152–162.
164. M. S. Shive, S. M. Hasan and J. M. Anderson, *J. Biomed. Mater. Res.*, 1999, **46**, 511–519.
165. D. P. Bakker, A. van der Mats, G. J. Verkerke, H. J. Busscher and H. C. van der Mei, *Appl. Environ. Microbiol.*, 2003, **69**, 6280–6287.
166. R. B. Dickinson and S. L. Cooper, *AIChE J.*, 1995, **41**, 2160–2174.
167. J. D. Patel, Y. Iwasaki, K. Ishihara and J. M. Anderson, *J. Biomed. Mater. Res., Part A*, 2005, **73**, 359–366.
168. C. J. Wright, M. K. Shah, L. C. Powell and I. Armstrong, *Scanning*, 2010, **32**, 134–149.
169. H. H. P. Fang, K. Y. Chan and L. C. Xu, *J. Microbiol. Methods*, 2000, **40**, 89–97.
170. Y. Chen, H. J. Busscher, H. C. van der Mei and W. Norde, *Appl. Environ. Microbiol.*, 2011, **77**, 5065–5070.
171. Y. L. Ong, A. Razatos, G. Georgiou and M. M. Sharma, *Langmuir*, 1999, **15**, 2719–2725.
172. A. P. Mosier, A. E. Kaloyeros and N. C. Cady, *J. Microbiol. Methods*, 2012, **91**, 198–204.
173. V. Dupres, D. Alsteens, G. Andre and Y. F. Dufrene, *Trends Microbiol.*, 2010, **18**, 397–405.
174. L. C. Xu and C. A. Siedlecki, in *Comprehensive Biomaterials*, ed. P. Ducheyen, K. E. Healy, D. W. Hutmacher, D. W. Grainger and C. J. Kirkpatrick, Elsevier, 2011, vol. 3, pp. 23–35.
175. L.-C. Xu and C. A. Siedlecki, *J. Biomed. Mater. Res., Part A*, 2010, **92**, 126–136.
176. R. D. Turner, N. H. Thomson, J. Kirkham and D. Devine, *J. Microsc.*, 2010, **238**, 102–110.
177. A. Méndez-Vilas, A. M. Gallardo-Moreno, M. L. González-Martín, R. Calzado-Montero, M. J. Nuevo, J. M. Bruque and C. Perez-Giraldo, *Appl. Surf. Sci.*, 2004, **238**, 18–23.
178. Y. F. Dufréne, *Nat. Rev. Microbiol.*, 2004, **2**, 451–460.
179. S. Scheuring and Y. F. Dufrêne, *Mol. Microbiol.*, 2010, **75**, 1327–1336.
180. A. Mendez-Vilas, A. M. Gallardo-Moreno and M. L. Gonzalez-Martin, *Antonie Van Leeuwenhoek International Journal of General and Molecular Microbiology*, 2006, **89**, 373–386.
181. Y. Liu, J. Strauss and T. A. Camesano, *Biomaterials*, 2008, **29**, 4374–4382.
182. X. X. Sheng, Y. P. Ting and S. O. Pehkonen, *J. Colloid Interface Sci.*, 2007, **310**, 661–669.
183. S. Kang and M. Elimelech, *Langmuir*, 2009, **25**, 9656–9659.
184. D. J. Muller, J. Helenius, D. Alsteens and Y. F. Dufrene, *Nat. Chem. Biol.*, 2009, **5**, 383–390.
185. Y. F. Dufrene, *Future Microbiol.*, 2006, **1**, 387–396.
186. V. Dupres, C. Verbelen and Y. F. Dufrene, *Biomaterials*, 2007, **28**, 2393–2402.

187. P. B. Chowdhury and P. F. Luckham, *Colloids Surf., A*, 1998, **143**, 53–57.
188. A. Agnihotri and C. A. Siedlecki, *Ultramicroscopy*, 2005, **102**, 257–268.
189. A. Touhami, M. H. Jericho and T. J. Beveridge, *Langmuir*, 2007, **23**, 2755–2760.
190. V. Dupres, F. D. Menozzi, C. Locht, B. H. Clare, N. L. Abbott, S. Cuenot, C. Bompard, D. Raze and Y. F. Dufrene, *Nat. Methods*, 2005, **2**, 515–520.
191. C. Verbelen and Y. F. Dufrene, *Integr. Biol.*, 2009, **1**, 296–300.
192. G. Francius, S. Lebeer, D. Alsteens, L. Wildling, H. J. Gruber, P. Hols, S. De Keersmaecker, J. Vanderleyden and Y. F. Dufrene, *ACS Nano*, 2008, **2**, 1921–1929.
193. M. Micic, D. H. Hu, Y. D. Suh, G. Newton, M. Romine and H. P. Lu, *Colloids Surf., B*, 2004, **34**, 205–212.
194. U. Schmidt, T. Dieing, K. Weishaupt, W. Liu and J. Yang, *Microsc. Microanal.*, 2012, **18**, 1504–1505.

Antimicrobial Interfaces

SABEEL P. VALAPPIL

Department of Health Services Research and School of Dentistry,
University of Liverpool, Liverpool, UK
Email: S.Valappil@liv.ac.uk

14.1 Introduction

The importance of understanding antimicrobial interfaces, where the anti-microbial agents interact with their target has increased dramatically in recent years. This is partly fuelled by the need for new antimicrobial agents because of the rise in antibiotic resistance among bacteria. Antimicrobial resistance can be defined as the ability of microorganisms that cause disease to resist actions of antimicrobial agents. The World Health Organization has acknowledged antimicrobial resistance as one of the three greatest threats to human health. Some of the recent figures show shocking numbers of people affected with antibiotic resistant bacteria; for example, such as carbapenem-resistant Enterobacteriaceae (CREs) can cause bladder, lung and blood in-fections that can potentially result in septic shock which endanger life. CREs evade the action of most of the antibiotics and they kill up to 50% of all patients who become infected with them.[1] Another antibiotic resistant or-ganism, methicillin-resistant *Staphylococcus aureus* (MRSA), is responsible for 100 000 infections annually in the United States alone with a mortality rate of almost 20%.[2] Currently, in certain parts of the world, previously potent drugs against tuberculosis have now become almost ineffective. This is a frightening development as multidrug-resistant *Mycobacterium tuberculosis* affects over 500 000 individuals worldwide.[2] A study from the United Kingdom forecast that antibiotic inefficiency may result in routine

RSC Smart Materials No. 10
Biointerfaces: Where Material Meets Biology
Edited by Dietmar Hutmacher and Wojciech Chrzanowski
© The Royal Society of Chemistry 2015
Published by the Royal Society of Chemistry, www.rsc.org

operations such as hip replacements leading to mortality for as many as one in six people.[3] In spite of these trends, only one new antibacterial drug, named linezolid, with a totally novel mechanism of action has been introduced in the past three decades, and the pipeline of new classes of antimicrobial pharmaceuticals is running dry.[4] Extensive use of antibiotics offers a potent selective force for antibiotic resistant bacteria and is consequently suggested to be responsible for the rise of resistance. In recent years, several countries have shown an increased alertness by implementing, or preparing to implement, regulations that cap the access to antibiotics. The rapid rise of the New Delhi metallo-β-lactamase-resistance gene among Enterobacteriaceae in India[5] and its quick spread to the United Kingdom guided the drug controller general of India to introduce new rules aimed at stopping the sale of antibiotics without a prescription.[2] Already the human and economic cost of antibiotic resistance is in a colossal scale. For example, in the year 2007, the number of infections by multidrug-resistant bacteria was 400 000, and there were 25 000 deaths associated with it in Europe alone. The number of extra hospital days associated with this stood at up to 2.5 million.[6] The expenditure linked with these infections in terms of extra hospital costs and efficiency losses exceeded 1.5 billion euros each year.[7] In the United States, antibiotic-resistant infections were blamed for $20 billion per year in excess health care costs, $35 billion per year in societal costs and 8 million additional hospital days per year.[8] Antimicrobial resistance is quickly becoming a key public health risk and is threatening to invalidate decades of advances in our ability to treat infections. It is also testing our perception of how to control communicable diseases. Antimicrobials, especially antibiotics, are important and valuable agents. Without them routine medical practices such as hip replacements and cancer chemotherapy would become much riskier and routine operations could become deadly if we lose the capacity to tackle infection. The chapter will discuss the bacterial interfaces with antibacterial materials, which is a key aspect of infection and potential starting point of drug resistance among bacteria.

14.2 Bacterial Interfaces

Bacteria can grow in two discrete forms of life and swap between them. These two forms include (1) a free-living existence, called planktonic growth, in which they are not adhered to a surface, and (2) a sessile state, called biofilm growth, which can form communities and is adhered to a surface. Bacteria growing as a biofilm, such as dental plaque, have been reported as being up to 1000 times more resistant to antimicrobial treatments compared to their planktonic counterparts[9] and are responsible for >65% of microbial infections in humans.[10] Diseases associated with bacterial biofilms include: lung infections of cystic fibrosis patients, colitis, urethritis, conjunctivitis, otitis media, infective endocarditis and periodontitis. Moreover, biofilm associated infections of indwelling medical devices are also of serious concern, because once the device is colonised by infective pathogens it become extremely difficult to eliminate them.[11] When bacteria inhabit surfaces, their gene expression model experiences a significant and fundamental shift to generate sole phenotype of biofilm which vary considerably from a bacterium in planktonic state. During some stages of biofilm development, as much as 50% of the proteome can be differentially produced compared with the same cells growing in planktonic culture.[12,13] Biofilms are normally characterised by a dense, extremely hydrated (water content up to 97%) group of bacterial cells. These structured but complex biofilm communities comprises of one or more species of microorganisms bonded to a substrate, to one another and embedded in an extracellular matrix consisting mainly of exopolysaccharide, proteins and extracellular DNA.[14–18] Bacteria that belong to either the Gram-positive or Gram-negative groups, classified based on their cell wall composition, could potentially generate biofilm configuration on almost all environmental surfaces that include inorganic substrates (such as minerals and shells of dead animals) as well as living organisms (plants as well as animals). In humans, bacteria colonise most sites which are either external (such as skin) or internal (such as mouth, nose and artificial devices) to the body.[19] The bacterial flora includes non-pathogenic commensal microorganisms which can become opportunistic pathogens especially in an immunosuppressed situation. Biofilm symbolises an evolved system for microbial survival which represents a sheltered means of growth that permit the cells to survive in adverse environments and also let them to break up to colonise new niches.[20] Biofilms are naturally resistant to antiseptics and antimicrobial agents that would typically eliminate their planktonic counterparts. Moreover, they are known to be more than 1000 times resistant to conventional antibiotics.[21]

Development of biofilm takes place as a result of a series of events: microbial surface attachment, cell proliferation, matrix production and detachment.[13] Among these, bacterial adhesion on a biomaterial surface is a key step in biofilm infections. The first step in bacterial adhesion is when bacteria approach a surface which may potentially be colonised. At this stage, weak attractive forces mediated transient attachment is assumed. But the majority of the bacteria split the bond with the surface with the minority remaining fixed for a longer period. It was reported that dissipation of the membrane potential ($\Delta\Psi$) totally opposes the change from transient to lasting attachment.[22] The transient attachment can be predisposed to electrical charges possessed by the bacteria. Although the exact type of interaction is disputed, both electrostatic attraction and van der Waals forces are implied.[11] It is also hypothesised that this shift from transient to lasting attachment is mediated by the signals from the environment. However, most of these signals are yet to be identified.

14.2.1 Quorum Sensing

Quorum sensing controls cell-to-cell communication signals (both at the intra-species and inter-species level) which have a pivotal part in the formation of multifaceted community structures of bacteria such as biofilm. Bacteria utilise those communication tracks to control a broad group of physiological functions (such as motility, virulence, antibiotic production and sporulation), which include biofilm formation.[23–25] Many of the bacteria that initiate infections at mammalian epithelial barriers are capable of intracellular invasion of the epithelial cells such as the putative periodontal pathogen *Porphyromonas gingivalis.*[26] Invasion of such bacteria is considered an important virulence factor, which protect them from the host immune system and contribute to tissue damage. Quorum sensing could act as a biofilm promoting signal which is crucial for the pathogenicity of a number of bacteria. The evolution of individual cells into biofilm is coordinated *via* a range of environmental and physiological signals, such as cell density of bacteria, availability of nutrients and cellular stress.[11] It was reported that bacteria within a biofilm can reach a much higher cell density (10^{11} CFU mL^{-1}) than do planktonic bacteria (10^8 CFU mL^{-1}).[27] A number of the quorum-sensing signals involved in biofilm formation have already been identified but many of them remain unidentified. In addition, biofilm development needs the activation and suppression of a wide array of genes. The cellular and transcriptional programs that guide biofilm formation is reviewed elsewhere.[28–30]

Quorum sensing is now recognised as a global regulatory mechanism for biofilm formation. The signals identified in biofilm formation include mechanical signals,[14,31–34] nutritional and metabolic signals,[35,36] presence of inorganic compounds[37–40] and host derived signals.[41,42] Bacterial cells may generate and sense signal molecules, termed autoinducers, that allow the entire population initiate concerted action once a critical concentration (matching to a specific population density) of the signal has been reached. The exposure of autoinducer concentration that meet the minimal threshold stimulatory level will trigger the change and synchronisation of the gene expression in a particular population of bacteria.

Bacteria display different quorum sensing systems and the well known ones include the *Lux-R-LuxI* system in Gram-negative bacteria that utilise an *N*-acyl-homoserine lactone as signal. Similarly, in Gram-positive bacteria the *agr* system, which uses a peptide thiolactone as signal and the RNAIII as effecter molecules.[43,44] Recently, bis-(3′,5′)-cyclic dimeric guanosine monophosphate (c-di-GMP), has emerged as a key player in bacterial signal transduction and regulation that control biofilm formation. The structural biology of c-di-GMP receptors, which is a fast developing scientific research field, holds promise for the targeted production of new therapeutic agents that might counter bacterial infections. This subject has been reviewed in detail recently.[45]

14.2.2   Production of Biofilm Matrix

The last phase in the permanent attachment on to a surface by bacterial cells is normally coupled with the fabrication of extracellular polymeric substances (EPS), which guides the creation of a mature biofilm structure. Most EPS are polyanionic owing to the occurrence of either uronic acids or

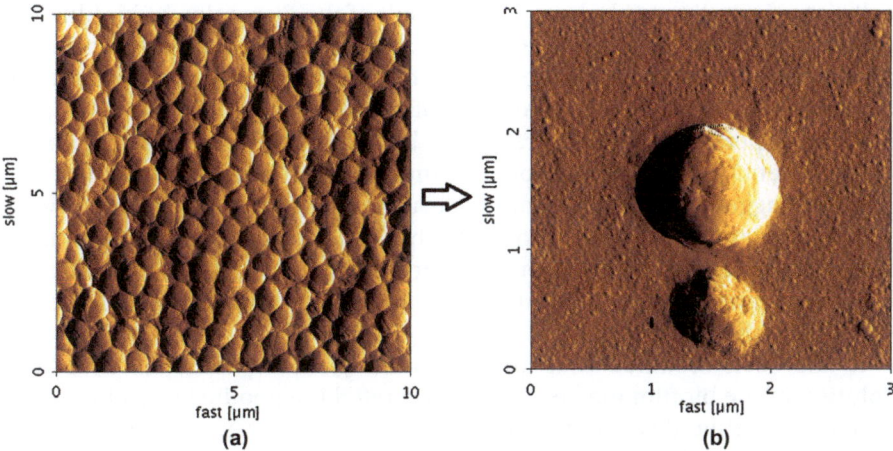

Figure 14.1 Atomic force microscopy images of a matured *Staphylococcus aureus* biofilm formed on a phosphate based glass substrate (a) and physically detached, using shear force, single planktonic *Staphylococcus aureus* cell (b).

ketal-linked pyruvate. Moreover, inorganic residues, such as phosphate or, less frequently, sulfate, could also result in polyanionic status. The composition of the EPS is known to be inconsistent from species to species but polyglycolic acid is the most frequent EPS component of the biofilm matrix. These are produced by bacteria such as *S. epidermidis*, *S. aureus* and *Escherichia coli.*[46] Cellulose-like and alginate-like EPS have also been found in *E. coli* and *Pseudomonas* spp.,[16] respectively.

The biofilm architecture is made up of EPS with embedded microcolonies of bacteria distributed throughout less dense area of matrix. In addition, the matrix is a porous network in which fluids run along channels and facilitate the movement of waste and nutrients. Every bacterium in a biofilm occupies a tailored microniche, in a multifaceted microbial community that has a primitive homeostasis, a primitive circulatory system and metabolic co-ordination. Every one of the sessile bacterial cells responds to such special environments in a way that is essentially divergent from their planktonic counterparts.[28] However, cell density dependant regulation of biofilm communities can yield free floating bacteria that may quickly scatter in to the surrounding environment (Figure 14.1).

14.2.3 Biofilm Dispersion

Dispersion is the change-over between biofilm and planktonic lifestyles. It was reported that biofilm dispersion can be described as a physiologically regulated process that occurs naturally during biofilm maturation and in the laboratory when flow was arrested in continuous culture biofilm reactors.[47] Dispersion is a synchronised disaggregation event that releases bacterial cells into the bulk liquid phase.[13] As a result of the dispersion of biofilm, cell clusters become hollow in the centre[13,48] appearing as doughnut-shaped structures under the microscope.[49] A substantial quantity of information is available regarding the regulation of biofilm adherence and maturation, yet, biofilm dispersion remains a relatively less well studied feature of biofilm development.[50] A hypothesis on the mechanisms leading to biofilm dispersion suggests that signalling molecules reaching a threshold concentration triggers the release of enzymes that cause cell cluster dismantling.[47] Another hypothesis suggests that dispersion is a response to two different environmental factors, accumulation of a metabolic product and depletion of a metabolic substrate.[51]

It has been assumed that quorum sensing is responsible for the dissolution of the biofilm matrix. Once an excreted metabolite (*e.g.*, a furanone) reaches a threshold concentration the release of bacterial cells into the bulk liquid is triggered.[52] It has also been proven that acyl-homoserine lactones are responsible for maintaining bacterial cells in a dispersed state both in *Rhodobacter sphaeroides* and *Pseudomonas aureofaciens*.[53,54] Moreover, in *P. aeruginosa*, the chemotaxis regulator BdIA has been hypothesised to be involved in the regulation of biofilm dispersion and cell adhesiveness by acting as a sensor protein in c-di-GMP signalling.[55]

14.3   Biofilm Infection

Antimicrobial interfaces can be best studied at an interface which involves development of biofilms on implant surfaces. The most frequently cultured microorganisms that are generally linked to implant surfaces include *Staphylococcus* spp., *Streptococcus* spp., *Enteroccoccus* spp. followed by Gram-negative bacilli (which include *Pseudomonas* spp. and *Enterobacter* spp.) and anaerobes (*Propionibacterium* spp.).[56] Apart from bacterial infections, a growing proportion of infections associated with indwelling devices, mainly the ones engaged with the urinary tract and the bloodstream, are fungal in origin, primarily *C. albicans*.[57,58] Bacterial colonisation of implants used in surgery does not automatically point towards infection; however, the growth of virulent bacteria typically points to infection. Low virulence bacteria, which are usually part of commensal flora of skin, might either become harmless contaminants or potent pathogens. In many cases, the objective of infection control is attained with the help of both surgical intervention and antibiotic therapy.

Bacteria involved in implant associated infection mainly include non-pathogen or opportunistic pathogens. But the species associated differ considerably based on the site of localisation and sorts of device employed.[28,59] *Staphylococcus* spp. (*S. epidermidis* and *S. aureus*) and other Gram-positive cocci usually infect arteriovenus shunts, sutures, vascular grafts, central venous catheters, sclera buckles and orthopaedic devices.[28,59] *Enterobacteriaceae* (frequently *E. coli*) and *Enterococcus* spp. colonise urinary catheters. The conversion of opportunistic or non-pathogenic bacteria into virulent sub-populations is activated by both phenotypic phase adaptation of bacteria (leading to the production of defensive biofilm structures) and local variation in host immune responses.[11] These two events are strongly inter-related as the intact host immune systems frequently get rid of transient bacterial contamination except when the inoculum load surpasses a distinct threshold. Although most of the medically used materials are inert and non-toxic they can often cause unfavourable physiological effect termed as a 'foreign body reaction' at the site of insertion, which induces local im-munosuppression.[60,61] Under such conditions, even low numbers of bacteria will not be eradicated by an appropriate response from the host immune system which then leads to a bacterial threshold that may cause an infection to decrease dramatically.[62–65]

14.4 Biofilm Antimicrobial Surface Interfaces

The biofilm matrix which constitutes EPS, proteins and DNA can trap anti-microbial agents by physicochemical interactions. This leads to antibacterial agents (such as antibacterial metal ions, antibiotics, disinfectants or ger-micides) being deactivated, delayed or stopped from penetrating into the biofilm interior through diffusion. It was reported that formation of a dead bacterial layer at the interface with the silver-releasing phosphate-based glasses caused re-emergence of viable bacteria after 24 h growth of *S. aureus*[66] suggesting the diffusion limitation of silver ions released from the phosphate-based glasses (Figure 14.2). Many reports studied the diffusion and interaction of antibacterial agents with that of the biofilm matrix ma-chinery.[67–70] The uptake of antibacterial agents differs significantly with their chemical composition, size, charge and hydrophilicity. For instance, aminoglycosides such as gentamicin and tobramycin adhere strongly to biofilm matrix of *Pseudomonas aeruginosa* made of alginate thus totally blocking its diffusion through to the interior.[71] However, β-lactam anti-biotics displayed a higher diffusion coefficient.[72] The phenotypic variation in growth, metabolism and genetic mutations undergone by biofilm associated species all add to the creation of drug resistance among such bacteria. Bacterial cells which are part of a biofilm are known to grow at a significantly reduced speed compared to their planktonic counterparts and possess a decreased metabolic activity. These reduced metabolic activities contribute to decreased sensitivity to antimicrobial compounds.[73,74] Furthermore, ele-vated rates of exchange of genetic material and mutations inside biofilms

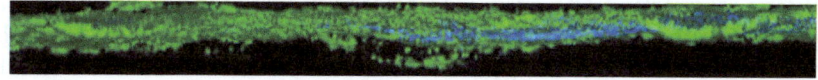

Figure 14.2 A cross-sectional view of confocal laser scanning microscopic image showing the *S. aureus* biofilm on silver-doped phosphate based glasses. Viable (green) and non-viable (blue) bacteria.

support the progress of survival mechanism such as expression of particular efflux pumps and up-regulation of proteins (such as porin proteins) that could quickly disseminate right through the biofilm population.[75]

Biofilm associated implant infections show a high rate of recurrence that leads to further aggressive diseases. The return of such infection is now attributed to the presence of a protected sub-population of persister cells.[76] Persister cells might have inactivated their automatic cell death mechanisms which help them to display resistance to the actions of antibiotics. More importantly, after earlier—seemingly successful—therapies, multi-drug tolerant progeny originating from persister cells could repopulate the community and be responsible for persistent infections. The mechanism of multi-drug tolerance among biofilm bacteria and the nature of persister cells are still under intense research. Multi-drug tolerance occurs due to elevated expression of toxin–antitoxin modules and related genes. This may hinder vital cell functions that include translation, a major target for certain antibiotics.[76] Due to such variation and yet to be elucidated resistant mechanisms possessed by biofilm, a single substance or strategy to alleviate the issues linked to implant infections appears to be a remote possibility.

14.5   Challenges of Antimicrobial Material Design

There is an underlying need to improve the properties of existing materials due to the incidence of healthcare associated infections, which often lead to revision surgery. Currently, prophylaxis in the form of systemically administered antibiotics serves as the main weapon against infection following implant surgery.[77] The first step in bacterial colonisation that could lead to a mature bacterial biofilm is the attachment of an individual bacterium to material surfaces and interfaces. The key steps in bacterial adhesion to surfaces can normally be summed up in a two-stage model. That includes an early reversible interaction between the material surface and the bacterial cell surface followed by a second stage which comprise of specific and nonspecific interactions between binding molecules on the material surface and adhesion proteins expressed on bacterial surface structures.[11] This latter stage is gradually reversible but frequently deemed to be irreversible.[78]

14.5.1   Anti-adhesive surfaces

Adhesion of bacteria to materials is a multifaceted interaction that involves varying means including inherent motility of bacteria, Brownian motion, diffusion, convection and gravity associated movement. When a material is implanted in the human body a thin film, termed a conditioning film, which comes from organic matter (mainly adsorbed proteins and blood platelets) present in the surrounding fluid, is deposited on the material surface. The composition of this organic matter affects the physicochemical and morphological properties of the surface. The conditioning film forms an

Table 14.1 Bacteria–material surface interactions and distances.[79]

Distance (nm)	Nature of the interactions
< 5	Short-range interactions: electrostatic forces, specific and hydrophobic groups
10–20	van der Waals forces and electrostatic forces
> 50	van der Waals forces

interconnecting layer between substrate and adhering bacteria and mediates their direct interactions.[79] Adhesion of bacteria onto static surfaces can be explained based on their specific as well as nonspecific interactions[79] (Table 14.1). Specific interactions normally function over short distances (about 5 nm) and are directional, while the non-specific interaction is the result of interactive forces between every molecules on the whole cell and substratum (which are normally long-range in nature, typically more than 50 nm). Once a bacterium gets in touch with a surface through these broad macroscopic interactive forces, interaction amongst the stereochemical cluster (short range and specific ones) will start to occur by means of significantly elevated attractive interaction energy.[80]

There are many features on the surface of materials which were projected to influence the primary stage in attachment of bacteria. Such factors comprise of material stiffness, surface charge, surface roughness, hydrophobicity, zeta potential, hydrogen bonding capacity,[81] Lewis acid–base character, specific receptor–adhesion interactions[82] and van der Waals forces. The involvement of so many factors in the colonisation of bacteria on an implant surface obscure the contribution made by each of these factors in bacterial attachment. Furthermore, it is difficult to devise universal conclusions for adhesion mechanisms of bacteria as the environmental conditions (which include concentrations of glucose and oxygen, temperature, flow conditions and presence of adsorbed proteins) in which bacterial strains proliferate vary considerably. Therefore environmental stimuli can influence the comparative significance of material surface characters and adhesion mechanisms even in the case of a single strain of bacteria and material surface. Thus no particular material characteristic or bacterial trait can wholly illustrate or manage bacterial adhesion.[78] In order to analyse familiar strategies and draw universal conclusions polymer based biomaterial and bacterial interfaces is predominantly discussed in the following sections.

Designs of antibacterial interfaces have focussed on limiting or preventing colonisation of bacteria to material surfaces. Different strategies are pursued which include exploiting material surface characteristics such as anti-adhesiveness, antimicrobial activity of material by contact and the controlled release of antibacterial agents.[83] One of the important factors that play a significant role in bacterial attachment is the overall surface properties of bacteria. The relative degree of hydrophobicity in relation to both bulk liquid and the substratum has been demonstrated to influence attachment.[84]

Differences in surface hydrophobicity are believed to result to form the net surface properties conferred upon the cell surface molecules such as proteins and lipids. The production of these molecules at the surface of the bacterial envelope is inducible in at least some instances. Bacteria might also be capable of adjusting their surface hydrophobicity in ways that would either enhance or diminish attachment to a substratum.

The diverse physicochemical tactics reported so far are rooted on detailed analyses of the interaction energy between surface characters and the microbial system. However, limitations in the systems design to study these interactions do not provide the overall picture. For example, cell surfaces of bacteria are structurally heterogeneous and possess an intricacy that is not routinely considered while contemplating physicochemical modelling at the molecular level. More importantly concepts such as distance between the bacterial surface to that of biomaterial surface partly lose its sense when taking into account that some of the bacterial cell surface appendages such as pili can become as long as 1 μm.[85] Adhesion of bacteria to surface of materials may be predicted as means of reassigning disorganised cells (bulk phase) to almost firmly connected status at the interface.[86] In order to forecast the behaviour of bacteria on various surfaces, physicochemical approaches such as the thermodynamic approach[87] and the Derjaguin–Landau–Verwey–Overbeek; (DVLO) theory[88] are being explored. It is envisaged that the mechanistic data on bacterial adhesion derived from the extended DVLO theory could provide developmental strategies for new surfaces displaying nominal bacterial adhesion. A comprehensive discussion on the subject is reported elsewhere.[85]

The fact that surface irregularities of material generally encourage attachment of bacteria, while ultra smooth surfaces inhibit biofilm formation was approved in many reports.[89–91] The initial bacterial adhesion is likely to favour locations where bacterial cells can be shielded from shear forces. Therefore increasing the surface roughness can lead to enhancement of the overall surface accessible for adhesion.[92] However, the concept that ultrasmooth surfaces prevent attachment of bacteria and subsequent biofilm formation has been challenged recently by the finding that bacteria could effectively colonise surfaces that possess very low (nanometre or subnanometre range) surface roughness (Ra).[93] Therefore new advances in techniques to modify the nanotopography of surfaces can be used as a valuable technique for the production of antibacterial materials. It was reported that physical parameters such as stiffness, in particular Young's modulus, alter the mechano-selective adhesion of bacteria such as *S. epidermidis* and *E. coli*.[94] An elevated level of Young's modulus of the material was found to be positively correlated with the attachment of *S. epidermidis* and *E. coli* in the range of 1 MPa to 100 MPa.[94]

Even though it is not widely included in antimicrobial investigations, the fundamental features, which include the evaluation of the biocompatibility of the antibacterial surfaces, require comprehensive analysis while designing such materials. Polymers (including naturally derived, synthetic or blended forms) are extensively utilised to produce antimicrobial surfaces on implant materials. These polymers could either be employed in isolation or used in conjunction with different materials or molecules in order to generate superior quality composite materials. Polymer-based materials are prepared by surface adsorption driven by electrostatic interactions. These polyelectrolyte multi-layers include, for example, poly(acrylic acid), poly-(methacrylic acid), poly(allylamine hydrochloride), sulfonated polystyrene,[95] polyethyleneimine,[96] chitosan[97] and hyaluronic acid.[98] Different studies have shown that the presence of cationic groups correlated with potent antimicrobial effects.[76] It was reported that surfaces coated with a cationic *N,N*-dimethyl-2-morpholinone are capable of inactivating bacteria in dry environment while a zwitterionic carboxybetaine help surfaces to defy bacterial adhesion in wet environments.[99] Surfaces were also prepared using polyelectrolyte that can stop the first attachment of cells thus avoiding steady bacterial attachment on material surface. One such polymer, poly-ethylene glycol (PEG), which has a strong affinity for water molecules, showed drastically reduced adsorption of *E. coli* cells when immobilised or grafted onto surfaces due to the development of a highly hydrated layer.[82] But PEG is reported to be susceptible to oxidation impairment which restricted lasting application of it in complex media.[100] Other developments include the anti-adhesive polymer heparin (which is negatively charged) embedded into the antibacterial biopolymer chitosan (which is positively charged) to generate surfaces that inhibit *E. coli* cells.[97] Likewise hyaluronan incorporated into a multi-layer showed reduced adhesion of *S. epidermidis* due to its hydrophilicity.[98]

14.5.2 Contact-active Materials

Contact active materials are known to cause death or impairment of bacteria that which firmly attached to its surfaces. A number of these materials are prepared based on the antibacterial properties of surface immobilised polycations.[76,101] It was hypothesised that the antibacterial mechanism is based on the destabilisation and permeation of bacteria resulted from the dislocation of charges on the cell membrane.[76] The antimicrobial peptide defensin was also incorporated into polyelectrolyte multi-layers that inactivated bacteria.[102] Other contact killing species including particles of titanium[103] and groups of transition metals (polyoxometalates)[104] were also exploited. The development of light-activated antimicrobial surfaces such as crystal violet and nanogold incorporated silicone which kill *S. epidermidis* and *E. coli* by the action of light have been developed recently.[105] Likewise it was reported that light activation led to a decrease in *S. aureus* attachment among certain Ni–Ti alloy surface materials covered with titanium dioxide.[106] This was suggested to be due to the photon accumulation in the surface and likely increase in defects which could have resulted in free oxygen. These approaches have potential applications in domestic and healthcare settings and the topic is reviewed elsewhere.[107]

14.5.3 Controlled-drug Delivery Agents

Controlled delivery of the antibacterial agents from the surface involves release-based antibacterial mechanisms. Medical implant coatings can include antibacterial agents (such as antibiotics or metal ions) and the benefits of such a local drug delivery approach compared to systemic drug therapy are published elsewhere.[108] The similarity in properties, such as hydrophobicity between surface polymer and the intended drug for delivery, was shown to produce homogenous distribution of the drug within the matrix.[11] Recent developments in this area of research include the use of biodegradable polymers (*e.g.*, polyhydroxyalkanoates[109]) for the controlled delivery of drugs. Biodegradable surface coatings could actively discharge high doses of antibiotics for a long time and the degradation end products are easily metabolised by the human body.[109,110] Immunotherapy, which involves the release of *ex situ* produced antibodies in a clinical setting, hold promise for highly efficient killing of bacteria. Likewise, leachable antimicrobial agents which are incorporated into polymers such as poly(β-amino esters) with a tuneable discharge of the aminoglycoside antibiotic gentamicin have been explored.[111] Moreover, antimicrobial peptides that can be released from polyelectrolyte multi-layers have also been reported.[112] Recently, emerging classes of polymers such as polynorbornenes or poly-(phenylene ethylene) derived from antimicrobial peptides (such as defensin or magainin) display excessive antibacterial activity but with low levels of cytotoxicity.[113–115] Polyelectrolyte multi-layers are reported to be used for the delivery of antimicrobial compounds such as silver including ionic and nanoparticulate forms.[116–120] It was reported that a coating based on silver nanoparticles fixed in a modified-chitosan layer deposited on methacrylic thermosets was found to inhibit bacterial growth.[121] Silver-nanoparticle-embedded antimicrobial paint based on vegetable oil which could have biomedical application was also reported.[122] Silver, photocatalytic TiO_2, nitric oxide releasing nanoparticles and metal oxides such as zinc and magnesium nanoparticles were also found to inhibit bacterial growth.[123,124] It was reported that the bactericidal actions of nanoparticles are derived from its electrostatic forces, basic character, oxidising power of halogens, generation of reactive oxygen species and accrual of nanoparticles near the cytoplasm which kills the cells.[123,125] Quaternary ammonium compounds, such as cetrimide, were incorporated into polyelectrolyte multi-layers as leaching agents to inactivate *S. aureus* and *E. coli*.[96] In spite of its antibacterial properties, it was reported that some bacteria are capable of developing resistance against surfaces coated with quaternary ammonium compounds.[126] Phosphate glasses offer potential alternatives to the existing means offered for the treatment of infections since it can be used as localised antibacterial delivery systems through the inclusion of ions known for its antibacterial effects such as copper, silver and gallium.[127] Those materials could be placed at a site of infection to release antibacterial ions as part of the glass degradation, which could be valuable in wound healing applications.

Their use could be extended to the prevention of implant or biomaterial related infections, which are one of the main causes of revision surgery and to augment or replace the current prophylaxis of systemically administered antibiotics.[77]

14.6 The Interface Between Veterinary and Human Antibiotic Use

Antimicrobial resistance happens when organisms are able to endure antimicrobial agents which are meant to damage them. This is a worrying trend as unregulated and widespread use of antibacterial agents, particularly antibiotics in veterinary applications, might have a link in the recent surge of resistance among bacteria. Non-human use of antibiotics could be an additional factor of antimicrobial resistance among humans (particularly with regard to food-borne diseases, such as salmonellosis and campylobacteriosis).[128] Antibiotics have been supplemented to livestock animals even in the absence of infection, as a means of prophylaxis and to promote growth. Almost 70% of the antibiotics used annually are consumed by farm animals[2] and hence it is major concern for the development of antimicrobial resistant strains of bacteria. It was reported that MRSA frequently colonises livestock animals and might act as an extended reservoir of MRSA in humans. Moreover, it is likely that the increased focus in research on animal MRSA might pave the way for discoveries of other strains with atypical characteristics that may also be riskier to human health.[129]

14.7 Strategies to Control Antimicrobial Resistance

As described in earlier paragraphs antimicrobial resistance is a global issue and controlling its spread could only be achieved by improving our understanding of antimicrobial resistance using data from surveillance and diagnostics. Furthermore, control of infection transmission between countries could be achieved by promoting good clinical practice in prescribing antibiotics. From this angle, maintaining the efficacy of existing antimicrobials is key. This can be achieved by applying stricter infection prevention and control practices, optimising prescribing practices and embedding antimicrobial stewardship programmes that ensure antibiotics are used only when needed and then in the right way, at the right dose and for the right duration. The second most important step towards antimicrobial resistance control is supporting the development of new antimicrobials and alternative treatments. This could be achieved by great international collaboration to stimulate the antibiotic pipeline by encouraging innovative approaches to develop diagnostics, new antibiotics and novel therapies. In this regard, the forecast that the market for antimicrobials is likely to be US$40.3 billion in 2015[2] should encourage investment in the discovery and development of novel drugs.

14.7.1   Role of Antibacterial Metal Ions

Novel therapeutic approach such as the gallium-based materials which act on multiple targets of the bacterial cells would generate very low chances of developing resistance.[130] Metabolism of iron (Fe) plays a vital role in the vulnerability of infecting bacteria because they need Fe for growth as well as the functioning of important enzymes; including those involved in DNA synthesis, electron transport and oxidative stress defences.[131] Consequently, efficient functioning of Fe metabolism is key in the pathogenesis of bacterial infections. The ionic radius of gallium (Ga^{3+}) is almost identical to that of Fe^{3+} and hence it can act as a 'Trojan horse' as many biological systems are unable to differentiate Ga^{3+} from Fe^{3+}.[132] More crucially, sequential oxidation and reduction are essential for a number of the biological functions of Fe^{3+}. Thus, insertion of Ga^{3+} can disrupt Fe^{3+}-dependent processes as it cannot be reduced under the same conditions.[132] Therefore Ga^{3+} has emerged as a new generation antibacterial ion that might be effective in treating and preventing localised infections.[130,133] Much research in this area focuses on the use of gallium complexes, including gallium maltolate,[134] desferrioxamine gallium,[135] and gallium salts [*e.g.*, Ga $(NO_3)_3$].[136] Gallium-doped phosphate-based glasses (PBGs) containing Ga^{3+} were reported as a potent antibacterial agents against both *P. aeruginosa* growing either planktonically or as biofilms.[130,137] Recently, it was shown that a combinational approach whereby different antibacterial agents with diverse antibacterial mechanisms such as silver (which inhibits DNA replication, protein inactivation and destabilisation of the intercellular adhesion forces within bacterial biofilms[138]) and gallium (which interfere with iron-dependent enzymes and hence could act concurrently on multiple targets[132]) could be a promising strategy in controlling the putative periodontal pathogen *P. gingivalis*.[139] Such strategies have huge potential as the mutation of a single intracellular drug target might not yield high-level resistance in bacteria undergoing such treatment. A protocol to evaluate cell–cell adhesion within *P. aeruginosa* biofilm exposed to antibacterial metal-doped glasses is shown in Figure 14.3. Furthermore, combined approaches were also reported that include incorporating silver releasing ability of the surfaces with contact-killing capabilities of quaternary ammonium salts,[118] titanium-doped iron, silver-doped titanium, silver-doped silica films, silver-doped phenyltriethoxysilane, silver-doped inorganic–organic hybrids, and silver-doped HA coatings.[140,141]

14.8   Future Directions

Despite recent significant progress in the understanding of the means by which infection develops, the design and manufacture of antibacterial surfaces remains a high public health and scientific research priority. The colonisation of surfaces by bacteria is well recognised to unfavourably affect the function of a number of specific interfaces, such as medical implants, yet

A flow diagram showing Constant Depth Film Fermentor (CDFF), media and connections

Key:
1. Constant-depth film fermentor (CDFF);
2. CDFF motor unit;
3. 0.3 μm filter;
4. Inoculation flask (*P. aeruginosa*);
5. Magnetic stirrer;
6. Peristaltic pump (0.7 mL min^{-1});
7. Growth medium (1%TSB) supply flask;
8. Peristaltic pump (0.1 mL min^{-1});
9. Grow-back trap;
10. Effluent flask;
V. Vent to atmosphere

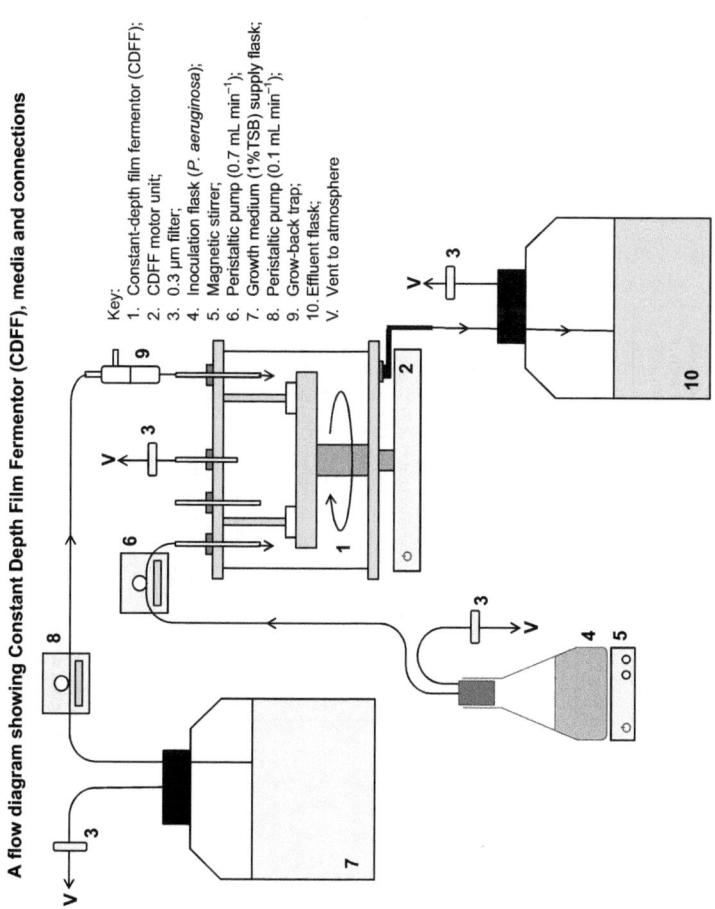

Figure 14.3

Protocol to measure cell-cell adhesion within a *P.aeruginosa* biofilm grown in a CDFF using Atomic Force Microscopy (AFM)

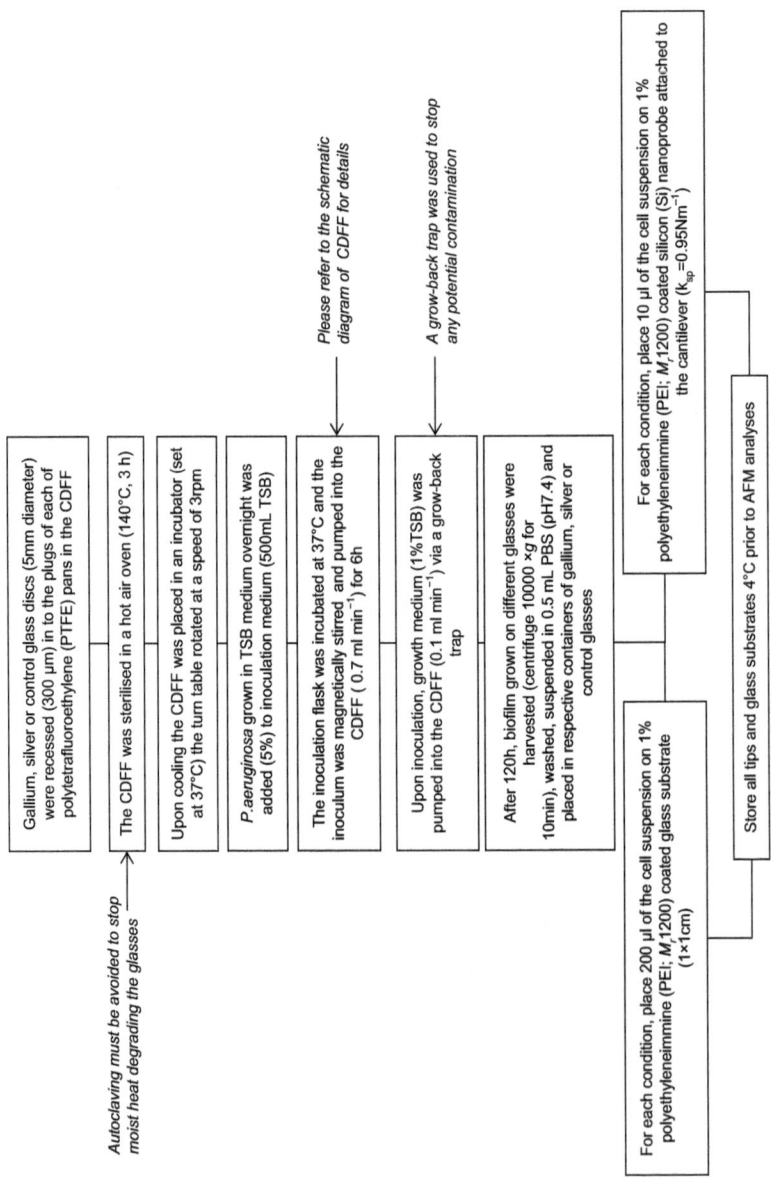

Gallium, silver or control glass discs (5mm diameter) were recessed (300 μm) in to the plugs of each of polytetrafluoroethylene (PTFE) pans in the CDFF

The CDFF was sterilised in a hot air oven (140°C; 3 h)

Autoclaving must be avoided to stop moist heat degrading the glasses

Upon cooling the CDFF was placed in an incubator (set at 37°C) the turn table rotated at a speed of 3rpm

P.aeruginosa grown in TSB medium overnight was added (5%) to inoculation medium (500mL TSB)

The inoculation flask was incubated at 37°C and the inoculum was magnetically stirred and pumped into the CDFF (0.7 ml min^{-1}) for 6h

Please refer to the schematic diagram of CDFF for details

Upon inoculation, growth medium (1%TSB) was pumped into the CDFF (0.1 ml min^{-1}) via a grow-back trap

A grow-back trap was used to stop any potential contamination

After 120h, biofilm grown on different glasses were harvested (centrifuge 10000 ×g for 10min), washed, suspended in 0.5 mL PBS (pH7.4) and placed in respective containers of gallium, silver or control glasses

For each condition, place 200 µl of the cell suspension on 1% polyethyleneimmine (PEI; M_r1200) coated glass substrate (1×1cm)

For each condition, place 10 µl of the cell suspension on 1% polyethyleneimmine (PEI; M_r1200) coated silicon (Si) nanoprobe attached to the cantilever (k_{sp}=0.95Nm^{-1})

Store all tips and glass substrates 4°C prior to AFM analyses

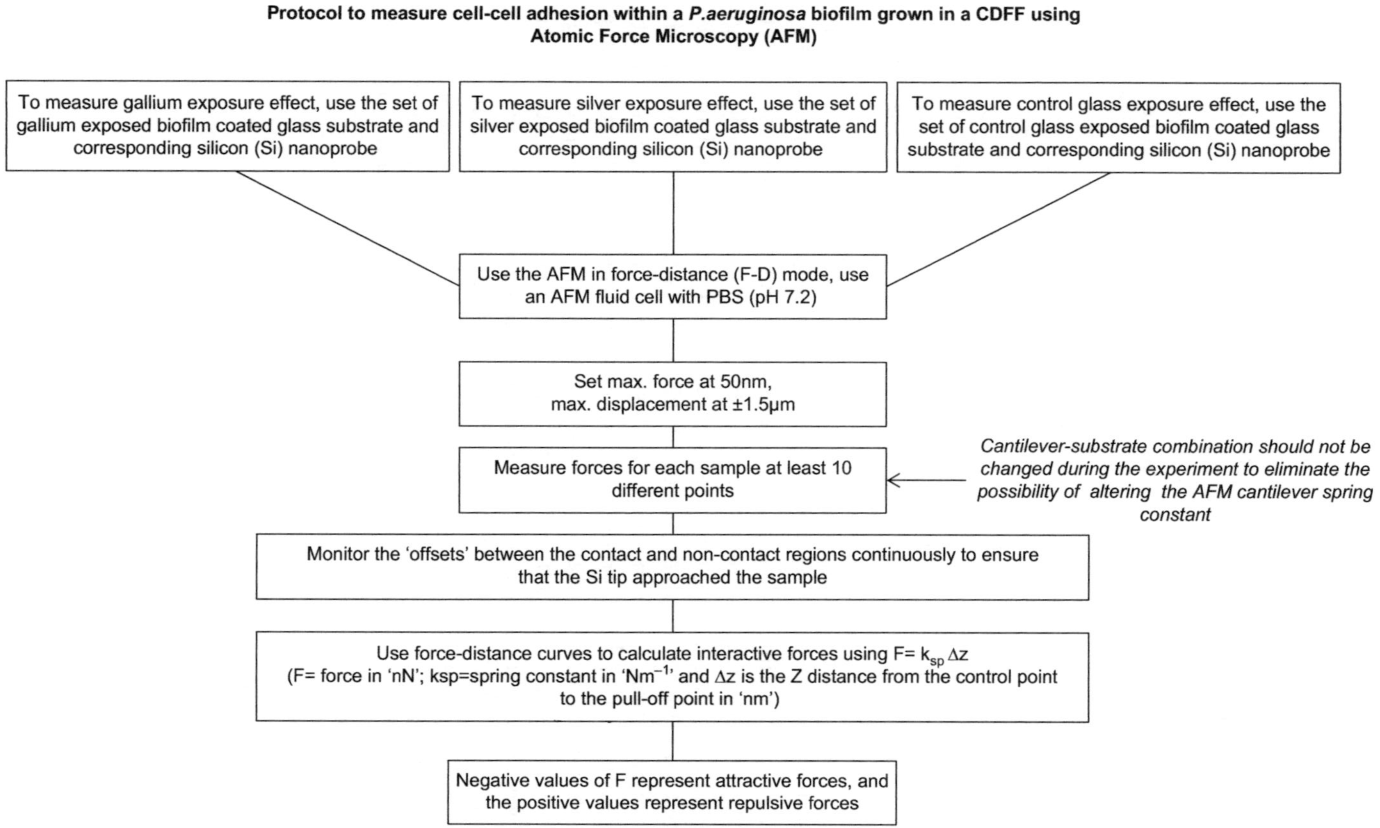

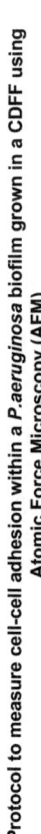

Protocol to measure cell-cell adhesion within a *P.aeruginosa* biofilm grown in a CDFF using Atomic Force Microscopy (AFM)

To measure gallium exposure effect, use the set of gallium exposed biofilm coated glass substrate and corresponding silicon (Si) nanoprobe

To measure silver exposure effect, use the set of silver exposed biofilm coated glass substrate and corresponding silicon (Si) nanoprobe

To measure control glass exposure effect, use the set of control glass exposed biofilm coated glass substrate and corresponding silicon (Si) nanoprobe

Use the AFM in force-distance (F-D) mode, use an AFM fluid cell with PBS (pH 7.2)

Set max. force at 50nm, max. displacement at ±1.5μm

Measure forces for each sample at least 10 different points

Cantilever-substrate combination should not be changed during the experiment to eliminate the possibility of altering the AFM cantilever spring constant

Monitor the 'offsets' between the contact and non-contact regions continuously to ensure that the Si tip approached the sample

Use force-distance curves to calculate interactive forces using F= $k_{sp} \Delta z$
(F= force in 'nN'; ksp=spring constant in 'Nm^{-1}' and Δz is the Z distance from the control point to the pull-off point in 'nm')

Negative values of F represent attractive forces, and the positive values represent repulsive forces

Figure 14.3 A protocol to evaluate cell–cell adhesion within *P. aeruginosa* biofilm exposed to antibacterial metal-doped glasses.

little is known about the conformational properties and interactions of the individual components of a bacterium in such an interface. Moreover, individual microbial cells can vary extensively from each other (with regards to their genetic composition, biochemistry, physiology and behaviour) and such heterogeneity can have a significant effect on several processes, including resistance to antibiotics. Recent developments in nanotechnology and tools such as atomic force microscopy offers significant breakthrough in evaluating single cells and molecules.[142,143] These advances should be utilised to probe the physical and chemical properties of bacterial cell surfaces and measure their interactions to understand the unfolding pathways of membrane proteins, cell–surface polymers (especially their conformational properties) and the surface receptors localisation.[144] Importance must also be given to studying the role that the nanoscale surface topography has in the design and manufacturing of an antibacterial surface. Factors that determine the biocompatibility of the antibacterial surface and/or their toxicological consequence need thorough analyses. Until now, only a few studies have been reported which carry out concurrent experiments to address these issues.

So far, the major drawback in most of the reported antibacterial materials is the longevity of their antibacterial activity. Surfaces which work by releasing antibacterial agents are self-limiting as they tend to release the active components for only a limited period. On the other hand, newly emerging surfaces with long-lasting contact-based antibacterial mechanisms are confronted with the accumulation of dead cells which causes weakening of their efficacy over time. From the material design perspective, a recent report on the production of biodegradable polymer-supported co-ordination of gallium in gallium carboxymethyl cellulose (Ga-CMC) looks very promising.[145] This is because of the ability of the Ga-CMC material to act as 'suicide pills' for susceptible pathogenic bacteria such as *P. aeruginosa* that are capable of breaking the CMC moiety and releasing the highly active gallium ions.[145] However, surfaces which possess broad spectrum antibacterial activity for prolonged periods will enhance the application of such materials. So the quest for biofilm prevention should continue on innovative ideas and strategies to produce one or more successful solutions.

Acknowledgements

I thank Dr Christopher K. Hope for his constructive comments during the preparation of this manuscript.

References

1. M. McKenna, *Nature*, 2013, **499**, 394–396.
2. Editorial., *Nat. Rev. Microbiol*, 2010, **8**, 836–836.
3. S. Richard and C. Joanna, *Br. Med. J.*, 2013, **346**, f1493.

4. S. R. Norrby, C. E. Nord and R. Finch, *Lancet Infect. Dis.*, 2005, **5**, 115–119.
5. K. K. Kumarasamy, M. A. Toleman, T. R. Walsh, J. Bagaria, F. Butt, R. Balakrishnan, U. Chaudhary, M. Doumith, C. G. Giske, S. Irfan, P. Krishnan, A. V. Kumar, S. Maharjan, S. Mushtaq, T. Noorie, D. L. Paterson, A. Pearson, C. Perry, R. Pike, B. Rao, U. Ray, J. B. Sarma, M. Sharma, E. Sheridan, M. A. Thirunarayan, J. Turton, S. Upadhyay, M. Warner, W. Welfare, D. M. Livermore and N. Woodford, *Lancet Infect. Dis.*, 2010, **10**, 597–602.
6. K. Bush, P. Courvalin, G. Dantas, J. Davies, B. Eisenstein, P. Huovinen, G. A. Jacoby, R. Kishony, B. N. Kreiswirth, E. Kutter, S. A. Lerner, S. Levy, K. Lewis, O. Lomovskaya, J. H. Miller, S. Mobashery, L. J. V. Piddock, S. Projan, C. M. Thomas, A. Tomasz, P. M. Tulkens, T. R. Walsh, J. D. Watson, J. Witkowski, W. Witte, G. Wright, P. Yeh and H. I. Zgurskaya, *Nat. Rev. Microbiol.*, 2011, **9**, 894–896.
7. ECDC/EMEA, *Joint technical report*, European Centre for Disease Prevention and Control (ECDC) & European Medicines Agency (EMEA). Stockholm, 2009.
8. R. R. Roberts, B. Hota, I. Ahmad, R. D. Scott 2nd, S. D. Foster, F. Abbasi, S. Schabowski, L. M. Kampe, G. G. Ciavarella, M. Supino, J. Naples, R. Cordell, S. B. Levy and R. A. Weinstein, *Clin. Infect. Dis.*, 2009, **49**, 1175–1184.
9. G. Rossi-Fedele and A. P. Roberts, *Int. Endod. J.*, 2007, **40**, 772–777.
10. C. Potera, *Science*, 1999, **283**, 1837–1839.
11. E. Marsich, A. Travan, I. Donati, G. Turco, F. Bellomo and S. Paoletti, *Antimicrobial Polymers*, John Wiley & Sons, Inc., 2011, pp. 379–428.
12. M. A. Patrauchan, S. A. Sarkisova and M. J. Franklin, *Microbiology*, 2007, **153**, 3838–3851.
13. K. Sauer, A. K. Camper, G. D. Ehrlich, J. W. Costerton and D. G. Davies, *J. Bacteriol.*, 2002, **184**, 1140–1154.
14. K. M. Blair, L. Turner, J. T. Winkelman, H. C. Berg and D. B. Kearns, *Science*, 2008, **320**, 1636–1638.
15. S. Jabbouri and I. Sadovskaya, *FEMS Immunol. Med. Microbiol.*, 2010, **59**, 280–291.
16. C. Ryder, M. Byrd and D. J. Wozniak, *Curr. Opin. Microbiol.*, 2007, **10**, 644–648.
17. C. B. Whitchurch, T. Tolker-Neilsen, P. C. Ragas and J. S. Mattick, *Science*, 2002, **295**, 1487.
18. I. W. Sutherland, *Trends Microbiol.*, 2001, **9**, 222–227.
19. M. Wilson, *Microbial Inhabitants of Humans*, Cambridge University Press, Cambridge, 2005.
20. L. Hall-Stoodley and P. Stoodley, *Trends Microbiol.*, 2005, **13**, 7–10.
21. T. B. Rasmussen and M. Givskov, *Int. J. Med. Microbiol.*, 2006, **296**, 149–161.
22. K. L. Van Dellen, L. Houot and P. I. Watnick, *J. Bacteriol.*, 2008, **190**, 8185–8196.

23. E. A. Novak, H. Shao, C. A. Daep and D. R. Demuth, *Infect. Immun.*, 2010, **78**, 2919–2926.
24. X. G. Wu, H. M. Duan, T. Tian, N. Yao, H. Y. Zhou and L. Q. Zhang, *FEMS Microbiol. Lett.*, 2010, **309**, 16–24.
25. T. Bjarnsholt, T. Tolker-Nielsen, N. Hoiby and M. Givskov, *Expert Rev. Mol. Med.*, 2010, **12**, e11.
26. R. J. Lamont, A. Chan, C. M. Belton, K. T. Izutsu, D. Vasel and A. Weinberg, *Infect. Immun.*, 1995, **63**, 3878–3885.
27. C. L. Quave, M. Estevez-Carmona, C. M. Compadre, G. Hobby, H. Hendrickson, K. E. Beenken and M. S. Smeltzer, *PLoS One*, 2012, 7, e28737.
28. J. W. Costerton, P. S. Stewart and E. P. Greenberg, *Science*, 1999, **284**, 1318–1322.
29. E. Karatan and P. Watnick, *Microbiol. Mol. Biol. Rev.*, 2009, 73, 310–347.
30. M. Moscoso, E. García and R. López, *Int. Microbiol.*, 2009, **12**, 77–85.
31. E. S. Garrett, D. Perlegas and D. J. Wozniak, *J. Bacteriol.*, 1999, **181**, 7401–7404.
32. M. Soriani and J. L. Telford, *Future Microbiol.*, 2010, **5**, 735–747.
33. A. Houry, R. Briandet, S. Aymerich and M. Gohar, *Microbiology*, 2010, **156**, 1009–1018.
34. K. P. Lemon, D. E. Higgins and R. Kolter, *J. Bacteriol.*, 2007, **189**, 4418–4424.
35. Y. Lim, M. Jana, T. T. Luong and C. Y. Lee, *J. Bacteriol.*, 2004, **186**, 722–729.
36. M. Shemesh, A. Tam and D. Steinberg, *J. Med. Microbiol.*, 2007, **56**, 1528–1535.
37. N. Bollinger, D. J. Hassett, B. H. Iglewski, J. W. Costerton and T. R. McDermott, *J. Bacteriol.*, 2001, **183**, 1990–1996.
38. G. M. Patriquin, E. Banin, C. Gilmour, R. Tuchman, E. P. Greenberg and K. Poole, *J. Bacteriol.*, 2008, **190**, 662–671.
39. Y. Wu and F. W. Outten, *J. Bacteriol.*, 2009, **191**, 1248–1257.
40. R. D. Monds, P. D. Newell, R. H. Gross and G. A. O'Toole, *Mol. Microbiol.*, 2007, **63**, 656–679.
41. K. Mathee, O. Ciofu, C. Sternberg, P. W. Lindum, J. I. Campbell, P. Jensen, A. H. Johnsen, M. Givskov, D. E. Ohman, S. Molin, N. Hoiby and A. Kharazmi, *Microbiology*, 1999, **145**, 1349–1357.
42. D. T. Hung, J. Zhu, D. Sturtevant and J. J. Mekalanos, *Mol. Microbiol.*, 2006, **59**, 193–201.
43. M. Boyer and F. Wisniewski-Dye, *FEMS Microbiol. Ecol.*, 2009, **70**, 1–19.
44. R. P. Novick and E. Geisinger, *Annu. Rev. Genet.*, 2008, **42**, 541–564.
45. P. V. Krasteva, K. M. Giglio and H. Sondermann, *Protein Sci.*, 2012, **21**, 929–948.
46. N. Cerca and K. K. Jefferson, *FEMS Microbiol. Lett.*, 2008, **283**, 36–41.
47. D. G. Davies, *Microbial extracellular polymeric substances*, Spinger-Verlag, Berlin, 1999.

48. B. Purevdorj-Gage, W. J. Costerton and P. Stoodley, *Microbiology*, 2005, **151**, 1569–1576.
49. A. P. Stapper, G. Narasimhan, D. E. Ohman, J. Barakat, M. Hentzer, S. Molin, A. Kharazmi, N. Hoiby and K. Mathee, *J. Med. Microbiol.*, 2004, **53**, 679–690.
50. G. O'Toole, H. B. Kaplan and R. Kolter, *Annu. Rev. Microbiol.*, 2000, **54**, 49–79.
51. S. M. Hunt, E. M. Werner, B. Huang, M. A. Hamilton and P. S. Stewart, *Appl. Environ. Microbiol.*, 2004, **70**, 7418–7425.
52. M. Hentzer, K. Riedel, T. B. Rasmussen, A. Heydorn, J. B. Andersen, M. R. Parsek, S. A. Rice, L. Eberl, S. Molin, N. Hoiby, S. Kjelleberg and M. Givskov, *Microbiology*, 2002, **148**, 87–102.
53. A. Puskas, E. P. Greenberg, S. Kaplan and A. L. Schaefer, *J. Bacteriol.*, 1997, **179**, 7530–7537.
54. S. Atkinson, J. P. Throup, G. S. Stewart and P. Williams, *Mol. Microbiol.*, 1999, **33**, 1267–1277.
55. R. Morgan, S. Kohn, S. H. Hwang, D. J. Hassett and K. Sauer, *J. Bacteriol.*, 2006, **188**, 7335–7343.
56. H. Segawa, D. T. Tsukayama, R. F. Kyle, D. A. Becker and R. B. Gustilo, *J. Bone Joint Surg.*, 1999, **81**, 1434–1445.
57. E. M. Kojic and R. O. Darouiche, *Clin. Microbiol. Rev.*, 2004, **17**, 255–267.
58. M. J. Richards, J. R. Edwards, D. H. Culver and R. P. Gaynes, *Infect. Control Hosp. Epidemiol.*, 2000, **21**, 510–515.
59. J. M. Schierholz and J. Beuth, *J. Hosp. Infect.*, 2001, **49**, 87–93.
60. W. J. Hu, J. W. Eaton, T. P. Ugarova and L. Tang, *Blood*, 2001, **98**, 1231–1238.
61. J. Y. Wang, B. H. Wicklund, R. B. Gustilo and D. T. Tsukayama, *J. Orthop. Res.*, 1997, **15**, 688–699.
62. M. Aviv, I. Berdicevsky and M. Zilberman, *J. Biomed. Mater. Res., Part A*, 2007, **83**, 10–19.
63. J. L. Del Pozo, M. S. Rouse, G. Euba, C. I. Kang, J. N. Mandrekar, J. M. Steckelberg and R. Patel, *Antimicrob. Agents Chemother.*, 2009, **53**, 4064–4068.
64. G. Lemperle and N. Gauthier-Hazan, *Plast. Reconstr. Surg.*, 2009, **123**, 1864–1873.
65. I. Uckay, D. Pittet, P. Vaudaux, H. Sax, D. Lew and F. Waldvogel, *Ann. Med.*, 2009, **41**, 109–119.
66. S. P. Valappil, D. M. Pickup, D. L. Carroll, C. K. Hope, J. Pratten, R. J. Newport, M. E. Smith, M. Wilson and J. C. Knowles, *Antimicrob. Agents Chemother.*, 2007, **51**, 4453–4461.
67. P. A. Suci, M. W. Mittelman, F. P. Yu and G. G. Geesey, *Antimicrob. Agents Chemother.*, 1994, **38**, 2125–2133.
68. I. G. Duguid, E. Evans, M. R. Brown and P. Gilbert, *J. Antimicrob. Chemother.*, 1992, **30**, 803–810.
69. N. Cerca, S. Martins, F. Cerca, K. K. Jefferson, G. B. Pier, R. Oliveira and J. Azeredo, *J. Antimicrob. Chemother.*, 2005, **56**, 331–336.

70. I. Sadovskaya, E. Vinogradov, J. Li, A. Hachani, K. Kowalska and A. Filloux, *Glycobiology*, 2010, **20**, 895–904.

71. R. A. Hatch and N. L. Schiller, *Antimicrob. Agents Chemother.*, 1998, **42**, 974–977.

72. C. A. Gordon, N. A. Hodges and C. Marriott, *J. Antimicrob. Chemother.*, 1988, **22**, 667–674.

73. I. G. Duguid, E. Evans, M. R. Brown and P. Gilbert, *J. Antimicrob. Chemother.*, 1992, **30**, 791–802.

74. D. J. Evans, D. G. Allison, M. R. Brown and P. Gilbert, *J. Antimicrob. Chemother.*, 1990, **26**, 473–478.

75. L. Zhang and T. F. Mah, *J. Bacteriol.*, 2008, **190**, 4447–4452.

76. K. Lewis, *Biochem (Mosc)*, 2005, **70**, 267–274.

77. A. G. Gristina, *Science*, 1987, **237**, 1588–1595.

78. J. A. Lichter, K. J. Van Vliet and M. F. Rubner, *Macromolecules*, 2009, **42**, 8573–8586.

79. H. J. Busscher, G. I. Doornbusch and H. C. Van der Mei, *J. Dent. Res.*, 1992, **71**, 491–500.

80. C. J. van Oss, *Colloids Surf., B*, 1995, **5**, 91–110.

81. N. I. Abu-Lail and T. A. Camesano, *Langmuir*, 2006, **22**, 7296–7301.

82. K. D. Park, Y. S. Kim, D. K. Han, Y. H. Kim, E. H. Lee, H. Suh and K. S. Choi, *Biomaterials*, 1998, **19**, 851–859.

83. A. Travan, E. Marsich, I. Donati and S. Paoletti, *Nanotechnologies for the Life Sciences*, Wiley-VCH Verlag GmbH & Co. KGaA, 2007.

84. M. O. Samuelsson and D. L. Kirchman, *Appl. Environ. Microbiol.*, 1990, **56**, 3643–3648.

85. R. Bos, H. C. van der Mei and H. J. Busscher, *FEMS Microbiol. Rev.*, 1999, **23**, 179–230.

86. M. Hermansson, *Colloids Surf., B*, 1999, **14**, 105–119.

87. D. R. Absolom, F. V. Lamberti, Z. Policova, W. Zingg, C. J. van Oss and A. W. Neumann, *Appl. Environ. Microbiol.*, 1983, **46**, 90–97.

88. C. J. Van Oss, R. J. Good and M. K. Chaudhury, *J. Colloid Interface Sci.*, 1986, **111**, 378–390.

89. G. Speranza, G. Gottardi, C. Pederzolli, L. Lunelli, R. Canteri, L. Pasquardini, E. Carli, A. Lui, D. Maniglio, M. Brugnara and M. Anderle, *Biomaterials*, 2004, **25**, 2029–2037.

90. T. R. Scheuerman, A. K. Camper and M. A. Hamilton, *J. Colloid Interface Sci.*, 1998, **208**, 23–33.

91. E. W. McAllister, L. C. Carey, P. G. Brady, R. Heller and S. G. Kovacs, *Gastrointest. Endosc.*, 1993, **39**, 422–425.

92. L. Mauclaire, E. Brombacher, J. D. Bunger and M. Zinn, *Colloids Surf., B*, 2010, **76**, 104–111.

93. V. K. Truong, R. Lapovok, Y. S. Estrin, S. Rundell, J. Y. Wang, C. J. Fluke, R. J. Crawford and E. P. Ivanova, *Biomaterials*, 2010, **31**, 3674–3683.

94. J. A. Lichter, M. T. Thompson, M. Delgadillo, T. Nishikawa, M. F. Rubner and K. J. Van Vliet, *Biomacromolecules*, 2008, **9**, 1571–1578.

95. J. D. Mendelsohn, S. Y. Yang, J. Hiller, A. I. Hochbaum and M. F. Rubner, *Biomacromolecules*, 2003, **4**, 96–106.
96. J. C. Grunlan, J. K. Choi and A. Lin, *Biomacromolecules*, 2005, **6**, 1149–1153.
97. J. Fu, J. Ji, W. Yuan and J. Shen, *Biomaterials*, 2005, **26**, 6684–6692.
98. C. Cassinelli, M. Morra, A. Pavesio and D. Renier, *J. Biomater. Sci., Polym. Ed.*, 2000, **11**, 961–977.
99. Z. Cao, L. Mi, J. Mendiola, J. R. Ella-Menye, L. Zhang and H. Xue, *Angew. Chem.*, 2012, **124**, 2656–2659.
100. G. Cheng, H. Xue, Z. Zhang, S. Chen and S. Jiang, *Angew. Chem.*, 2008, **47**, 8831–8834.
101. H. Murata, R. R. Koepsel, K. Matyjaszewski and A. J. Russell, *Biomaterials*, 2007, **28**, 4870–4879.
102. O. Etienne, C. Picart, C. Taddei, Y. Haikel, J. L. Dimarcq, P. Schaaf, J. C. Voegel, J. A. Ogier and C. Egles, *Antimicrob. Agents Chemother.*, 2004, **48**, 3662–3669.
103. W. Yuan, J. Ji, J. Fu and J. Shen, *J. Biomed. Mater. Res., Part B*, 2008, **85**, 556–563.
104. Y. Feng, Z. Han, J. Peng, J. Lu, B. Xue, L. Li, H. Ma and E. Wang, *Mater. Lett.*, 2006, **60**, 1588–1593.
105. S. Noimark, M. Bovis, A. J. MacRobert, A. Correia, E. Allan, M. Wilson and I. P. Parkin, *RSC Adv.*, 2013, **3**, 18383–18394.
106. W. Chrzanowski, S. P. Valappil, C. W. Dunnill, E. A. Abou Neel, K. Lee, I. P. Parkin, M. Wilson, D. A. Armitage and J. C. Knowles, *Mater. Sci. Eng., C*, 2010, **30**, 225–234.
107. S. Noimark, C. W. Dunnill and I. P. Parkin, *Adv. Drug Delivery Rev.*, 2013, **65**, 570–580.
108. P. Wu and D. W. Grainger, *Biomaterials*, 2006, **27**, 2450–2467.
109. S. P. Valappil, S. K. Misra, A. R. Boccaccini and I. Roy, *Expert Rev. Med. Devices*, 2006, **3**, 853–868.
110. J. S. Price, A. F. Tencer, D. M. Arm and G. A. Bohach, *J. Biomed. Mater. Res.*, 1996, **30**, 281–286.
111. H. F. Chuang, R. C. Smith and P. T. Hammond, *Biomacromolecules*, 2008, **9**, 1660–1668.
112. A. Shukla, K. E. Fleming, H. F. Chuang, T. M. Chau, C. R. Loose, G. N. Stephanopoulos and P. T. Hammond, *Biomaterials*, 2010, **31**, 2348–2357.
113. L. Timofeeva and N. Kleshcheva, *Appl. Microbiol. Biotechnol.*, 2011, **89**, 475–492.
114. F. Siedenbiedel and J. C. Tiller, *Polymers*, 2012, **4**, 46–71.
115. A. Muñoz-Bonilla and M. Fernández-García, *Prog. Polym. Sci.*, 2012, **37**, 281–339.
116. D. Lee, R. E. Cohen and M. F. Rubner, *Langmuir*, 2005, **21**, 9651–9659.
117. M. Malcher, D. Volodkin, B. Heurtault, P. Andre, P. Schaaf, H. Mohwald, J. C. Voegel, A. Sokolowski, V. Ball, F. Boulmedais and B. Frisch, *Langmuir*, 2008, **24**, 10209–10215.

118. Z. Li, D. Lee, X. Sheng, R. E. Cohen and M. F. Rubner, *Langmuir*, 2006, **22**, 9820–9823.
119. Z. Shi, K. G. Neoh, S. P. Zhong, L. Y. Yung, E. T. Kang and W. Wang, *J. Biomed. Mater. Res., Part A*, 2006, **76**, 826–834.
120. J. Fu, J. Ji, D. Fan and J. Shen, *J. Biomed. Mater. Res., Part A*, 2006, **79**, 665–674.
121. A. Travan, E. Marsich, I. Donati, M. Benincasa, M. Giazzon, L. Felisari and S. Paoletti, *Acta Biomater.*, 2011, **7**, 337–346.
122. A. Kumar, P. K. Vemula, P. M. Ajayan and G. John, *Nat. Mater.*, 2008, **7**, 236–241.
123. U. K. Parashar, V. Kumar, T. Bera, P. S. Saxena, G. Nath, S. K. Srivastava, R. Giri and A. Srivastava, *Nanotechnology*, 2011, **22**, 415104.
124. A. W. Carpenter, D. L. Slomberg, K. S. Rao and M. H. Schoenfisch, *ACS Nano*, 2011, **5**, 7235–7244.
125. M. J. Hajipour, K. M. Fromm, A. Akbar Ashkarran, D. Jimenez de Aberasturi, I. R. D. Larramendi, T. Rojo, V. Serpooshan, W. J. Parak and M. Mahmoudi, *Trends Biotechnol.*, 2012, **30**, 499–511.
126. S. Takenaka, T. Tonoki, K. Taira, S. Murakami and K. Aoki, *Appl. Environ. Microbiol.*, 2007, **73**, 1799–1802.
127. E. A. Abou Neel, D. M. Pickup, S. P. Valappil, R. J. Newport and J. C. Knowles, *J. Mater. Chem.*, 2009, **19**, 690–701.
128. T. R. Shryock and A. Richwine, *Ann. N. Y. Acad. Sci.*, 2010, **1213**, 92–105.
129. A. Pantosti, *Front. Microbiol.*, 2012, **3**, 127.
130. S. P. Valappil, D. Ready, E. A. Abou Neel, D. M. Pickup, W. Chrzanowski, L. A. O'Dell, R. J. Newport, M. E. Smith, M. Wilson and J. C. Knowles, *Adv. Funct. Mater.*, 2008, **18**, 732–741.
131. J. J. Bullen, H. J. Rogers, P. B. Spalding and C. G. Ward, *FEMS Immunol. Med. Microbiol.*, 2005, **43**, 325–330.
132. C. R. Chitambar and J. Narasimhan, *Pathobiology*, 1991, **59**, 3–10.
133. Y. Kaneko, M. Thoendel, O. Olakanmi, B. E. Britigan and P. K. Singh, *J. Clin. Invest.*, 2007, **117**, 877–888.
134. L. R. Bernstein, T. Tanner, C. Godfrey and B. Noll, *Met. Based Drugs*, 2000, **7**, 33–47.
135. E. Banin, A. Lozinski, K. M. Brady, E. Berenshtein, P. W. Butterfield, M. Moshe, M. Chevion, E. P. Greenberg and E. Banin, *Proc. Natl. Acad. Sci. U. S. A.*, 2008, **105**, 16761–16766.
136. J. E. Moore, A. Murphy, B. C. Millar, A. Loughrey, P. J. Rooney, J. S. Elborn and C. E. Goldsmith, *J. Microbiol. Methods*, 2009, **76**, 201–203.
137. S. P. Valappil, D. Ready, E. A. Abou Neel, D. M. Pickup, L. A. O'Dell, W. Chrzanowski, J. Pratten, R. J. Newport, M. E. Smith, M. Wilson and J. C. Knowles, *Acta Biomater.*, 2009, **5**, 1198–1210.
138. K. C. Chaw, M. Manimaran and F. E. H. Tay, *Antimicrob. Agents Chemother.*, 2005, **49**, 4853–4859.
139. S. P. Valappil, M. Coombes, L. Wright, G. J. Owens, R. J. M. Lynch, C. K. Hope and S. M. Higham, *Acta Biomater.*, 2012, **8**, 1957–1965.

140. M. Marini, S. De Niederhausern, R. Iseppi, M. Bondi, C. Sabia, M. Toselli and F. Pilati, *Biomacromolecules*, 2007, **8**, 1246–1254.
141. Y. Yin and C. Wang, *J. Sol-Gel Sci. Technol.*, 2011, **59**, 36–42.
142. D. J. Muller and Y. F. Dufrene, *Nat. Nanotechnol.*, 2008, **3**, 261–269.
143. Y. F. Dufrene, *Nat. Rev. Microbiol.*, 2004, **2**, 451–460.
144. Y. F. Dufrene, *Nat. Rev. Microbiol.*, 2008, **6**, 674–680.
145. S. P. Valappil, H. H. Yiu, L. Bouffier, C. K. Hope, G. Evans, J. B. Claridge, S. M. Higham and M. J. Rosseinsky, *Dalton Trans.*, 2013, **42**, 1778–1786.

Subject Index